Lecture Notes in Networks and Systems

Volume 176

The series "Lecture Notes in Networks and Systems" publishes the latest developments in Networks and Systems—quickly, informally and with high quality. Original research reported in proceedings and post-proceedings represents the core of LNNS.

Volumes published in LNNS embrace all aspects and subfields of, as well as new challenges in, Networks and Systems.

The series contains proceedings and edited volumes in systems and networks, spanning the areas of Cyber-Physical Systems, Autonomous Systems, Sensor Networks, Control Systems, Energy Systems, Automotive Systems, Biological Systems, Vehicular Networking and Connected Vehicles, Aerospace Systems, Automation, Manufacturing, Smart Grids, Nonlinear Systems, Power Systems, Robotics, Social Systems, Economic Systems and other. Of particular value to both the contributors and the readership are the short publication timeframe and the world-wide distribution and exposure which enable both a wide and rapid dissemination of research output.

The series covers the theory, applications, and perspectives on the state of the art and future developments relevant to systems and networks, decision making, control, complex processes and related areas, as embedded in the fields of interdisciplinary and applied sciences, engineering, computer science, physics, economics, social, and life sciences, as well as the paradigms and methodologies behind them.

Indexed by SCOPUS, INSPEC, WTI Frankfurt eG, zbMATH, SCImago.

All books published in the series are submitted for consideration in Web of Science.

More information about this series at http://www.springer.com/series/15179

Subarna Shakya · Valentina Emilia Balas ·
Wang Haoxiang · Zubair Baig
Editors

Proceedings of International Conference on Sustainable Expert Systems

ICSES 2020

Springer

Editors
Subarna Shakya
Institute of Engineering
Tribhuvan University
Pulchowk, Nepal

Valentina Emilia Balas 🆔
Intelligent Systems Research Centre
Aurel Vlaicu University of Arad
Arad, Romania

Wang Haoxiang
Go Perception Laboratory
Cornell University
Ithaca, NY, USA

Zubair Baig
School of Information Technology
Deakin University
Geelong, VIC, Australia

ISSN 2367-3370   ISSN 2367-3389   (electronic)
Lecture Notes in Networks and Systems
ISBN 978-981-33-4357-3   ISBN 978-981-33-4355-9   (eBook)
https://doi.org/10.1007/978-981-33-4355-9

This Springer imprint is published by the registered company Springer Nature Singapore Pte Ltd.
The registered company address is: 152 Beach Road, #21-01/04 Gateway East, Singapore 189721,
Singapore

We are honored to dedicate this book to all the technical program committee members, editors, and authors of ICSES 2020.

Preface

It is my great pleasure to welcome you all to the International Conference on Sustainable Expert Systems (ICSES 2020) held in Nepal during September 28–29, 2020.

The main thrust of ICSES 2020 is to deal with the design, implementation, development, testing, and management of intelligent and sustainable expert systems and also to provide both theoretical and practical guidelines for the deployment of these systems. Due to the well-established track record of intelligent and networked systems, it mandates the implementation of sustainability aspect to become more accessible for the network users. This makes ICSES 2020 an excellent venue for exploring the sustainable computing foundations for the challenging communication network issues.

The success of ICSES 2020 completely depends on the efforts made by the researchers in the field of artificial intelligence and sustainable computing, who have written and submitted research articles on various research topics.

The program committee members and external reviewers deserve sincere accolades, as they have invested more time in assessing and analyzing multiple papers submitted on different research domains to significantly hold and maintain high quality for this conference event. Additional thanks are given to the Springer publications. We hope that you will take advantage of many research insights into intelligent expert systems.

Pulchowk, Nepal

Arad, Romania

Ithaca, USA

Geelong, Australia

Dr. Subarna Shakya

Prof. Dr. Valentina Emilia Balas

Dr. Wang Haoxiang

Dr. Zubair Baig

Acknowledgements

We ICSES 2020 are very honored to have the most prominent keynote speakers, who have significant research expertise in their respective specialty. They are Prof. Dr. Shashidhar Ram Joshi, Dean, Institute of Engineering, Tribhuvan University, Nepal, and Dr. Joy Iong Zong Chen, Da-Yeh University, Changhua County, Taiwan.

Further, we would like to thank the technical conference committee and external reviewers, who have worked very hard in reviewing the papers and making valuable suggestions to enhance the research works submitted by authors. We express our sincere gratitude to the authors for contributing their innovative research results to the conference. Special thanks go to Springer publications.

Last but not least, we wish to express our thanks to all the members of the ICSES 2020 committee for all their unprecedented efforts in making ICSES 2020 as a successful conference. Finally, our thanks go to the Tribhuvan University (TU), Nepal. It would have been impossible to organize the conference without their moral support. We really hope that all the participants of ICSES 2020 will benefit tremendously from the conference event. Finally, we would like to wish all success in their future research direction.

About the Conference

This proceedings includes the papers presented in the International Conference on Sustainable Expert Systems (ICSES 2020), Organized by Tribhuvan University (TU), Nepal, with a primary focus on the research information related to artificial intelligence (AI), sustainability, and expert systems applied in almost all the areas of industries, government sectors, and academia worldwide. ICSES 2020 will provide an opportunity for the researchers to present their research results and examine the advanced applications in artificial intelligence (AI) and the expert systems field. Nevertheless, this proceedings will promote novel techniques to extend the frontiers of this fascinating research field.

Advancement in both the sustainability and intelligent systems disciplines requires an exchange of thoughts and ideas of audiences from different parts of the world. The 2020 International Conference on Sustainable Expert Systems was potentially designed to encourage the advancement and application of sustainability and artificial intelligence models in the existing computing systems. ICSES 2020 received a total of 191 manuscripts from different countries, wherein the submitted papers have been peer-reviewed by at least three reviewers drawn from the technical conference committee, external reviewers, and also editorial board depending on the research domain of the papers. Through a rigorous peer-review process, the submitted papers are selected based on the research novelty, clarity, and significance. Out of these, 52 papers were selected for publication in ICSES 2020 proceedings. It covers papers from several significant thematic areas, namely data science, wireless communication, intelligent systems, social media, and image processing.

To conclude, this proceedings will provide a written research synergy that already exists between the intelligent expert systems and network-enabled communities and represents a progressive framework from which new research interaction will result in the near future. I look forward to the evolution of this conference theme over time and to learning more on the network sustainability and continue to pave the way for applications of network sustainability in the emerging intelligent expert systems.

Conference Chair
Prof. Dr. Subarna Shakya
Professor
Department of ECE
Pulchowk Campus
Institute of Engineering
Tribhuvan University
Pulchowk, Nepal

Contents

Editors and Contributors

About the Editors

Dr. Prof. Subarna Shakya is currently Professor of Computer Engineering, Department of Electronics and Computer Engineering, Central Campus, Institute of Engineering, Pulchowk, Tribhuvan University, Coordinator (IOE) and LEADER Project (Links in Europe and Asia for engineering, eDucation, Enterprise and Research exchanges), ERASMUS MUNDUS. She received M.Sc. and Ph.D. degrees in Computer Engineering from the Lviv Polytechnic National University, Ukraine, 1996 and 2000, respectively. Her research area includes government system, computer systems & simulation, distributed & cloud computing, software engineering & information system, computer architecture, information security for e-government and multimedia system.

Prof. Valentina Emilia Balas is currently Full Professor in the Department of Automatics and Applied Software at the Faculty of Engineering, "Aurel Vlaicu" University of Arad, Romania. She holds a Ph.D. in Applied Electronics and Telecommunications from Polytechnic University of Timisoara. Dr. Balas is the author of more than 250 research papers in refereed journals and international conferences. Her research interests are in intelligent systems, fuzzy control, soft computing, smart sensors, information fusion, modeling and simulation. She is Editor-in-Chief to the International Journal of Advanced Intelligence Paradigms (IJAIP) and to the International Journal of Computational Systems Engineering (IJCSysE), an Editorial Board member of several national and international journals and an evaluator expert for national, international projects and Ph.D. thesis. Dr. Balas is the director of Intelligent Systems Research Centre in Aurel Vlaicu University of Arad.

Wang Haoxiang is currently the director and a lead executive faculty member of GoPerception Laboratory, NY, USA. His research interests include multimedia information processing, pattern recognition and machine learning, remote sensing image processing and data-driven business intelligence. He has co-authored over 60 journals and conference papers in these fields on journals such as Springer MTAP, Cluster Computing, SIVP; IEEE TII, Communications Magazine; Elsevier Computers & Electrical Engineering, Computers, Environment and Urban Systems, Optik, Sustainable Computing: Informatics and Systems, Journal of Computational Science, Pattern Recognition Letters, Information Sciences, Computers in Industry, Future Generation Computer Systems; Taylor & Francis International Journal of Computers and Applications and conferences such as IEEE SMC, ICPR, ICTAI, ICICI, CCIS and ICACI. He is Guest Editor for IEEE Transactions on Industrial Informatics, IEEE Consumer Electronics Magazine, Multimedia Tools and Applications, MDPI Sustainability, International Journal of Information and Computer Security and Journal of Medical Imaging and Health Informatics, Concurrency and Computation: Practice and Experience.

Dr. Zubair Baig received the B.S. degree in Computer Engineering from the King Fahd University of Petroleum & Minerals, Dhahran, Saudi Arabia, in 2002, M.S. degree in Electrical Engineering from the University of Maryland, College Park, USA, in 2003, and the Ph.D. degree in Computer Science from Monash University, Melbourne, Victoria, Australia, in 2008. Currently, he is working in the division of Cyber Security from the School of Information Technology, Faculty of Sci Eng & Built Env, Deakin University, Australia. He has authored over 55 journal and conference articles and book chapters.

Contributors

Sohaib N. Abdullah Northern Technical University, Mosul, Iraq

M. B. Abhishek Department of Electronics and Communication, Dayananda Sagar College of Engineering, Bengaluru, India

Nanda Bikram Adhikari Department of Electronics and Computer Engineering, Pulchowk Campus, Institute of Engineering, Tribhuvan University, Lalitpur, Nepal

Rajneesh Agrawal Comp-Tel Consultancy, Mentor, Jabalpur, India

M. K. A. Ahamed Khan Faculty of Engineering, UCSI University, Kuala Lumpur, Malaysia

Tanvir Ahmed Department of Computer Science and Engineering, Ahsanullah University of Science and Technology (AUST), Dhaka, Bangladesh

Al Amin Department of Computer Science and Engineering, Southeast University, Dhaka, Bangladesh

Chun Kit Ang Faculty of Engineering, UCSI University, Kuala Lumpur, Malaysia

Milos Antonijevic Singidunum University, Belgrade, Serbia

M. Ashwin Department of Electronics & Communication Engineering, Kumaraguru College of Technology, Coimbatore, India

Galanis Athanasios Department of Civil Engineering, International Hellenic University, Terma Magnesias, Serres, Greece

Nebojsa Bacanin Singidunum University, Belgrade, Serbia

Vikramjit Banerjee Hope Foundation's, International Institute of Information Technology, Pune, India

Shahana Bano Department of CSE, Koneru Lakshmaiah Education Foundation, Vaddeswaram, India;
Department of Computer Science and Engineering, Koneru Lakshmaiah Education Foundation, Vaddeswaram, Andhra Pradesh, India

S. Basheer Ahamed Department of Electronics and Communication Engineering, Kumaraguru College of Technology, Coimbatore, India

Pranav Bathija Thadomal Shahani Engineering College, Mumbai, India

Ezzohra Belkadi Laboratoire de Recherche en Management, Information et Gouvernances, Faculté Des Sciences Juridiques Économiques et Sociales Ain-Sebaa, Hassan II University of Casablanca, Casablanca, Maroc

Kishore Bhamidipati School of Electrical Engineering and Computer Science, Oregon State University, Corvallis, USA

Sagar Bhandari Department of Electronics and Computer Engineering, Pulchowk Campus, Institute of Engineering, Tribhuvan University, Lalitpur, Nepal

Shiva Bhandari Department of Electronics and Computer Engineering, Pulchowk Campus, Institute of Engineering, Tribhuvan University, Lalitpur, Nepal

Nirdesh Bhattarai Department of Electronics and Computer Engineering, Pulchowk Campus, Institute of Engineering, Tribhuvan University, Lalitpur, Nepal

Girish P Bhole Department of Computer Engineering & Information Technology, Veermata Jijabai Technological Institute, Matunga Mumbai, Maharastra, India

Pushkar Bhuse Dwarkadas J. Sanghvi College of Engineering, Mumbai, India

Anupkumar M. Bongale Symbiosis Institute of Technology, Symbiosis International (Deemed University), Pune, Maharashtra, India

Arunkumar M. Bongale Symbiosis Institute of Technology, Symbiosis International (Deemed University), Pune, Maharashtra, India

W. M. M. Botejue Faculty of Information Technology, University of Moratuwa, Moratuwa, Sri Lanka

Pornpimol Chaiwuttisak Department of Statistics, Faculty of Science, King Mongkut's Institute of Technology Ladkrabang, Bangkok, Thailand

Ganesh Chandrasekaran Department of ECE, Karunya Institute of Technology and Sciences, Coimbatore, India

Joseph Charles Department of Physical Sciences and Technology, Faculty of Applied Sciences, Sabaragamuwa University of Sri Lanka, Belihuloya, Sri Lanka

Chandrashekhar Choudhary Department of Computer Science Engineering and Information Technology, Jaypee University of Information Technology, Waknaghat, India

Randilu de Zoysa Faculty of Information Technology, University of Moratuwa, Katubedda, Sri Lanka

J. Deny ECE Department, Kalasalingam Academy of Research and Education, Krishnankoil, Tamil Nadu, India

J. R. Dinesh Kumar Department of Electronics & Communication Engineering, Sri Krishna College of Engineering & Technology, Coimbatore, Tamil Nadu, India

Aleksandar Djordjevic Singidunum University, Belgrade, Serbia

T. G. I. Fernando Department of Computer Science, University of Sri Jayewardenepura, Gangodawila, Nugegoda, Sri Lanka

Prashant Gadakh Department of Computer Engineering, Savitribai Phule Pune University, Pune, India

Sharvari Gadiwan Hope Foundation's, International Institute of Information Technology, Pune, India

Upeksha Ganegoda Faculty of Information Technology, University of Moratuwa, Katubedda, Sri Lanka

M. Gayathri Department of Computer Science and Engineering, SRM Institute of Science and Technology, Chennai, India

Aparna George Applied Electronics and Instrumentation, Rajagiri School of Engineering and Technology, Kochi, India

Botzoris George School of Civil Engineering, Democritus University of Thrace, Xanthi, Greece

Deepika Ghanta Department of CSE, Koneru Lakshmaiah Education Foundation, Vaddeswaram, India

R. A. R. C. Gopura Department of Mechanical Engineering, University of Moratuwa, Moratuwa, Sri Lanka

Anil Gupta Department of Computer Science and Engineering, MBM Engineering College, Jai Narain Vyas University, Jodhpur, Rajasthan, India

Gaurav Hajela Department of Computer Science and Engineering, MANIT, Bhopal, India

Mustafa W. Hamadalla Northern Technical University, Mosul, Iraq

D. Jude Hemanth Department of ECE, Karunya Institute of Technology and Sciences, Coimbatore, India

Lim Wei Hong Faculty of Engineering, UCSI University, Kuala Lumpur, Malaysia

Mohammad Ashraful Hoque Department of Computer Science and Engineering, Southeast University, Dhaka, Bangladesh

Ehssan M. A. Ibrahim Northern Technical University, Mosul, Iraq

Thouhidul Islam Department of Computer Science and Engineering, Southeast University, Dhaka, Bangladesh

Shrushti Jagtap Hope Foundation's, International Institute of Information Technology, Pune, India

Anku Jaiswal Department of Electronics and Computer Engineering, IOE, Pulchowk Campus, Patan, India

M. I. M. Jayarathna Department of Statistics and Computer Science, University of Kelaniya, Kelaniya, Sri Lanka

A. M. Imanthi C. K. Jayathilake Department of Statistics and Computer Science, Faculty of Science, University of Peradeniya, Peradeniya, Sri Lanka

Jony Guru Jambheshwar University of Science and Techonology, Hisar, Haryana, India

Omar Mahmood Jumaah Northern Technical University, Mosul, Iraq

Rohan Kalra Department of Computer Science and Engineering, SRM Institute of Science and Technology, Chennai, India

Atharva Kawade Hope Foundation's, International Institute of Information Technology, Pune, India

Rohit Kawari Department of Electronics and Computer Engineering, Pulchowk Campus, Institute of Engineering, Tribhuvan University, Lalitpur, Nepal

Sunkara Venkata Krishna Department of CSE, SRM University, Amaravati, AP, India

Satish Kumar Symbiosis Institute of Technology, Symbiosis International (Deemed University), Pune, Maharashtra, India

Sunil Kumar Guru Jambheshwar University of Science and Techonology, Hisar, Haryana, India

Oumayma Labti Laboratoire de Recherche en Management, Information et Gouvernances, Faculté Des Sciences Juridiques Économiques et Sociales Ain-Sebaa, Hassan II University of Casablanca, Casablanca, Maroc

Y. Lakshmi Pranthi Department of Computer Science and Engineering, Koneru Lakshmaiah Education Foundation, Vaddeswaram, Andhra Pradesh, India

Sugeeswari Lekamge Department of Computing and Information Systems, Faculty of Applied Sciences, Sabaragamuwa University of Sri Lanka, Belihuloya, Sri Lanka

Leandro L. Lorente-Leyva SDAS Research Group, Ibarra, Ecuador

C. Malathy Department of Computer Science and Engineering, SRM Institute of Science and Technology, Chennai, India

R. Manasa Department of Electronics and Communication, Dayananda Sagar College of Engineering, Bengaluru, India

Alex Mathew Department of Cybersecurity, Bethany College, Bethany, USA

Karam Hashim Mohammed Northern Technical University, Mosul, Iraq

D. Mohan ECM Department, Sreenidhi Institute of Science and Technology, Hyderabad, India

Sangeet Mohanty SureIT Solutions Inc, Hyderabad, India;
Sreenidhi Institute of Science and Technology, Hyderabad, India

G. Muneeswari Department of Computer Science and Engineering, School of Engineering and Technology, CHRIST (Deemed to be University), Bangaluru, India

Naresh Babu Muppalaneni National Institute of Technology, Silchar, India

A. Murugan Department of Computer Science, Dr. Ambedkar Government Arts College (Autonomous), Affiliated to University of Madras, Chennai, India;
PG & Research Department of Computer Science, Dr. Ambedkar Government Arts College, Chennai, India

Sangeetha Muthiah Department of Computational Logistics, Alagappa University, Karaikudi, Tamil Nadu, India

P. N. P. S. Nagarathne Division of Orthodontics, Faculty of Dental Sciences, University of Peradeniya, Peradeniya, Sri Lanka

S. Naren Department of Electronics & Communication Engineering, Kumaraguru College of Technology, Coimbatore, India

Lakshika S. Nawarathna Department of Statistics and Computer Science, Faculty of Science, University of Peradeniya, Peradeniya, Sri Lanka

Ruwan D. Nawarathna Department of Statistics and Computer Science, Faculty of Science, University of Peradeniya, Peradeniya, Sri Lanka

Kriti Nemkul Central Department of Computer Science and Information Technology, Tribhuvan University, Kirtipur, Nepal

Gorsa Lakshmi Niharika Department of CSE, Koneru Lakshmaiah Education Foundation, Vaddeswaram, India

Eliou Nikolaos Department of Civil Engineering, University of Thessaly, Volos, Greece

Lemonakis Panagiotis Department of Civil Engineering, University of Thessaly, Volos, Greece

Shishir Panta Department of Electronics and Computer Engineering, Pulchowk Campus, Institute of Engineering, Tribhuvan University, Lalitpur, Nepal

Gajanan Shankarrao Patange Department of Mechanical Engineering, Faculty of Technology and Engineering, Charotar University of Science and Technology, Changa Ta, Petlad Dist Anand, Gujarat, India

Delio R. Patiño-Alarcón Academia del Conocimiento, Ibarra, Ecuador

Fernando A. Patiño-Alarcón Academia del Conocimiento, Ibarra, Ecuador

Amit Paudyal Department of Electronics and Computer Engineering, Pulchowk Campus, Institute of Engineering, Tribhuvan University, Lalitpur, Nepal

P. Pavan Kalyan Department of Computer Science and Engineering, Koneru Lakshmaiah Education Foundation, Vaddeswaram, Andhra Pradesh, India

Diego H. Peluffo-Ordóñez SDAS Research Group, Ibarra, Ecuador;
Yachay Tech University, Urcuquí, Ecuador;
Coorporación Universitaria Autónoma de Nariño, Pasto, Colombia

Rakesh Chandra Prajapati ORION Space, Madhyapur Thimi, Nepal

Yerramreddy Lakshmi Pranathi Department of CSE, Koneru Lakshmaiah Education Foundation, Vaddeswaram, India

Prakash Chandra Prasad Department of Electronics and Computer Engineering, IOE, Pulchowk Campus, Patan, India

Ch. Prathima Sree Vidyanikethan Engineering College, Tirupati, India

S. Preethi Department of Electronics & Communication Engineering, Kumaraguru College of Technology, Coimbatore, India

I. A. Premaratne Department of Electrical and Computer Engineering, The Open University of Sri Lanka, Nawala, Nugegoda, Sri Lanka

S. Priya Applied Electronics and Instrumentation, Rajagiri School of Engineering and Technology, Kochi, India

K. Priyadharsini Department of Electronics & Communication Engineering, Sri Krishna College of Engineering & Technology, Coimbatore, Tamil Nadu, India

K. Priyanka Department of Computer Science and Engineering, Koneru Lakshmaiah Education Foundation, Vaddeswaram, Andhra Pradesh, India

K. Priyankan Department of Computer Science, University of Sri Jayewardenepura, Gangodawila, Nugegoda, Sri Lanka

K. Punitha Department of Computer Science, AgurchandManmull Jain College (Shift II), Chennai, India

Antony Puthussery Department of Computational Sciences, School of Science, CHRIST (Deemed to be University), New Delhi, India

Nabin Rai Department of Electronics and Computer Engineering, Pulchowk Campus, Institute of Engineering, Tribhuvan University, Lalitpur, Nepal

Thilina C. Rajapakse Postgraduate Institute of Science, University of Peradeniya, Peradeniya, Sri Lanka

Manickam Ramasamy Faculty of Engineering, UCSI University, Kuala Lumpur, Malaysia

Rajeev Ramesh Hope Foundation's, International Institute of Information Technology, Pune, India

L. Ranathunga Faculty of Information Technology, University of Moratuwa, Moratuwa, Sri Lanka

Tarik A. Rashid Computer Science and Engineering Department, University of Kurdistan Hewler, Erbil, KRG, Iraq

Akhtar Rasool Department of Computer Science and Engineering, MANIT, Bhopal, India

H. U. W. Ratnayake Department of Electrical and Computer Engineering, The Open University of Sri Lanka, Nawala, Nugegoda, Sri Lanka

Sahil Ratra Cognizant Technology Solutions India, Pune, Maharashtra, India

Nazmus Sakib Ribhu Faculty of Engineering, UCSI University, Kuala Lumpur, Malaysia

Ram Prasad Rimal Ramlaxman Innovations, Kathmandu, Nepal

Shamim H. Ripon Department of Computer Science and Engineering, East West University, Dhaka, Bangladesh

W. U. D. Rodrigo Department of Electrical and Computer Engineering, The Open University of Sri Lanka, Nawala, Nugegoda, Sri Lanka

P. Sai Shreyashi Department of Computer Science and Engineering, SRM Institute of Science and Technology, Chennai, India

M. Sakthimohan ECE Department, Kalasalingam Academy of Research and Education, Krishnankoil, Tamil Nadu, India

Hrithik Sanyal Department of Electronics and Telecommunications, Bharati Vidyapeeth College of Engineering, Pune, Pune, India;
Department of Electronics & Telecommunications, Bharati Vidyapeeth College of Engineering Pune, Pune, India

Priyanka Saxena Department of Computer Science, Rajiv Gandhi Proudyogiki Vishwavidyalaya, Bhopal, India

Jarashanth Selvarajah Postgraduate Institute of Science, University of Peradeniya, Peradeniya, Sri Lanka

A. Senthilrajan Department of Computational Logistics, Alagappa University, Karaikudi, Tamil Nadu, India

Nooriya Shahul Department of Electronics & Communication Engineering, Mar Athanasius College of Engineering, Kothamangalam, Kerala, India

Subarna Shakya Institute of Engineering, Tribhuvan University, Kirtipur, Nepal; Department of Electronics and Computer Engineering, IOE, Pulchowk Campus, Patan, India

M. Shanmugapriya Applied Electronics and Instrumentation, Rajagiri School of Engineering and Technology, Kochi, India

Rishav Mani Sharma Department of Electronics and Computer Engineering, Pulchowk Campus, Institute of Engineering, Tribhuvan University, Lalitpur, Nepal

K. Anitha Sheela ECE Department, JNTUH College of Engineering, Kukatpally, Hyderabad, India

Siddharth Shelly Department of Electronics & Communication Engineering, Mar Athanasius College of Engineering, Kothamangalam, Kerala, India

Mir Moynuddin Ahmed Shibly Department of Computer Science and Engineering, East West University, Dhaka, Bangladesh

Aman Shinde Hope Foundation's, International Institute of Information Technology, Pune, India

Kavita Pankaj Shirsat Department of Computer Engineering & Information Technology, Veermata Jijabai Technological Institute, Matunga Mumbai, Maharastra, India

Sailesh Singh Department of Electronics and Computer Engineering, IOE, Pulchowk Campus, Patan, India

Prattyush Sinha Department of Computer Science Engineering and Information Technology, Jaypee University of Information Technology, Waknaghat, India

Zankhan Sonara Department of Mechanical Engineering, Faculty of Technology and Engineering, Charotar University of Science and Technology, Changa Ta, Petlad Dist Anand, Gujarat, India

Gorsa Datta Sai Sreya Department of CSE, Pragati Engineering College, Surampalem, India

Sridevi Thiagarajar College of Engineering, Madurai, India

Ivana Strumberger Singidunum University, Belgrade, Serbia

P. Sudhakar ECE Department, JNTUH College of Engineering, Kukatpally, Hyderabad, India;
ECM Department, Sreenidhi Institute of Science and Technology, Hyderabad, India

S. Tejashree Department of Electronics and Communication, Dayananda Sagar College of Engineering, Bengaluru, India

Susara S. Thenuwara Department of Computational Mathematics, University of Moratuwa, Moratuwa, Sri Lanka

Guru Sree Ram Tholeti Department of CSE, Koneru Lakshmaiah Education Foundation, Vaddeswaram, India

Tahmina Akter Tisha Department of Computer Science and Engineering, East West University, Dhaka, Bangladesh

Prakhar Tiwari Ingenious E-Brain Solutions, Gurugram, Haryana, India

Duc Chung Tran FPT University, Ho Chi Minh City, Vietnam

Aeshana Udadeniya Faculty of Information Technology, University of Moratuwa, Katubedda, Sri Lanka

T. P. Umadevi Department of Computer Science, JBAS College for Women (Autonomous), Chennai, India

Aruna Varanasi SureIT Solutions Inc, Hyderabad, India;
Sreenidhi Institute of Science and Technology, Hyderabad, India

Jyoti Vashishtha Guru Jambheshwar University of Science and Techonology, Hisar, Haryana, India

T. G. Vibha Department of Electronics and Communication, Dayananda Sagar College of Engineering, Bengaluru, India

V. Vimal kumar Applied Electronics and Instrumentation, Rajagiri School of Engineering and Technology, Kochi, India

Sunil Kumar Vishwakarma Department of Computer Science and Engineering, MANIT, Bhopal, India

Nisal Waduge Faculty of Information Technology, University of Moratuwa, Katubedda, Sri Lanka

W. A. C. Weerakoon Department of Statistics and Computer Science, University of Kelaniya, Kelaniya, Sri Lanka

Shakthi Weerasinghe Faculty of Information Technology, University of Moratuwa, Katubedda, Sri Lanka

Akkem Yaganteeswarudu SureIT Solutions Inc, Hyderabad, India; Sreenidhi Institute of Science and Technology, Hyderabad, India

Mohammed N. Yousif Northern Technical University, Mosul, Iraq

Miodrag Zivkovic Singidunum University, Belgrade, Serbia

Captcha-Based Defense Mechanism to Prevent DoS Attacks

G. Muneeswari and Antony Puthussery

Abstract The denial of service (DoS) attack, in the current scenario, is more vulnerable to the banking system and online transactions. Conventional mechanism of DoS attacks consumes a lot of bandwidth, and there will always be performance degradation with respect to the traffic in any of the communication networks. As there is an advent over the network bandwidth, in the current era, DoS attacks have been moved from the network to servers and API. An idea has been proposed which is CAPTCHA-based defense, a purely system-based approach. In the normal case, the protection strategy for DDoS attacks can be achieved with the help of many session schedulers. The main advantage is to efficiently avoid the DoS attacks and increase the server speed as well as to avoid congestion and data loss. This is majorly concerned in a wired network to reduce the delays and to avoid congestion during attacks.

Keywords DoS attack · Captcha-based mechanism · Security threat · Network topology · Authentication

1 Introduction

Denial of service attacks always in the next generation attacks category to destroy the entire network built for commercial Web sites. Most of the hackers in cyber-world steal the information through DoS attacks. The main concern with DoS attacks is many computational devices that make up the IoT world are forced to close the target commercial Web sites due to DoS attacks. Mostly, a DoS attack uses a single

G. Muneeswari (✉)
Department of Computer Science and Engineering, School of Engineering and Technology,
CHRIST (Deemed to be University), Bangaluru, India
e-mail: muneeswari.g@christuniversity.in

A. Puthussery
Department of Computational Sciences, School of Science, CHRIST (Deemed to be University),
New Delhi, India
e-mail: frantony@christuniversity.in

S. Shakya et al. (eds.), *Proceedings of International Conference on Sustainable Expert Systems*, Lecture Notes in Networks and Systems 176,
https://doi.org/10.1007/978-981-33-4355-9_1

1

computer and a single IP address to attack its target Web sites; thus, it is easy for us to protect against this. In the modern world, the DoS attack is rejecting the service from the customers, and it is a very big threat to commercial transactions over the communication network. Conventional threat degrades the performance of the system in terms of the quality of service, and the bandwidth around the cluster of computers also gets affected. To detect these types of attacks, many security measures had been imposed over the network of systems. Such mechanisms identify the attacks by some traffic control implementations and frequency of patterns during the data transfer through routers held between the source and the destination. Earlier DoS attacks were targeted only in the networks making it easy to design a protocol for the network, but currently, the server resources and certain applications are vulnerable to the hackers. Pertaining to the application-based DoS attacks concerned, since it is mainly causing problems at the end system, whatever network monitoring system implemented at the router level does not give an advantage over the threat. In this paper, two mechanisms are implemented to overcome the DoS attacks: First mechanism deals with every activity in the network incorporated with session validation, and the second mechanism is designed with captcha-based defense mechanism. In session scheduler, every user is authenticated by solving some puzzles, and these schedulers are designed as part of the DoS shield. The main disadvantage is the overhead incurred during every session validation. When the network traffic is tremendously high, then the captcha-based defense mechanism will introduce additional delays and more overhead through interactive-based systems. Many of the online transactions with a specific application system do not employ a captcha-based security system which is simple to implement and does not require major changes in the existing system.

The actual organization of the paper is elaborated as follows. Section 2 deals with the existing work with a literature survey, and Sect. 3 throws limelight on the system methodology adopted. Experimental results are described in Sect. 4, and the conclusion has been given in Sect. 5.

2 Literature Survey

Whenever a new application is created, designed, and implemented, there should be a mechanism that must be incorporated to handle [1] the effect of denial of service attack. Considering the application layer of the network, the work in this paper [2] has given all the possibilities over the protocol-related issues, and the system is integrated with non-intrusive complaint. The attacks reduce the speeding of the transfer of data and thus introduce network delay affecting the overall performance of the system. The idea [3] of snort detection has proposed a wide knowledge in preventing DoS attacks which handles a lot of requests to the servers. Captcha-based system is indicated here that differentiates the original client and intruder system with the help of the IP addresses. Here [4], a novel approach called live baiting based on group testing theory was proposed to manage DoS attacks. This novel mechanism can be implemented in an online environment where millions of requests sent from a client and that can be in differentiable from the attacker's request. While considering the novel attacks in the current scenario, even though a lot of research [5, 11, 13] has been done, it is very difficult to distinguish malicious traffic with the normal network flow. A confidence-based filtering method [9] is a new novel mechanism to prevent the threat in cloud computing.

In this research [6, 8, 10], some set of taxonomies are proposed foreseeing the different type of attacks that could occur in the network and provided a possible solution for that. In [7], F. Kargl et al. provided a mechanism for protecting a huge number of Web servers from the conventional DoS attacks, and with the help of unique coding technique [12], provided a solution for DoS attacks using IP Traceback. In [14], Soft computing-based autonomous low rate DDOS attack detection and security for cloud computing is proposed which gives a better result with respect to vulnerability in cloud-based systems. The solution provided in [15] is based on machine learning algorithms for DDOS attack detection in a telecommunication network which has the training data set obtained from the real-time data set.

3 Working Methodology

The proposed methodology adopted in this paper is to detect DoS attack using captcha-based technique. In the simulation, around ten clients had been defined along with the server definition. The server is implemented with the full functional server monitoring along with the captcha verification. Every client node is illustrated with several parameters including an IP address, the current status of the client, and the port number defined for a specific topology. Mutual authentication with the help of conventional authentication algorithm like MD5 has been incorporated in the client–server communication. Once the client is authenticated, it is allowed to initiate the data transfer either to the destination node or to the server node. The flexibility of scalability makes the node addition and deletion most easily.

Throughout the system design, a centralized database is maintained by the server system in the network. It contains all the details about the client nodes including the major authentication and encryption protocols details. Every node can be identified in the centralized database by the server with the help of an IP address. Each time when the client sends the request as shown in Fig. 1, it is verified by the server in the database, and the request is responded consecutively.

Since the protocol is built on top of the transport layer, the corresponding information has to be added in every packet transferred by the client node. The underlying protocol once receives the packet gets the header and forwards the packet to the upper layers for further investigations. In the second module of the proposed methodology, all the client activities are monitored by the server, and it is created by the normal IP address identification as shown in Fig. 2. Every network topology is pre-initialized with a traffic threshold and the pattern of traffic.

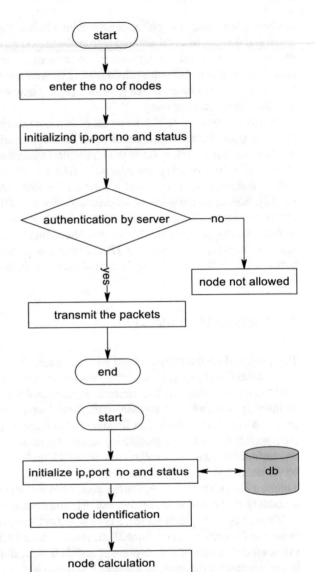

Fig. 1 Creation of client nodes

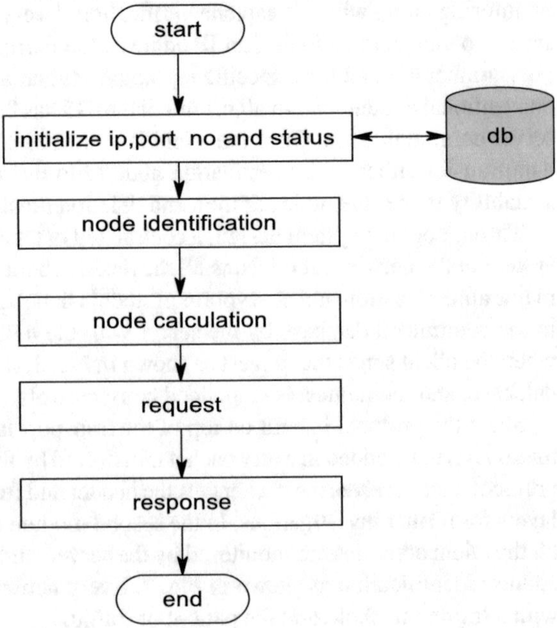

Fig. 2 Creation of server node

Whenever the system detects a change in the threshold value due to the DoS attack, then the corresponding client node is removed from the network topology, and all the relevant information about the node is deleted from the centralized database by the server. Initially, the packet will be discarded and later removal of the node from the network will take place. This ensures the detection of DoS attacks from the respective client machine as well as from the network in a simple way.

Further to the enhancement over the server monitoring as shown in Fig. 3, a more efficient captcha method is also incorporated by the server. According to this mechanism, captcha is generated with the corresponding node id, process id. The server checks whether this captcha is within the threshold or not. If it does not fall under the threshold, then it identifies that the client node is trying to make a DoS attack. Further, the node will be removed from the centralized database. The monitoring process log will often be maintained in the database on the server side.

Predominantly, every node in the network topology that is intended to transfer the data has to generate a distinct captcha depicted in Fig. 4. As mentioned earlier, every unique captcha consists of two different components such as node id and process id. Node id can be the combination of IP address and port number. For the identification of the DoS attack, it is needed to find out the node along with the type of process which caused the threat.

Fig. 3 Server monitoring

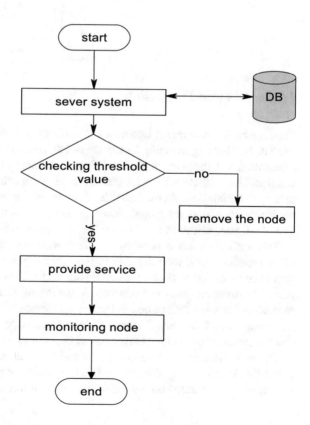

Fig. 4 Captcha generation

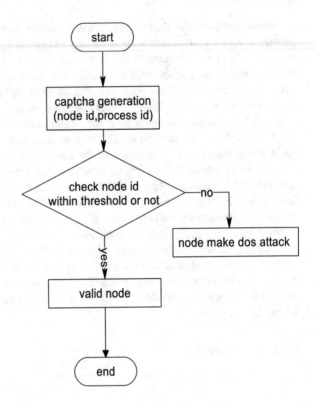

4   Results and Discussion

The captcha mechanism for DoS attack is implemented with the help of a simulation system. This is designed using Java swing and SQL Server 2000. Initially, all the node registration with the given network topology is initiated with the server-side database creation. The database is the centralized repository with all the details about the client nodes are maintained. At the initial level, node, IP address, and port number of all the legitimate clients are stored. When the node wants to initiate any data transfer first of all, authentication certification has to be generated by the server.

This authentication is done by verifying with the centralized database for the client identity. Later, with the help of the captcha generation, DoS attacks can be prevented by checking the frequency of patterns in the network transfer. Every time client transmission status is maintained in the log for future investigation. Figure 5 represents the node registration in the network topology, and server maintains this information in the database. Figure 6 represents the node registration in the network, and the corresponding client is recognized by the server node.

Figure 7 indicates captcha generation and verification procedures. The captcha generation is initiated by the client when it is ready for transmission of data. Once the captcha is generated on the legitimate client, it has to be subsequently verified

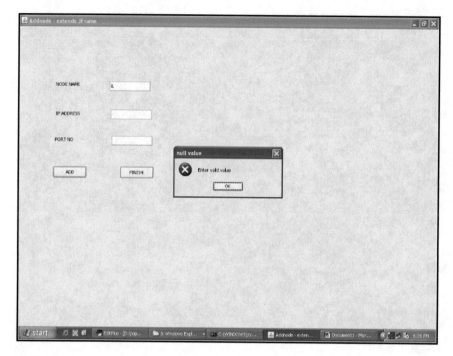

Fig. 5 Node registration

by the server. This verification finally authenticates the client by the servers on the network. For the simulation evaluation, Table 1 parameters and values are taken into account, and the results are obtained for TCP packets. The packet size is taken as 1024 bytes, and the payload could be up to 500 bytes. The network traffic is measured every 100 s for the client–server communication.

The comparison of attack rate versus average detection time (ADT) is depicted in Fig. 8, and it shows that captcha-based detection mechanism time is 32% lesser compared to the existing OTP based protocols and machine learning-based algorithms.

Similarly, the performance metric like Accuracy, Precision, Recall, and F-measure are compared for the same three algorithms, and captcha-based mechanism outperforms the other existing algorithms as shown in Fig. 9.

5 Conclusion

In the current scenario, most of the users are trying to access the Internet for commercial transaction purpose. As a result, every access over the network is vulnerable to a security threat, and the idea proposed in the paper is to give a solution for preventing DoS attacks. Since malicious transmission of data is difficult to be differentiated from

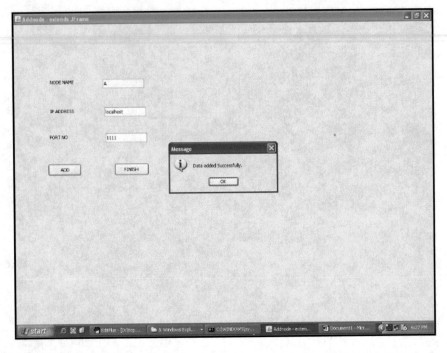

Fig. 6 Successful registration of clients

Fig. 7 Captcha generation
and verification

Name

Michelle

Email

michelle@domain.com

Subject

Inquiry about pricing

Message

Hi, I am intersted in …

Retype the characters from the picture

Wrong code. Try again please.

Submit

Table 1 Simulation parameters and values for TCP packets

S. No.	Parameter	Value
1	Type of packet	TCP
2	Traffic flow in the network	1000
3	Data size	500 Bytes
4	Size of each packet window	100
5	Packet size	1024 KB

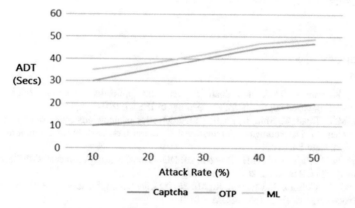

Fig. 8 Attack rate versus ADT comparison

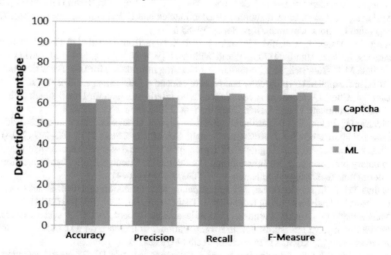

Fig. 9 DoS attack detection percentage comparison

the normal data transfer, the security mechanism is implemented on the authentication side. The captcha-based mechanism incorporated on the client and server side not only provides mutual authentication but also launches a defense mechanism over the DoS attacks. The novelty in this paper is mainly incorporated over the double-level hierarchical security mechanism wherein with the normal mutual authentication with OTP and user verification in commercial Web sites, a more advanced level of captcha has been implemented which ensures much more security than the existing mechanisms.

References

1. Xuan, Y., Shin, I., Thai, M.T, Znati, T.: Detecting application denial-of-service attacks: a group-testing-based approach. In: Proceedings of IEEE (2010)
2. Thai, M.T., Xuan, Y., Shin, I., Znati, T.: On detection of malicious users using group testing techniques. In: Proceedings of International Conference on Distributed Computing Systems (ICDCS), New York (2018)
3. Lanke, N.M., Raja Jacob, C.H.: Detection of DDOS attacks using snort detection international. J. Emerg. Eng. Res. Tech. 2(9), 13–17 (2014)
4. Khattab, S., Gobriel, S., Melhem, R., Mosse, D.: Live baiting for service-level DoS attackers, In: Proceedings of IEEE INFOCOM (2018)
5. Saxena, R., Dey, S.: Cloud shield: effective solution for DDoS in cloud. In: IDCS 2015, Lecture Notes in Computer Science (LNCS), vol. 9258, pp. 3–10. Springer, Berlin Heidelberg (2015)
6. Mirkovic, J., Reiher, P.: A taxonomy of DDoS attack and DDoS defense mechanisms. ACM Sigcomm Comput. Commun. Rev. 34(2), 39–53 (2004)
7. Kargl, F., Maier, J.,Weber,M,: Protecting Web Servers from Distributed Denial of Service Attacks, In:Proc. 10th Int'l Conf. World Wide Web (WWW '01), pp. 514–524, (2011)
8. Atallah, M.J., Goodrich, M.T.,Tamassia, R.: Indexing Information for Data Forensics, In:Proc. Int'l Conf. Applied Cryptography and Network Security (ACNS), pp. 206–221, (2005)
9. Dou, W., Chen, Q., Chen, J.: A confidence-based filtering method for DDoS attack defense in cloud environment. Future Gener. Comput. Syst. 29(7), 1838–1850 (2012)
10. Meena, D., Jadon, R.S.: Distributed denial of service attacks and their suggested defense remedial approaches. Int. J. Adv. Res. Comput. Sci. Manage. Stud. 2(4), 183–197 (2014)
11. Katkar, V.D., Bhatia, D.S.: Lightweight approach for the denial of service attacks using numeric to binary preprocessing, In: International Conference on Circuit System Communication and Information Technology Application (CSCITA) in IEEE (2014)
12. Sathya Priya, J., Ramkrishnan, M., Rajagopalan, S.P.: Detection of DDoS attacks using IP traceback and network coding technique. J. Theor. Appl. Inf. Tech. 99–106 (2014)
13. Muneeswari, G., Antony Puthussery, Multilevel security and dual OTP system for online transaction against attacks. In: IEEE 3rd International Conference on IOT in Social, Mobile, Analytics and Cloud, SCAD Institute of Technology (2019)
14. Mugunthan, S.R.: Soft computing based autonomous low rate DDOS attack detection and security for cloud computing. J. Soft Comput. Parad. (JSCP) 1(02), 80–90 (2019)
15. Smys, S.: DDOS attack detection in telecommunication network using machine learning. J. Ubiquit. Comput. Commun. Tech. (UCCT) 1(01), 33–44 (2019)

Prognostic of Depression Levels Due to Pandemic Using LSTM

Shahana Bano, Yerramreddy Lakshmi Pranathi, Gorsa Lakshmi Niharika, and Gorsa Datta Sai Sreya

Abstract Depression is a medical illness that affects the way you think and how you react. It is a serious medical issue that impacts the stability of the mind. Depression occurs at many stages and situations. With the help of classification, the stage of depression the person is in can be tried to categorize. Nowadays, many users are sharing their views on social media, and it became a platform for knowing people around us. From the data that is shared on social media, the depressing posts are being classified using machine learning techniques. With these reports collected, the depressed person might be helped from making any sudden decisions. So, in our research study, the large datasets of the people in depression during the COVID-19 pandemic situations is analyzed and not in pandemic situations. Here to analyze the data, the neural networks have been trained with the current pandemic analysis report, and it has given a prediction that the people are less likely to get depressed when they are not in a pandemic situation like COVID-19.

Keywords Depression · Long short-term memory · Recurrent neural network · Preprocessing · Feature extraction · Visualization

S. Bano · Y. L. Pranathi · G. L. Niharika (✉)
Department of CSE, Koneru Lakshmaiah Education Foundation, Vaddeswaram, India
e-mail: niharikagorsa2000@gmail.com

S. Bano
e-mail: shahanabano@icloud.com

Y. L. Pranathi
e-mail: pranathi.yr@gmail.com

G. D. S. Sreya
Department of CSE, Pragati Engineering College, Surampalem, India
e-mail: dattasaisreya@gmail.com

© The Editor(s) (if applicable) and The Author(s), under exclusive license
to Springer Nature Singapore Pte Ltd. 2021
S. Shakya et al. (eds.), *Proceedings of International Conference on Sustainable Expert Systems*, Lecture Notes in Networks and Systems 176,
https://doi.org/10.1007/978-981-33-4355-9_2

1   Introduction

Depression is a common mental disorder. It is seen in all ages nowadays. According to the World Health Organization (WHO), 264 million people of all ages suffer from depression. Depression affects the functioning of the nervous system. It lasts very long. The major problem that comes with depression is the lack of proper diagnosis for this. There are different kinds and stages of depression. Some people did not identify the correct stage or suggest the correct type. The extreme stage of depression can lead the person to attempt suicide. In a study conducted, it is known that 7.2% people are affected with this disease starting at the age of 12. So, there is no age limitation for the depression. But it is considerably high in medieval people. The major depressive period is the highest among individuals between the age of 18 and 25 according to a study conducted. Women are said to have 8.7% depression rate while men are having a depression rate of 5.3%. This study shows that women are getting more depressed than men due to many factors.

Depression first affects the mood of the person [1], and it makes the person feel sad or depressed. It might affect the body as the person changes the appetite, losses interest in activities, and has trouble in sleeping. This may also lead to suicidal thoughts. Depression is different from having sadness and grief. Depression is the second most common illness in the world said by the World Health Organization (WHO). It is more prevalent in the world. It can be tested through direct physical interaction with the psychiatrist and analysis provided by them. It is believed that depression is caused due to chemical imbalance caused in the brain, but it is not caused due to that [2]. Many other factors can cause depression. It can be caused due to the stressful life, mental illness, and medications which add on lead to depression.

For this extreme point of the study, people interactions gave us a report of reasons for the cause of the depression which is joblessness, loss of loved ones, health problems, money issues, etc. The ultimate stage of their depression is that they start questioning their existence in life and starts comparing their life with others.

Many researchers made a strategical analysis by taking social media posts [3, 4] and measuring the depression levels by different machine learning approaches [5]. This related work on the depression and getting an analysis of the data is the main approach of our research work. By comparing the report of depression levels in pandemic days and predicting that the depression levels would be less if there is no pandemic.

So, for all this analysis, the neurons have been trained with the current report of depression levels at the pandemic stage COVID-19. Based on those training, our model predicts that the depression rates will be less if there is no pandemic situation. In the final stage of predictions, the output is got in the form of graph model like a visualization report. Here, long short-term memory (LSTM) algorithm is used for training and building our model. This model can be used for prediction of stock prices [6] which recurrently checks the real price of stocks.

The LSTM RNN makes the classification of timely changing factors like the psychological series [7] which changes according to the reason of depression that they are in. After an in-depth survey on LSTM [8, 9] RNN methods, it is found to implement this technique in providing statistics of depression levels by using the count obtained through the examination of depression.

2 Procedure

2.1 Importing All the Packages

To process our model, the required packages are needed to import for training the data in the LSTM algorithm. Here, in this model, NumPy has been imported for calculating high mathematical functions in making the data values into matrices and also for multidimensional array. The pandas, matplotlib, and sklearn for making regression and clustering algorithms have also been imported for the representation of the data in a graphical way. LSTM have been imported for this model for training the previous dataset sequentially to predict the next dataset within the provided constraints.

2.2 Importing Datasets

For making sure of our model, the dataset have been imported which have been got from the recent survey done with some individuals during the pandemic situations. This survey gave the data rate which contains the depression levels and their reasons for the depression. The dataset taken contains depression reason with the depression level value. The survey was taken during the pandemic situations like COVID-19.

2.3 Preprocessing and Feature Extraction

After importing packages and datasets, need to do scaling, centering, and also normalize the raw data in the dataset. The whole dataset is not needed for making predictions of depression levels if there is no pandemic situation. So, the data will be preprocessed by localization of depression levels of individuals and to scale it for feature extraction of raw data. Since this preprocessing and feature extraction role is the keen step for the entire model to progress the raw data.

2.4 Working Principle

When started to import all the necessary packages, datasets, preprocessing is the crucial stage for the extraction of the data and to allocate for sequential training of the available dataset. LSTM algorithm is a part of the recurrent neural network (RNN) [10]. This algorithm solves our daily ways of approaches by predicting the previous memory. For example, take cricket world cup matches which are held every year, having all the records of previous matches, if trying to predict which team is going to win this year's match, can be predicted it by the performance of that country in previous matches. It all happens in the gap while thinking of the winning possibility of that country. LSTM model here checks sequentially with every previous data.

The model is made to train with the dataset which selects the depression value as an attribute. This code of LSTM is trained with the dataset along with its reason for depression. So for further analysis of data, it keeps on training the data with depression value and reason for depression.

Here, in our model of code, the training of the previous report is made on depression levels of the individuals during the pandemic situation. It stores the data of the particular person at that particular time. It is nothing like the history of depression levels which are trained in every iteration. For the LSTM [11] model, it stores the long-term memory, and it only happens because of their default nature. RNN makes the recurrent analysis of each neuron module with the data using every single layer [12]. The main point in LSTM is the cell state C_{S-1} which adds or removes the data if needed. The depression levels will not be high in the future if there is no pandemic situation. The possibility is that it can be high or cannot be high, but it will not be like '0' or '1.' So, these acts as gates for the process whether the data can be added as depression values at that period of time. This is like predicting the situation with no pandemic. So, if the model is trained with the depression values during pandemic days, then when tested with no pandemic days, the model forgets the old depression value and checks whether it can be the new value of depression or not, and finally adds that value as a new cell state C_S [13].

$$\text{forget layer}_s = \sigma\left(W_f.\left[h_{s-1}, x_s\right] + b_f\right) \tag{1}$$

Here, LSTM has the interactive layers where continuous or recurrent analysis is made with depression values. The state value is got by training the depression values of the pandemic situation based on their reason. The first step of the LSTM model is to decide whether to take the value for future prediction of data. So through Eq. 1, the model is being made to get into a decision either to keep the value for prediction of future or to neglect it. It is also called a decision taking layer as h_{s-1}, x_s together makes with sigmoid function [13].

Equation 2 acted as the input layer for taking data. This layer only takes values that are to be added to the data. However, in forget layer, activation vector is multiplied to the cell state and can set values to zero.

Equations 1, 2, and 3 used b_f, b_i, b_c which is continuously learned, and initialized with random numbers by neurons training biases. However, W_f, W_i, W_c are considered as neurons weights matrices to train the data with initializing random numbers.

$$i_s = \sigma\left(W_i.[h_{s-1}, x_s] + b_i\right) \qquad (2)$$

The tanh layer in Eq. 3 is creating a function to add the value taken from the input layer to add in a new variable with a new state. For the final upgrading of a new state value, the new updated state value is being calculated through forget and input layer values (Eq. 4)

$$\widetilde{C}_s = \tanh\left(W_C.[h_{s-1}, x_s] + b_C\right) \qquad (3)$$

$$C_s = f_s * C_{s-1} + i_s * \widetilde{C}_s \qquad (4)$$

Fundamental analysis, statistics, linear regression, and all types of analysis can also be used, but they did not give us the correct idea on daily changes of mental health. So, this LSTM algorithm uses the recurrent approach of each neuron and predicts the next value to be placed. By this, the human changes recurrently based on their thoughts if the recurrent neural network approach is used to train every neuron with all the previously available data. If the person is more depressive in a pandemic situation and it tries to predict the depression rates when there is no pandemic situation, the basic motto of prediction is to identify the person's depressive levels in no pandemic situation based on the reason that they are into depression at pandemic days.

Step 1   Start.
Step 2   Import all the necessary packages, and the dataset is taken from a recent survey.
Step 3   Preprocessing and feature extraction of the dataset.
Step 4   Building the LSTM algorithm model.
Step 5   Training the model by passing through each neuron on the given data.
Step 6   Test the model with depression values when there is no pandemic situation.
Step 7   Displays the graph of predicting depression values after an analysis was done using the LSTM model.
Step 8   Prediction is done.
Step 9   Repeat the steps 1, 2, 3, 4, 5, 6, 7, and 8 if you want to see a random change in the depression values by passing the updated reports or datasets.
Step 10   End.

3 Flowchart

To visualize our process of approach, a flowchart has been given (Fig. 1). For the overview of our model, the packages and datasets required to train the LSTM model have been imported. Importing these packages in Python helps to do all types of high-level mathematical functions. After this, scaling, centering the dataset for feature extraction, will be performed, and also this stage is named for preprocessing the dataset. Now, the main stage of implementation takes place that is building the

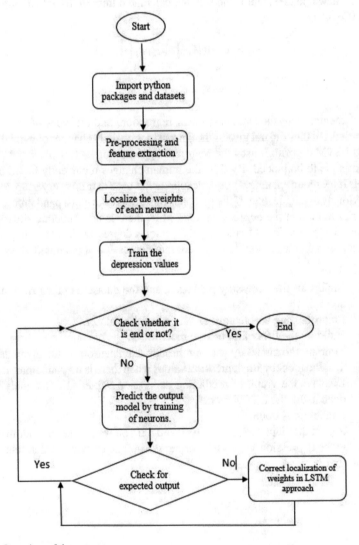

Fig. 1 Overview of the process

LSTM model. It localizes the weights of each neuron that needed to be trained for each recurrent iteration. A recurrent neural network predicts the possible output for the next stage by comparing it with all the previous data. If it does not identify the depression levels of a person, it corrects the weight of neurons and starts predicting the depression value again. This recurrent approach of the model gives the correct predictions by calculating the long-term memory of the model. Finally, it gives the prediction of depression value will be less in no pandemic situations like COVID-19.

4 Results

Figure 2 is a snapshot of our dataset that is taken to train our model. Here, we have trained our model that has been trained on the reason for the depression. Based upon their reason for depression during pandemic days, we have trained our model; if the reason does not have a long-lasting end, then the person's depression value will be the same irrespective of the pandemic situation. Predicting that if the reason is solved in the future, then depression values may decrease by taking all the possibilities in the approach of predicting depression values in the future.

Figure 3 shows preprocessing the dataset, and it is scaled for the training of dataset in the next stage of the LSTM model of approach. Figure 4 shows training each neuron with the passed dataset along with the reasonable value of depression and compares it with test data in a later stage. Then, it gives predictions for the depression levels when there is no pandemic situation like COVID-19 (Fig. 5).

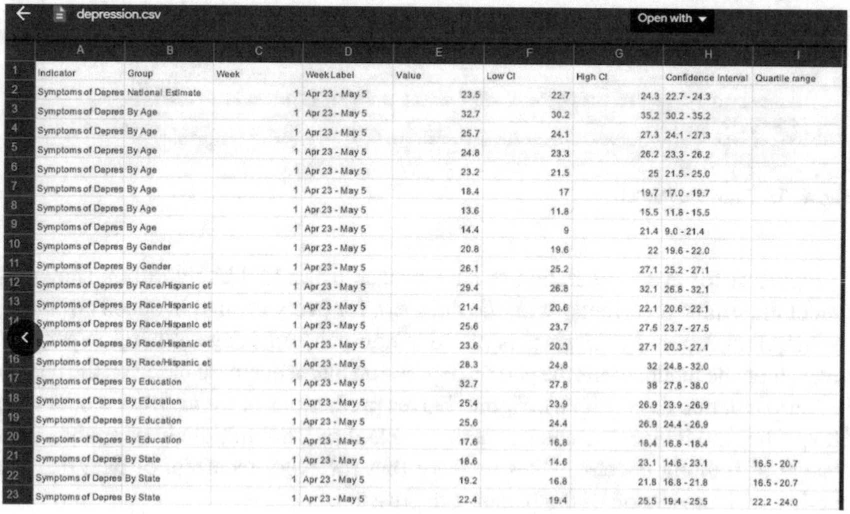

	Indicator	Group	Week	Week Label	Value	Low CI	High CI	Confidence Interval	Quartile range
		A	B	C	D	E	F	G	H	I
1	Indicator	Group	Week	Week Label	Value	Low CI	High CI	Confidence Interval	Quartile range
2	Symptoms of Depres	National Estimate	1	Apr 23 - May 5	23.5	22.7	24.3	22.7 - 24.3	
3	Symptoms of Depres	By Age	1	Apr 23 - May 5	32.7	30.2	35.2	30.2 - 35.2	
4	Symptoms of Depres	By Age	1	Apr 23 - May 5	25.7	24.1	27.3	24.1 - 27.3	
5	Symptoms of Depres	By Age	1	Apr 23 - May 5	24.8	23.3	26.2	23.3 - 26.2	
6	Symptoms of Depres	By Age	1	Apr 23 - May 5	23.2	21.5	25	21.5 - 25.0	
7	Symptoms of Depres	By Age	1	Apr 23 - May 5	18.4	17	19.7	17.0 - 19.7	
8	Symptoms of Depres	By Age	1	Apr 23 - May 5	13.6	11.8	15.5	11.8 - 15.5	
9	Symptoms of Depres	By Age	1	Apr 23 - May 5	14.4	9	21.4	9.0 - 21.4	
10	Symptoms of Depres	By Gender	1	Apr 23 - May 5	20.8	19.6	22	19.6 - 22.0	
11	Symptoms of Depres	By Gender	1	Apr 23 - May 5	26.1	25.2	27.1	25.2 - 27.1	
12	Symptoms of Depres	By Race/Hispanic et	1	Apr 23 - May 5	29.4	26.8	32.1	26.8 - 32.1	
13	Symptoms of Depres	By Race/Hispanic et	1	Apr 23 - May 5	21.4	20.6	22.1	20.6 - 22.1	
14	Symptoms of Depres	By Race/Hispanic et	1	Apr 23 - May 5	25.6	23.7	27.5	23.7 - 27.5	
	Symptoms of Depres	By Race/Hispanic et	1	Apr 23 - May 5	23.6	20.3	27.1	20.3 - 27.1	
16	Symptoms of Depres	By Race/Hispanic et	1	Apr 23 - May 5	28.3	24.8	32	24.8 - 32.0	
17	Symptoms of Depres	By Education	1	Apr 23 - May 5	32.7	27.8	38	27.8 - 38.0	
18	Symptoms of Depres	By Education	1	Apr 23 - May 5	25.4	23.9	26.9	23.9 - 26.9	
19	Symptoms of Depres	By Education	1	Apr 23 - May 5	25.6	24.4	26.9	24.4 - 26.9	
20	Symptoms of Depres	By Education	1	Apr 23 - May 5	17.6	16.8	18.4	16.8 - 18.4	
21	Symptoms of Depres	By State	1	Apr 23 - May 5	18.6	14.6	23.1	14.6 - 23.1	16.5 - 20.7
22	Symptoms of Depres	By State	1	Apr 23 - May 5	19.2	16.8	21.8	16.8 - 21.8	16.5 - 20.7
23	Symptoms of Depres	By State	1	Apr 23 - May 5	22.4	19.4	25.5	19.4 - 25.5	22.2 - 24.0

Fig. 2 Snapshot of our trained dataset

Fig. 3 Preprocessing and
feature extraction snapshot

```
[ ]  training_scaled

 ⌑  array([[0.35555556],
           [0.58271605],
           [0.40987654],
           ...,
           [0.57037037],
           [0.42962963],
           [0.58024691]])
```

```
[ ]  regressor.fit(x_train,y_train,epochs = 100, batch_size = 32)
```

```
 ⌑  Epoch 1/100
    1199/1199 [==============================] - 5s 4ms/step - loss: 0.0532
    Epoch 2/100
    1199/1199 [==============================] - 4s 3ms/step - loss: 0.0221
    Epoch 3/100
    1199/1199 [==============================] - 4s 3ms/step - loss: 0.0239
    Epoch 4/100
    1199/1199 [==============================] - 4s 3ms/step - loss: 0.0215
    Epoch 5/100
    1199/1199 [==============================] - 4s 3ms/step - loss: 0.0211
    Epoch 6/100
    1199/1199 [==============================] - 4s 3ms/step - loss: 0.0214
    Epoch 7/100
    1199/1199 [==============================] - 4s 3ms/step - loss: 0.0214
    Epoch 8/100
    1199/1199 [==============================] - 4s 3ms/step - loss: 0.0217
    Epoch 9/100
    1199/1199 [==============================] - 4s 3ms/step - loss: 0.0210
    Epoch 10/100
    1199/1199 [==============================] - 4s 3ms/step - loss: 0.0212
    Epoch 11/100
    1199/1199 [==============================] - 4s 3ms/step - loss: 0.0212
    Epoch 12/100
    1199/1199 [==============================] - 4s 3ms/step - loss: 0.0200
    Epoch 13/100
    1199/1199 [==============================] - 4s 3ms/step - loss: 0.0207
    Epoch 14/100
    1199/1199 [==============================] - 4s 3ms/step - loss: 0.0206
    Epoch 15/100
    1199/1199 [==============================] - 4s 3ms/step - loss: 0.0215
```

Fig. 4 Training the model

Figure 5, for every prediction of an approach using LSTM algorithm, shows its
model that trains recurrently with the data. So, for every change in epochs training,
the output of depression levels gets varied in non-pandemic days. When 25 epochs is
given, then the model trains the person depression might decrease based on the reason
in non-pandemic days (Fig. 8). Figure 7 gives predictions of training the model with
50 epochs, and it also fluctuates data in raise and fall. So, this study of analytics on
change of training epochs makes us learn that for every recurrent of data that are
varying with the reason of the person's depression.

Hereby, changes in the prediction of data have been given with training data for
every change of epochs like 100 (Fig. 5), 75 (Fig. 6), 50 (Fig. 7), and 25 (Fig. 8).

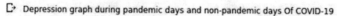

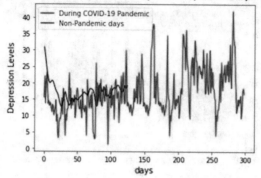

Fig. 5 Prediction of depression levels in non-pandemic days with 100 epochs

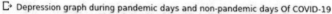

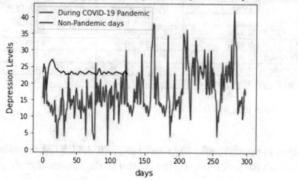

Fig. 6 Prediction of depression levels in non-pandemic days with 75 epochs

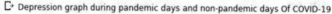

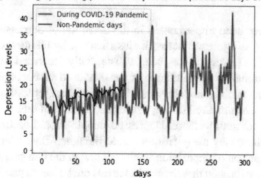

Fig. 7 Prediction of depression levels in non-pandemic days with 50 epochs

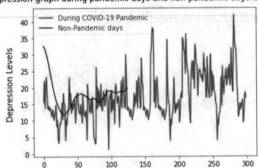

Fig. 8 Prediction of depression levels in non-pandemic days with 25 epochs

Type	Output shape	Number of parameters
Input	359, 1	0
LSTM(Forward)	125, 60	1470
LSTM(Backward)	125, 60	1470
Dropout	50	0
Epochs	100, 75, 50, 25	0
Fully connected (Sigmoid)	1	51

Fig. 9 Representation analytics of model

These changes in epochs make the model to train the LSTM algorithm and to learn for new data with each iteration on the reason of depression that they are in (Fig. 9).

5   Conclusion

Many methods have been implemented in the detection of depression. Depression can be detected by using neural network classification by taking the data available from social media. Depression has been substantially increased over a period of time. From the outbreak of the pandemic, it has increased exponentially than before, due to many factors such as hopelessness, loneliness, and joblessness. The data of depression have been compared during pandemic days. With this, it can be said that depressed people are getting more. Depressed people should be treated and taken care of to make them stay away from suicidal thoughts. Based on the reason for depression, the situation of depression might get low in the non-pandemic situation. From the study, it can be said that depression levels are higher in pandemic days than compared to normal days.

6 Future Scope

By considering this scenario of research, an app can also be built for further improvement of the research work. The depression levels of data from the social media posts [14–16] can be used to analyze the depression levels of the person. Through the analysis of the levels of depression that the person is in, the person can be helped and prevented from taking any serious action, and suggested something that can help the person.

References

1. Ramalingam, D., Sharma, V., Zar, P.: Study of depression analysis using machine learning techniques. Int. J. Innov. Tech. Explor. Eng. (IJITEE) **8**, 188–191 (2019)
2. Patel, M.J., Khalaf, A., Aizenstein, H.J.: Studying depression using imaging and machine learning methods. NeuroImage Clin. **10**, 115–123 (2016)
3. Shuai, H., et al.: A comprehensive study on social network mental disorders detection via online social media mining. IEEE Trans. Knowl. Data Eng. **30**(7), 1212–1225 (2018)
4. Sonawane, N., Padmane, M., Suralkar, V., Wable, S., Date, P.: Predicting depression level using social media posts. Int. J. Innov. Res. Sci. Eng. Tech. **7**(5), 6016–6019 (2018)
5. Sridharan, S., Akila Banu, M., Bakkiyalakshmi, A., Buvana, P.: Detection and diagnosis on online social network mental disorders using convolutional neural network. Int. J. Eng. Sci. Comput. (IJESC) **8**, 16146–16150 (2018)
6. Pang, X., Zhou, Y., Wang, P., et al.: An innovative neural network approach for stock market prediction. J. Supercomput. **76**, 2098–2118 (2020)
7. Karim, F., Majumdar, S., Darabi, H., Chen, S.: lstm fully convolutional networks for time series classification. IEEE Access **6**, 1662–1669 (2018). https://doi.org/10.1109/ACCESS.2017.2779939
8. Liu, Y., Wang, Y., Yang, X., Zhang. L.: Short-term travel time prediction by deep learning: a comparison of different LSTM-DNN models. In: 2017 IEEE 20th International Conference on Intelligent Transportation Systems (ITSC), Yokohama, pp. 1–8 (2017). https://doi.org/10.1109/itsc.2017.8317886
9. Duan, Y., Lv Y, Wang, F.: Travel time prediction with LSTM neural network. In: 2016 IEEE 19th International Conference on Intelligent Transportation Systems (ITSC), Rio de Janeiro, pp. 1053–1058 (2016)
10. Zhu, X., Sobihani, P., Guo, H.: Long short-term memory over recursive structures. Int. Conf. Mach. Leatn. **37**, 1604–1612 (2015)
11. Hochreiter, S., Schmindhuber, J.: Long short-term memory. Neur. Comput. **9**(8), 1735–1780 (1997)
12. Gers, F.A., Schraudolph, N.N., Schmidhuber, J.: Learning precise timing with LSTM recurrent networks. J. Mach. Learn. Res. **3**, 115–143 (2003)
13. Jozefowicz, R., Zaremba, W., Sutskever, I.: An empirical exploration of recurrent network architectures. Int. Conf. Mach. Learn. (ICML) **37**, 2342–2350 (2015)
14. Li, A., Jiao, D., Zhu, T.: Detecting depression stigma on social media: a linguistic analysis. J. Affect. Disord. **232**, 358–362 (2018)
15. Aldarwish, M.M., Ahmad, H.F.: Predicting depression levels using social media posts. In: 2017 IEEE 13th International Symposium on Autonomous Decentralized System (ISADS), Bangkok, pp. 277–280 (2017)

16. Roshini, T., Sireesha, P.V., Parasa, D., Bano, S.: Social media survey using decision tree and Naive Bayes classification. In: 2019 2nd International Conference on Intelligent Communication and Computational Techniques (ICCT), Jaipur, India, pp. 265–270 (2019). https://doi.org/10.1109/icct46177.2019.8969058

An Efficient Design of 8 * 8 Wallace Tree Multiplier Using 2 and 3-Bit Adders

M. Sakthimohan and J. Deny

Abstract In VLSI, hardware architecture requires the multiplier unit as one of the important parts for arithmetic operation. A multiplier is a major component in many hardware architectures, so various experts are focusing their research in multiplier design to accomplish compact area, delay, and power. Numerous case studies were done for many architectures, in that the increased speed and low area are achieved through a reduction of partial products. One and only of the finest methods is Wallace tree multiplier (WTM). In this research article, Wallace tree 8 * 8 multiplier architecture is proposed, and it produces optimized area and delay. Our work targets structuring and execution of Wallace tree 8 * 8 multiplier utilizing VHDL language. Using limiting quantity of partial products, 2-bit and 3-bit adders are utilized in the 8-bit multiplier. In this work, 8 * 8 Wallace tree multiplier development is inspected and reproduced in XILINX Integrated Software Environment tool. In this 8-bit Wallace tree multiplier circuit, our primary objectives are to diminish the area of multiplier circuit and speed up multiplier routine.

Keywords Carry look-ahead adder · Wallace tree multiplier · 2-bit and 3-bit adders · VLSI · VHDL · XILINX

1 Introduction

Signal handling is an essential advance in sight and sound correspondence frameworks. Numerous application frameworks dependent on DSP, particularly the ongoing innovative optical correspondence frameworks, require amazingly quick handling of an enormous measure of computerized information. In signal processing applications, multiplication is a fundamental activity. By comparison with addition

M. Sakthimohan (✉) · J. Deny
ECE Department, Kalasalingam Academy of Research and Education, Krishnankoil, Tamil Nadu, India
e-mail: sakthimohan.phd@gmail.com

J. Deny
e-mail: j.deny@klu.ac.in

S. Shakya et al. (eds.), *Proceedings of International Conference on Sustainable Expert Systems*, Lecture Notes in Networks and Systems 176, https://doi.org/10.1007/978-981-33-4355-9_3

23

and subtraction computation, multiplication has a huge delay. To overcome this problem, Wallace multiplier gives an effective solution. In 1964, Wallace derived quick multiplication with the mixture of 2-bit and 3-bit adders.

Execution of the circuit relies upon the exhibition of the little pieces of the circuit. Along these lines, it is a troublesome and testing task the architect to structure the elite circuit. A wide range of parameters is considered in the exhibition of the circuit like area, power dissemination, engendering delay, and so on. Conservative area, low power scattering, and rapid are a primary worry of any circuit designer. A multiplier comprises three phases' initially generation of the partial product then the addition of the same and afterwards last phase addition. This paper focuses on the decrease of power utilization and latency of Wallace 8 * 8 tree multiplier. It is formed with the use of 2-bit, 3-bit adders and look-ahead carry adder. The main usage is to multiply whole numbers (unsigned). By using AND gate, it is possible to generate partial products after that adding generated partial products into carry look-ahead adder circuit. In the last phase, the overall totaling process is performed with carry propagate adder.

2   Related Work

Wang et al. [1] proposed non-volatile logic implemented using memristor-based MIG logic. A full viper of rapid is modified, in light of which a multiplier is given the selection of Wallace tree calculation. This approach has given a chance to investigate progressed processing designs contrasted and the old-style von Neumann design. Contrasted and the conventional multiplier was developed by the CMOS unit, and this plan can successfully understand the coordination of capacity and calculation, what is more, halfway tackle memory bottleneck issue.

Dr. Karuppusamy [2] proposed a fast and a low force multiplier with a diminished force utilization and upgraded speed of the activity, the paper proposes the Baugh Wooley multiplier with the altered circuit utilizing the rhythm virtuoso. The structure is mimicked utilizing the rhythm Specter by starting a simple plan condition. The outcome acquired demonstrates the ability of the Baugh Wooley multiplier.

Navin Kumar et al. [3] described calculation capacities are completed utilizing the multiplier, where it is seen as more force devouring part in the electronic circuits. The significant multiplier design utilizing KSA is recognized dependent on the investigation of execution with the multiplier engineering made out of changed surmised full viper. With the imminent advances in future, the significant multiplier might be upgraded and can be utilized at different places, for example, picture handling, video conferencing, and DSP.

Sundhar et al. [4] proposed 16 × 16 bit Wallace tree multiplier utilizing 154 blower engineering that has structured and integrated force analysis of 16 × 16 Wallace tree multiplier utilizing on Spartan 3 XC3S100E board and recreated in Xilinx ISE 14.5. The presentation of proposed multiplier with Kogge–Stone viper is contrasted and similar engineering of multiplier utilizing equal viper. It very well may be surmised

Fig. 1 Process
flow—multiplier

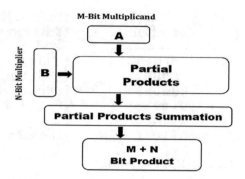

that 16×16 multiplier design utilizing 15–4 blower with Kogge–Stone viper is quicker contrasted with a multiplier with an equal viper.

Mukherjee et al. [5] presented the proposed counter-based GDI Wallace tree multiplier that shows better outcomes in low force utilization, region concern, and defer execution. The improvement will be more for higher no. of bits. The Wallace tree structure can be used for changed Booth Wallace tree multiplier circuit for further improvement in marked piece increases. Such on-chip multipliers can be executed distinctive convenient small-scale frameworks and MEMS processor units.

3 Basic Multiplication Practice

In 2-bit binary numbers, the multiplication process is carried out, first digit is called multiplicand, and the subsequent one is multiplier [6]. The multiplication procedure comprises the partial product of multiplier and multiplicand at the initial stage. In partial product on the off chance that the multiplier bit is "0," at that point, the multiplication result will be zero, and in the event that the multiplier bit is "1" at that point, the multiplication result will be same as multiplicand. After creation of partial product, following task in multiplication is addition. Square graph of multiplier (N-Bit) and multiplicand (M-Bit) is shown in Fig. 1.

A case of a multiplier and multiplicand in 4-bit and a product of two positive (unsigned) binary numbers radix 2 is shown in Fig. 2.

4 Proposed Work

4.1 Wallace Tree Multiplier

Speed of multiplier is relied upon the entire time taken for summation in partial products. Array multiplier is slower related to Wallace tree multiplier. Researcher

```
          1010              (This binary number represents 10 in decimal)
       x  1001              (This binary number represents 9 in decimal)
    -------------------
          1 0 1 0           (1010 x 1)
          0 0 0 0 x         (1010 x 0, shifted one position to left)
          0 0 0 0 x x       (1010 x 0, shifted two position to left)
       +1 0 1 0 x x x       (1010 x 1, shifted three position to left)

    -------------------
        1 0 1 1 0 1 0       (This binary number represents 90 in decimal)
```

Fig. 2 Multiplication process

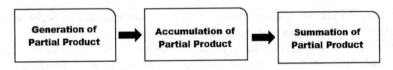

Fig. 3 Process flow—WTM

Wallace Chris (1964) presented a simple and straightforward scheme for adding the partial product bits with the equal utilizing tree of the look-ahead carry adder, which is known as "Wallace tree." Wallace tree multiplier utilizes look-ahead carry addition calculation (Fig. 3).

Figure 3 shows, Wallace tree multiplier consists of three important procedures. At the initial period, the partial product is generated by multiplier * multiplicand. In the second step, the 2-bit and 3-bit adder help in minimizing the generated partial products. At the third step, the final addition was done [7].

4.2 Wallace Tree 4 * 4 Multiplier

Figure 4 indicates two 4-bit multiplication. The numeric is specified as A and B and is written as $a_0 \ldots a_3$ and $b_0 \ldots b_3$. The LSB is denoted as a_0 and MSB as a_3. Likewise in B term, it is also denoted like A term. For getting partial products, multiply A and B terms of four bits, and finally, it is producing a new variable P as $p_0 \ldots p_7$. The p_0 is taken as LSB, and p_7 is considered as MSB.

The two 4-bit numbers are multiplied to produce the middle terms that were,

$$a_3 * b_3, \quad a_2 * b_3, \quad a_1 * b_3, \quad a_0 * b_3,$$
$$a_3 * b_2, \quad a_2 * b_2, \quad a_1 * b_2, \quad a_0 * b_2,$$
$$a_3 * b_1, \quad a_2 * b_1, \quad a_1 * b_1, \quad a_0 * b_1,$$
$$a_3 * b_0, \quad a_2 * b_0, \quad a_1 * b_0, \quad a_0 * b_0,$$

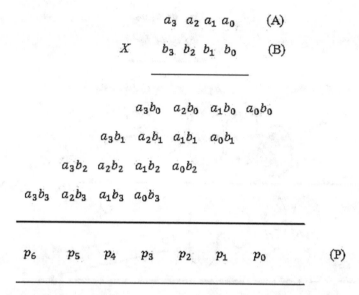

Fig. 4 Multiplication arrangement of 4 × 4 quantities

Two intermediate terms in a single section are included utilizing a 2-bit adder, and multiple terms in a single segment are included utilizing 3-bit adder as clarified in Fig. 5. Sum got afterward Si represents every addition process, where i shifts from 1 to 10. So also, C_j means carries where j shifts from 1 …10 and next carry signified by Qk, where k differs from 0 …3.

Figure 6 spectacles the physical portrayal of 4 × 4 multiplier utilizing full and half adders [8]. After the multiplication of the two digits, the support of full and half adders expand the intermediate term. Here, product exposed how the product of each bit demonstrated. $R_0, R_1…R_{14}, R_{15}$ is taken as different product terms which are related from the first phase of multiplication. The product P is released from the first bit of p_0 which is presented in $a_0 * b_0$ which is mentioned in $R_0…R_2$. Then half adder produced two output carry C_1 and sum S_1. Now, product R_0 is considered as $a_0 * b_0$. Thus, various notations from R_1 to R_{15} speak to the next product terms. The sum S_1 is mentioned as P (bit of product) that is indicated by p_1. $R_3…R_5$ which is turned as of full adder input bits given as output for carry C_2 and sum S_2. The older C_1 and S_2, which is acted as next half adder as input which produces two output, i.e., S_6 and C_6. The third bit of product P is sum S_6, and it is considered as P_2. The remaining product bits P, for example, $P_3…P_8$, individually were gotten similarly as clarified previously.

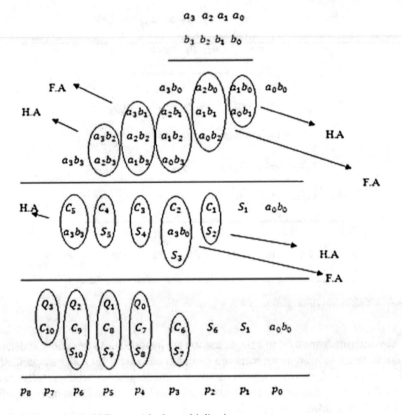

Fig. 5 Adders usage (middle terms) in the multiplication process

4.3 Wallace Tree 8 * 8 Multiplier

The major limitations of 4 * 4 Wallace tree multiplier are that it cannot able to give more number of bits as an input, and also it occupies more area and delays for smaller amount of inputs compared to 8 * 8 Wallace tree multiplier. Because of the minimum bits of inputs in 4 * 4 multiplier, structure had a scheduled pipelined structure, and it will consume huge power [9].

Figure 7 shows the essential design and the means associated with Wallace tree 8 * 8 multiplier. The same strategy is followed as like Wallace tree 4 * 4 multiplier in the multiplication method [10] (Fig. 8).

From Fig. 8, The WTM is quicker than an array. Subsequently, architects frequently maintain a strategic distance from Wallace trees, while structure multi-faceted nature is a worry to them. 8-bit multiplier dependent on Wallace Tree is progressively productive as far as power and consistency with low latency and area. 8 * 8 bit multiplier utilizing Wallace tree circuit comprises AND array and likewise comprise half, full adders, and look-ahead carry adder. In that, AND gates are utilized for the production of partial products in parallel. There will be a decrease in power

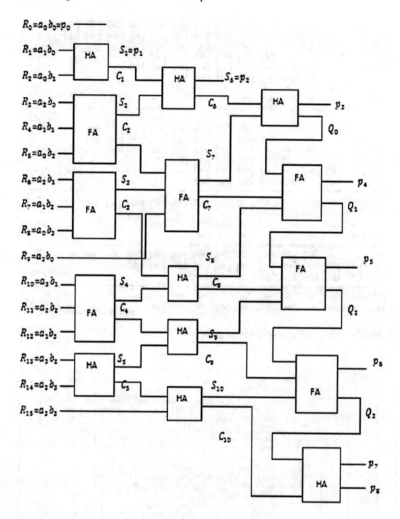

Fig. 6 Physical illustration of 4×4 multiplier

utilization since the power is given uniquely to the level that is engaged with the calculation [11].

4.4 2-Bit Adder

From Fig. 9, Half adder means that it performs addition of two bits in the combinational circuit. It requires two binary input and gives two binary output. One of the input variables designates the augend, and the other designates the addend. The output produces the sum and the carry [12] (Fig. 9).

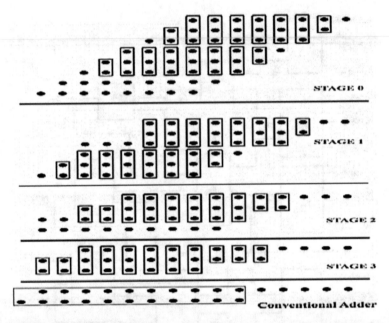

Fig. 7 Multiplication structure of 8 * 8 numbers

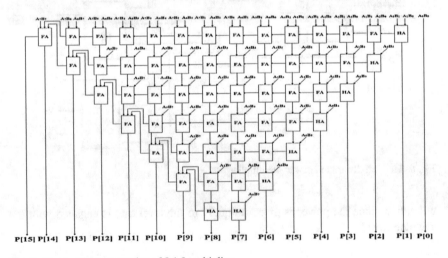

Fig. 8 Structural representation of 8 * 8 multiplier

4.5 3-Bit Adder

Full adder defines the addition of more than 2-bit numbers. Figure 10 explains that a, b, and carry-in were taken as inputs. For implementing FA, the following gates are

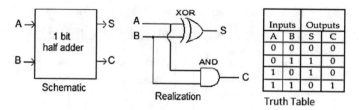

Fig. 9 2-bit adder

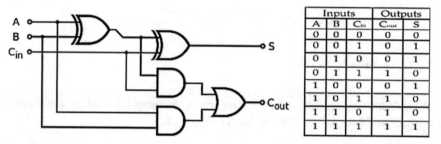

Fig. 10 3-bit adder

required. That is two exclusive OR gates, three AND gates, and a single OR gates. It produces the output as S and C_{out} [13] (Fig. 10).

4.6 Ripple Carry Adder (RCA)

Adder unit is the most dominant part in multiplier regarding power and speeds accomplishable. The 4-bit RCA involves four individual units of full adders connected in an exceeding chain as shown in Fig. 11. During this case, the carryout of full adder unit is the carry-input of the subsequent FA unit. To undertake and expand the speediness

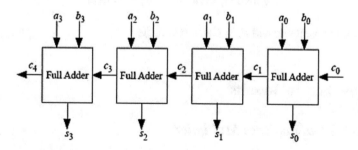

Fig. 11 Ripple carry adder

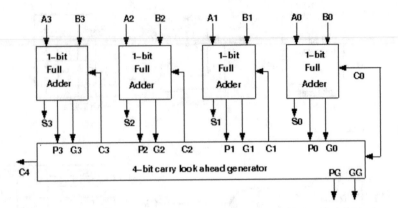

Fig. 12 Look-ahead carry adder

of the adder, consequently enhancing the speed, the carry-select adder additionally enforced and compared to the first RCA [14] (Fig. 11).

4.7 Look-Ahead Carry Adder

From Fig. 12, Look-ahead carry adder (CLA) regulates the carry delay dispute by totaling the carry flags in advance of time, in light of inputs [15]. It is because a carry sign produced in two occurrences: The first one is ai, bi bits that were 1, and the second one is ai or bi that is 1. Hence, one compose (Fig. 12),

$$C_{i+1} = a_i \cdot b_i + (a_i \oplus b_i) \cdot c_i \qquad S_i = (a_i \oplus b_i) \cdot \oplus c_i$$

With the inclusion of P_i, G_i, the above two equations are rewritten as

$$C_{i+1} = G_i + P_i \cdot c_i \qquad S_i = P_i \oplus c_i$$
$$\text{where } G_i = a_i \cdot b_i \qquad P_i = a_i \oplus b_i$$

G_i = Carry generate and P_i = Carry propagate.

5 Experimental Results

5.1 8 * 8 Wallace Tree Multiplier

The whole structure realized utilizing Xilinx Integrated Software Environment Design Suite 14.7 and the focused on a group of FPGA is Vertex 5—xc5vsx50t. This

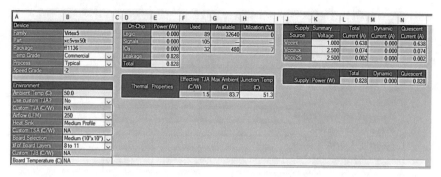

Fig. 13 Power utilization summary for Wallace tree 8 * 8 multiplier

segment manages power, area, and delay utilization of 8 * 8 Wallace tree multiplier. It additionally communicates RTL schematic, technology schematic, and package view on the Wallace tree multiplier equipment plan.

5.2 Power

Figure 13 shows that dynamic power utilization is 0.00 W and quiescent power utilization is 0.828 W. Subsequently, the total power utilization is 0.828 W for the Wallace tree 8 * 8 multiplier.

5.3 Area

From Fig. 14, it has indicated a device usage outline that is area utilization. It has indicated the utilization of slice registers, flip flops, LUTs, involved slices, LUT flip flop combinations, bonded IOBs, and average fan-out. A flip flop is major building blocks of every sequential circuit, which is used to store one bit of binary information. It is used to store the previous set of input values for future references. A LUT flip flop pair for this architecture speaks to one LUT combined with one flip flop inside a cut. A control set is an interesting mix of a clock, reset, set, and empower signals for an enrolled component. The slice logic distribution report is not important if the structure is over-planned for a non-cut asset or if placement comes up short. Overmapping of BRAM assets ought to be overlooked if the plan is over-planned for a non-BRAM asset or if arrangement falls flat.

Device Utilization Summary			
Slice Logic Utilization	Used	Available	Utilization
Number of Slice LUTs	89	32,640	1%
Number used as logic	89	32,640	1%
Number using O6 output only	89		
Number of occupied Slices	42	8,160	1%
Number of LUT Flip Flop pairs used	89		
Number with an unused Flip Flop	89	89	100%
Number with an unused LUT	0	89	0%
Number of fully used LUT-FF pairs	0	89	0%
Number of slice register sites lost to control set restrictions	0	32,640	0%
Number of bonded IOBs	32	480	6%
Average Fanout of Non-Clock Nets	4.67		

Fig. 14 Device utilization summary Wallace tree 8 * 8 multiplier

```
===============================================
Total REAL time to Xst completion: 23.00 secs
Total CPU time to Xst completion: 23.28 secs
```

Fig. 15 Time utilization for 8 * 8 Wallace tree multiplier

5.4 Delay

From Fig. 15, it has demonstrated that total real-time completion is 23 s, and total CPU time completion is 23.28 s.

5.5 RTL Schematic View

Figure 16 shows that RTL schematic, for the Wallace tree 8 * 8 multiplier equipment plan. RTL View is a register-transfer level graphical portrayal of this work. This portrayal (.ngr record delivered by Xilinx Synthesis Technology (XST)) is produced by the synthesis tool at prior phases of a synthesis procedure when innovation planning is not yet finished. The objective of this view is to be as close as conceivable to the first HDL code. In the RTL view, the structure is spoken to as far as large-scale squares, for example, adders, multipliers, and registers. The standard combinatorial rationale is planned onto rationale entryways, for example, AND, NAND, and additionally.

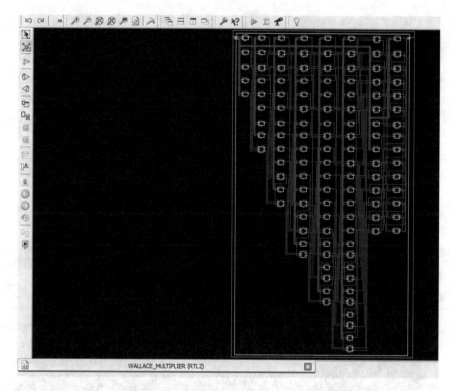

Fig. 16 RTL schematic of Wallace tree 8 * 8 multiplier

5.6 Technology Schematic View

Figure 17 shows that technology schematic view for the Wallace tree 8 * 8 multiplier equipment plan. After the advancement and innovation focusing on the period of the amalgamation procedure, the plan shows a schematic portrayal of your integrated source document. This schematic shows a portrayal of the plan as far as rationale components enhanced to the objective Xilinx gadget or "innovation," for instance, as far as of LUTs, convey rationale, I/O cushions, and other innovation explicit segments. Review of this schematic permits you to see an innovation level portrayal of your HDL advanced for a particular Xilinx engineering, which may assist you with finding configuration that gives right off the bat in the plan procedure.

5.7 Package View

Figure 18 shows that package view for the Wallace tree 8 * 8 multiplier equipment plan.

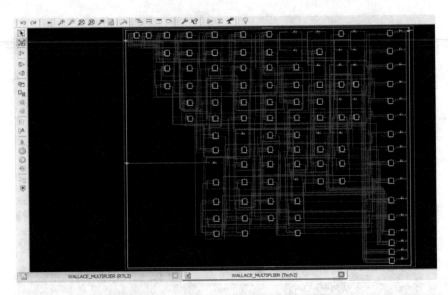

Fig. 17 Technology schematic of Wallace tree 8 * 8 multiplier

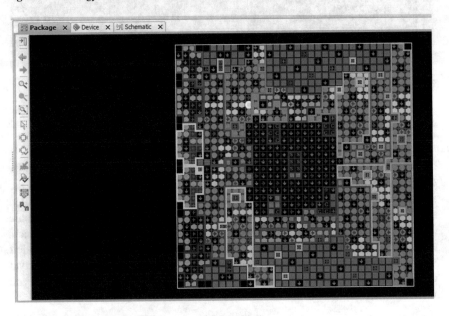

Fig. 18 Package schematic of Wallace tree 8 * 8 multiplier

Table 1 Investigation of Wallace tree 8 * 8 multiplier

Parameters	Wallace tree multiplier
Number of slice LUTs	89
Number of occupied slices	42
Number of fully used LUT-FF pairs	0
Number of bonded IOBs	32
Total power consumption	0.828 W
Junction temperature	51.3 C
Average fan-out of non-clock nets	4.67
Total REAL time for completion	23.00 s
Total CPU time for completion	23.28 s

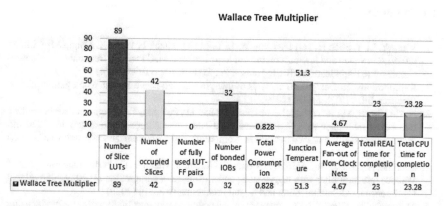

Fig. 19 Analysis of 8 * 8 Wallace tree multiplier

5.8 Investigation of Wallace Tree 8 * 8 Multiplier

Table 1 and Fig. 19 display the investigation of Wallace tree 8 * 8 multiplier as far as total power consumption, junction temperature, slice register numbers, scope of LUTs, scope of involved slices, scope of completely utilized LUT-FF sets, scope of ensured IOBs, average fan-out of non-clock nets, total real time for completion, and total C.P.U. time for completion.

6 Conclusion

In this article, the execution and investigation of a WTM design are projected. It tends to be inferred that Wallace Multiplier is prevalent in the whole regard like area, speed, delay, multifaceted nature, and utilization of power. Nonetheless, array multiplier necessitates extra utilization of power and provides an ideal number of segments needed, yet multiplier delay is bigger. Henceforth, for low power necessity and fewer delay prerequisite, WTM is proposed. The code is actualized on Xilinx ISE 14.7. The computational path delay for proposed 8 * 8 Wallace tree multiplier is seen as 23 s. In addition, the power utilization for proposed 8 * 8 Wallace tree multiplier is seen as 0.828 Watts. Contrasted with the conventional array multiplier, Wallace tree multiplier is profoundly effective as far as execution time (speed), power and area optimization. Finally from the entire work, power and area optimization achieved very well, but delay optimization is somewhat lacking compared to other multipliers. In future work, the delay optimization will be achieved better by replacing the look-ahead carry adder through Kogge–Stone adder.

References

1. Wang Y, Li Y, Shen H, Fan D, Wang W, Li L, Liu Q, Zhang F, Wang X, Chang M-F, Liu M (2019) A few-step and low-cost memristor logic based on MIG logic for frequent-off instant-on circuits in IoT applications. IEEE Trans. Circ. Syst. II: Expr. Briefs **66**(4)
2. Karuppusamy, P.: Design and analysis of low-power, high-speed Baugh Wooley multiplier. J. Electron. Inf. **01**(02), 60–70 (2019)
3. Navin Kumar, M.; Adithyaa, R.S., Bhavan Kumar, D., Pavithra, T.: Design analysis of wallace tree based multiplier using approximate full adder and Kogge stone adder. In: 2020 6th International Conference on Advanced Computing and Communication Systems (ICACCS) (2020)
4. Sundhar, A., Deva Tharshini, S., Priyanka, G., Ragul, S., Saranya, C.: Performance analysis of Wallace tree multiplier with Kogge stone adder using 15–4 compressor. In: 2019 International Conference on Communication and Signal Processing (ICCSP) (2019)
5. Mukherjee, B., Ghosal, A.: Counter based low power, low latency Wallace tree multiplier using GDI technique for on-chip digital filter applications. IEEE Conf. **EC-13**, 14–17 (2019)
6. Nagaraj, S., Thyagarajan, K., Srihari, D., Gopi, K.: Design and analysis of Wallace tree multiplier for CMOS and CPL logic. IEEE Conf. **EC-25**, 345–348 (2018)
7. Waters, R.S., Swartzlander, E.E.: A reduced complexity Wallace multiplier reduction. IEEE Trans. Comput. **59**(8), 1134–1137 (2010)
8. Sakthimohan, M., Deny, J.: An optimistic design of 16-Tap FIR filter with Radix-4 booth multiplier using improved booth recoding algorithm, Microprocessors and Microsystems (2020). https://doi.org/10.1016/j.micpro.2020.103453
9. Murugeswari, S., Mohideen, S.K.: Design of area efficient and low power multipliers using multiplexer based full adder. In: 2nd International Conference on Current Trends in Engineering and technology (ICCTET), Enathi, India, pp. 388–392, July 2014
10. Rao, M.J., Dubey, S.: A high speed and area efficient booth recoded Wallace tree multiplier for fast arithmetic circuits. In: Asia Pacific Conference on Postgraduate Research in Microelectronics and Electronics (PrimeAsia), Hyderabad, India, pp. 220–223 57 Dec 2012
11. Jiang, Y., Al-Sheraidah, A., Wang, Y., Sha, E., Chung, J.-G.: A novel multiplexer -based low-power full adder. IEEE Trans. Circ. Syst. **51**(7), 345–348 (2004)
12. Sakthimohan, M., Deny, J.: An enhanced 8x8 vedic multiplier design by applying Urdhva-Tiryakbhyam Sutra. Int. J. Adv. Sci. and Technol. **29**(05), 3348–3358 (2020). Retrieved from http://sersc.org/journals/index.php/IJAST/article/view/12015
13. Oskuii, S.T., Kjeldsberg, P.G.: Power optimized partial product reduction interconnect ordering in parallel multipliers. NORCHIP, Aalborg **19–20**, 1–6 (2007)
14. Kumar, M.V.P., Sivanantham, S., Balamurugan, S., Mallick, P.S.: Low power reconfigurable multiplier with reordering of partial products. In: International Conference on Signal

Processing, Communication, Computing and Networking Technologies (ICSCCN), Thuckafay, pp. 532-536, 21–22 July 2011
15. Anitha, P., Ramanathan, P.:A new hybrid multiplier using Dadda and Wallace method. In: International Conference on Electronics and Communication Systems (ICECS), Coimbatore, India, pp. 1-4, 13–14 Feb 2014

Application of Rules and Authorization Key for Secured Online Training—A Survey

Priyanka Saxena, Hrithik Sanyal, and Rajneesh Agrawal

Abstract Online training has been present for over a decade, and its importance is increasing every day. Today, it has become very important due to ongoing COVID-19 pandemic. As it has been accepted widely by the educational systems; nowadays, challenges like hardware resources, network resources, software resource, and security have become more demanding. Security threats among all challenges require more researches to develop rigid systems where data of all stakeholders remain secured. ML has a proven track record to solve such problems. In terms of security, ML continuously learns by analyzing data to find patterns so unauthorized access to encrypted traffic is detected better and find insider threats to keep information safe. Here, a new system is being developed using an improved algorithm, described in proposed work. Using this new algorithm, machines are trained to identify unauthorized access attempts and stop them from stealing data even if, they are authenticated.

Keywords Machine learning · Artificial intelligence · Security · Authorization · Role-based access control · Classification · Decision tree algorithm

1 Introduction

Cyber security incidents will cause significant financial and reputation impacts on enterprises. In the world, numerous organizations are requiring to implement high

P. Saxena (✉)
Department of Computer Science, Rajiv Gandhi Proudyogiki Vishwavidyalaya, Bhopal, India
e-mail: anusaxena1812@gmail.com

H. Sanyal
Department of Electronics and Telecommunications, Bharati Vidyapeeth College of Engineering, Pune, Pune, India
e-mail: hrithiksanyal14@gmail.com

R. Agrawal
Comp-Tel Consultancy, Mentor, Jabalpur, India
e-mail: rajneeshag@gmail.com

S. Shakya et al. (eds.), *Proceedings of International Conference on Sustainable Expert Systems*, Lecture Notes in Networks and Systems 176, https://doi.org/10.1007/978-981-33-4355-9_4

41

security tools as digital assault events have increased to a great extent. These organizations require a powerful and adaptable machine learning-based threat detection and prevention as a vital piece for their security monitoring portfolio. Similarly, online education is need for the hour and for the future; hence, they are also in a desperate need of highly secured online training system and recently come across the similar issue faced by unacademy, where about 22 million student's data is been hacked; the proposed system can be implemented to such e-learning platform to identify such unauthorized access and make platforms secure. According to CSO India, (CSO provides news, analysis and research on security and risk management) [1] from IDG has provided data that very big companies have been affected by security breaches in recent past such as Adobe, eBay, LinkedIn, MySpace, NetEase, and Yahoo. It has been estimated that around 3 million encrypted credit card details including 38 million active user accounts were breached from Adobe, whereas from eBay 145 million user accounts were hacked. In a similar manner, around 360 million user accounts of MySpace, 235 million user accounts of NetEase were breached. Yahoo's 3 billion user accounts were compromised and were considered as the biggest data breach of history.

In secured systems, three aspects of security are having, viz. authentication, encryption, and authorization. All of these three aspects are to be considered as an important mechanism for security implementation. As by applying password, it can have authentication of the users to stop the non-authentic entry in system and by encryption data transfer and storage can be made secure. But imagine what if someone hacked or get a password and enters into the system, then proper authorization becomes crucial.

So, building a system, which applies machine learning concept to train machines for providing authorizations to the users of the system, is very much required. Machine learning will help identify the unauthorized persons and stop them to access data even if they somehow authenticate themselves.

In Sect. 1, introductions of machine learning and its use in security are elaborated. In Sect. 2, explanation of why and how security in machine learning is done; in Sect. 3, existing systems have been discussed along with classification and different types of classification algorithms have been explored with machine learning techniques. In Sect. 4, the existing access control mechanism has been explored. In Sect. 5, elaborate proposed work. Section 6 shows the expected outcome obtained. In Sect. 7, conclusions are discussed.

2 Machine Learning

2.1 Overview

The main purpose of machine learning is to understand the format of data in a dataset and accommodate that data into models which are better understood and utilized by people. Machine learning is used anywhere from automating mundane tasks to offering intelligent insights, industries in every sector try to benefit from it. Machine learning can help businesses to better analyze threats and respond to attacks and security incidents. Machine learning in security is a fast-growing trend. Cyber-attacks have and hence drawing increased attention to the vulnerabilities of cyber systems and the need to increase their security. Role-based access control enables access management to the innermost level. Using RBAC, the level of access can be granted only that users need to perform their work.

2.2 Security in Machine Learning

Machine learning detects threats by continuously monitoring network for anomalies. ML is revolutionizing almost every field of science because of its unique properties like adaptability and ability to handle unknown challenges. Machine learning is the most common approach to describe its application in security. ML is used to collect data to predict fraudulent activity which in turn helps the security team to address the liability before it evolves into a costly data theft. Many cybersecurity tasks can be made more efficient with the implementation of ML algorithms.

2.3 Authorization

Authorization is whether the individual is allowed to have an asset or not. This is typically controlled by seeing whether that individual has a piece of a specific key or if that individual has taken paid confirmation or has a specific degree of trusted status. Access control components figure out which tasks the client can or can't do by contrasting the client's personality with an access control list (ACL). Access control is a strategy for ensuring that clients are who they state they are, and that they have suitable access to organization information.

2.4 Classification

In supervised machine learning, an approach called classification is mostly used in which the system learns from the data fed to it and make new observations or classifications or results. Classification is a type of supervised learning where input data is provided. In classification, a given set of data is categorized into classes. Classification can be accessed on both structured and unstructured types of data. The process of classification begins with the prediction of the class in which the datasets are fed. These classes are also called as targets or labels or categories. Classification also referred to as predictive modelling uses a process of approximation of mapping functions from input dataset variables and gives us discrete output variables. The main goal of it is to judge whether the new dataset fed in, should belong to which class or category.

Classification algorithms in machine learning are as follows:

- Logistic Regression
- Decision Tree
- Random Forest
- Naive Bayes

Logistic Regression: Logistic regression is a type of classification algorithm in machine learning which makes use of one or more independent input variables to come up with an outcome/output. The output is measured with a dichotomous variable dataset hinting that it will only have two possible outputs. The main idea of logistic regression is to find the best connection between dependent and a set of independent variable datasets. Logistic regression is more useful than any other form of binary classification algorithms.

Some of the pros and cons of logistic regressions are:

- Logistic regression is specifically meant for classification, and it is useful in understanding how a set of independent variables affect the outcome of the dependent variable.
- The main disadvantage of the logistic regression algorithm is that it only works when the predicted variable is binary. It assumes that the data is free of missing values and assumes that the predictors are independent of each other.

Naive Bayes Classifier: It is another type of classification algorithm which is solely based on Bayes algorithm in which it assumes the presence of a particular feature present in a class which is not related to any other feature's presence. Its model is easy to make and mostly used when having very large datasets. Although its approach is very simpleton, Naïve Bayes is known to perform far better than the rest of the classification algorithms in ML.

Albeit, the features depend on each other, all of their properties independently contribute to the probability. Its model is easy to make and mostly used when having very large datasets. Some of the pros and cons of logistic regressions are:

- The naive Bayes classifier requires a very small amount of training dataset to give an estimation of the necessary parameters for obtaining the results.
- They are extremely fast in nature when compared to other classifiers.
- One big disadvantage is that they are bad estimator.

Decision Tree: The decision tree algorithm is another type of classification algorithm which makes use of a tree structure. It uses the "if-then" rule which is mutually exclusive in classification.

The process of the decision tree begins on by breaking down the data into smaller chunks/structures and consequently associating it with a decision tree which will be incremental in nature. The final structure looks like a tree with nodes and sub-nodes.

This also includes using training datasets for learning of rules sequentially one at a time. After each rule is learned, the rule gets removes. This process goes on and continues on the training datasets until the culmination point is reached, i.e., the termination point. Some of the pros and cons of the decision tree are:

- A decision tree is very simple to understand and imagine and visualize.
- It requires very little datasets for preparation.
- One disadvantage of a decision tree is that it can create complex trees that may become inefficient later on.
- Another disadvantage is that it can be quite unstable, thus hindering the whole structure of the decision tree.

Random Forest: Random decision trees or random forest is another type of classification algorithm. It works by construction multiple decision trees while at training time and giving outputs of the class that also happens to be the mode of that class or classification or even mean regression of the individual decision trees. It is like a meta-estimator that fits a number of decision trees from various datasets and then uses an approximation to improve the accuracy of the model's predictive nature. The training size of the datasets should always be the same as of the original input variable size but replacements done on these datasets can often be seen. Some of the pros and cons of the decision tree are:

- An advantage of random forest is that it is more accurate than the conventional decision tree due to the reduction in the size of fittings.
- One disadvantage of random forest classifiers is that it is quite complex in implementation and gets pretty slow in real-time prediction.

3 Existing Systems

Machine learning is used for many tasks and hence catches public eyes. It is used for prediction and classification, and therefore securing sensitive data which is involved in training and testing computation is important. Security is improved, and risk is reduced by applying encryption and decryption. This encryption and decryption are done in both the client and server machine in the proposed system architecture. In the existing engine and framework, a module is built by using python and loaded in this module encryption and then decryption is done. Encryption is done by using CP-ABE, and after this, machine learning computation gets started [2].

User-centric machine learning system provides a complete framework and solution to risky user detection for enterprise SOC. An ML system is used between soc workflow, SIEM alerts and soc analysis the result and generates a user risk score for soc. And therefore now soc analyst uses these risk score to prioritize their investigation and hence improves their efficiency, and as a result, security is enhanced [3].

Taxonomy of ML attacks is designed for the identification of different types of attacks on ML. These hypotheses of representation and ML risk are done as well as architectural risk analysis is also done. Since cyber threats have increased exponentially, so the better data mining technique is required for analyzing security loss and thus ensuring effective and automated cyber threat detection. ML is an emerging trend to minimize operational overhead and false positive. For this, the optimal ML algorithm and models are being used [4].

A security flaw in access control of a data control system is explained while using CP-ABE and attribute-based signature scheme by the authors. It has been discussed that CP-ABE is exposed to collusion attack, and as a result, the unauthorized user can decrypt the cipher-text. The exposure to collusion attack in CP-ABE scheme is due to the improper bonding among user secret key elements. Collusion attack should be handled as its resistance is one of the most important properties to make a system to be fully secured. When CP-ABE scheme is modified, the cryptanalysis result shows that the system's security assurance does not hold [5].

In enterprise security of the accesses of different resources is ascertained by the use of a high-level entry control system which applied a guard at each entry point. Enterprise-level high-tech entry control systems automate many decision functions. These functions put into higher tiered infrastructure. Care of the other functions like authorized and unauthorized access to the system is taken by a guard. Guard is guided by the risk managers, and his major functions include authentication and controlled authorization. Such a system will have multi-layer security and expected performance issues [6].

In this paper, it is explained that normal web-based systems are completely different from e-learning systems. Authors have proposed to apply a role-based access control scheme that is successfully used in the development of a complete e-learning system. Two rules are explained here for complete security: One is called access rule and other is called an access matrix. Access rules are defined to authentic users who are authorized to access web documents and services, whereas the other rule is denial rule, which does not allow a person to access web documents. Scope of further improvement is also mentioned like access and denial matrix, along with the scope of improvement can be a web-based interface, built to manage the matrices in the database, which allows for managing access to the integrated e-learning system [7].

4 Access Control

Access control is a strategy to ensure that clients are who they say or pretend they are and that they have suitable access to organization information. Without validation and approval, there is no information security. Validation is possible by ID and secret key, and approval can be given or done for controlling purpose.

Role-based access (RBAC) controls the limit on dependency of an individual's role in an association. It has got one of the primary strategies for getting cutting control. It guarantees that clients just approach the data or records that are pertinent to their present position or undertaking. In associations that have significant divisions, instituting a job-based access control framework is fundamental in moderating information misfortune and for protection and security reason. It includes setting authorizations and benefits to empower access to approved clients. Most enormous associations use job-based access control to give their workers shifting degrees of access dependent on their jobs and duties and to secure important information.

5 Proposed Work

In security, machine learning causes continuous learning of the machine by analyzing data to find patterns so it can better detect accesses which are unauthorized during access request from encrypted traffic and can also find threats to keep information stored in them safe from unauthorized users. In this paper, a new system is being proposed to be built by training machines to identify the unauthorized accesses and stop data theft or mal-processed if authentication of such users has already been done by different means.

Further, in this proposed work, after authentication, the authorization mechanism is applied by using modified role-based access control.

Step 1   Users data with their roles will be availed or created.
Step 2   All user's transactions shall be encrypted
Step 3   Users will be provided with an encrypted key of sixteen characters, which will be used by them to get authorization.
Step 4   Authorization key shall be generated randomly which will be applied using different rules enlisted here for Modified RBAC algorithm.

RULES

- Forever for particular roles module-wise
- Forever for particular roles for a complete system
- Duration based (to be taken from module db) for a particular module
- Duration based (to be taken from module db) for a complete system
- Forever/Duration-Based access for particular pages of the module/complete System
- Forever/Duration-Based access for particular action applicable on-page/module/complete system
- Guest User roles (for a limited duration and limited page/module/action-wise access)

KEY:

A random key generation is also proposed in this system for authorization of the users who are authenticated, and system access has been allowed to them. The

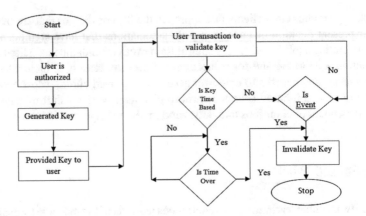

Fig. 1 Flowchart of the authorization process in the proposed system

proposed key shall be of 32 alpha-numeric characters which shall be not only generated randomly but will be shuffled to avoid prediction of a key by the application of man in the middle attack. The key is a major source of continuous access of the resource for any user as it will be validated every time user accesses the system and in random time interval if the access is continued for long.

IMP: Fig. 3 at the end of paper depicts the application of the above rules with the overall time sequencing of the various use case environment.

Step 5   The key provided shall be valid for a specific duration either by event or by time decided by administrators of the system

Step 6   Following factors shall be used to assign keys to different users:

- Role
- Module
- Page (Submodules)
- Action (add/delete/update/view/download/share, etc.)

Step 7   Key management system shall be implemented as per the following scenario:

- Key (which key)
- User (who owns it)
- Module(s) (for which module it is applicable)
- Duration (How long will be useful either forever or for a particular duration decided by the administrator of the system (Figs. 1, 2 and 3).

6   Expected Outcome

The proposed system is expected to provide high security over the e-learning portal where thousands of students and teachers can register. There are various parameters as

Fig. 2 Flowchart of the
proposed system

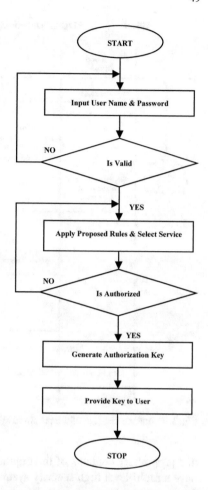

discussed in proposed work section will be expected to have only well authenticated and authorized users to get into the system. It is also expected that the performance compromise is not applicable in such a system as the parameters will be used hierarchically, i.e., Time → Module → Page → Action with the roles as categorization of the users resulting in tree-based time complexities which are in the proportion of log (n) time.

7 Conclusion

The need for security is paramount for the web-based systems as they are open for the whole world and particularly when the systems are having secured and non-sharable information the criticality of the secured system is very high. E-learning is also having data of students, teachers, and financial information of the organization

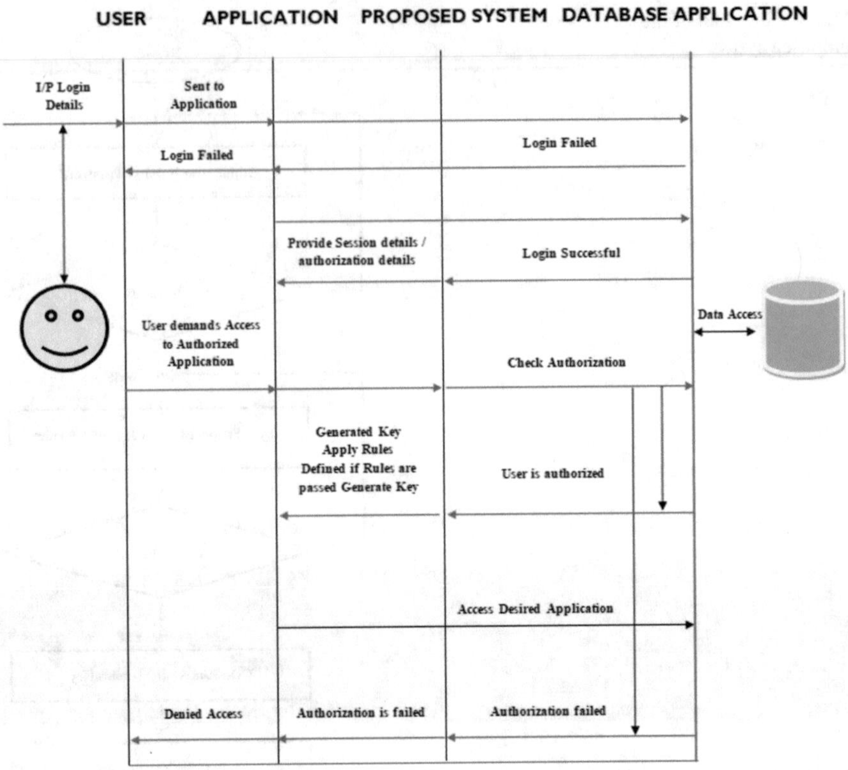

USER   APPLICATION   PROPOSED SYSTEM   DATABASE APPLICATION

Fig. 3 Sequence diagram of the proposed system

and proprietary contents of the organizations. The proposed system is expected to have a multi-level high security system in which authentication will not guarantee for the users to have access to the whole or partial system. Users must have properly assigned access rights and authorization key to get into the system.

References

1. SANS Technology Institute: The 6 Categories of Critical Log Information (2013)
2. Kurniawan, A., Kyas, M.: Securing machine learning engines in IoT applications with attribute-based encryption. In: 2019 IEEE International Conference on Intelligence and Security Informatics (ISI), Shenzhen, China, pp. 30–34 (2019). https://doi.org/10.1109/isi.2019.8823199
3. Feng, C., Wu, S., Liu, N.: A user-centric machine learning framework for cyber security operations center. In: 2017 IEEE International Conference on Intelligence and Security Informatics (ISI), Beijing, pp. 173–175 (2017). https://doi.org/10.1109/isi.2017.8004902
4. Farooq, H.M., Otaibi, N.M.: Optimal machine learning algorithms for cyber threat detection. In: 2018 UKSim-AMSS 20th International Conference on Computer Modelling and Simulation (UKSim), Cambridge, pp. 32–37 (2018). https://doi.org/10.1109/uksim.2018.00018
5. Tan, S.: Comment on secure data access control with ciphertext update and computation outsourcing in fog computing for internet of things. IEEE Access **6**, 22464–22465 (2018).

https://doi.org/10.1109/ACCESS.2018.2827698

6. Shane, D.: Managing the training guards in the operation of a high-tech facility access control system. In: 2013 47th International Carnahan Conference on Security Technology (ICCST), Medellin, pp. 1–4 (2013). https://doi.org/10.1109/ccst.2013.6922074

7. Wang, H.: An access control scheme for web-based E-learning systems. In: 2006 7th International Conference on Information Technology Based Higher Education and Training, Ultimo, NSW, pp. 447–453 (2006). https://doi.org/10.1109/ithet.2006.339796

Detecting Semantically Equivalent Questions Using Transformer-Based Encodings

Pushkar Bhuse and Pranav Bathija

Abstract Due to the increasing influx of users on various Q and A forums like Quora, Stack Overflow, etc., answers to questions get fragmented across different versions of the same question due to this redundancy of questions. However, if questions which are lexically or semantically identical could be grouped, the search results for a given question could yield the assimilation of answers provided for all versions of a question. In this paper, an ensemble of four separate models is created using four different word-embedding techniques in each. The vectorized data from each of the embedding layers is passed through a custom-made architecture using transformer-based encodings. This architecture is used to determine if a pair of questions have the same semantic meaning or not. Finally, on testing the model on a subset of the provided data, the experiments show that the proposed model achieves an $F1$ score of 84.63%

Keywords Artificial intelligence · Deep learning · Transformer · Duplicate text detection · Semantic equivalence · Word embeddings

1 Introduction

In natural language processing (NLP), semantic similarity plays an important role in many NLP applications like sentiment analysis, neural machine translation, natural language understanding and many others as mentioned by Majumder et al. [1]. It is the process of encompassing the underlying meaning of two pieces of text and then comparing how similar they are. This technique is used in various Q and A forums to tackle the problem of question duplication.

P. Bhuse (✉)
Dwarkadas J. Sanghvi College of Engineering, Mumbai, India
e-mail: pushkar.s.bhuse@gmail.com

P. Bathija
Thadomal Shahani Engineering College, Mumbai, India
e-mail: pranavb30@gmail.com

S. Shakya et al. (eds.), *Proceedings of International Conference on Sustainable Expert Systems*, Lecture Notes in Networks and Systems 176,
https://doi.org/10.1007/978-981-33-4355-9_5

However, due to a massive increase of users on such forums, it is quite likely for a subset of these users to ask questions which semantically have the same meaning. According to Bogdonova et al. [2], two questions can be called semantically equivalent if they can be answered by the same answer. These answers get fragmented across different versions of the same question due to the redundancy of questions in these forums. Thus, at platforms like Quora, identifying duplicate or semantically similar questions and merging them makes knowledge sharing more efficient and effective in many ways. This way, users can get answers to all the questions on a single thread and writers do not need to write the same answer in different locations for the same question. At the time, Quora used the random forest model to identify duplicate questions. Due to the inaccuracy of this method attributed to its simplistic architecture which leads to an unsatisfactory performance for the given task and the rapid advancements in the field of machine learning and deep learning, Quora released a dataset of more than 400,000 pairs of questions with labels indicating if a given pair could be considered duplicate or no. This dataset was then set out to be harnessed in a Kaggle competition. Participants were challenged to tackle this natural language processing problem by applying advanced techniques to classify whether question pairs are duplicates or not.

To tackle the problem of detecting text equivalence, a bunch of methods were experimented with. The most successful architecture included the creation of a setup similar to a Siamese network, using transformer-based encodings. An ensemble of this architecture was created which used the combination of four different types of word2vec models, i.e., FastText crawl, FastText crawl subword, Google news vector and FastText Wiki News Embeddings. This ensemble was then harnessed to predict the semantic similarity between a pair of questions.

It is assumed that questions marked as duplicates in the Quora dataset are semantically equivalent since Quora's duplicate question policy [3] concurs with this definition of semantic equivalence [4].

In Table 1, a few examples of semantically equivalent and nonequivalent questions are shown. In the first example, since the question is asked about two inherently different things (Pokemons and Pokemon Eggs) in the same game (Pokemon GO),

Table 1 Description of the Quora dataset

S. No.	Question 1	Question 2	Is duplicate?
1	How do I find Pokémon in Pokémon GO?	How do I find Pokémon Eggs in Pokémon GO?	NO
2	Why do worrying so much when worrying doesn't bring any difference at all is known?	Why do worrying?	NO
3	How can someone control their anger?	How can a person control anger?	YES
4	What are your views on the Netaji Subhash Chandra Bose's mystery death case?	What is secret of Subhash Chandra Bose's death?	YES

they cannot be considered equivalent. In the second example, the first question points to a more specific question while the second one is more general, because of which they are again not semantically equivalent. However in example 3 and 4, each pair of questions, although in different tones and approaches, are directed towards the same purpose and expect the same answer. Due to this reason, the pairs stated in example 3 and 4 are considered semantically equivalent to each other.

The rest of the paper is structured as follows. Section 2 describes the current research done related to this work. In Sect. 3, the dataset and the preprocessing techniques applied to it are explained. Section 4 introduces the proposed architecture of the proposed model and the underlying deep learning concepts used in the design of it. Section 5 presents the results obtained in this research using the proposed model. The paper ends with a conclusion and direction for future work.

2 Related Work

The task of detecting duplicate texts on a semantic rather than lexical level involves understanding the underlying ideas the pairs of texts convey. The complete meaning of a text can change the insertion or deletion of a few words so even though a pair of text may be lexically similar, their semantic meaning might be very different. To understand the meaning of the text, various sequence models were adopted. Multiple methods were researched [5] to determine the similarity of texts by calculating the distance between the vectorized representations of the pair of texts. The initial methods for detecting duplicate questions included the use of a siamese network [6]. This method included using parallelly stacked character-based BI-LSTM. Cosine similarity was used as a distance metric for job-title normalization. Dadashov et al. in [7] used a Siamese Ma-LSTM architecture was created for the task of detecting duplicate questions on the Quora Question Pairs datasets. This model achieved an $F - 1$ score of 79.5% and accuracy of 83.8%.

In machine learning models, the words need to be converted to some numerical representation before they can be processed by the model. Traditionally, BoW and TFIDF have been very popular but in recent times, pre-trained word embeddings have been very instrumental in models which have produced a state-of-the-art results in NLP tasks. These embeddings have been better at capturing the meaning of words in a sentence.

Ensemble models helped in harnessing the capabilities of multiple word embeddings into a single model. Imtiaz and et al. used the combination of three-word embeddings for the task of duplicate question detection [8].

However as the sequence length increases the training time for LSTMs, RNNs and comparable models increase drastically. This kind of behavior is also noticed with a considerable increase in training data. Due to the underlying complexity, LSTMs and RNNs are not considered hardware friendly and take a considerably large amount of time to generate a trained model.

With the introduction of transformers [9], a new method to machine translation was created. It contains an encoder and decoder to convert sequential data from one form to another. Ostendroff et al. used a Vanilla Transformer (BERT) for the task of text similarity [10]. This architecture also used deep neural networks for computing the semantic distance between two pieces of text which provided a change from the earlier methods like Manhattan distance, cosine similarity, etc. Transformer encodings can be used for a multitude of applications like text classification [11], question-answering, neural machine translation, etc. Due to the ubiquity of the research done on this topic using the Quora dataset, it was decided to use it as the primary dataset for the research. In the next section, an in-depth analysis of the Quora dataset is done along with a description of how the data was preprocessed to suit the purpose of the proposed architecture.

In conclusion, there is an increased need to propose an architecture which not only improves the accuracy as well as effectivity of the given use case but also considerably reduces the training time and model complexity. Siamese networks have proven to be highly effective in detecting equivalent content. However, its variant using LSTM suffers from the problems of increased training time. This is one aspect where transformers hold the upper hand. This motivates the proposed architecture, where integration of transformers and Siamese networks is created. The performance of this system is then further improved by creating an ensemble of word embeddings.

3   Dataset and Preprocessing

The dataset used for this research is the Quora Question Pairs dataset downloaded from Kaggle [4]. Each record has a pair of questions and a target class that represents whether the questions are duplicate or not. The description of dataset columns is shown in Table 1. In the following subsection, an in-depth description of the dataset and an explanation about the preprocessing techniques used on this dataset are provided.

A. Dataset

Quora released a public dataset that consists of 404,351 question pairs in January 2017. The question pairs are from various domains including technology, entertainment, politics, culture and philosophy. This dataset is downloaded from Kaggle [4]. The data is split randomly into approximately 376 k train examples, 11 k dev examples and 15 k test examples. In the given dataset, duplicated question pairs only make up 37% of the training data, indicating that the dataset is heavily imbalanced with many more non-duplicated question pairs than duplicated ones. In addition, 33% of questions appear multiple times in different question pairs

B. Preprocessing

Data preprocessing is a technique that is used to convert the raw data into a clean dataset. Various preprocessing steps were performed on the Quora dataset. NLP techniques are used such as the conversion of text to lower case, lemmatization and tokenization, with the help of freely available libraries such as Keras. Since the length of the questions was small, it is preferable to abstain from using functions provided by the above mentioned libraries since it included negative words like "no" and "not" as a part of stopwords. Since these words are quite important when considering the semantic meaning of the question, stopwords are not dropped

After the preprocessing stage, the data is converted in a format suitable to be fed to the embedding layer and only contains information relevant for prediction. Using the help of tokenizers, raw textual data is converted into a vectorized representation of words. This vector of words is then assigned unique token numbers for respective words. Once the data is converted to a numeric format, a post-padding step to set the maximum length of all the text to 30 is performed. Questions with lengths larger than 30 will be cut short and the ones with length less than 30 will be zero-padded. Once the numerical data is padded, it is fed to the embedding layer where all the four-word embeddings (i.e., FastText Wiki News, FastText crawl, FastText subwords crawl,

Google news vector) are applied individually and then processed further. The preprocessed data was made suitable to be used in the architecture explained in Sect. 4. The following section provides a comprehensive understanding of the various components of the setup and the architecture as a whole.

4 Proposed Methodology

The proposed model is inspired by the transformer architecture [9, 12]. It uses the transformer encoder which consists of positional encodings, multi-head attention and a feed-forward network which consists of layer normalization [13] and two fully connected layers. Unlike RNNs and LSTMs models in which parallelization is not possible because of their sequential nature, the transformer can be parallelized leading to faster training times. The function of the encoding layer is to find encoding for each word of the sentence by taking every other word of the sentence into consideration which is achieved by the self-attention mechanism which draws information from every other word and weighs their relevance to the current word before calculating the encoding for that word.

In Fig. 1, the complete architecture of the Transformer is shown. Its contains the internal components which help in the encoding of word embeddings to be further passed to the deep neural network. This transformer architecture is applied in the Siamese network as shown in Fig. 2, which serves an overall view of the complete system architecture. Raw data is passed as input to this architecture and predicted responses are given back as output.

The proposed model, shown in Fig. 2, consists of an embedding layer, the transformer encoding layer and the feed-forward network. The word embedding layer uses four distinct word embedding models to vectorize data. The Transformer encoding layer converts this vectorized data from both sentences into representations that are compared to each other by flattening and concatenating them before passing them

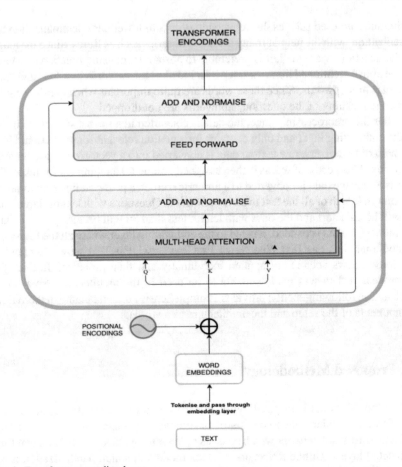

Fig. 1 Transformer encoding layer

through a series of fully connected layers with ReLU and dropouts [14]. The prediction of the question pairs is found by passing the output of the fully connected layers to a softmax classifier.

Multiple models have been created using different pre-trained embeddings namely FastText Wiki News, FastText crawl, FastText subwords crawl, Google news vector [15–17] and a model which is a combination of all the models. After each of the single models makes their predictions, the ensemble model takes an average of the prediction probabilities of all the models which give us the class probabilities for the ensemble model. The probabilities obtained are then used to make the final class predictions. The preprocessed data is first passed through a series of layers explained below and shown in Fig. 1. These layers are interconnected to form the transformer encoding layer .

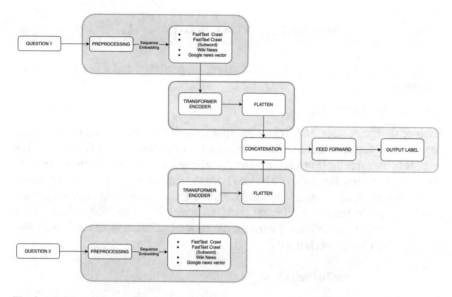

Fig. 2 Model architecture diagram

(1) Positional encodings

The position and order of words in a sentence created the desired meaning which makes it very important. Since LSTMs and RNNs analyze the sentence word by word, it implicitly takes the position of words into account. In transformers since there is no such mechanism, a way must found to pass this information to the model. Positional encodings [18] offer a way to solve this problem. Out of the various possible choices for positional encodings, sine and cosine functions are used as used in [8].

$$PE_{(pos,2i)} = \sin\left(\left(\frac{pos}{10000}\right)^{\frac{2i}{dmodel}}\right). \tag{1}$$

$$PE_{(pos,2i+1)} = \cos\left(\left(\frac{pos}{10000}\right)^{\frac{2i}{dmodel}}\right) \tag{2}$$

where pos is the position of the word, i is the dimension, and dmodel is the dimensions of the model.

(2) Scaled dot product attention

In the proposed model, scaled dot product attention is used in which three vectors from each of the input vectors are created; it creates the query, keys and values vectors each having the same dimension d. The dot product of the query and key vectors is computed before scaling it by dividing it by $\sqrt{d}$ and passing it to a softmax function to get the self-attention weights. These self-attention weights are then multiplied by the value vectors to get the scaled dot product attention.

$$\text{Attention}(Q, K, V) = \text{softmax}\left(\frac{QK^{\text{T}}}{\sqrt{d}}\right)V \qquad (3)$$

where Q, K and V are query, keys and values, respectively, and d is the dimension of the model.

(3) Multi-head attention

Each of the query, keys and, values are split into multiple heads and each of these heads is then passed to the scaled dot product attention where they run in parallel. The outputs of these multiple heads are then concatenated and are linearly transformed to obtain the expected dimensions. The idea is to combine the knowledge from multiple heads. According to the [9], multi-head attention allows combining knowledge from different representations at different positions which is not possible with a single head. For the model, different values for the number of heads were tried and it was found that ten heads worked the best.

$$\text{MultiHead}(Q, K, V) = [\text{head}_1; \ldots; \text{head}_h]W^O \qquad (4)$$

$$\text{where head}_i = \text{Attention}\left(QW_i^Q, KW_i^K, VW_i^V\right) \qquad (5)$$

where Q, K and V are query, keys and values, respectively, and W_i^Q, KW_i^K, VW_i^V are their trained parameters.

(4) Feed-forward networks

The output from the multi-head attention is passed to a fully connected feed-forward network consisting of two linear transformations with ReLU activation in between them. The dimension of the input and output of the model is 300 and the dimension of the feed-forward layer is 512. The output of the multi-head attention is added with the original input before it is normalized and passed to the feed-forward network where linear transformation takes place. This followed by layer normalization where the input to it is $x + \text{sub-layer}(x)$ where x is the input to the feed-forward block and sub-layer(x) is the output of the feed-forward block

$$\text{FFN}(x) = \max(0, xW_1 + b_1)W_2 + b_2 \qquad (6)$$

The designed model was then trained, validated and tested on the Quora dataset, and in Sect. 5, the results obtained by various components of the model as well as the ensemble are discussed.

5 Results and Discussion

The model has been trained on various pre-trained embeddings (FastText Wiki News, FastText crawl, FastText crawl subword and Google news vector) individually and tried various combinations of them. The ensemble of FastText Wiki News, FastText crawl, FastText crawl subword and Google news vector provided the best results. The model was trained on 376 k question pairs and tested them on 15 k question pairs. For the proposed model, this research used a single transformer and the dimensions of the feed-forward network that provided the best results was 1024. This research used a dropout rate of 0.1 while training the model and L2 regularization with a regularization parameter of 1e-5. Additionally, label smoothing with a smoothing of value 0.2 was used [19]. Adding label smoothing increased the accuracy by around 1–2% and improved the $F1$ score slightly. A custom averaging layer which considered the original length of the sentence before adding any padding was also experimented with, global average pooling, global max pooling and a flatten as the layer after the transformer and observed that flattening the transformer output provided the best results. After concatenating the output from both transformers, it was tried to directly pass it to the softmax, passing it to a feed-forward network with tanh activation and passing it to a feed-forward network with ReLU activation. The feed-forward network with ReLU activation provided the best results. As evident from Table 2, FastText Wiki News, FastText crawl, FastText crawl subword and Google news vector provided accuracies of 83.36%, 83.85%, 83.32% and 83.23% and $F1$ scores of 81.85%, 82.55%, 81.72% and 81.72%, respectively, and the ensemble model had an accuracy of 86% and an $F1$ score of 84.63%. The details of models with each type of word embedding as well the ensemble are shown in Table 2 where its precision, recall and accuracy are shown as well. The performance of each of the models is shown diagrammatically in the form of confusion matrices as shown in Fig. 3.

To understand how different embeddings and the combination are working, the results for five pairs for each model have been extracted as shown in Table 3. In the first question, all models have predicted the same label as the true label. In the second question, the model using Google news vector has predicted the wrong label but as

Table 2 Description of the Quora dataset

Attribute	Description
id	ID of a question pair.
qid1	ID of question 1
qid2	ID of question 2
question 1	Full content of question two
question 2	Full content of question two
is_duplicate	The target label. When set to 1, it means that the pair of questions are semantically equivalent. When set to 0, it means that the pair of questions is not equivalent

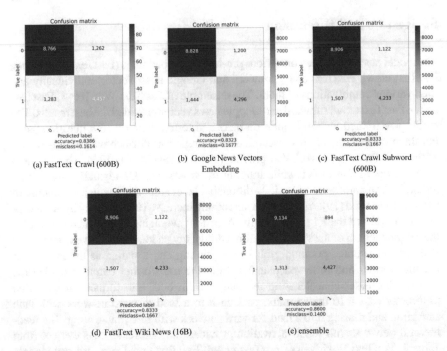

Fig. 3 Confusion matrix of different word embeddings on test data

Table 3 Results using different word embeddings

Word embedding	Accuracy (%)	Precision (%)	Recall (%)	F1-Score (%)
FastText Crawl (600B)	83.85	82.58	82.53	82.55
FastText Crawl Subword (600B)	83.32	82.28	81.27	81.72
FastText Wiki News (16B)	83.36	82.21	81.54	81.85
Google News Vectors	83.23	82.05	81.43	81.72
Ensemble of 4 word embeddings	86.0	85.31	84.1	84.63

all other models have predicted the label correctly, the combination model predicts the label correctly. In the third and fifth questions, the pattern of the results is similar to the second question but instead of the model with Google news vector, the model with FastText subword crawl predicts the wrong class. In the fourth question, two models predict the wrong class and two predict the right class so the prediction of the combination model depends on the probability with which each model predicts the class.

In Figs. 4 and 5, a graphical description on the variation of the accuracy and loss (in Figs. 4 and 5, respectively) is shown for all of the embeddings used in the ensemble. Each of the embedding's performance is analyzed during its respective training phase.

Fig. 4 Training and validation accuracies for individual word embeddings

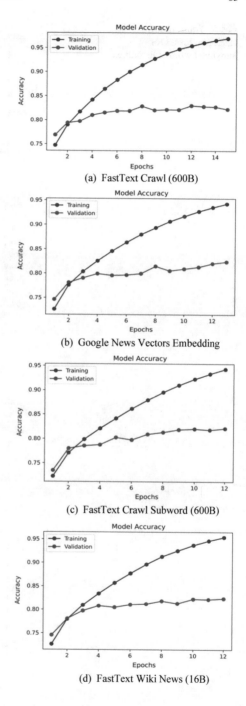

(a) FastText Crawl (600B)

(b) Google News Vectors Embedding

(c) FastText Crawl Subword (600B)

(d) FastText Wiki News (16B)

Fig. 5 Training and
validation losses for
individual word embeddings

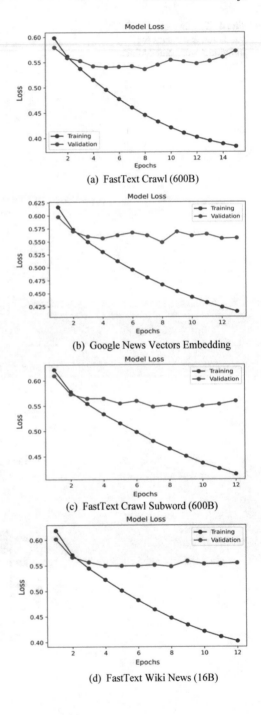

(a) FastText Crawl (600B)

(b) Google News Vectors Embedding

(c) FastText Crawl Subword (600B)

(d) FastText Wiki News (16B)

The metrics that are used to evaluate this model's performance are accuracy, model loss and $F-1$ score. Accuracy simply helps in understanding that out of all tested examples, how many are predicted correctly. The proposed model uses a cross-entropy loss. $F1$ Score is used in cases where the distribution of all labels are unequal. $F1$ Score is the weighted average of precision and recal. Therefore, this score takes both false positives and false negatives into account. A true positive is a correctly predicted positive class; similarly, a false positive is a positive class predicted incorrectly, a true negative is a correctly predicted negative class and a false negative is a negative class predicted incorrectly (Table 4).

$$\text{Precision} = \frac{TP}{TP + FP} \tag{7}$$

$$\text{Recall} = \frac{TP}{TP + FN} \tag{8}$$

$$F1 \text{ score} = 2 \times \frac{\text{Precison} \times \text{Recall}}{\text{Precision} + \text{Recall}} \tag{9}$$

The successful performance of this architecture opens doors to new methodologies and techniques for this use case. The future scope and concluding remarks on this research have been discussed in the next section.

6 Conclusion and Future Work

In this paper, transformers have been used as the primary source of encoding text to extrapolate its semantic meaning (feature extraction). It is found that the performance by coupling of transformer-based encodings and deep neural networks for similarity calculation performed comparable and sometimes better results than existing methods depending on the preprocessing and word embeddings used. To improve performance further, a combination of the results predicted by four different word embedding models (Google news vector, FastText crawl, FastText crawl subwords and FastText Wiki News) has been used. This combination helped us achieve a performance superior to the one achieved when their embeddings were used individually. This research managed to achieve an accuracy of 86% and an $F1$ of 84.63% with this ensemble model.

As a continuation of this research, various data augmentation techniques can be used to counter overfitting of the proposed models during training phase after a given number of iterations. The use of different distance metrics can be used to calculate the similarity of the transformer encodings, and at the same time, different transformers like BERT, ALBERT, T5-11B, which have proved most effective for the process of text similarity can be used in this architecture in order to attempt to produce more accurate feature extraction.

Table 4 Question pair classification using different word embeddings and their combination

Question 1	Question 2	True Label	Google News Vectors	FastText Wiki News (16B)	FastText crawl subword (600B)	FastText crawl (600B)	Ensemble of 4 word embeddings
Is Kellyanne Conway annoying in your opinion?	Did Kellyanne Conway really imply that attention should not be paid to the words that come out of Donald Trump's mouth?	0	0	0	0	0	0
What are your views on the Netaji Subhash Chandra Bose's mystery death case?	What is secret of Subhash Chandra Bose's death?	1	0	1	1	1	1
How do I play Pokemon GO in Korea?	How do I play Pokemon GO in China?	0	0	0	1	0	0
What is the probability that a leap year has 53 Sundays?	What are some unknown facts about leap year and 29th February?	0	0	1	0	1	1
Why is US mainstream media biased against Trump?	Why mainstream media hates Donald Trump?	1	1	1	0	1	1

References

1. Majumder, G., Pakray, P., Gelbukh, A.F., Pinto, D.: Semantic textual similarity methods, tools, and applications: a survey. Computación y Sistemas **20**(4), 647–665 (2016)
2. Bogdanova, D., Santos, C., Barbosa L., Zadrozny, B.: Detecting semantically equivalent questions. In: Online user forums. Proceedings of the 19th Conference on Computational Natural Language Learning, pp. 123–131 (2015)
3. Quora. What's Quora's policy on merging questions? https://help.quora.com/hc/en-us/art icles/360000470663-What-s-Quora-s-policy-on-merging-questions (2019). Online Accessed 20 May 2020
4. Quora. Quora question pairs, version 1. https://www.kaggle.com/c/quora-questionpairs/data (2017, November). Online Accessed 20 May 2020
5. Gomma, W.H., Fahmy, A.A.: A survey of text similarity approaches. Int. J. Comput. Appl. **68**(13), 13 (2013)
6. Neculoiu, P., Versteegh, M., Rotaru M.: Learning text similarity with siamese recurrent networks. In: *Rep4NLP@ACL* (2016)
7. Dadashov, E., Sakshuwong, S., Yu, K.: Quora question duplication, Stanford University (2017)
8. Imtiaz, Z., Umer, M., Ahmad, M., Ullah, S., Choi, G.S., Mehmood, A.: Duplicate questions pair detection using siamese MaLSTM. IEEE Access **8**, 21932–21942 (2020). https://doi.org/10.1109/ACCESS.2020.2969041
9. Vaswani, A.,Shazeer, N., Parmar, N., Uszkoreit, J., Jones, L., Gomez, A., Kaiser, L., Polosukhin, I.: Attention Is All You Need (2017)
10. Ostendorff, M., Ruas, T., Schubotz, M., Rehm, G., Gipp, B.: Pairwise multi-class document classification for semantic relations between wikipedia articles. In: Proceedings of the ACM/IEEE Joint Conference on Digital Libraries (JCDL) (2020)
11. Yüksel, A.E., Türkmen, Y.A., Ozgür, A., Altınel, A.B.: Turkish tweet classification with transformer encoder. In: Proceedings of Recent Advances in Natural Language Processing, pp. 1380–1387, Varna, Bulgaria, Sep 2–4, pp. 1380–1387 (2019)
12. Alammar, J.: The Illustrated Transformer (2018). https://jalammar.github.io/illustrated-transf ormer/
13. Ba, J, Kiros, J., Hinton, G.: Layer normalization. arXiv preprint arXiv:1607.06450 (2016)
14. Srivastava, N., Hinton, G., Krizhevsky, A., Sutskever, I., Salakhutdinov, R.: Dropout: a simple way to prevent neural networks from overfitting. J. Mach. Learn. Res. **15**(1), 1929–1958 (2014)
15. Mihaltz, M.: word2vec-googlenews-vectors. https://github.com/mmihaltz/word2vecGoogleN ewsvectors
16. Joulin, A., Grave, E., Bojanowski, P., Douze, M., Jégou, H., Mikolov, T.: Fasttext.zip: compressing text classification models. In: CoRR, abs/1612.03651 (2016)
17. Bojanowski, P., Grave, E., Joulin, A., Mikolov, T.: Enriching word vectors with subword information. In: CoRR, abs/1607.04606 (2016)
18. Kazemnejad, A.: Transformer Architecture: The Positional Encoding. https://kazemnejad.com/blog/transformer_architecture_positional_encoding/
19. Muller R., Kornblith S., Hinton, G.: When does label smoothing help In: arXiv preprint arXiv: 1906.02629 (2019)

Detection and Monitoring of Alzheimer's Disease Using Serious Games—A Study

Shrushti Jagtap, Atharva Kawade, Vikramjit Banerjee, Sharvari Gadiwan, Rajeev Ramesh, Aman Shinde, and Prashant Gadakh

Abstract Alzheimer's disease is an intensifying disorder attacking the neurons of the brain which causes hampering of memory skills and language skills as well as a change in behaviour of the patient affected. Alzheimer's has no definite cure but the patient's life can be prolonged with the help of certain treatments. Serious games have proven to be a source with high potential for the improvement of cognitive abilities. The paper suggests a single application that can be accessed anywhere, free of cost to single-handedly detect and identify the symptoms of Alzheimer's using a series of tests that can be conducted anywhere and predict if the user has Alzheimer's or not along with corresponding stages using machine learning algorithms.

Keywords Alzheimer's disease · Serious games · Machine learning · Cost-effective

S. Jagtap · A. Kawade · V. Banerjee · S. Gadiwan · R. Ramesh (✉) · A. Shinde
Hope Foundation's, International Institute of Information Technology, Pune, India
e-mail: rajeevramesh21@gmail.com

S. Jagtap
e-mail: shrushti.jagtap1999@gmail.com

A. Kawade
e-mail: atharvakawade@protonmail.com

V. Banerjee
e-mail: vikramjit1999@gmail.com

S. Gadiwan
e-mail: 1999sharvari@gmail.com

A. Shinde
e-mail: amanshinde.i2it@gmail.com

P. Gadakh
Department of Computer Engineering, Savitribai Phule Pune University, Pune, India
e-mail: prashantg@isquareit.edu.in

© The Editor(s) (if applicable) and The Author(s), under exclusive license
to Springer Nature Singapore Pte Ltd. 2021
S. Shakya et al. (eds.), *Proceedings of International Conference on Sustainable Expert Systems*, Lecture Notes in Networks and Systems 176,
https://doi.org/10.1007/978-981-33-4355-9_6

69

1 Introduction and Motivation

Alzheimer's is a neurodegenerative illness, the cause of which is still unknown. It is the most common cause of dementia. It is a generic term for cognitive impairment that is very severe, causes hindrance in day-to-day life. The elderly population is most prone to get affected by Alzheimer's. Early detection of AD is important so that preventative measures can be taken. Alzheimer's disease is incurable, but certain treatments can be used for slowing down the advancement of the disease [1, 2].

All over the world, an estimated 44 million people suffer from Alzheimer's disease. It is said that every 4 s, a new case of dementia is diagnosed [1]. Among old people, it is the third leading cause of death after heart diseases and cancer (Refer Fig. 1) [3, 4]. There are over 4 million people suffering from Alzheimer's and other forms of dementia in India. 47.5% of people in India live with it without even knowing about it. In the USA, 5.5 million people of an estimated age have Alzheimer's disease (Refer Fig. 2).

Serious games have a primary purpose other than amusement or relaxation. Under the commercial and educational categories, some serious games have been widely used that improve cognitive skills such as attention, visual–spatial abilities and memory skills among people [4]. There are a lot of serious games available worldwide but they are not used to detect Alzheimer's or any other form of dementia. The currently available serious games are developed for the sole purpose of entertainment. People without any cognitive impairment can also improve their cognitive skills by playing these serious games [6]. Instead, they can be used for healthcare purposes, after some updates in them so that they can be used for detection of dementia, especially Alzheimer's [7].

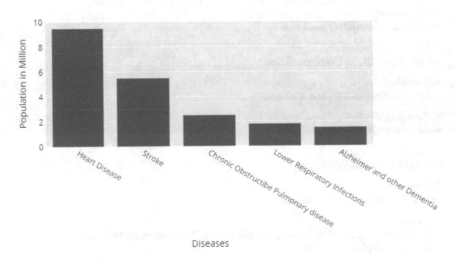

Fig. 1 Global cause of death in 2016 [5]

Fig. 2 Alzheimer's age-wise
patient distribution in the
USA among 5.8 million [5]

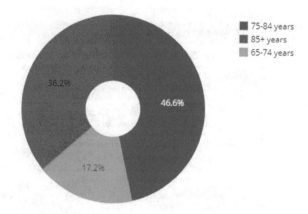

2 Findings and Methodology

The disease is categorized into seven different stages, and each of the stages has few
different symptoms. The diagram (Fig. 3) shows the stages of Alzheimer's along
with their symptoms briefly classified.

The proposed method aims at detecting Alzheimer's disease (AD) at the near stage
so that the patient will have a brief idea of her/his mental health before consulting
a doctor [6]. The system uses serious games which correspond to the symptoms of
each stage of AD. Furthermore, after seeking medical consultation, the user can opt

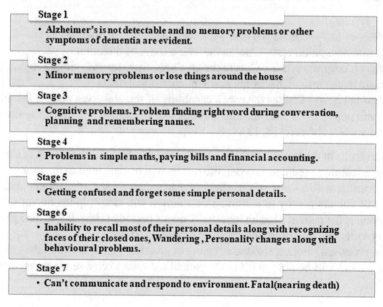

Fig. 3 Stages of Alzheimer's disease

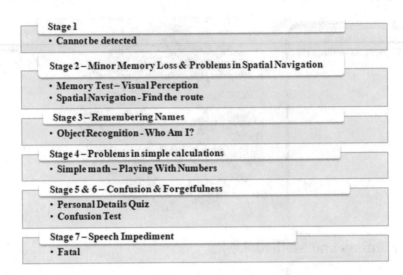

Fig. 4 Proposed serious games corresponding to each stage of Alzheimer's disease

to continue using serious games to monitor her/his mental health and consult their doctor for continuous evaluation. In the proposed solution, the symptoms which the patient has will be identified through the game (Refer Fig. 4) [8, 9].

Each level of this game is designed in accordance with the stages of Alzheimer's which will help in the identification of the problems the user is dealing with [10].

The games along with their basic functionality are as follows (Refer Figs. 4 and 5):

1. Memory test—Checks the user's immediate/short-term memory [11].
2. Spatial navigation test—Checks the user's sense of direction.
3. Object recognition test—Checks the user's ability to identify simple objects [12].
4. Basic maths test—Checks the user's mathematical skills.
5. Personal quiz—Checks the user's capability of remembering personal details.
6. Confusion test—Checks the user's ability to stay focused in a confusing situation and answer correctly [8, 13].

The results of these games will be stored in the database. With the help of a proper machine learning prediction model, the results will be delivered to the user [14].

The result will consist of the stage of Alzheimer's user, precautions and also details of doctors nearby to consult. This solution will also help the doctor to track and monitor the patient's health repeatedly over a period of time. It will also give the patient a clear picture of her/his medical progress [10, 15].

The proposed solution also has a feature of guardian panel, in which the guardian of the patient will be able to fill in the patient's details to ensure the correctness of data.

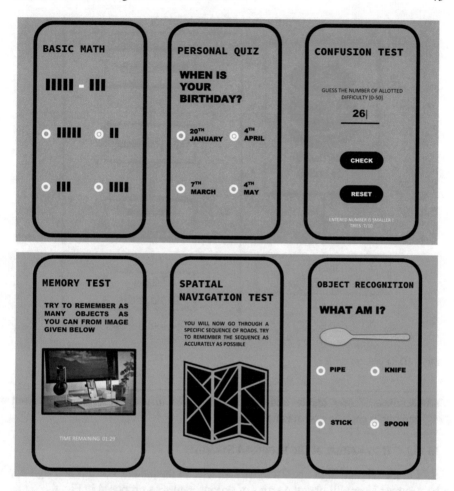

Fig. 5 UI of games showing basic functionality

In Fig. 6(Data Process Cycle):

a. Data Extraction: Initially, the test results and personal details filled by user/guardian will be stored in a database.

b. Data Processing: All stored details are fetched, processed and analysed. The raw information is processed into meaningful data.

c. Predictive Model: The result obtained from the games will be passed to the predictive model. With the help of machine learning and the existing database, the model will predict if a user has AD or not. Classification algorithms like decision tree, neural networks, logistic regression, Naïve Bayes, etc., can be used to classify into seven stages of Alzheimer's.

d. Model Deployment: Deploys model on any cloud infrastructure for 24×7 uptime and real-time changes.

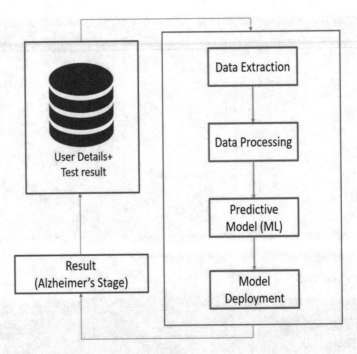

Fig. 6 Data process cycle

e. Database: Stores results and user details. Used to share results with relevant personnel such as doctors and guardians.

In Fig. 7 (Flow chart of the Proposed System):

a. Fill Details and Sign-up: User/guardian will fill the patient's details and sign-up.
b. Games: User will play the series of serious games as proposed in Fig. 4
c. Game Results: The game results are stored in the database in a suitable data format which can be used for predictive modelling.
d. Predictive Model: The results of the games are passed to the predictive model as mentioned in Fig. 6. The model uses machine learning algorithms to predict if the user has Alzheimer's disease (AD) or not. If AD is detected, the stage which the user is facing currently will be displayed and doctors will also be suggested
e. Result: Stage of Alzheimer's: If Alzheimer's is detected, the stage and the severity are predicted
f. Doctor/Guardian: The patient's result related to Alzheimer's is shared with the guardian or doctor for further diagnosis.

Fig. 7 Flow chart of the
proposed system

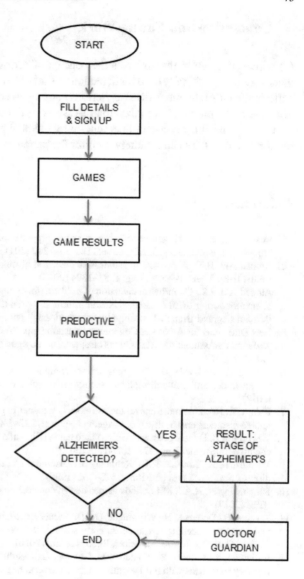

3   Results and Discussions

The results obtained from the games will be given as input to the pre-trained machine
learning predictive model. The model will predict the stage of Alzheimer's the patient
is suffering, along with the precautions to be taken and nearby expert doctors.

4 Conclusion and Future Work

Cognitive stimulation therapy uses the concept of serious games. Existing serious games are not specifically used for detection of Alzheimer's or as a tool to help people suffering from Alzheimer's disease. Serious games powered by machine learning are effective in the prediction of Alzheimer's disease at the early stages.

Thus, the need for constant intervention by medical personnel could be reduced or eliminated as they can remotely monitor the patient, saving time and resources.

References

1. Rosen, A., et al.: Cognitive training changes hippocampal function in mild cognitive impairment: a pilot study. J. Alzheimers Dis. 234–240 (2011)
2. Ahmad Akl, B.T.: Autonomous unobtrusive detection of mild cognitive impairment in older adults. IEEE Trans. Biomed. Eng. 1383–1394 (2015)
3. Ahmad Akl, J.S.: Unobtrusive detection of mild cognitive impairment in older adults through home monitoring IEEE J. Biomed. Health Inform. 224–234 (2015)
4. Barbara Legrand Brandt, J.A.:Cognitive apprenticeship approach to helping adults learn. In: New Directions for Adult and Continuing Education, pp. 22–29 (1993)
5. Alzheimer's Association "Alzheimer's disease facts and figures". Alzheimer's Dementia 15(3), 321–387 (2019)
6. CharikliaTziraki, R.B.-D.: Designing serious computer games for people with moderate and advanced dementia: interdisciplinary theory-driven pilot study. JMIR Serious Games 34–47 (2017)
7. Robert, P.: Recommendations for the use of serious games in people with Alzheimer's disease, related disorders and frailty. Front. Ageing Neurosci. 222–240 (2014)
8. Weybright, E.:Effects of an interactive video game (Nintendo Wii) on older women with mild cognitive impairment, pp. 271–287. IEEE (2010)
9. Mendez, M.F., Tomsak, R.L., Remular, B.: Disorders of the visual system in Alzheimer's disease. J. Clin. Neuroophthalmol. 320–332 (1990)
10. Samarasinghe, H.A.S.M.: Serious games design considerations for people with Alzheimer's. IEEE (2017)
11. Takechi, H., Dodge, H.H., Wijesinghe, D.P.D.: Scenery picture memory test: a new type of quick and effective screening test to detect early stage Alzheimer's disease patients, pp. 345–356. Geriatr. Gerontol. Int. ACM, Petras, Rhodes Island (2010)
12. Belleville, S., Gilbert, B., Fontaine, F., Gagnon, L., Gauthier, S.: Improvement of episodic memory in persons with mild cognitive impairment and healthy older adults: evidence from a cognitive intervention program. Dementia Geriatr. Cogn. Disorders 22(5–6), 486–499 (2006)
13. Yamaguchi, H.: Rehabilitation for dementia using enjoyable video-sports games. Int. Psychogeriatrics 989–994 (2010)
14. Austin, J., Dodge, H.H., Riley, T., Jacobs, P.G., Thielke, S., Kaye, J.: A smart-home system to unobtrusively and continuously assess loneliness in older adults. IEEE J. Transl. Eng. Health Med. 1–11 (2016)
15. Ian McCarthy, T.S.: Detection and localisation of hesitant steps in people with Alzheimer's disease navigating routes of varying complexity. Healthc. Technol. Lett. 220 (2019)

Factors Affecting the Online Travel Purchasing Decision: An Integration of Fuzzy Logic Theory

Oumayma Labti and Ezzohra Belkadi

Abstract Understanding consumer's characteristics should be a crucial question in studying consumer behavior. In this context, investigation is done on how consumers' characteristics affect their intention to purchase travel online. This research aims to examine the effect of extraversion and personal innovativeness on consumers' travel online purchase intention, by using the fuzzy logic technique. The implementation of fuzzy logic enables us to identify the level of these factors' effect on online purchase intention, and it helps to identify the link that exists between input variables and output variables. To do this, 102 sample surveys were collected from customers with experience in purchasing travel online. Results show that extraversion and personal innovativeness have both a meaningful influence on consumers' travel online purchase intention

Keywords Online purchase intention · Travel industry · Personality characteristics · Extraversion · Personal innovativeness

1 Introduction

Over recent years, information technology has emerged as an increasingly important part of our personal and professional lives and has remarkably influenced them. Nowadays, no one can argue the prominent relevance of technology in society as it affects more and more people's lives. Technology has influenced both the way services are offered and the nature of the services provided [1]. This substantial

O. Labti (✉) · E. Belkadi
Laboratoire de Recherche en Management, Information et Gouvernances, Faculté Des Sciences Juridiques Économiques et Sociales Ain-Sebaa, Hassan II University of Casablanca, BP 2634 Route des Chaux et Ciments Beausite, Casablanca, Maroc
e-mail: Labtioumayma6@gmail.com

E. Belkadi
e-mail: zo.belkadi@yahoo.fr

S. Shakya et al. (eds.), *Proceedings of International Conference on Sustainable Expert Systems*, Lecture Notes in Networks and Systems 176, https://doi.org/10.1007/978-981-33-4355-9_7

77

progress in technology, boosted by the Internet, has brought changes to the business environment in many markets. The Internet permits consumers to look for information and purchase products/services at lower costs when compared with offline shopping. Consumers can find product's information through a fast search process on the Internet helping to save time and money [2]. In addition to this, it has been a medium through which commercial communications, sales transactions and logistics have been facilitated [3]. Therefore, the Internet is considered as a central marketing tool for many products/services [2], with the dramatic expansion of Internet users. Similarly, the Internet has led to a revolution in business operations in all service sectors, including online services [4].

According to the survey on the online shopping behaviors conducted in 2014, led by MasterCard, Moroccan consumers are increasingly making their purchases on the Internet. Almost half of those surveyed reported making online purchases over the Internet, with more than 90% saying they are very satisfied with their online shopping experience. Consumers reported spending the majority of their money on airlines, travel and hotels. The travel industry is just as much a part of this revolution. Many travel consumers use the Internet either to book airline tickets, make hotel reservations or other online purchases instead of going to travel agencies to do it for them [5]. This has been supported in various studies with the requirement that the Internet offers to travel consumers a greater choice of products, extra information and often competitive prices than they can get in travel agencies [6]. In this context, it is vitally important to understand consumers' intention to purchase travel online. Most research has been interested in testing existing theories of consumer behavior. The technology acceptance model (TAM) [7] has been heavily utilized to investigate online purchase behavior in the travel context [8]. The unified theory of acceptance and use of technology (UTAUT) [9] have also been employed to examine various drivers of online airline ticket purchasing behavior [10]. One further study [11] has used the theory of planned behavior (TPB) [12], to examine travelers' intention to purchase travel online by including trust and risk. Another research has focused on determinants and results of consumer trust toward online travel Web sites [13]. The authors of those previous papers have investigated attitude models that seem to be a fairly sustainable foundation. Another key driver in the acceptance of Internet shopping is personality because individual differences affect the decision customers make. Few research has focused on the consequence of personal characteristics on the consumer's decision [14]. It is, therefore, necessary to examine the influence of personality characteristics on travel online purchase intention and that makes the current study urgent and topical.

This study attempts to determine how travelers' online purchase intention varies depending on their personal innovativeness and on their extraversion degree. The remainder of the paper is structured into five sections: Sect. 2 briefly presents literature associated with the subject of the study, including the definition of constructs. Methodological details are then given in Sect. 3 including an introduction of the concept of fuzzy inference system. In Sect. 4, the results and discussions are provided. Section 5 outlines the conclusion, limitations and directions for future research.

2 Literature Review

The theory of reasoned action [15] is able to describe human behavior drivers. In that respect, humans are rational and will look at the outcomes of their acts when deciding to perform a particular behavior [16]. This theory is considered as a pioneer model for understanding individuals' behavior toward technology adoption [17]. According to this theory, customers' purchase behavior is determinable from customers' purchase intention [16], and this concept is one of the most studied variables in marketing and management research [18]. Therefore, online purchase intention is used in this study as a dependent variable. Online purchase intention can be explained by many factors [19]. Huang [20] and Lan [21] observed that emotional and psychological attributes could be used to measure online purchase intention. Demographic factors can also be viewed as parameters that influence the online purchase intention. According to Brown et al. [22], gender can affect the intention to purchase products online. Monica and Mark [23] have worked on gender factors and shoppers backgrounds and their consequences on the online purchase intention. In addition to demographic characteristics, consumers' shopping motivations have been reached predicting online purchase intention [24]. Nevertheless, little research has been done to evaluate the impact of personality characteristics on customers' online purchase intention. Hence, the relevance of examining the impact of personality characteristics on online purchase intention.

Personal innovativeness is a very important dimension in order to study individual behavior in the context of innovation. Several researchers have tried to define this concept. Hurt et al. [25] have considered innovativeness as a personality characteristic that reflects "The willingness to change." Consumer innovation corresponds to individual's response to new experiences and new stimuli [26]. Consumers' capacity for innovation is described as the tendency to shop new and various goods instead of following traditional choices and consumption habits [27]. In particular, this construct was identified in the context of information technology "the openness of an individual to experiment emerging information technology" [28], in other words, it means that highly innovative users are likely to adopt computer innovations rather than others that are not very innovative [29]. It is generally accepted that people who are highly innovative embrace new technologies very easily [30]. Bommer and Jalajas [31] have linked personal innovativeness to risk tolerance, meaning that individuals tend to innovate if they are willing to take risks, particularly in the area of Internet shopping, as online shopping involves risks and uncertainties [32]. This concept has been studied in various fields of research, such as office automation [28], web retailing [29], online shopping [33] and online travel purchasing [34]. Kanthawongs [35] have found that personal innovativeness favorably affects the intention to purchase online clothing. However, Lestari [36] has rejected the hypothesis that assumes that consumers' innovativeness has no influence on the intention to use an online shopping forum.

Recently, a limited but emerging body of research literature on personality traits in terms of consumer behavior has appeared [37]. Earlier studies found that shoppers' character is a substantial factor for the success of e-vendors [38]. Personality can be defined as a group of characteristics or traits relatively durable, which distinguish individuals between them and trigger certain reactions to stimuli coming from their environment [39]. Over the last many years, the big five models have been considered as a supported model of personality [40]. According to this model, five aspects are needed to introduce human personality in terms of traits: neuroticism, extraversion, agreeableness, openness and conscientiousness [37]. In an online shopping context, Tsao and Chang [41] have observed that the five factors could affect the shoppers' buying motivations. Due to the lack of research on the extraversion trait, we will focus on this concept and its impact on the intention to purchase online. Based on the literature review, Robu [42] has considered that extroverts are communicative, dynamic, act in a controlling manner and are seeking for sensation. For those having a limited interest in communication, online shopping may be the preferred choice [43]. Nevertheless, consumers with a high desire to interact with others will be prone to buy from offline stores [44]. Saleem et al. [45] have shown that there is a favorable relationship between extraversion and computer use. At the same time, Khare et al. [46] found that extroverts tend to assume risks as they are experimenting with innovations; as a result, they are inclined to adopt online banking. Lissitsa and Kol [47] have found that extraversion is positively correlated with m-shopping intention. Hence, not only consumers' innovativeness may determine their purchase intention, but it may also depend on the consumers' extraversion degree.

The analysis of the literature reveals that personal characteristics, especially personal innovativeness, can affect the adoption of technological innovations in general and the online purchase intention in particular. Extraversion was found to significantly enhance purchasing behavior. Extraversion can positively affect purchase intention in a digital context. Following this line of research, this paper focuses on two main aspects of personal characteristics as determinants of travel online purchase intention, namely extraversion and personal innovativeness.

In the existing consumer behavior literature, many studies are employing statistical techniques to examine consumer purchasing behavior. However, these traditional statistical approaches are not capable of explaining such phenomena. Indeed, the results of these techniques do not capture the imprecision, uncertainty and vagueness of the participants' choices. Therefore, fuzzy logic technique has been used in this paper.

Fuzzy logic applies set theory to examine the relationship, which exists of an independent variable (called input variable) with a dependent variable (called output variable), viewing the input or combinations of inputs as an adequate condition for the output. Fuzzy logic can lead to a more precise comprehension of the difficult reality behind the intention to buy online. Fuzzy logic may provide an alternative approach to investigate how an individual factor in a system may lead to a positive or negative influence on the online purchase intention. In other words, fuzzy logic provides a clearer explanation of how perceptual and personal factors interact to shape online purchase intention. Fuzzy logic has been first introduced by Zadeh [48]. This method has been developed in several works as a decision support instrument [49]. It is relevant to mention that fuzzy logic is an effective method for depicting uncertain, vague and imprecise data. In addition, fuzzy logic helps to provide more accurate insights into customer preferences and to improve the understanding of such a topic. This research helps to support the literature relative to consumer purchasing behavior.

3 Methodology

3.1 Fuzzy Inference System

A fuzzy logic inference system is used in this paper. It is a system that allows transferring input variables to output vectors starting from 0 to 1 following defined rules. Fuzzy logic inference system includes four-step process: they are fuzzification, rule-based inference, aggregation and defuzzification. (The design of the fuzzy logic system is illustrated by Fig. 1.)

(1) Fuzzification: It is the first step in the fuzzy inference system that transfers the system's input values into fuzzy sets with their corresponding ranges of values where they are determined.
(2) Rule-based inference: Defining the rules is a crucial phase in the process of developing fuzzy systems. Rules are made by a collection of fuzzy "If–Then" syntax that links between one or more antecedents clause and one or more consequences clause. The formation of fuzzy rules, in general, is derived by using questionnaires, panel techniques or from experts.
(3) Aggregation: In this step, the subsets are aggregated to form a single set and to calculate the fuzzy output.
(4) Defuzzification: It is the unit where the fuzzy output is converted to a crisp output.

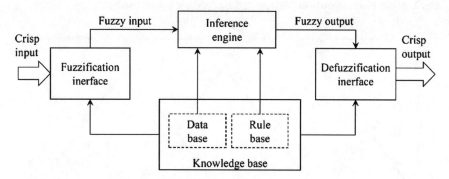

Fig. 1 Typical fuzzy logic controller structure

3.2 Sampling and Procedures

The fuzzy mathematical model presented in the current study intends to analyze online purchase intention as an output variable by using extraversion and personal innovativeness as input variables (Fig. 2). In this case, the first phase consists of data collection. A questionnaire was distributed online to collect data. The questionnaire included three constructs. A five-point Likert scale (from 1—strongly disagree to 5 strongly agree) was used for all these constructs. The second step in the fuzzy logic within our model is to associate input variables and output variables with membership functions. A triangular membership function is going to be employed in this study. In this context, it is relevant to indicate that the triangular membership function is commonly utilized in the practice. Triangular membership functions are made from straight lines. These straight lines are simple to work with and easy to compute. Extraversion (EX) and personal innovativeness (PI) are the main input variables. The corresponding fuzzy values for the first input extraversion (EX) are very low, low, medium, high and very high (Fig. 3). Similarly, for the second input variable personal innovativeness (PI), the corresponding fuzzy values are defined to be very low, low,

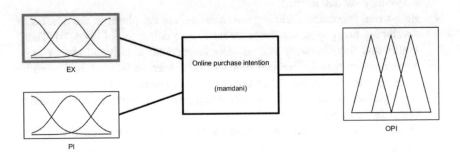

Fig. 2 Fuzzy logic system with two input variables and one output variable

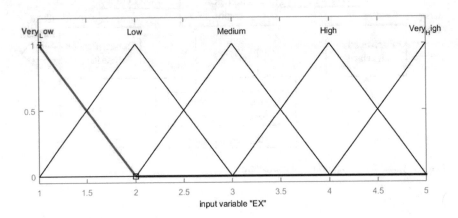

Fig. 3 Membership function plots for the input variable "extraversion"

medium, high and very high (Fig. 4). Finally, the output of the fuzzy model is the online purchase intention (OPI) denoted as very low, low, medium, high and very high (Fig. 5). The fuzzy rules have been extracted from the questionnaire responses. In this study, a total of 15 rules are defined (Fig. 6). In this contribution, the fuzzy Mamdani inference model [50] is applied. This tool allows adding rules via logical operators OR/AND. These rules are described through a series of fuzzy "If–Then" rules in which the antecedents/consequent implies linguistic variables. This series of rules explain the behavior of the system. After defining the rules, the gravity center or centroid method shall be implemented to get the crisp output. It is one of the most widely employed techniques for defuzzification [50]. To implement this method, fuzzy logic toolbox for the Mamdani system in MATLAB has been applied.

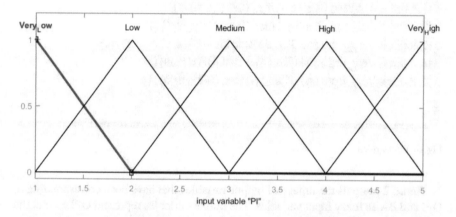

Fig. 4 Membership function plots for the input variable "personal innovativeness"

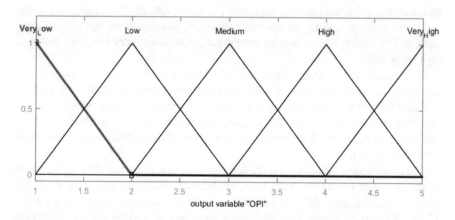

Fig. 5 Membership function plots for the output variable "online purchase intention"

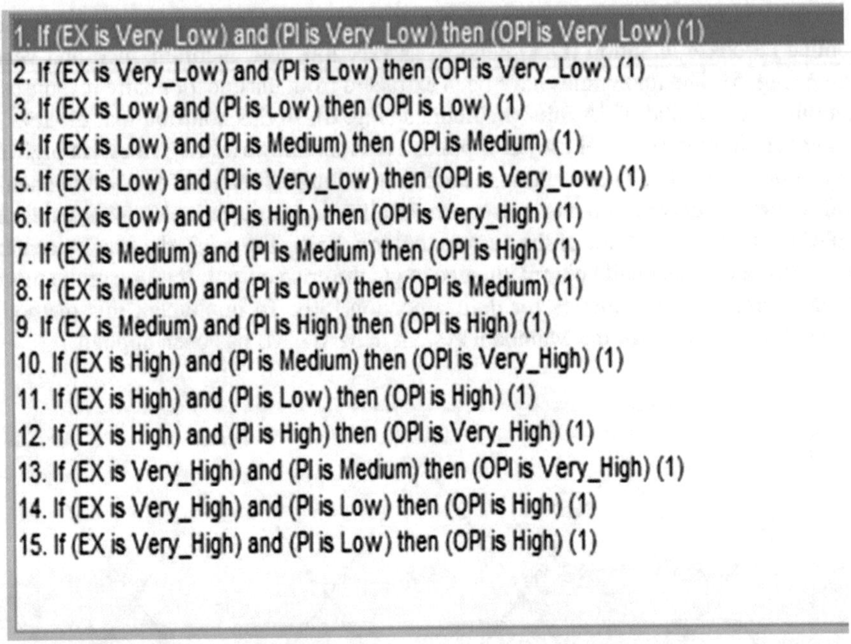

1. If (EX is Very_Low) and (PI is Very_Low) then (OPI is Very_Low) (1)

2. If (EX is Very_Low) and (PI is Low) then (OPI is Very_Low) (1)

3. If (EX is Low) and (PI is Low) then (OPI is Low) (1)

4. If (EX is Low) and (PI is Medium) then (OPI is Medium) (1)

5. If (EX is Low) and (PI is Very_Low) then (OPI is Very_Low) (1)

6. If (EX is Low) and (PI is High) then (OPI is Very_High) (1)

7. If (EX is Medium) and (PI is Medium) then (OPI is High) (1)

8. If (EX is Medium) and (PI is Low) then (OPI is Medium) (1)

9. If (EX is Medium) and (PI is High) then (OPI is High) (1)

10. If (EX is High) and (PI is Medium) then (OPI is Very_High) (1)

11. If (EX is High) and (PI is Low) then (OPI is High) (1)

12. If (EX is High) and (PI is High) then (OPI is Very_High) (1)

13. If (EX is Very_High) and (PI is Medium) then (OPI is Very_High) (1)

14. If (EX is Very_High) and (PI is Low) then (OPI is High) (1)

15. If (EX is Very_High) and (PI is Low) then (OPI is High) (1)

Fig. 6 Fuzzy rules

Figure 2 presents the input and output variables that have been corresponding to OPI and EX as fuzzy input variables. The output variables represent OPI on that the fuzzy sets have been determined.

Figure 3 shows the membership functions defined for the first input variable "EX," with five triangular membership functions (very low, low, medium, high, very high).

Figure 4 illustrates the membership functions developed for the second input variable "PI," using five triangular membership functions (very low, low, medium, high, very high).

The membership functions identified in Fig. 5 indicate five triangular membership functions for the output variable "OPI" (very low, low, medium, high, very high).

Once the membership functions for the fuzzy variables were established, the rules that the inference system will employ to generate the ultimate result have been outlined as seen in Fig. 6. Based on the responses of the questionnaire, those rules have been derived.

The questionnaire starts with some demographic questions, and then the measures of the different constructs have been established. The sample consisted of 102 Moroccan consumers. All the participants have already purchased travel online. 97.1% of the participants have received a university or college degree, with more females (51%) than males (49%). Approximately, 32.4% were <25 years old, 63.8% were aged 25–44, and 3.9% were 45 and older.

3.3 Measures

The reliability of the constructs in this research was evaluated through Cronbach alpha indicator. All used measures have been found reliable since all Cronbach's alpha values have exceeded 0.7. Extraversion (four indicators) with scale reliability Cronbach's $\alpha = 0,789$, personal innovativeness (four indicators) with scale reliability Cronbach's $\alpha = 0.713$ and online purchase intention (three indicators) with scale reliability Cronbach's $\alpha = 0.931$ (Table 1).

Table 1 Reliability of constructs

	Construct and indicator	Cronbach alpha	Scales origin
Input variables	*Extraversion* I am the life of the party I am skilled in handling social situations I make friends easily I know how to captivate people	0.789	Adapted from Buchanan [51]
	Personal innovativeness If I heard about a new method of buying travel, I would look for ways to experiment with it Among my peers, I am the first to explore online travel purchasing I like to experiment with new online travel purchasing techniques In general, I am hesitant to try to purchase travel online	0.713	Adapted from Agarwal and Prasad [28]
Output variables	*Online purchase intention* I will use the Internet to purchase travel regularly in the future I will frequently use the Internet in the future to purchase travel I will strongly recommend others to use the Internet to purchase travel	0.931	Adapted from Moon and Kim [52]

4 Results and Discussion

4.1 Fuzzy Inference System Results

The results of the inference system that have been developed in MATLAB are provided in Figs. 7, 8, 9, 10 and 11. The rule viewer and the surface viewer are presented, respectively. Once the membership functions for the fuzzy variables had been established, a definition of the rules that were used by the inference engines to obtain the final result has been made as given in Fig. 6. The rules were extracted from the questionnaire. Figs. 7, 8, 9 and 10 depict the rule viewer, and Fig. 11 illustrates the surface viewer of the fuzzy set.

A fuzzy set of rules linking the input variables and the output are constructed as illustrated by Fig. 7. The model is used to test the intention to purchase travel online using two inputs, namely extraversion and personal innovativeness. The result of the model is relevant on a scale of 1–5, as shown in Fig. 7. The developed model is evaluated by testing all input coefficients equal to 3, providing an output of 4, meaning a high level of purchasing travel online.

As shown in Fig. 8, the model is designed to be operated with inputs equal to 2. The fuzzy logic output of the model is 2, representing a low intention to purchase travel online.

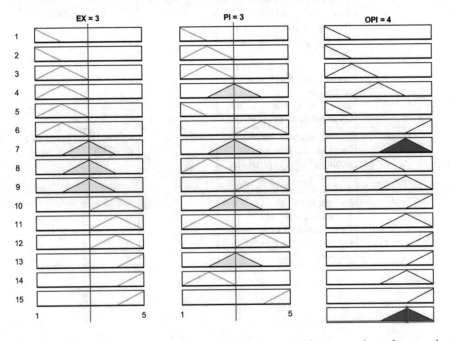

Fig. 7 Summary of results using the Mamdani's inference method for extraversion = 3, personal innovativeness = 3 and online purchase intention = 4

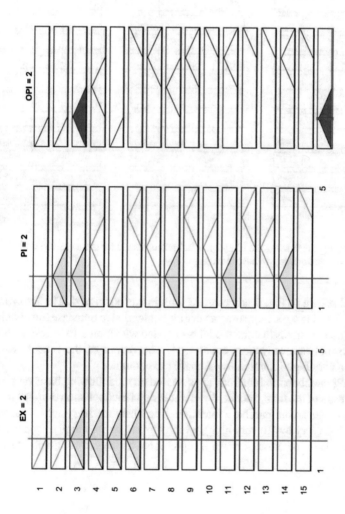

Fig. 8 Summary of results using the Mamdani's inference method for extraversion = 2, personal innovativeness = 2 and online purchase intention = 2

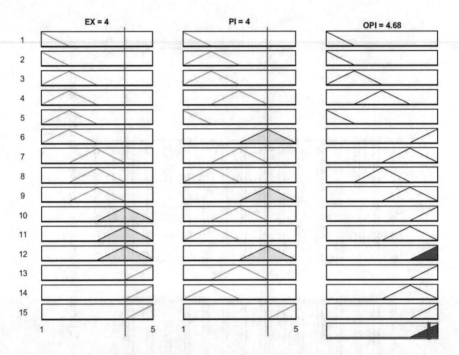

Fig. 9 Summary of results using the Mamdani's inference method for extraversion = 4, personal innovativeness = 4 and online purchase intention = 4.68

The model shown in Fig. 9 is measured with inputs equal to 4. The fuzzy logic output of the model is 4.68, indicating a very high intention to purchase travel online.

Figure 10 demonstrates that the model is operated with inputs, extraversion equal to 2 and personal innovativeness equal to 4. The output of the model is 4.68, representing a very high intention to purchase travel online.

EX and OPI membership functions are assessed on the following basis on a scale of 1–5. As a result of the fuzzy model, the membership functions as an output variable are based on travel online purchase intention. The fuzzy model result is measured on a scale of 1–5 (very low to very high), which turns to be very high as one nears 5. The online purchase intention of travel is represented by the surface graph as shown in Fig. 11.

4.2 Discussion

Figure 9 depicts a possible combination in which the results of the input variables extraversion, personal innovativeness would be EX = 4PI = 4, and the output variable online purchase intention equals to 4.68. Hence, Fig. 9 shows the relationship between extraversion, personal innovativeness and the online purchase intention.

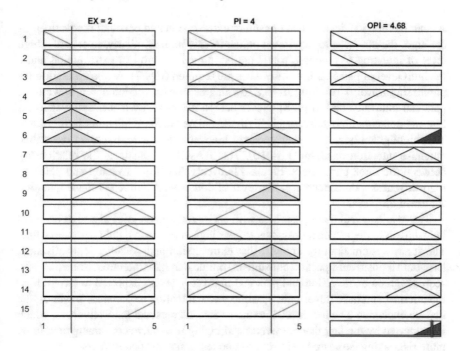

Fig. 10 Summary of results for extraversion = 2, personal innovativeness = 4 and online purchase intention = 4.68

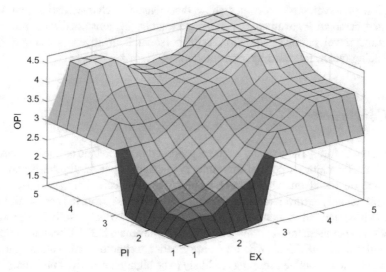

Fig. 11 Surface viewer of the relationship between the extraversion (EX), personal innovativeness (PI) and online purchase intention (OPI)

As can be seen in Fig. 9, when consumer extraversion and personal innovativeness are high, the intention to buy travel online is very high. Similarly, to ensure a high level of intention to purchase travel online, a medium level of extraversion and a medium level of personal innovativeness are required (Fig. 7). A weak intention for purchasing travel is resulting from a low degree of extraversion and a low rate of personal innovativeness (Fig. 8). Nevertheless, it was found that if extraversion is low and personal innovativeness is high, the online purchase intention is very high. This highlighted how important personal innovativeness is to assure a high online purchase intention of travel (Fig. 10). The 3D surface illustrates the relationship between the three parameters. Figure 11 reveals that if extraversion and personal innovativeness values decrease, the level of consumers' online purchase intention for travel decreases.

The results reveal that consumers with high extraversion level and high personal innovativeness degree are more closer to get a high intention to purchase travel online. The fuzzy system findings support that extraversion and personal innovativeness have both a positive impact on consumers' online purchase intention. Someone who scores high on extraversion and personal innovativeness is expected to have a high intention to purchase travel online, while a low intention to purchase travel online may be explained by a low level of extraversion and personal innovativeness, which is supported by the fact that extroverts and highly innovative consumers tend to be more risk-taking compared to introverts and less innovative consumers.

This paper introduces a new model in an e-commerce travel environment and has explored the relationship that exists between two factors and the online purchase intention of travel. The outcomes show that consumer characteristics (extraversion and personal innovativeness) have successfully demonstrated their influence on online travel purchase intention, with personal innovativeness being a high influencing factor, followed by extraversion.

5 Conclusion

In this contribution, a new model has been proposed to understand online consumers' behavior. The major purpose of this study is to examine personality traits and their influence on travel online purchase intention in Morocco. This study tests two concepts, namely extraversion and personal innovativeness, from the perspective that more extroverted individuals tend to assume more risks which results in a growing intention to purchase travel online. Personal innovativeness has also been included in our model since individuals who adopt information technology innovations particularly in the area of online shopping are likely to be highly innovative consumers. The fuzzy logic specificity consists of identifying the impact of the two inputs proposed in our model at once on the output. The findings revealed that both input variables have a positive impact on the output variable. With these findings, online service managers will be able to implement more appropriate strategies for consumers who tend to be less extroverts and also for consumers who are considered less innovative. Certain limitations of this study should be mentioned. Primarily, this work is limited by the sample size $N = 102$. Secondly, the respondents are all Moroccans, the necessity to generalize our findings in future studies is then required. In this paper, only two variables are considered as personality characteristics that could affect consumers' intention, but if other personality traits are going to be included, results ban be varied. Finally, as previously mentioned, fuzzy logic approach provides results of both input variables at the same time, therefore, it will be interesting to combine fuzzy logic with another analysis method as structural equation modeling (SEM) to investigate the effect of each input variable on the output variable.

Acknowledgements This article was supported by the "Moroccan national center of research" CNRST, Morocco.

References

1. Barnett, J.E., Scheetz, K.: Technological advances and telehealth: ethics, law, and the practice of psychotherapy. Psychotherapy Theory Res. Pract. Train. **40**(1–2), 86–93 (2003). https://doi.org/10.1037/0033-3204.40.1-2.86

2. McGaughey, R.E., Mason, K.H.: The internet as a marketing tool. J. Market. Theory Pract. Summer 1–11 (1998). https://doi.org/10.1080/10696679.1998.11501800

3. Burke, R.R.: Do you see what I see? The Future of virtual shopping. J. Acad. Mark. Sci. **25**, 352–360 (1997)

4. Parasuraman, A., Colby, C.L.: An updated and streamlined technology readiness index: TRI 2.0. J. Serv. Res. **18**(1), 59–74 (2015). https://doi.org/10.1177/1094670514539730

5. Morrison, A.M., Jing, S., O'Leary, J.T., Cai, L.A.: Predicting usage of the internet for travel bookings: an exploratory study. Inf. Technol. Tour. **4**(1), 15–30 (2001)

6. Xiang, Z., Gretzel, U.: Role of social media in online travel information search. Tour. Manage. **31**(2), 179–188 (2010). https://doi.org/10.1016/j.tourman.2009.02.016

7. Davis, F.D.: Perceived usefulness, perceived ease of use, and user acceptance of information technology. MIS Quart. **13**(3), 319–340 (1989)

8. Amaro, S., Duarte, P.: An integrative model of consumers' intentions to purchase travel online. Tour. Manage. **46**, 64–79 (2015). https://doi.org/10.1016/j.tourman.2014.06.006

9. Venkatesh, V., Thong, J.I.L., Xu, X.: Consumer acceptance and use of information technology: extending the unified theory of acceptance and use of technology. MIS Quart. **36**(1), 157–178 (2012)

10. Escobar-Rodríguez, T., Carvajal-Trujillo, E.: Online drivers of consumer purchase of website airline tickets. J. Air Transp. Manage. **32**, 58–64 (2013). https://doi.org/10.1016/j.jairtraman.2013.06.018

11. Suzanne, A., Paulo, D.: Travellers' intention to purchase travel online: integrating trust and risk to the theory of planned behavior. Anatolia (2016). https://doi.org/10.1080/13032917.2016.1191771

12. Ajzen, I.: The theory of planned behavior. Organ. Behav. Hum. Decis. Process. **50**(2), 179–211 (1991)

13. Agag, G., El-Masry, A.: Why do consumers trust online travel websites? Drivers and outcomes of consumer trust towards online travel websites. J. Travel Res. **56** (2016).https://doi.org/10.1177/0047287516643185

14. Labay, D.G., Kinnear, T.C.: Exploring the consumer decision process in the adoption of solar energy systems. J. Consumer Res. **24**(2), 271–278 (1997)

15. Ajzen, I., Fishbein, M.: Belief, attitude, intention, and behavior: an introduction to theory and research. Addison-Wesley, Boston, MA (1975)
16. Ajzen, I., Fishbein, M.: Understanding attitudes and predicting social behavior. Prentice-Hall, Englewood Cliffs, NJ (1980)
17. Venkatesh, V., Morris, M.G., Davis, G.B., Davis, F.D.: User acceptance of information technology: towards a unified view. MIS Quart. **27**(3) (2003)
18. Zeithaml, V.A., Berry, L.L., Parasuraman, A.: The behavioral consequences of service quality. J. Market. **60**(2), 31–46 (1996)
19. Pavlou, P.: Consumer acceptance of electronic commerce: integrating trust and risk with the technology acceptance model. Int. J. Electron. Commerce **7**(3), 69–103 (2003)
20. Huang, M.H.: Modeling virtual exploratory and shopping dynamics: an environmental psychology approach. Inf. Manag. **41**(1), 39–47 (2003)
21. Lan, X.: Affect as information: the role of affect in consumer online behaviors. In: NA— Advances in Consumer Research, 29 (2002)
22. Brown, M., Pope, N., Voges, K.: Buying or browsing? An exploration of shopping orientations and online purchase intention. Eur. J. Mark. **37**(11), 1666–1684 (2003)
23. Monica, L., Mark, N.: Age and gender differences: understanding mature online users with the online purchase intention model. J. Global Sch. Market. Sci. **26**(3), 248–269 (2016). https://doi.org/10.1080/21639159.2016.1174540
24. Wolfinbarger, M., Gilly, M.: Shopping online for freedom, control and fun. Calif. Manage. Rev. **43**(2), 34–56 (2001); Rogers, E.M.: Diffusion of Innovations (4th ed). The Free Press, NY (1995)
25. Hurt, H.T., Joseph, K., Cooed, C.D.: Scales for the measurement of innovativeness. Human Comm. Res.**4**, 58–65 (1977)
26. Goldsmith, R.E.: Convergent validity of four innovativeness scales. Educ. Psychol. Measur. **46**, 81–87 (1986)
27. Steenkamp, J.-B.E.M., ter Hofstede, F., Wedel, M.: A cross-national investigation into the individual and national cultural antecedents of consumer innovativeness. J. Market. **63**, 55–69 (1999)
28. Agarwal, R., Prasad, J.: A conceptual and operational definition of personal innovativeness in the domain of information technology. Inf. Syst. Res. **9**(2), 204–215 (1998)
29. O'Cass, A., Fenech, T.: web retailing adoption: exploring the nature of internet users web retailing behavior. J. Retail. Consumer Serv. **10**(2), 81–94 (2003)
30. Grace, P., Osman, Z., Cheuk, C.: Young adult Malaysian consumers' intention to shop via mobile shopping apps. Asian J. Bus. Res. **8**, 18–37 (2018)
31. Bommer, M., Jalajas, D.S.: The threat of organizational downsizing on the innovative propensity of R&D professionals. R&D Manage. **29**(1), 27–34 (1999)
32. Kwak, H., Fox, R.J., Zinkhan, G.M.: What products can be successfully promoted and sold via the internet? J. Advert. Res. **42**(1), 23–38 (2002)
33. Im, S., Bayus, B.L., Mason, C.H.: An empirical study of innate consumer innovativeness, personal characteristics, and new-product adoption behavior. J. Acad. Market. Sci. **31**(1), 61–73 (2003)
34. Sigala, M.: Reviewing the profile and behavior of internet users research directions and opportunities in tourism and hospitality. J. Travel Tour. Market. **17**(2–3), 93–102 (2004)
35. Kanthawongs, P.: Factors positively affecting intention to purchase of online clothing shoppers (2019)
36. Lestari, D.: Measuring e-commerce adoption behavior among gen-Z in Jakarta, Indonesia. Econ. Anal. Pol.**64**, 103–115 (2019). https://doi.org/10.1016/j.eap.2019.08.004
37. Luchs, M.G., Mooradian, T.A.: Sex, personality, and sustainable consumer behavior: elucidating the gender effect. J. Consumer Pol. **35**, 127–144 (2012)
38. Barkhi, R., Wallace, L.: The impact of personality type on purchasing decisions in virtual stores. Inf. Technol. Manage. **8**, 313–330 (2007). https://doi.org/10.1007/s10799-007-0021-y
39. Kassarjian, H.H., Sheffet, M.J.: Personality and consumer behavior: an update. In: Kassarjian, H.H., Robertson, T. (eds.), Perspectives in Consumer Behavior, pp. 281–303. Englewood Cliffs, NJ, Prentice-Hall (1991)

40. McCrae, R.R., Costa, P.T., Jr.: Validation of the five-factor model of personality across instruments and observers. J. Pers. Soc. Psychol. **52**(1), 81–90 (1987)
41. Tsao, W.C., Chang, H.R.: Exploring the impact of personality traits on online shopping behavior. Afr. J. Bus. Manage. **4**(9), 1800–1812 (2010)
42. Robu, V.: Un posibil model de integrare a rezultatelor la inventarul NEO PI-R în cadrul demersului de evacuare specific selecției de personal. Revista de psihologie și științele educației **1**(3) (2007)
43. Dabholkar, P., Bagozzi, R.: An attitudinal model of technology-based self-service: moderating effects of consumer traits and situational factors. J. Acad. Market. Sci. J ACAD MARK SCI. **30**, 184–201 (2002). https://doi.org/10.1177/0092070302303001
44. Monsuwé, T., Dellaert, B., De ruyter, K.: What drives consumers to shop online? a literature review. Int. J. Serv. Ind. Manage. **15**, 102–121 (2004). https://doi.org/10.1108/095642304105 23358
45. Saleem, H., Beaudry, A., Croteau, A.M.: Antecedents of computer self-efficacy: a study of the role of personality and gender. Comput. Hum. Behav. **27**(5), 1922–1936 (2011)
46. Khare, A., Khare, A., Singh, S.: Role of consumer personality in determining preference for online banking in India. J. Database Market. Cust. Strat. Manage. **17**, 174–187 (2010)
47. Lissitsa, S., Kol, O.: Four generational cohorts and hedonic m-shopping : association between personality traits and purchase intention. Electron. Comm. Res. (2019). https://doi.org/10.1007/s10660-019-09381-4
48. Zadeh, L.A.: Fuzzy sets. Inf. Control **8**, 338–353 (1965)
49. Darney, P.E., Jacob, I.J.: Performance enhancements of cognitive radio networks using the improved fuzzy logic. J. Soft Comput. Paradigm (JSCP) **1**(02), 57–68 (2019)
50. Mamdani, E.H.: Application of fuzzy algorithms for control of simple dynamic plant. Proc. Inst. Electric. Eng. **121**(12), 1585 (1974). https://doi.org/10.1049/piee.1974.0328
51. Buchanan, T.: Online implementation of an IPIP. Five Factor Personality Inventory (2001)
52. Moon, Y.W., Kim, Y.G.: Extending the TAM for a world-wide-web context. Inf. Manage. **38**(4), 217–230 (2001)

An Empirical Study on the Occupancy Detection Techniques Based on Context-Aware IoT System

Kavita Pankaj Shirsat and Girish P Bhole

Abstract Occupancy detection and behavior in buildings has a huge impact on cooling, heating, ventilation demand, building controls, and energy consumption in lighting appliances. The human factor is an important factor in real-time occupancy information and building energy management systems that offer great potential for maximizing energy efficiency and assessing energy flexibility. The occupancy predictive strategy provided a better quality of service and energy savings performance than reactive strategies. In this research paper, 20 papers based on context-aware IoT systems for occupancy detection are reviewed. The research works are categorized into the sensor, sensor fusion, Wi-Fi, LAN, radio frequency (RF) signals, machine learning, and so on. The research gaps and the challenges faced during the occupancy detection are listed for further enhancement in the occupancy detection methods. The research work is analyzed based on the performance metrics, classification methods, and the publication year. The analysis shows that the most frequently used performance metrics is accuracy, the most commonly used classification technique is the sensor, whereas most of the research papers are published in the year 2018.

Keywords Occupancy detection · Internet of things · Sensor · Network · Automation system

1 Introduction

Internet of things (IoT) is becoming a reality in everyday life due to recent findings and commercial products. Sensors are used for the transmission, acquisition, and data analysis, which is then coupled with cheap and low-power transceivers and micro-controllers for digital communications that enable the user for accessing numerous

K. P. Shirsat (✉) · G. P. Bhole
Department of Computer Engineering & Information Technology, Veermata Jijabai Technological Institute, Matunga Mumbai, Maharastra, India
e-mail: kpshirsat_p16@ce.vjti.ac.in

© The Editor(s) (if applicable) and The Author(s), under exclusive license to Springer Nature Singapore Pte Ltd. 2021
S. Shakya et al. (eds.), *Proceedings of International Conference on Sustainable Expert Systems*, Lecture Notes in Networks and Systems 176, https://doi.org/10.1007/978-981-33-4355-9_8

applications in various domains, like mobile health care, home automation, energy management, and other applications. The IoT is foreseen as an important technology in the Smart City concept as it fits perfectly in urban scenarios. The smart cities tend to increase the life quality of the citizen along with the reduction in the operation costs of public administrations [1]. Smart buildings and IoT technology required the need for smart devices to operate without human intervention. The basic application of the home automation system is the detection of occupancy. The automation systems are built by incorporating IoT technologies in them. For instance, the occupancy tracking can help during a fire or natural disaster for evacuating the survivors from the building. The occupancy detection techniques are also used in the intrusion detection that detected the activities at abnormal times [2].

Recent studies have shown that the identification of occupancy patterns in building and smart home applications had lower-energy consumption when compared to the approaches that assume usage patterns and fixed occupancy. The detection of occupancy in buildings is an expensive and difficult process as it needs to overcome the issues, like false detection and the intrusive nature of visual sensors [3]. The devices that are used for detection of occupancy are based on audible sensors, radio frequency identification (RFID)-based systems, video cameras, passive infrared (PIR) technologies, and microwave [4]. PIR sensors detected the body movements using the infrared image for the detection of occupancy. Ultrasonic and the microwave sensors used the change in the pattern of the reflected wave for the detection of occupancy. The drawback of the motion sensor in occupancy detection is the inability to detect the presence of the user if he/she remains idle, and it resulted in uncomfortable situations, like turning OFF fans and lights during the presence of the user [5]. The audible sound sensors are not able to distinguish the non-human and human noises, and they are prone to false alarms. To overcome these limitations, the occupancy detectors are used in the place of motion detector [2].

The accurate occupancy detection is obtained by blending a multi-sensor data, like motion sensors, sound, humidity, temperature, and CO_2. Accurate detection of the occupancy assists in developing context-driven control techniques, in which sensing and actuation tasks are done based on the contextual changes. Moreover, with recent advances in wireless sensors networks, most of the industries and researchers have confirmed the prospective of IoT [6, 7] as an enabler to the development of context-aware and intelligent services and applications. These services energetically respond to the environment changes and users' preferences. Figure 1 shows the block diagram of the occupancy detection system.

The objective of this research is to facilitate a detailed survey of occupancy detection systems.

The organization of the paper is as follows: Sect. 1 describes the introduction to occupancy detection, Sect. 2 elaborated the literature review of the occupancy detection methods, Sect. 3 describes the research gaps and issues, Sect. 4 depicts the analysis of the researches, and finally, Sect. 5 concludes the paper.

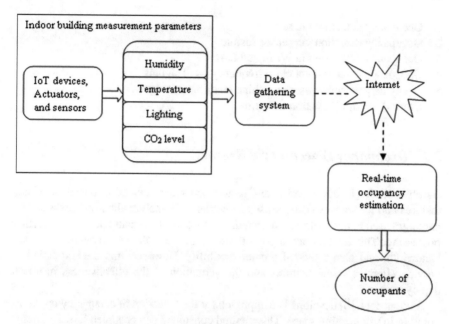

Fig. 1 Block diagram of the occupancy detection system

2 Categorization of Occupancy Detection Techniques

This section explains various research papers for occupancy detection. The occupancy detection techniques are categorized into the sensor, sensor fusion, Wi-Fi, LAN, radio frequency (RF) signals, machine learning techniques, and so on. Figure 2 shows the categorization of occupancy detection techniques. The existing occupancy detection techniques are categorized into five groups based on the way, in which, they detect the occupancies.

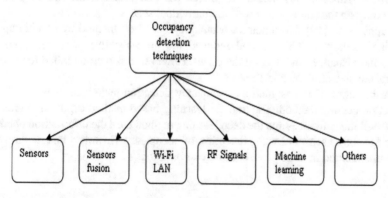

Fig. 2 Categorization of occupancy detection techniques

1. Occupancy detection via sensors
2. Occupancy detection via sensor fusion
3. Occupancy detection via Wi-Fi and LAN
4. Occupancy detection via radio frequency (RF) signals
5. Occupancy detection using machine learning techniques
6. Other types of occupation detection.

2.1 Occupancy Detection via Sensors

Baroffio et al. [1] developed a distributed occupancy detection based on visual features and low-power visual sensor networks. The analyze-then-compress (ATC) paradigm used in this method required the features of the sensing devices for further processing. This method had a good trade-off between the accuracy, transmission bandwidth, and also provided greater flexibility. However, this method failed to provide effective visual features and the validation of the effectiveness in a real deployment.

Akkaya et al. [8] developed an approach for the detection of occupancy by incorporating IoT in meeting space. This method consumed and occupied fewer amounts of energy and space in the room. This method provided the status of real-time room occupancy, generated notifications, room management, and booking and the use of remote location for controlling the environment of the room, but it failed to fuse the sensors for achieving better accuracy.

Jeon et al. [9] modeled a particulate matter (PM) concentration for IoT-based occupancy detection system. This method detected the occupancy using triangular shape extraction and point extraction algorithm. This method was applied for the occupancy detection for the patients and the elderly patients, who were living alone, but it failed to evaluate the spatial distribution of particulate matter.

Ji et al. [4] designed an IoT environment sensor for occupancy detection. This method detected the occupancy based on the information on temperature, CO_2, and humidity. Although this method improved the performance, the accuracy varied depending on the characteristics of the individual spaces.

Luppe et al. [10] developed an occupancy detection method by combining the sensing modalities. This method reduced the false-negative rates, false-positive rates, and visualized the data on the cloud. However, this method failed to estimate occupancy detection on a large scale.

Tushar et al. [11] designed an occupancy detection technique using IoT-based signal processing method. The transfer learning-based method captured the images at the building entrances and the deep learning method used the information obtained from the sound sensors for the detection of occupancy. However, this method was not implemented in large-scale commercial building.

2.2 Occupancy Detection via Sensor Fusion

Akbar et al. [2] designed a non-intrusive approach for the detection of occupancy. This method used the electricity consumption data that detected the occupancy along with the contextual information. It also dealt with the situations that occur when the user leaves the seat for short intervals and the requirement of extra equipment for the detection of occupancy. Although this method detected occupancy with high efficiency, it failed to fuse the possibility of other sensors along with the energy consumption data.

Nesa and Banerjee et al. [12] modeled a Dempster–Shafer evidence theory for the detection of occupancy in the room. The Dempster–Shafer evidence theory fused the information that was collected from the heterogeneous sensors and derived a conclusion by performing a mass combination. The probability density functions were considered for calculating the mass assignments. This method provided high accuracy, precision, and specificity in the detection of occupancy. However, this method failed to estimate the number of occupants.

Javed et al. [13] developed an occupancy detection method using the RNN-based intelligent controller. The intelligent controller integrated the IoT with cloud computing. Mean vote-based set points and the number of occupants were calculated by learning the user preferences for controlling the heating, cooling, and ventilation. This method provided accurate occupancy estimation and better power consumption. However, this method had a high computational complexity.

Roselyn et al. [14] designed an automation system for the detection of occupancy. The slave controllers, like image processing algorithms and thermal sensor algorithms, were combined using the sensor fusion model. The dynamic and the statistical changes in the environment were detected based on the ROI segregation model and background subtraction-based model using the image processing algorithm. This method provided highly secured storage and low latency, but it failed to generate measurable ROIs within the initial phase.

2.3 Occupancy Detection via Wi-Fi and LAN

Zou et al. [15] developed a Wi-Fi-enabled Internet of things (IoT) devices for device-free occupancy detection. The information fusion and the transfer kernel learning were done using crowd counting classifiers for the environmental and temporal disparities. The difference between the target and the source distributions was minimized by constructing domain-invariant kernel. Although this method provided better crowd counting accuracy and occupancy detection accuracy, it had a high computational cost.

Sadhukhan [16] modeled a prototype for the E-parking system. This method provided information regarding the availability of parking space. The occupancy duration of the parking lot and the improper parking were detected using the integrated component known as the parking meter, which was deployed at every parking lot. This method reserved the parking lot and gathered information through a suitable reservation-based parking management facility.

Huang et al. [17] developed the IoT prototype for the detection of occupancy of rooms in energy-efficient buildings. The prototype had Raspberry Pi modules and Lattice iCE40-HX1K stick FPGA boards. The prototype was installed at the door frame, and the infrared streams were blocked when a person entered the door frame. The direction of the movement of humans was obtained by comparing the time instances of the obstructive events and updating the quantity of the occupancy of a thermal zone. The room occupancy information was acquired by building the automation system or by allowing the anonymous users in the open-source application user interface. This method had better occupancy counting accuracy and a low failure rate.

2.4 Occupancy Detection via Radio Frequency (RF) Signals

Baldini et al. [18] developed an occupancy detection technique using plug and play solution. The location of the smart objects was derived using the anchor nodes which was estimated by determining the weighted distance matrix. In this method, the smart objects were identified automatically using the threshold weighted matrix. However, it failed to determine the robustness of the approach.

Ng and She [19] developed a device-free occupancy detection using denoising-contractive autoencoder (DCAE). This method constructed the fingerprint vector by appending the temporal difference between subsequent RSS measurements with time average received signal strength. The DCAE method dealt with the common issues of RF fingerprint methods, like sparsity and noise. Although this method provided occupancy detection with better accuracy, it failed to optimize the detection performance.

Ng et al. [20] designed a DCAE method for the detection of occupancy in the sub-room level. This method encoded the meaningful hidden representation even if the size of the input data was larger. Even though the input data size was larger, the meaningful hidden representations were encoded. This method was robust in environmental noise and RSS variations, but it failed to consider the dimensionality of the encoded feature.

2.5 Occupation Detection Using Machine Learning Techniques

Adeogun et al. [21] designed a machine learning technique for the detection of occupancy. An augmented data set was formed by combining the multi-sensor measurements and door status. A two-layer feed-forward neural network with sigmoid output neurons was applied to the data. This method provided better accuracy for multi and binary-class problems. However, this method failed to improve classification performance.

Ling et al. [22] designed machine learning techniques for developing a parking space based on parked vehicle positions. The occupation detection pipeline along with the clustering-based learning method identified the parking spaces correctly and determined the occupancy without specifying the parking locations manually. Although this method provided high accuracy in the detection of parking space, it had high computational complexity.

Elkhoukhi et al. [23] developed a holistic platform by combining big data and IoT technologies for occupancy detection. The machine learning algorithms were integrated for the detection of the presence of the occupant. The mining of the big data streams was done by integrating with the Scalable Advanced Massive Online Analysis (SAMOA) platform. However, this method had a high computational complexity.

2.6 Other Types of Occupation Detection

Casado-Mansilla et al. [24] designed a context-aware and a human-centric architecture for occupancy detection. The energy consumption was reduced by cooperating the devices with eco-aware users. The interactions between the energy-consuming assets and the occupants were enhanced in this method. The necessary understanding was provided by the socioeconomic behavioral model that transformed the energy-consuming device into active pro-sustainability agents. Although this method had reduced energy consumption, it failed to get rid of the intermediate devices that performed forecasting algorithms.

Paganelli et al. [25] designed an ontology-based context model for handling and monitoring patient chronic conditions. This method supported the operators by developing a prototype and integrating the prototype with the service platform in home-based care networks. However, this method required modification in the alarm threshold values and reasoning rules.

Forkan et al. [26] developed a BDCaM for context-aware computing. This method distinguished the normal from the emergency conditions accurately. Although this method predicted the abnormal conditions accurately, it required proper training for predicting the abnormal conditions for large samples.

3   Research Gaps and Issues

Internet of things (IoT) faces challenges during the development of green IoT technologies, employment of artificial intelligence methods during the creation of intelligent things of smart objects, development of context-aware IoT middleware solutions, integration of IoT solutions with social networking and while combining cloud computing and IoT [13]. In [9], the occupancy was detected using particle concentration, but the main challenge lies in the consideration of the spatial distribution of particulate matter and the effects of particulate matter on other factors. Occupancy detection using visual features provided higher flexibility, but the challenge lies during the real-time deployment of the method [1]. Occupancy detection using Dempster–Shafer evidence theory faces challenges during the estimation of the number of occupants [12]. In [14], the occupancy detection using the automation system provided low latency and highly secured detection, but the main challenge lies in the generation of measurable ROIs in the initial phase.

4   Result and Discussion

This section describes the analysis of different techniques of occupancy detection based on the publication year, performance metrics, and classification techniques.

4.1   Analysis Based on Publication Year

This section describes the analysis of the occupancy detection technique based on the year of publication. Figure 3 shows the analysis based on the publication year. From the analysis, it is concluded that more research papers were published in the year 2018.

Fig. 3 Analysis based on publication year

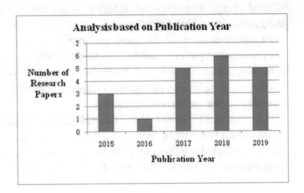

Table 1 Analysis based on performance metrics

Performance	Number of research papers
Accuracy	[1, 4, 8, 12, 17, 19–23]
Success rate	[18]
Training time	[20]
Recall, precision, F-measure	[2]
Humidity, temperature	[18]
Load power	[1]
Energy consumption	[13, 22]
Control decision delay	[13]
Energy efficiency	[24]

4.2 Analysis Based Performance Metrics

This section depicts the analysis based on the performance metrics. The performance metrics used for the analysis are accuracy, success rate, training time, recall, precision, F-measure, load power, energy consumption, control decision delay, and energy efficiency. The most commonly used performance metrics is accuracy. Table 1 shows an analysis based on performance metrics.

4.3 Analysis Based on Classification Techniques

This section describes the analysis of the occupancy detection methods based on classification techniques. The analysis shows that the most frequently used classification methods are the sensor. Table 2 shows the analysis based on classification techniques.

Table 2 Analysis based on classification techniques

Classification techniques	Number of research papers
Sensor	[1, 4, 8, 18, 22]
Sensor fusi	[2, 12, 13, 23]
Wifi, LAN	[16, 17]
Radio frequency (RF) signals	[18–20]
Machine learning techniques	[21–23]
Others	[24, 26]

5   Conclusion

Occupancy detection is the process of detection of occupancy by overcoming issues like false detection, energy issues, and intrusive nature of sensors. In this research, the detailed survey of context-aware IoT system for occupancy detection is discussed, and for the analysis, 20 research papers are considered moreover; the methods are categorized as sensor, sensor fusion, Wi-Fi, LAN, radio frequency (RF) signals, machine learning, and so on. The research papers are collected from Google Scholar, IEEE, Elsevier, Science Direct, and so on. The research gaps and challenges faced during occupancy detection are elaborated. The research works are analyzed based on performance metrics, classification methods, accuracy, and the year of publication. From the analysis, it is concluded that the most commonly used classification technique is the sensor. Most of the research works are published in the year 2018. Future enhancement in occupancy detection can be done by including more coding methods and visual features in the system.

References

1. Baroffio, L., Bondi, L., Cesana, M., Redondi, A.E., Tagliasacchi, M.: A visual sensor network for parking lot occupancy detection in smart cities. In: 2nd World Forum on Internet of Things (WF-IoT). IEEE (2015)
2. Akbar, A., Nati, M., Carrez, F., Moessner, K.: Contextual occupancy detection for smart office by pattern recognition of electricity consumption data. In: International Conference on Communications (ICC). IEEE (2015)
3. Nguyen, T.A., Aiello, M.: Energy intelligent buildings based on user activity: a survey. Energy Build. **56**, 244–257 (2013)
4. Ji, Y., Ok, K., Choi, W.S.: Occupancy detection technology in the building based on IoT environment sensors. In: Proceedings of the 8th International Conference on the Internet of Things (2018)
5. Abowd, G.D., Dey, A.K., Brown, P.J., Davies, N., Smith, M., Steggles, P.: Towards a better understanding of context and context-awareness. In: Computer ScienceLecture Notes, pp. 304–307(1999)
6. Kumar, T.S.: Efficient resource allocation and Qos enhancements of IoT with fog network. J. ISMAC **1**, 21–30 (2019)
7. Sivaganesan, D.: Design and development AI-enabled edge computing for intelligent-iot applications. J. Trends Comput. Sci. Smart Technol. (TCSST) **1**, 84–94 (2019)
8. Patel, J., Panchal, G.: An IoT-based portable smart meeting space with real-time room occupancy. In: Networks and Systems Lecture Notes, pp.35–42. Springer (2017)
9. Jeon, Y., Cho, C., Seo, J., Kwon, K., Park, H., Oh, S., Chung: IoT-based occupancy detection system in indoor residential environments. In: Building and Environment, vol. 132, pp. 181–204 (2018)
10. Luppe, C., Shabani, A.: Towards reliable intelligent occupancy detection for smart building applications. In: 30th Canadian Conference on Electrical and Computer Engineering (CCECE). IEEE (2017)
11. Tushar, W., Wijerathne, N., Li, W.T., Yuen, C., Poor, H.V., Saha, T.K., Wood, K.L.: Iot for green building management (2018). arXiv:1805.10635
12. Nesa, N., Banerjee, I.: IoT-based sensor data fusion for occupancy sensing using Dempster–Shafer evidence theory for smart buildings. IoT J. **4**(5), 1563–1570. IEEE (2017)

13. Javed, A., Larijani, H., Ahmadinia, A., Gibson, D.: Smart random neural network controller for HVAC using cloud computing technology. IEEE Trans. Industr. Inf. **13**(1), 351–360 (2017)
14. Roselyn, J.P., Uthra, R.A., Raj, A., Devaraj, D., Bharadwaj, P., Kaki, S.V.D.: Development and implementation of novel sensor fusion algorithm for occupancy detection and automation in energy efficient buildings. In: Sustainable Cities and Society, vol. 44, pp. 85–98 (2019)
15. Zou, H., Zhou, Y., Yang, J., Spanos, C.J.: Device-free occupancy detection and crowd counting in smart buildings with WiFi-enabled IoT. In: Energy and Buildings, vol. 174, pp. 309–322 (2018)
16. Sadhukhan, P.: An IoT-based E-parking system for smart cities. In: International Conference on Advances in Computing, Communications and Informatics (2017)
17. Huang, Q., Rodriguez, K., Whetstone, N., Habel, S.: Rapid internet of things (IoT) prototype for accurate people counting towards energy efficient buildings. In: IT Conference, pp. 1–13 (2019)
18. Baldini, A., Ciabattoni, L., Felicetti, R., Ferracuti, F., Longhi, S., Monteriu, A., Freddi, A.: Room occupancy detection: combining RSS analysis and fuzzy logic. In: 6th International Conference on Consumer Electronics Berlin (ICCE-Berlin). IEEE (2016)
19. Ng, P.C., She, J.: Denoising-contractive autoencoder for robust device-free occupancy detection. IoT J. IEEE (2019)
20. Ng, P.C., She, J., Ran, R.: Towards sub-room level occupancy detection with denoising-contractive autoencoder. In: International Conference on Communications. IEEE (2019)
21. Adeogun, R., Rodriguez, I., Razzaghpour, M., Berardinelli, G., Christensen, P.H., Mogensen, P.E.: Indoor occupancy detection and estimation using machine learning and measurements from an IoT LoRa-based monitoring system. In: Global IoT Summit (2019)
22. Ling, X., Sheng, J., Baiocchi, O., Liu, X., Tolentino, M.E.: Identifying parking spaces and detecting occupancy using vision-based IoT devices. In: Global Internet of Things Summit (GIoTS) (2017)
23. Mansilla, D.C., Moschos, I., Esteban, O.K., Tsolakis, A., DeIpina C.E.: A human-centric and context-aware IoT framework for enhancing energy efficiency in buildings of public use. IEEE Access **6**, 31444–31456 (2018)
24. Elkhoukhi, H., NaitMalek, Y., Berouine, A., Bakhouya, M., Elouadghiri, D., Essaaidi, M.: Towards a real-time occupancy detection approach for smart buildings. Proc. Comput. Sci. **134**, 114–120 (2018)
25. Paganelli, F., Giuli, D.: An ontology-based system for context-aware and configurable services to support home-based continuous care. IEEE Trans. Inf. Technol. Biomed. **15**(2), 324–333 (2010)
26. Forkan, A.R.M., Khalil, I., Ibaida, A and Tari, Z.: BDCaM: big data for context-aware monitoring a personalized knowledge discovery framework for assisted healthcare. IEEE Trans. Cloud Comput. **5**, 628–641 (2015)

Study of Holoportation: Using Network Errors for Improving Accuracy and Efficiency

Hrithik Sanyal and Rajneesh Agrawal

Abstract Holoportation is a technique in which two persons communicate with each with anyone's virtual presence in front of the other. There have been efforts to make it as much smoother and real as possible by the researchers in the recent past. But the challenges are many in this field not only due to unavailability of software resources but due to hardware constraints as well. Major hardware constraints are based on the transmission of a lot of data being collected by the camera and audio devices which require good data transfer rates between the communicating devices. Reason of challenge is viewed in two faces, i.e., one is slow data transfer speed and the second is huge amount of data transfer. Slow data transfer speed of resources is being tackled, and a good data transfer rate has been reached to but still not suffice and unavailable in all areas around the world. A huge amount of data transfer may also suffer from network lag spikes and dropouts of the signals which will lead to disruption in reproduced Holoportation on the receiver's end. In this paper, the focus is on proposing a buffering and correction mechanism which will require to be applied on both sender and receiver ends. The system will produce high accuracy and will not increase network latency and hence the smoothness of service. The system will leverage normal human behaviour and persistence of vision delays to provide better accuracies.

Keywords Holoportation · Data transport · Human bond communication · Real-time processing · Virtual reality · Augmented reality · Intelligent system architecture

H. Sanyal (✉)
Department of Electronics & Telecommunications, Bharati Vidyapeeth College of Engineering Pune, Pune, India
e-mail: hrithiksanyal14@gmail.com

R. Agrawal
Comp-Tel Consultancy, Mentor, Jabalpur, India
e-mail: rajneeshag@gmail.com

S. Shakya et al. (eds.), *Proceedings of International Conference on Sustainable Expert Systems*, Lecture Notes in Networks and Systems 176,
https://doi.org/10.1007/978-981-33-4355-9_9

1 Introduction

Holoportation is a vivid and creative type of innovation that has incredible potential in changing human life. Major work in Holoportation is being done by Microsoft which characterizes Holoportation as a 3D capture technology spectacle the high-resolution 3D models of people that are to be renewed, smoothed and broadcasted anywhere on the planet. They are putting efforts to develop hardware and software for Holoportation such as 3D catch innovation and reality headsets (HoloLens). Holo-portation requires clients must be in a room furnished with the innovation camera, which reproduces and transmits the client as a 3D model. They are further elabo-rating on concepts of augmented reality (AR) and virtual reality (VR) for identical purposes.

Holoportation with its enhancement is being envisioned to be immensely helpful for mankind, e.g., remote meetings, training, gaming, research, medical sciences, etc. Figure 1 below, gives a step by step processing in Holoportation.

Advancements in virtual reality and augmented reality techniques are becoming a boon for holoportation. Virtual reality is used to create an artificial environment to

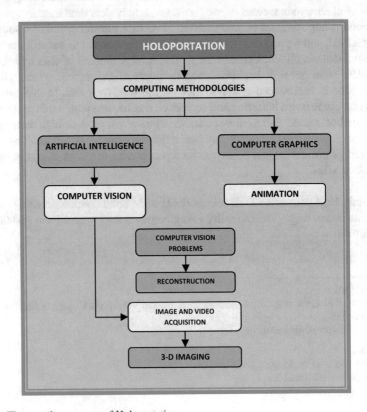

Fig. 1 The complete process of Holoportation

be in which users feel that they are part of it, whereas augmented reality produces the images which are seen by any user and portray the same in the 3D vision for users. This is the greed of human to communicate with others around the world virtually which also has been leveraged in developing applications for communication of text, images, audio, video, etc., in the 2D environment and now shifting to the 3D transmission which will make the communication to be most effective and real experience for the users of the system.

All technologies used together will make it possible to implement Holoportation and providing the most revolutionary communication system for the human. Holoportation is employable in all parts of human communication and will save a lot of human efforts being applied otherwise.

In this paper, Sect. 1 introduces and discusses more on the term Holoportation and also about its history and technologies. Sect. 2 details about techniques of virtual reality (VR) and augmented reality (AR), and Sect. 3 discusses more on human bond communication. Section 4 describes the network errors, and Sect. 5 enlists the details about the existing systems of Holoportation. In Sect. 6, the limitations of the existing models have been discussed on, Sect. 7 enlists all the errors and shortcomings and what has been proposed to overcome them, and in Sect. 8, the future scope has been discussed along with the conclusion.

2 Virtual Reality and Augmented Reality

2.1 Virtual Reality (VR)

Humans know the world through senses and perception systems which are taste, contact, smell, sight and hearing. Out of these for a few senses, electronic sensors have been developed and work is under research for others too. VR uses these sensors to create a virtual human eye-oriented vision, and the user feels embedded within to feel it as a real situation. A computerized simulated environment is created in VR using the sensors which can be transmitted to a user remotely situated, and special headsets can make the remote user feel like a part of the simulated environment. Though it seems to be pretty real, a lot of the work is yet to be carried on VR, as a human only has 180° of vision, and similarly, other senses make it more susceptible to be away from reality. Still, some of the wide varieties of applications for virtual reality include

1. Architecture
2. Sport
3. Medicine
4. The arts
5. Entertainment.

2.2 *Augmented Reality (AR)*

AR is a combination of real-time processing, computer-generated, real and accurate 3D generation of the superimposed real and virtual objects. Augmented reality (AR) creates a camera vision image which is transmitted to a remote user which is sensed using similar sensors as used in VR. VR uses the human eye vision to create near reality, whereas AR uses sensors to see the genuine condition directly before the user. AR produces computer visions fetched from sensors and software algorithms and superimposes the same in human view to make a realistic view for the users. Apart from gaming, AR is being applied in innumerable human areas such as architectural design, commerce, archaeology and manufacturing.

3 Human Bond Communication

In past, the human had been communicating electronically with the remote users using multiple different media such as text, images, audio and video. This not only shows the greed of human bonding but also depicts the gradual enhancement of electronic communication systems. But these communications are 2D communication, and user on the remote end views these only. The remote user cannot become a part of it actually or virtually. Growth in communication technologies, networking and sensor technologies has made a communication to the virtual reality (VR), augmented reality (AR) and Internet of things (IoT)-based communication. But, except for aural and optical media, challenges in the transmission of the sensory features, namely gustatory, olfactory and tactile are quite remote from reality.

HBC is a notion of understanding the transmission of data that involves all the five sensory features such as the gustatory, tactile and olfactory ((five human body sensations (smell, sight, touch, taste and sound)). This human body sense is significant to have the data exchange through communication methods for human sentiment-centric communication that ranges from the digital to broadcast and replicate at the receiver's end to permit the data transfer between the human beings and in peculiar events between the machines and human beings (M2H)/Internet of things (IoT).

Some of its applications are virtual and augmented presence, augmented reality, virtual reality and gaming. These applications attain an advantage with the assistance of the medical field. At present, there are innovative models of social networking applications that are available based on HBC technology, merely the application would have to undergo radical changes to meet them.

Some technologies that go hand in hand to make the process of Holoportation smooth and success are as follows:

- Augmented reality/virtual reality
- Haptics
- Real-time human bond communication human avatar creation/human body model acquisition based on 3D

- Gaze-aware facial reenactment and real-time facial identification
- Real-time animation of the avatar—photorealistic.

4 Network Error

When large data is transferred over the network, there are lots of chances because of which network errors may occur. The biggest cause may be manipulations of bits due to network errors. Network errors may change the important bits and will mutilate the actual Holoportation image, and output may be confusing and distorted. Another important network issue is network latency, i.e., when packets are transferred over the network, then it flows through various routes some of which might cause huge latency. This causes the Holoportation to be non-smooth and will cause the users to be in viewing delayed information which may lead to Holoportation to be very much annoying. Yet another reason may be due to security flaws over the network leading to producing unexpected Holoportation images. Making data to be secure over the network, application of encryption and compression may be applied which further may lead to latency in a reproduction of the images resulting in distorted Holoportation (Fig. 2).

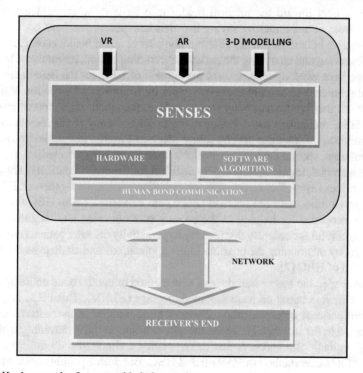

Fig. 2 Hardware and software used in holoportation

5 Existing System

The author talks about the immersive AR/VR telepresence is a technology which people could enjoy and experience around the world. The 3D telepresence system addresses the eye contact between the participants and also paves the way for gaze perception and the sense of having the difference between the orator ("me") and the presenter's whereabouts ("here") or "spatial faithfulness". With the expanded accessibility and refinement of the buyer VR/AR glasses and the RGB-D sensors, the current extent of 3D telepresence systems is still limited. Further, the skeleton data is extricated from the intensity level of the sensor that aids to animate a 3D model of the human, and also it incorporates rigging face to facial expression. Though thought to be encouraging at the beginning (no latency, no loss of data), it was seen that the sensors used did not allow the transfer of complicated hand movements and many users with real-time telepresence. The system works by instigating a 3D capturing and pipeline remodelling which would generate and distribute 3D user pictures in the real-time which will be very realistic [1]. Table 1 gives works of the authors who have contributed a lot in the field of Holoportation.

The author in this article starts the abstract talking about the recent advancements of technologies in the grounds of sensor technologies and wireless networks. This technology helps to monitor remote patients constantly. Henceforth, the author's concern is bent toward the medical field. Since he is more concerned about the medical area, he continues by saying that these advancements have opened scope for innovative boundaries in the different domains of smart health care, particularly when observing and analyzing the patients. According to him, researches and scientists have been working more toward the domain of HBC for the betterment of the medical science field, mainly suggested while transmitting and detection of data by utilizing all the five human body senses such as the smell, sight, touch, taste and sound. HBC recognizes these senses to duplicate and copy at the remote location and also allows to monitor several minor diseases through diagnosis. To make all these possible, special devices are available that would help the medical experts in making a crucial, important and timely decision for the critical patients. The author then describes saying that he presents an aesthetic/artistic investigation on HBC and all the possibilities between medical health care and HBC. So an HBC framework model is proposed for helping in the process of monitoring and diagnosis which would be helpful for courtesy discrepancy hyperactivity disorder patients and finally concludes by mentioning the possible future applications and challenges might face in the field of HBC [2].

In this paper, the author introduces a new method of human bond communication (HBC), which is based on head-mounted displays (HMDs). This helps in consecutive bidirectional communication among a large number of users irrespective of their access to the Internet. However, the author brings in the disadvantages of these existing technologies and says that these existing technologies do limit the possibilities of this new method of HMD-based HBC. But with a positive connotation, it brings out the suspense that a technology known as optical camera communication

Table 1 Comparison of existing researches

S. No.	Authors	Paper title	Year	Methodology	Achievements
1	Agata Manolova, Nikolay Neshov, Krasimir Tonchev, Pavlina Koleva, Vladimir Poulkov [1]	Challenges for real-time long-distance Holoportation to enable human bond communication	2019	Transmission of dynamic movements of dataset	• Successful HBC by extraction of skeleton data by depth sensors for animating facial expression • 3D capture and pipeline reconstruction which would generate and distribute 3D user pictures in real-time.
2	Tayaba Iftikhar, Hasan Ali Khattak, Zoobia Ameer, Munam Ali Shah, Faisal Fayyaz Qureshi and Muhammad Zeeshan Shakir [2]	Human bond communications: architectures, challenges and possibilities	2019	Study of HBC	• Making smart health care • The five human body senses serve as the framework to detect and transmit the data for all the major and minor diseases
3	Md. Tanvir Hossan, Mostafa Zaman Chowdhury, Md. Shahjalal and Yeong Min Jang [3]	Human bond communication with head-mounted displays: scope, challenges, solutions and applications	2019	Study of optical camera communication (OCC)-based head-mounted displays (HMD)	• A new method of optical camera communication (OCC)-based head-mounted displays (HMD) in the field of HBC • Challenges faced, advantages, architecture
4	Sudhir Dixit, Seshadri Mohan, Ramjee Prasad, Hiroshi Harada [4]	Multi-sensory human bond communication	2019	Study of multi-sensory structures	• The sensory structures such as gustatory (taste), tactile (touch) and olfactory (smell) are integrated to overwhelm the challenges faced during the breakdown of transmission

(continued)

Table 1 (continued)

S. No.	Authors	Paper title	Year	Methodology	Achievements
5	Christoph Müller, Matthias Braun, Thomas Ertl [5]	Optimized molecular graphics on the HoloLens	2019	Study of microsoft HoloLens technology	• Improvement of rendering speed of atom-based molecules • Conservative depth output
6	Ziyang Wang Wei Liao [6]	Study on non-direct signal transmission and characteristic of human body communication	2019	Study of human body characteristics for non-direct transmission	• Penetration of a body region over the network system in the medical field • Comparison between the surface and non-surface link • Surface link plays a decisive role and has delays
7	Jingzhen Li, Zedong Nie, Yuhang Liu and Lei Wang [7]	Modeling and characterization of different channels based on human body communication	2017	Study of HBC in characterization of channels	• Use of electrical signal in HBC • The finite difference time domain (FDTD) method aids to assist the investigations on the on-body to in-body (OB-IB), on-body to on-body (OB-OB), in-body to in-body (IB-IB) and in-body to on-body (IB-OB) channels
8	Ramjee Prasad [8]	Human bond communication	2015	Study of HBC	• Insight into the HBC • Need for improvements in technology and algorithms

(OCC) would help to break the barriers of the existing technologies, thus are being introduced with virtual applications with the approach of visual communications. In this process, the OCC could be jumble sale as a recipient, but proximate infrared radiation (IR), or infrared light source has to be further integrated to the HMD so that it could work as a transmitter. And the writer then accomplishes talking about the future scope, future applications, effective architectures and the challenges might face in the forthcoming future and the solutions to these challenges when working on OCC-based HMD for HBC [3].

The author in this paper discusses firstly on the swift progress of the communication and telecommunication technologies, ranging from image, speech, text and video, and the corresponding manner of communication has procrastinated to the machine-to-machine (M2M), Internet of things (IoT) and machine-to-human (M2H) communication. Then the author explained the challenges that still fail to overcome including the three sensory structures, namely olfactory (smell), gustatory (taste) and tactile (touch) and integrating them. HBC is a concept that includes all the five sensible features of data ranging from identifying digital to broadcast and allows to have the extra attractive, expressive, holistic and representative to transport all the data between people and also between the machine-to-human (M2H) or IoT. Other sorts of applications are augmented reality, virtual reality, gaming and virtual presence. HBC-based social schmoozing applications also exist with the new framework and would undergo extreme changes to compete with other technologies [4].

Due to the advancements in recent technologies, communication enables the individuals to have interconnections using the media (optical) and speech (aural). These communications are established through their auditory and optical senses. The challenges faced by the sensory features are far away from reality. HBC helps to stabilize the transmission of data through tactile, olfactory and gustatory. It also aids to communicate through the communication techniques for more of the human sentiment-centric data. In the future, innovation ideas or concepts promote the strategy of holistic communication [5].

6   Limitation of Existing Models

As per the studies of the various Holoportation techniques and models, the challenges are many in this field not only due to unavailability of software resources but due to hardware constraints as well. Major hardware constraints are based on the transmission of a lot of data being collected by the camera and audio devices which requires good data transfer rates between the communicating devices. Reason of challenge is viewed in two faces, i.e., one is slow data transfer speed and second huge amount of data transfer. Slow data transfer speed of resources is being tackled and to a good data transfer rate have been reached but still not suffice and unavailable in all areas around the world. A huge amount of data transfer may also suffer from network lag spikes and dropouts of the signals which will lead to disruption in reproduced Holoportation on the receiver's end.

7 Proposed System

In the past few years, researches on Holoportation are increasing slowly. Although it is in its infantry state, the results are promising and encouraging. Despite many challenges related to hardware, software, network bandwidth, network speed and network errors, the researchers are overcoming them in steps. In this work, the network error which might occur during the data transfer is being addressed. The error on the network may occur due to many reasons but they can lead to any of the two discrete states, viz. "SPIKE" or "DROPOUT". The complete proposed system will be required to be applied on both source and receiver ends as follows:

1. **Sender's End**:

    a. The camera or other capturing devices will be collecting information to be used for Holoportation.
    b. The collected information will be sent to a device where the information shall be divided into chunks and stored in a data queue for re-transmission if receiver demands it again.
    c. The transmitter will send data chunks over the network toward a receiver.

2. **Receiver's End**:

    a. Received chunks of data shall be stored in a data queue.
    b. The preprocessor unit will fetch the data chunks for filtering and 3D modelling one by one.
    c. The preprocessor will analyze the data chunks and will mark them as "CORRECT", "SPIKE" or "DROPOUT".
    d. The preprocessor will send the marked chunks to the filter.
    e. The filter will send data chunks to the 3D modeller for creating 3D models at the receiver which are marked "CORRECT" by the preprocessor.
    f. The filter will drop any chunks marked as "SPIKE" or "DROPOUT" and will proceed to work on next chunks of data (Fig. 3).

The preprocessor will use the algorithm to mark the chunks to be "CORRECT", "SPIKE" and "DROPOUT" based on the following:

a. The preprocessor will analyze "n" data chunks from the data queue.
b. It will apply parameters for analysis based on the previous "$n/2$" chunks and next "$n/2$" chunks for chunk under preprocessing.
c. Parameters will be decided based on the bit patterns.
d. If the bit patterns are very different from previous and next chunk bit patterns and most of them are negative, then chunk will be marked as "DROPOUT".
e. If the bit patterns are very different from previous and next chunk bit patterns and most of them are positive, then chunk will be marked as "SPIKE".
f. If the bit patterns are having similarity with previous and next chunk bit patterns, then chunk will be marked as "CORRECT".

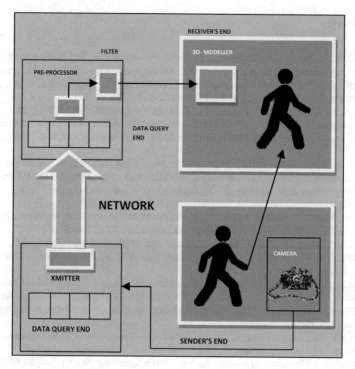

Fig. 3 System flow of the proposed work

8 Conclusion & Future Scope

Holoportation technique is currently in its infantry state but is very much promising for the future of the communication systems where human will be sitting virtually, participating in different events, attending meetings and many more. This has several challenges which include hardware, software and networking-based challenges. Researchers are working hard to tackling problems. The proposed work in this paper has been focused on network errors-related problems and by taking the advantage of human behaviour and different persisting nature of senses of the human, problems related with the network can be handled and will affect accuracy and performance of the Holoportation. The system is expected to have high accuracy and low latency as it is applied end to end no extra burden on the network has been imposed.

The work can be enhanced in future to include more parameters in the preprocessing state for marking the data chunks; variation of chunk sizes can also be tested. Actual implementation can further lead to add some fundamental changes in the process and parameters.

References

1. Manolova, A., Neshov, N., Tonchev, K., Koleva, P. and Poulkov, V.: Challenges for real time long distance holoportation to enable human bond communication. In: 2019 42nd International Conference on Telecommunications and Signal Processing (TSP), Budapest, Hungary, pp. 529–534 (2019). https://doi.org/10.1109/TSP.2019.8769053
2. Iftikhar, T., Khattak, H.A., Ameer, Z., Shah, M.A., Qureshi, F.F., Shakir, M.Z.: Human bond communications: architectures, challenges, and possibilities. In: IEEE Communications Magazine, vol. 57, no. 2, pp. 19–25, February 2019. https://doi.org/10.1109/MCOM.2018.1800531
3. Hossan, M.T., Chowdhury, M.Z., Shahjalal, M., Jang, Y.M.: Human bond communication with head-mounted displays: scope, challenges, solutions, and applications. In: IEEE Communications Magazine, vol. 57, no. 2, pp. 26–32, February 2019. https://doi.org/10.1109/MCOM.2018.1800527
4. Dixit, S., Mohan, S., Prasad, R., Harada, H.: Multi-sensory human bond communication. IEEE Commun. Mag. **57**(2), 18–18 (2019). https://doi.org/10.1109/MCOM.2019.8647105
5. Müller, C., Braun, M., Ertl, T.: Optimised molecular graphics on the hololens. In: 2019 IEEE Conference on Virtual Reality and 3D User Interfaces (VR), pp. 97–102. Osaka, Japan (2019). https://doi.org/10.1109/VR.2019.8798111
6. Wang, Z., Liao, W.: Study on non-direct signal transmission and characteristic of human body communication. In: 2019 3rd International Conference on Electronic Information Technology and Computer Engineering (EITCE), pp. 510–513. Xiamen, China (2019). https://doi.org/10.1109/EITCE47263.2019.9094819
7. Li, J., Nie, Z., Liu, Y., Wang, L.: Modeling and characterization of different channels based on human body communication. In: 2017 39th Annual International Conference of the IEEE Engineering in Medicine and Biology Society (EMBC), pp. 702–705. Seogwipo (2017). https://doi.org/10.1109/EMBC.2017.8036921
8. Prasad, R.: Human bond communication. Wirel. Pers. Commun. **87**(3), 619–627 (2016)

Sixth-Gen Wireless Tech with Optical Wireless Communication

Alex Mathew

Abstract Fifth-generation (5G) communication is about to arrive. It has too many features than existing Fourth-generation (4G) communication. And then later Sixth-generation (6G) communication will arrive having the features of artificial intelligence in it. In 5G, many things need to be upgraded and improved like more system capacity, higher data speed, and quality of services. This paper is about Sixth-generation (6G) technology with wireless communication. It contains the discussion of new technologies like artificial intelligence, optical wireless technology, and also the required technology for 6G communication and challenges to achieve this target.

Keywords 5G · AI · Sixth generation · Optical wireless communication

1 Introduction

Today, everything is connected with the Internet known as the Internet of things (IoT). So, with the development of applications like IoT, artificial intelligence (AI), and virtual reality, there is a huge volume of traffic. It has increased from 7.462 to 5016 EB/Month in 20 years [1]. It shows the importance of improved communication systems. It has affected every sector of society, such as health, industry, education, road, smart life, and many other things. Hence, to meet these requirements and support these applications, 5G was introduced. It has a high data rate with reliable connectivity. It has many new features like the millimeter wave and the optical spectra. 5G communication is the next and very advanced level of 4G communication. But the technology is growing at a very faster rate. Certain devices need beyond 5G as they require higher data rate virtual reality which is one of them. It is believed that 5G will reach its goal in 2030. And then, 6G will be introduced with much higher

A. Mathew (✉)
Department of Cybersecurity, Bethany College, Bethany, USA
e-mail: amathew@bethanywv.edu

S. Shakya et al. (eds.), *Proceedings of International Conference on Sustainable Expert Systems*, Lecture Notes in Networks and Systems 176,
https://doi.org/10.1007/978-981-33-4355-9_10

data rates and many new features. The main key factor of 6G communication is that it will support all past features and high reliability.

2 Discussion

Sixth-generation (6G) communication is the future in the communication sector. It supports all the existing features of other communication such as high reliability, less energy utilization, and high connectivity. It will also add new technologies such as AI, smart devices, autonomous vehicles, sensing, and 3D mapping. The most important features of 6G are its high data rate (Figs. 1 and 2).

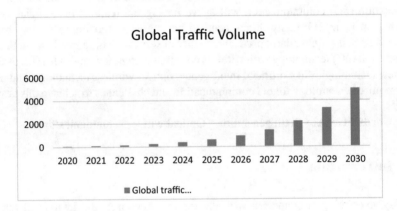

Fig. 1 Global traffic volume [2]

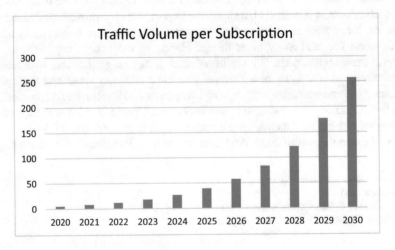

Fig. 2 Traffic volume per subscription [2]

Table 1 Reference [14]

Issue	2010	2020 (Predicted)	2030	Unit
Mobile subscriptions	5.32	10.7	17.1	Billion
Smartphones subscriptions	0.645	1.3	5.0	Billion
M2M subscriptions	0.213	7.0	97	Billion
Traffic volume	7.462	62	5016	EB/month
M2M traffic volume	0.256	5	622	EB/month
Traffic per subscription	1.35	10.3	257.1	GB/month

The 6G will improve the quality of services as well as it will also protect the data. It is estimated to be 1 Tb/s of data rate [2]. It is designed to provide wireless connectivity which will be 1000 times greater than 5G [3]. The most exciting feature of 6G is AI support terahertz band (THz), optical wireless communication (OWC), 3D networking, and wireless transfer.

The growing trend of mobile communication is shown below. It shows the new generation of communication, and it compares the use of mobile in connectivity in 2010, 2020, and 2030. It is believed that the mobile to mobile subscription will increase 33 times in 2020 and 455 times in 2030 as compared to 2010. It is also expected that the global traffic volume will also increase 670 times in 2030 compared to 2010 (Table 1).

Till the end of 2030, it is believed that 5G will not be able to meet the requirements of the market. So, 6G will be introduced to fill the gap between 5G and market demands. The main objective of the 6G system is high data rates, high connected devices, global connectivity, low energy use, high reliability, and AI. It is estimated that 6G will be integrated with satellites.

The 6G will have many technologies in it. Some of them are:

Artificial intelligence 4G does not support AI and 5G supports only partial AI. However, 6G will support AI fully. The introduction of AI will improve efficiency and reduce the communication delay [4–8]. AI will also play an important role in interaction with machines. AI based communication needs lots of supported metamaterials, smart structure, and networks.

Optical Wireless Technology It is one of the main features of 6G communications. It will enable all the devices to access networks. OWC technologies are practices in 4G communication and are more likely to meet the demand of 6G communications. OWC technologies such as light fidelity, visible light, and optical camera communication are based on optical band technologies [9–12]. It is on the progress to enhance these technologies. Communication in optical wireless technologies will be more secure and safer. It will also provide high data rates with low latency. One of its main technologies is LiDAR. It is based on the optical band. It is used for 3D mapping in high resolution in 6G communication.

2.1   *Advantages of Optical Wireless Communication*

There are many advantages of OWC in 6G communications. Some of them are:

- It does not require any spectrum license to use the band of 380 and 780 nm.
- It has a huge bandwidth of 400 THz. So, the data rates in it will be very high up to 1 Gbps.
- It has a more secure system for indoor, and the optical wave cannot pass through the walls. So, the signal cannot be heard by others.
- It does not require any extra components. The transmission and reception are available at cheap cost.
- It can also be used for illuminations.

2.2   *Indoor Applications*

Optical wireless technology supports a wide range of indoor applications. These applications have, but are not bound to,

Indoor networks: This application is used to connect OWC to devices inside an office, homes, hospitals, hotels, and many more things. It supports devices having less distances between transmitter and receiver. It can also be used to transfer the data of a patient to any storing devices. This application requires high security.

2.3   *Outdoor Applications*

Optical wireless technology supports a wide range of outdoor applications. These applications have and are not bound to,

- Wireless sensor networks (WSN): OWC can be used to recharge the battery of the sensors in WSN. These transmitters are small in size and can be fixed in any sensors. It can also be used to increase the transmit power and thus increase its reliability of the WSN.
- OWC also provides communication between wide ranges of vehicles. This comes into effect by using the front and taillight of the vehicle. So, it requires high network coordination.
- Due to high-speed rate in OWC, it can be used to connect with satellites in space. OWC has a data rate of 400Tbps.

3 Challenges and Future of 6G

6G communications are the demand of the future. But to implement this 6G completely, many technical problems are needed to be solved. Some of the possible concerns are:

High absorption of THz in the atmosphere: More the frequencies, more will be data rates. However, it has a major challenge of long-distance data transfer because of atmospheric absorption. So, to overcome this challenge, the receiver should be able to operate at high frequencies. It should also be able to use the complete bandwidths, so it must have THz band antennas. Safety of people is also important in THz band communications.

3D networking: Since the 3D networking has extended in upward direction, a new dimension was added. Moreover, a new technique was also introducing for resource management and optimization for mobile support, routing protocol, and many other essentials. So, it needed a new network design.

Heterogeneous hardware constraints: It includes many types of a communication system such as frequency bands, communication topologies, and so on. These all are involving in 6G communication. Different devices have different hardware configurations. So, a more complex structure is required which will also complicate the communication protocol and algorithm design. It will be a major challenge to integrate everything into a single platform.

Autonomous wireless system: The 6G system supports automation system like an autonomous car, UAVs on AI [13]. To execute these services, many subsystems are also required such as machine learning and machines of system. So, it becomes a challenging part.

Modeling of frequencies: Frequencies have few characteristics due to which absorption and dispersion effects are seen. Therefore, it has a complex channel band, and also it does not have any channel model.

Device support system: 6G communications will have lots of features, so devices must have the capability to support those features. These new features mainly include AI, XR, and sensing. So, it is believed that the cost of new devices will be high as compared to current devices. To overcome this challenge, devices which are used in 5G communications should also be compatible with 6G technology.

Backhaul connectivity: 6G has a very high density of access networks. It supports high data rate connectivity. Backhaul networks be ingused to connect the access network to a core network for a large amount of data. The optical fiber and FSO are used for backhaul connectivity.

Spectrum and interference: The resources in spectrum and interference are very low. So, it is very important to manage 6G spectrum and its technique. Then only, maximum resource utilization would be possible.

Beam management: Beam forming does support high data rate communications. However, in the THz band, it is a major concern because of its propagation characteristics. Therefore, beam management in unfavorable condition would be a challenging task for massive MIMO systems.

4 Conclusion

Every generation of communication has brought a revolution. The 5G communication has also come up with many new and exciting features which will meet the demand of the market for next ten years. However, 6G will be active then to meet the further demand of the market. It is still in the study phase. This paper has shown the lights to the future of 6G communications. The possible challenges and OWC are shown in it. Besides working theoretically on 6G, many technologies have been introduced that could be used for 6G communications.

References

1. ITU-R M.2370-0, IMT traffic estimates for the years 2020–2030, July 2015
2. David, K., Berndt, H.: 6G vision and requirements: is there any need for beyond 5G? IEEE Veh. Technol. Mag. **13**(3), 72–80 (2018)
3. Tariq, F., et al.: A Speculative Study on 6G, arXiv:1902.06700
4. Stoica, R.-A., Abreu, G.T.F.: 6G: the wireless communications network for collaborative and AI applications, arXiv:1904.03413
5. Loven, L., et al.: Edge AI: a vision for distributed, edge-native artificial intelligence in future 6G networks, 6G Wireless Summit, Levi, Finland (2019)
6. Clazzer, F., et al.: From 5G to 6G: has the time for modern random access come? arXiv:1903.03063
7. Mahmood, N.H., et al.: Six key enablers for machine type communication in 6G, arXiv:1903.05406
8. Zhao, J.: A survey of reconfigurable intelligent surfaces: towards 6G wireless communication networks with massive MIMO 2.0, arXiv:1907.04789
9. Chowdhury, M.Z., Hossan, M.T., Islam, A., Min Jang, Y.: A comparative survey of optical wireless technologies: architectures and applications. IEEE Access **6**, 9819–10220 (2018)
10. Chowdhury, M.Z., Hossan, M.T., Hasan, M.K., Jang, Y.M.: Integrated RF/optical wireless networks for improving QoS in indoor and transportation applications. Wirel. Pers. Commun. **107**(3), 1401–1430 (2019)
11. Hossan, M.T., et al.: A new vehicle localization scheme based on combined optical camera communication and photogrammetry. Mob. Inf. Syst. (2018)
12. Hossan, M.T., Chowdhury, M.Z., Shahjalal, M., Jang, Y.M.: Human bond communication with head-mounted displays: scope, challenges, solutions, and applications. IEEE Commun. Mag. **57**(2), 26–32 (2019)
13. Mathew, A.: Research article open access artificial intelligence for intent based networking. Int. J. Comput. Sci. Trends Technol. (IJCST) **8**(2), 13–17 (2020)
14. Elliott, D., Keen, W., Miao, L.: Recent advances in connected and automated vehicles. J. Traffic Transp. Eng. **6**(2), 109–131 (2019)

Intuitive and Impulsive Pet (IIP) Feeder System for Monitoring the Farm Using WoT

K. Priyadharsini, J. R. Dinesh Kumar, S. Naren, M. Ashwin, S. Preethi, and S. Basheer Ahamed

Abstract In this rapidly changing environment, caring for the pet is emerging as the prime need of PCS. Generally, the pets rely on their owners for both food and shelter. In the absence of their caretakers, their caring will be worst. The root cause for this problem is the lack of care on pets and believing that the pet feeding device will compensate this challenge. Pet feeding (PCS) occupies the major part of this paper, and the controlling of the device was done via the voice comment, which may generate using Google Assistance. This project will perform a research on the possibilities for deploying novel technologies to communicate and control and make the interaction between pet and owner by using the Internet of Things [IoT] model. The recent innovation in feeding system has leveraged advancement in the pet utilization. This major constrain is controlling the pet through virtual mode. This may uplift by using the fixing vision-based systems, which could help the pet to see their owners and follow their comments. In addition, this project has the robot design at the base. The feeding section is placed on the top of the robot in order to sense the movement of the pet in the house and timely monitoring with the food delivery. This system creates the texture by extracting the type of the pet and controls the metallic gallon in which the food has been placed. Once the food level is low, a notification will be sent to the owners via MQTT protocol, and through this M-app, user can identify the rate of changes in the taking nourishment to the pet via graphical representation. The graph variation could also help to diagnose the problem associated with the pet's health. Henceforth, this system becomes more suitable for the farmhouse, where large no.

K. Priyadharsini (✉) · J. R. Dinesh Kumar
Department of Electronics & Communication Engineering, Sri Krishna College of Engineering & Technology, Coimbatore, Tamil Nadu, India
e-mail: priyadharsinik@skcet.ac.in

J. R. Dinesh Kumar
e-mail: dineshkumarjr@skcet.ac.in

S. Naren · M. Ashwin · S. Preethi · S. Basheer Ahamed
Department of Electronics & Communication Engineering, Kumaraguru College of Technology, Coimbatore, India

© The Editor(s) (if applicable) and The Author(s), under exclusive license to Springer Nature Singapore Pte Ltd. 2021
S. Shakya et al. (eds.), *Proceedings of International Conference on Sustainable Expert Systems*, Lecture Notes in Networks and Systems 176,
https://doi.org/10.1007/978-981-33-4355-9_11

of the pets should be taken care at the same time and also to monitor them, which are at a different level of food consumption.

Keywords WoT · IoT · Pet feeders · MQTT · Google assistance

1 Introduction

The modern era of technology changes our life as much faster [1] and every new invention supports the different live beings of the earth [2, 3]. This paper is focused on caring of pet animals with automatic alert on feeding system to them based on WoT. This paper will give the ideal solutions for employees who have left their lovable pet animals in their home. Spending a separate time to feed pets on this busy life is difficult to overcome that problem [4], and our project helps to minimize the pressure faced by owner in feeding their pets on time. This pet feeding system is completely equipped for caring for the pet to the health concern process [5]. The process of this project is divided into modules for monitoring the pet with the automated system on feeding, which could be done with the Google Assistance (Fig. 2).

Catching convenient data, utilizing setting data, and communicating straightforwardly by a physical item are the fundamental solicitation in this modem world with the versatile customer. That is the principal use of the Web of things (WoT). That is, the IoT gives guidelines and strategies to operate the object in reality. There are a few methodologies for the arrangement of users, which make such associations as could reasonably be expected. A few researchers and people groups state that when setting off to a recreation center toward the end of the week, there are more individuals walking pets and taking their children to park. Designing the autonomous system for the pet feeding gets significance as the market, and the associated business with this is tremendously high. One of the research surveys predicts the US market on pet service (PSC) and care unit took nearly $4.68 Billion between 2014 and 2016,

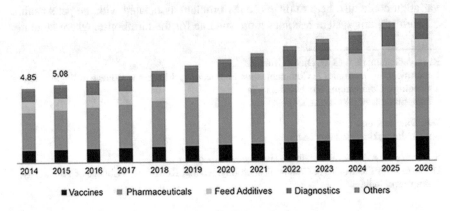

Fig. 1 Research survey on pet care service in the U.S. market

and it was 363 Million in 2005–2016 (Fig. 1). With a short period, the market rate has got a peak, and in future, the overall amount on spending toward pet care service increases by 10%; the probability of the growth rate of PCS is shown in (Fig. 1). In the US, not only almost all the inner and outer part of the world is looking for the PCS unit and ready to spend a huge amount. Especially, Europe and Asian area are a more promising and potential place for PCS.

Henceforth, this trend is accounted for by the pet care services (PCS). This paper aims to build the strong relationship between the pet and its owner by virtually. Even in the absence of the owner, the pet should be taken with utmost care. The modern technologies have different modulus [5, 6] on supporting segregated parts of them. One of the objectives is to cumulative the different process on PCS and convert them as more optimized one. It means making them as independent, relay on the other PCS support, consumes lower power, energy-optimized one, easy GUI [7], sensor nodes connected, easy access of data, and control. The evergreen trend on communication is controlling the device remotely by the Personal Assistant Devices (PAD). The PAD has a wide variety based on their configurations and the communicating network, and mostly this device will follow the adhoc-based networks which could communicate with base stations frequently. Therefore, our proposed model follows the principle of communication through WSN [8, 9], and all the devices are closely connected with owners, and also the signal integrity problem and handoff problems on the traditional devices can be overcome by this. This IoT–WoT based setup will eliminate the major supports of PCS [10], in which master will give the comment, the device associated with them will act as a slave, and it is like controlling the pet virtually. Almost the sensitivity part of this model design is using the camera [11] and type of material for delivering the food and carrying the water dispenser, which could not affect the pet and also could not get damaged by the pets. Upcoming sections describe the existing technology used in PCS and followed by the system architecture explanation with the experimental setup.

2 Study of Existing Methodology

(a) **RFID Technology**: One of the familiar and ancient technologies in tracking people in the indoor environment is RFID, since its low cost. This RFID [3, 11] based systems will use the RF tags to capture the information about the object movement in a defined limit. This system overcomes the drawback of IR based sensor for finding the dynamic movement in a door entrance. Initially, this looks good, and even different tags could be connected with the pet to track the movement inside the home. However, due to electromagnetic interference, it could fail to track. Another approach has been developed to come across this bug via EPC global connected architecture [11]. It uplifts the process of sharing the information with the pet's owners. The system was mainly used to realize collaboration between participants in terms of creating communication between them with the fullest cooperation and coordination. The EPC based models help to connect different adhoc pet owners to a single point of access (Fig. 2).

This could create a problem on the overlapping of the signal that could lead to the wrong window on an exchange, and the RF tags are limited to the range. So, the user always has to see their surrounding limits where they can communicate with the device without interference, and this makes the oriented figure of pattern to improve the performance and seamless information flows. Another technology was implementing pet feeders equipping with a camera, which is functions as allowing the users to monitor the behavior of the pet and its activity in a home from the workplace or somewhere out of home locations with uninterrupted streaming video in smart mobile. By using the monitoring device, the user can observe the behavior of pets only in the direction where the camera is focused. But, these machines were stable, so the pet movement cannot be tracked beyond the camera angle. Henceforth, it gives less precision on tracking the behavior of pet in the home. Another article on automated feeding process is based on the monitoring. The group of researchers develops it. The objective is to feed pet [11] anytime by using the special software application. It requires a huge investment, and also it sets the pet's mealtime. For example, it can be set up to 09 feed for the pets [11]; the pour food settings allow users to feed their animals any time since it uses program-based feeding system at the fixed time; it failed to read the food requirement of pet since the feed time are frequently altered. The taking care of cat likewise speeds down as indicated by breed type to define the eating speed. The feeder works with sauce food and oats. Shockingly, the level of the feeder does not show the sum of food left within the bowl.

(b) **Pet net Smart Feeder**: It requires a wireless router of 2.6 GHz [11, 12] frequency. Use the Pet net software to say about what breed, age, height, weight, and activeness of pet, and it will recommend specific set off level to feed the pet in the desired format. A specialized smartphone application is available in the play store [12] and app store which can be downloaded. One owner can control only a single animal at a time.

(c) **Petzi Treat Cam**: A feeder which can connect with the owner and feed our pet from all over at intervals the world is shown in Fig. 3. The Petzi Treat Cam [11] helps you connect at the side of your pets get through. Usually often not a wise feeder for meals, except dispensing treats to our pets throughout the day. Through the mobile app [13], the user can interact with their pet even in their absence. The recording is spot on with this. It has the camera, so all the movements can be captured by this and can be viewed by the user.

(d) **Pet cube Pet recorder**: Feed pet from anywhere [12]. Feed your pet from any distance in the world, or set automatic feeding from the Pet cube software application. The camera [11] used in this product uses the ultra-wide angle to record the video with the high definition. It has three stages as (i) Controller— for the feeder which can be controlled only through the Pet cube software and Amazon AI. (ii) Food throwing—the food is thrown at a distant to make the pets

Fig. 2 **a** Pet safe, **b** pet net smart feeder, **c** petzi treat cam, and **d** Arf pets automatic feeder

Fig. 3 Block diagram of system

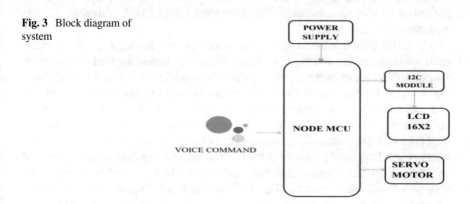

active and playful with their owner. (iii) Amount of food—the food is limited by the size and shape to make it sufficient enough on all day. The imitation of this device is communicating [14] with user, and the food delivery is limited which cannot be adjusted once it is fixed.

(e) **Arf Pets Automatic Feeder**: Programmed pet feeder makes [11] taking care of issue free for pets to get their sufficient nourishments to make them fit and sound in Fig. 3b. Arf Pets Automatic Feeder are intended to set the hour of six times each day [11, 15]. The pet feeder likewise has different other extraordinary highlights to set the time and set the degree of food that should be filled in the bowl, and the distributor additionally is uniquely intended to change the degree of an outlet with the goal that the measure of food poured can be controlled.

The different technology [11] has been implemented to feed the pet in an effective manner, but the main objective of the pet care is not satisfied, and these models were found effective for an animal, and it is not suited for more than two pets; hence, these models are not suitable for the farmhouses. Our system design was defined in next section to overcome the various drawbacks quoted in these sections.

3 Proposed Methodology

The proposed scheme has a two objective as designing a remote-control system and movement is based on [5, 14] wheel locomotion robot which is furnished with a vision-based camera along with the auto-feeding system for the required level of food and water. This setup makes the users receive the image captured by the vision system via a smart mobile, yet additionally to control its gesture through MQTT [10, 14] to achieve the basic need of the pet feed when it is controlled via remote location, also it could be developed for water supply alone. The scheme of pet feeder care (PFC) connection is shown in Fig. 3. It defines the way how the basic components of the pet feeder circuits are connected. The main blocks are LCD, I2C module, Google Assistant, and servo motor.

The I2C is used to communicate with the devices and has high speed communication of data transfer, and the servo motor section controlled the feed section of the system. In this, servo motor is highly précised one with the control of food delivery to the particular per. The vision-based system is used to identify the type of pet. The backend support of image processing is used to extract the information about the particular animal, and it performs the computations [16, 17] to show the level of food to be feed to the animal. According to this commend from the Arduino, the servo motor is acted to provide the food in load cell chamber, and simultaneously, the food weighted is calculated and sent to the control unit. The more understanding of the system is developed as shown in Fig. 4. The entire block diagram is defined for the communication between Arduino board which was controlling the motor section. The wheel mechanism control section is connected to the wheels of the robot, in which the mechanism is defined for a robot to move in different surface without any flaws. Another block is Node MCU ESP 8266 [5]-based system which is connected to the cloud for the data communication. Another sensor is associated with the setup which is laser range finder. The laser range finder sensor is used to identify and measure any obstacle or object that is between the feed robot and real-time environment. By using this sensor, our robot gets easily deviated from hitting the obstacle in the path. The relay switch is for transferring the type and level of food to be delivered to the different types of pet and relay-based switches.

If any pet lovers wish to bring up more than two pet in the same place, this switch will be helpful to identify the type of animal and provide different feeding to them and monitor the health condition [18]. The driver mechanism is based on the wheel locomotion type. One of the most familiar mechanisms used in the robotic industry is based on the wheels only. The wheels are driven by the driver IC LP293D [5]. This base robot has more dynamic stability and can able to hold the max of 2–2.5 kg of food. The level of stability is ensured by the four different wheels where each wheel is communicated with Arduino via motor driver IC [19, 20]. The standard

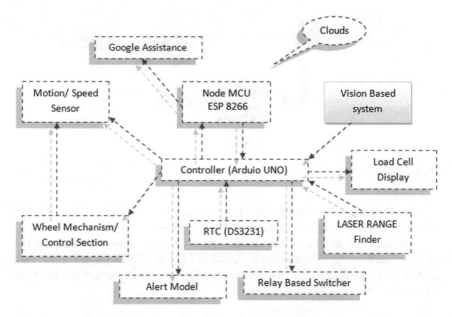

Fig. 4 Connective block setup

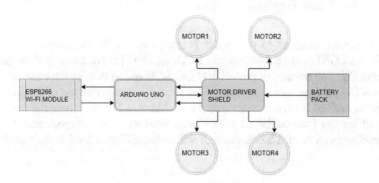

Fig. 5 Block diagram of robot design

fixed wheel [8] is used rather than Swedish wheel because of the environment where this robot is being used that is based on the smooth surface, and especially the friction loss and wheel dynamic losses are reduced when the system wheel is connected in a synchronized way. The front wheel sections are steering type, and the steering type is controlled by the signal generated from the wheel speed sensor which could be communicated with the wheel via controller (Fig. 5).

There are numerous approaches to actualize a pet feeder, you can set it to top off the bowl at a specific time, and you can order it to fill up whenever the bowl gets vacant, or possibly to give your canine food after they follow a lot of requests that

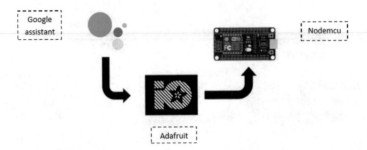

Fig. 6 Google assistant

you instructed them to be trained. In this particular task, the command instruction is given by the user through voice command by the platform Google.

In addition, it is also decided to add the option for user defined to feed the pet at a specific time interval. The combination of MQTT [7] and IFTTT [8] in Fig. 6 is used to integrate a seamless connection between user and pet feeder. After the instruction, the received the servo motor is activated.

3.1 Hardware Implementation

The schematic diagram of pet feeder is shown in Fig. 7. The common pins such as V_{cc} and GND pin of servo motor, LCD, and I²C [5] are connected to the node D1 and D2. The remaining pins of I²C like SCL and SDA pins are connected with NodeMCU receiver and data transmitter sections. The prototype of the pet feeder model is defined for the two or more pet is shown in Fig. 8. Where the metal body is used for the food container with a food bowl to store the food, and it has the internal separation of a block through which the different food is delivered to the

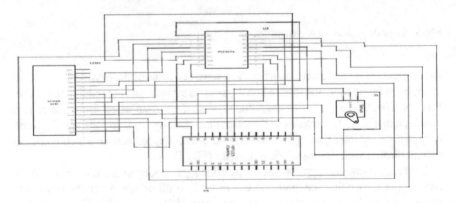

Fig. 7 Schematic of pet feeder

Fig. 8 Portable pet feeder model and level of display

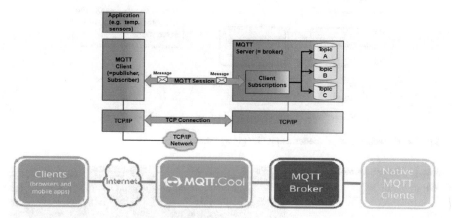

Fig. 9 Hub system and MQTT connections

corresponding pet, and this task are done based on the type of features taken from the vision-based systems (Fig. 9).

The servo motor is directly attached to the outlet of the food container. The display and controller placed on the board and display are shown below.

3.2 System Implementation

The pet feeder contains various components such as Google Assistant and servo motor which are used to send the data and receive data from the cloud through MQTT [7] and IFTTT. All the necessary commands and actions are stored by Adafruit IO server to make the working process easier as predicted in Fig. 10. Figure 11 shows the voice command given by the user which is stored in the cloud to make the actions work. Out of several protocols proposed for MOM/IoT applications, the MQTT [14]

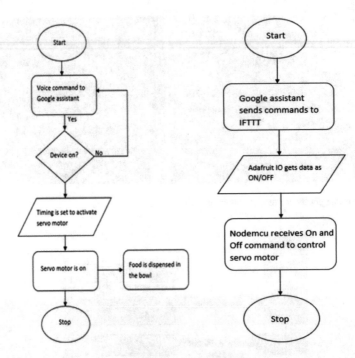

Fig. 10 Flowchart for pet feeder using Google assistance

Fig. 11 Screen on mobile app

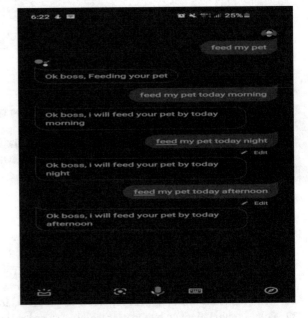

and CoAP are widely used because of its error handling capacity. MQTT is a weight free and using broker protocol which is beside the TCP/IP for M2M connectivity with 8-bit and 256 RAM controller. Usage of MQTT is provided with low bandwidth [8, 14].

MQTT Client: The client will send and receive from any microcontroller system. The client of MQTT can produce or collect the telemetry data by connecting to the server. When it is declared as publisher, the data can be viewed. For example, left food in the tray is 32.86 g or current feeding amount is 22.15 g. The reason behind the implementation of MQTT for this project is security, and it has better efficiency in WoT devices and machine-to-machine communications; also it is easily adopted to real-time signal [21] absorption and quick response to the sensory unit changes.

Corresponding library files are installed to make the use of the communication between client and server via the intermediate hub as shown in Fig. 10. A broker is the central device that plays a role of transferring the message from the client to the respective subscriber [13]. When information in a text format is iterated from the client, the broker could search for the topic on handout, and this can even handle up to 10,000 messages simultaneously. The user name and password given by the MQTT client will be the first process by the broker. After checking, the broker will initiate permission for the client whether to restrict or publish the data. For further communication, TLS and SLS encryption are used. With the publish/subscribe pattern, MQTT [8, 14] is used in various communication. The broker has also the additional capability of queuing the message when the client is not connected. Then, the client message will be released when the subscriber is ready. Figure 10 flowchart explains the working of the pet feeder. The process starts with the voice command given by the user to Google Assistant as "Ok Google, feed my pet." Then, the Wi-Fi checks whether the device is on or not. If the device is on, the further process goes as the timing inside the microcontroller is being activated and makes the servo motor open and close in the specific time to dispense the food, Table 1 shows the average amount of the meal that the dog can consume and the left out in the tray. Table 2 represents the meal consumed by the pet. The Google Assistant sends the word phrase which is being created in IFTTT as an applet where the certain keywords are triggered to make some specific action. The phrases are sent to Adafruit IO server as on/off data to process the task. The on/off is made into binary as 1 and 0 which is received by the NodeMCU to control the servo motor to dispense the food in the bowl.

The prototype model is tested, and the results were tabulated with the level of food delivered and at different time slots. Initially, for the prototype model, the limit is specified as 2 on initial days, and day 3 is meant for the pet owner who is out of the station for more than one day, it will automatically change its feeding type as 4 per day. However, this function is done based on the comment received from the user through a GUI or mobile app. The level of food consumed by the different species is listed below with the classification for adult and kid.

Google Assistant can move information to MQTT [8, 14] convention by a cloud worker setup by Adafruit IO. This cloud worker, the End client application, can create a correspondence between the cloud as the consequence of an associate procedure.

Table 1 Food amount per meal

No.	Prediction date and hour	Taking nourishment per meal (g)			
		Pattern hour	Craving food amount	Halted food in box(80)	Current taking nourishment amount
1	08/04/2020	2.00 am	65.00	15.00	81.25
2	08/04/2020	8.00 pm	45.50	35.50	56.875
3	09/04/2020	1.00 pm	60.77	20.23	75.9625
4	09/04/2020	7.00 pm	40.64	40.36	50.8
5	10/04/2020	8.00 am	40.00	40.00	50.2
6	10/04/2020	1.00 pm	10.00	30.00	12.5
7	10/04/2020	4.00 pm	20.00	10.32	25.3
8	10/04/2020	9.00 pm	40.55	40.45	50.6875
9	11/04/2020	10.00 am	75.20	5.35	93.75
10	11/04/2020	6.00 pm	60.00	20.00	75.3

Table 2 Amount of food eaten by different pets

S. No.	Type of pet	Average food consumption per day for each pet (g)	
		Adult	Kid
1	Dog	200–800	50–200
2	Cat	150–450	50–100
3	Rabbit	200–400	20–100
4	Hamster	100–200	50–100
5	Guinea Pig	150–250	10–100

For that reason, an applet is required that permits the IFTTT [7] to make activities and results in bringing certain procedure or work. Figure 12 shows the feed record. Consequently, this venture IF THEM THAT to make another activity where the client says a particular voice state is said Google Assistant. Pet feeder has built up an association with the Web worker of ESP8266 NodeMCU module of around 12 pins. The esp8266 accompanies NodeMCU module. NodeMCU will host and control the correspondence with MQTT convention.

This Wi-Fi module likewise underpins correspondence with encryption. Therefore, it switches the modes from "0" (which sets up correspondence without encryption) to "1" (made sure about correspondence). NodeMCU gets order from Web cloud and sends it to servo engine by SCL. This order is then scrambled by servo engine, so it can work the specific activity characterized by the client. In taking care of the procedure, the client provides the voice ordering through Google associate. The Google associate imparts the order to MQTTT [10] and IFTTT to trigger the

Fig. 12 Feed record

particular activities which are now coordinated in the worker to play out the assignment. Sometimes, the amount of food feeding exceed the amount of food which is defined by the user, so this is one of the disadvantages, Where the code inside the microcontroller has set that the measure to open and close of the servo motor can drop around 75 g in a single open. There are odds of getting cereals blocked in the pipeline prompting disappointment in the feeding process. In down to bowl, the servo motor can naturally diminish the blockage of food stuck such halfway open and close.

This pet feed system can be implemented in a farmhouse to support the farmer to monitor and feed their animals in the farmhouse as shown in Fig. 13. The most

Fig. 13 Automated pet feed systems in farmhouse

tedious process in the farm for owners is in maintaining the hygienic and equal food delivery to the large area of them. So, by implementing this pet feeder system in the farmhouse, it will help to identify the deficiency of food delivered to the pet and behavior of the pet, and isolate the affected pet before it starts to spread which will help to improve the yield and their economic with low investment at initially.

4 Conclusion

Pet feeding framework is regularly handled by transfer information to Google partner with cloud and NodeMCU, and that they move the data and orders with MQTT convention. For secure and quicker transmission, Adafruit IO was utilized all together that the correspondence between Google aid to cloud and NodeMCU is consistent and precise. Pet feeder is often made bigger using longer chute and tank to reserve the pet food/cereals. This app could further progress with features, like a free-rolling camera on 360°, implementation of AI for smarter feeding system, and water dispenser for brand-spanking devices. By implementing these devices in the farmhouse, economic support to the farmers can be provided which help to develop the nation's production and export in animal and boost the research associated with them. The feature of smart farming depends on the way technology it is using, and this project has encountered the problem of signal interference when it is implemented for the larger farm areas. Henceforth, this area has to be focused in future to utilize this technology which is well-defined manner to build the strong nation in farm–agriculture.

References

1. Dinesh Kumar, J.R., Ganesh Babu, C., Balaji, V.R.: Analysis of effectiveness of power on refined numerical models of floating point arithmetic unit for biomedical applications. IOP Conf. Ser. (2020)
2. Priyadharsini, K., Yazhini, V.: A novel methodology for smart soil management and health assessment system to support the contemporary cultivation using AI on mobile app environment, vol. 7s (29). IJAST (2020)
3. Priyadharsini, K., Nanthini, N., Soundari, D.V., Manikandan, R.: Design and implementation of cardiac pacemaker using CMOS technology. J. Adv. Res. Dyn. Control Syst. 10(12-Special Issue) (2018). ISSN 1943-023X
4. Own, C.-M.: For the pet care appliance of location aware infrastructure on cyber physical system. Int. J. Distrib. Sensor Netw. 8 (2012). https://doi.org/10.1155/2012/421259
5. Kranz, M., Holleis, P., et al.: Embedded interaction interacting with the internet of things. IEEE Internet Comput. 14(2), 46–53 (2010). https://doi.org/10.1109/MIC.2009.141
6. Dinesh Kumar, J.R., Ganesh Babu, C.: Performance investigation multiplier for computing and control applications. J. Adv. Res. Dyn. Control Syst. (2019)
7. Hunkeler, U., Stanford-Clark, A., Truong, H.L.: MQTT-S—a publish/subscribe protocol for wireless sensor networks. In: Proceedings of the COMSWARE '08

8. Wagle, S.: Semantic data extraction over MQTT for IoT centric wireless sensor networks. In: 2016 International Conference on Internet of Things and Applications (IOTA) Maharashtra Institute of Technology, Pune, India 22 Jan–24 Jan (2016)

9. Dinesh Kumar, J.R., Priyadharsini, K., Yazhini, V.: Performance Metric Interpretation on Handwritten Digit Recognition Using Neural Networks, vol. 7s (29). IJAST (2020)

10. Light, R.A.: Mosquitto: server and client implementation of the MQTT protocol. J. Open Source Softw. **2**(13) (2017)

11. Berhan, T.G., Ahemed, W.T., Birhan, T.Z.: PPF—Programmable pet feeder. Int. J. Sci. Eng. Res. (IJSER) (2015)

12. Kim, S.: Smart pet care system using internet of things. Int. J. Smart Home (2016)

13. Nanthini, N., Soundari, D.V., Priyadharsini, K.: Accident detection and alert system using Arduino. J. Adv. Res. Dyn. Control Syst. **10**(12-Special Issue) (2018)

14. Truong, H.L., Stanford-Clark: MQTT For Sensor Networks (MQTT-SN) Protocol Specification, MQTT Documents (2013)

15. Soundari, D.V., Padmapriya, R., Thirumariselvi, C., Nanthini, N., Priyadharsini, K.: Detection of breast cancer using machine learning support vector machine algorithm. J. Comput. Theor. Nanosci. **16**, 441–444 (2019)

16. Priyadharsini, R., Mr. Mahesh Kumar, H.: Effective power optimization of CMS scheme. Int. J. Adv. Res. Comput. Commun. Eng. **3**(2), February (2014)

17. Kumar, J.R.D., Babu, C.G., K.S.P.: Performance analysis of 16 bit adders in high speed computing applications. In: 2019 International Conference on Advances in Computing and Communication Engineering (ICACCE), Sathyamangalam, Tamil Nadu, India, pp. 1–7 (2019). https://doi.org/10.1109/ICACCE46606.2019.9079985

18. Dinesh Kumar, J.R, Priyadharsini, K., Ganesh Babu, C., Soundari, D.V. Karthi, S.P.: A novel system design for intravenous infusion system monitoring for betterment of health monitoring system using ML- AI. J. Int. J. Innovative Technol. Exploring Eng. (IJITEE) **9**(3), 2278–3075 (2020)

19. Parvin, J., Rejina, S., Gokul Kumar, A., Elakya, K., Priyadharsini, Sowmya, R.: Nickel material based battery life and vehicle safety management system for automobiles. Mater. Today: Proceedings (2020)

20. Priyadharsini, K., Little Judy, A., Mohan, K., Divya, R.: Design of an integrated circuit for electrical vehicle with multiple outputs. Int. J. Pure Appl. Math. **118**(20), 4887–4901 (2018)

21. Maheshkumar, H., Priyadharsini, K.: Self time regenerator for signaling scheme authors. J. Int. J. Adv. Res. Comput Commun. Eng. **3**(2)

22. Al-Fuqaha, M., Guizani, M., Mohammadi, M., Aledhari, Ayyash, M.: Internet of things: a survey on enabling technologies, protocols and applications. IEEE Commun. Surv. Tutor. **17**(4), 2347–2376 (2015)

23. Yick, J., Mukherjee, B., Ghosal, D.: Wireless sensor network survey. Comput. Netw. **52**(12), 229, 230 (2008).https://doi.org/10.1016/j.comnet.2008.04.002

24. Karyono, V.K., Nugroho, I.H.T.:Smart dog feeder design using wireless communication, MQTT and Android client. In: 2016 International Conference on Computer, Control, Informatics and its Applications (IC3INA), pp. 191–196. Tangerang (2016). https://doi.org/10.1109/IC3INA.2016.7863048

25. Priyadharsini, K., Dinesh Kumar, J.R., Ganesh Babu, C., Surendiran, P., Sankarshnan, S., Saranraj, R.: An experimental investigation on communication interference and mitigation during disaster using lifi technology. In: 2020 International Conference on Smart Electronics and Communication (ICOSEC), Trichy, India. pp. 794–800 (2020). https://doi.org/10.1109/ICOSEC49089.2020.9215280

26. Dinesh Kumar, J.R., Ganesh Babu, C., Balaji, V.R., Priyadharsini, K., Karthi, S.P.: Performance investigation of various SRAM cells for IoT based wearable biomedical devices. In: Ranganathan, G., Chen, J., Rocha, Á. (eds.) Inventive Communication and Computational Technologies. Lecture Notes in Networks and Systems, vol. 145. Springer, Singapore (2021). http://doi-org-443.webvpn.fjmu.edu.cn/10.1007/978-981-15-7345-3_49

27. Kumar, J.R.D., Dakshinavarthini, N.: Analysis and design of double tail dynamic comparator in analog to digital converter. IJPCSC **7**(2) (2015)

Machine Learning Algorithms for Prediction of Credit Card Defaulters—A Comparative Study

Sunil Kumar Vishwakarma, Akhtar Rasool, and Gaurav Hajela

Abstract The prime goal of this study is to predict the accuracy of the classifiers predicting the default credit card customers in Taiwan. Since the last few years, the transactional companies are providing loans to their customers on their credibility, but they suffer from the default customers' payments. It is difficult to predict the accuracy of real credit card customers who are going to be the next default. So, various classification methods including boosting methods and some other methods to predict the probability of default are studied. The accuracy of different classifiers is calculated by using the confusion matrix and area under the curve (AUC) and compared with other classification techniques.

Keywords Banking · Data mining · Classification · Machine learning

1 Introduction

In the late 90s, for the expansion of business and to increase the market share, the Taiwan banks started new bussiness-credit cards and cash cards [1]. To engage the more customers, banks lowered the requirement for getting credit card approvals, and the young peoples of Taiwan became the target, having not enough income to repay the loans at time. Due to this incautious credit card lending, banks faced bad debt, and in February 2006, the debt reached $268 billion USD, and more than half billion people were not able to pay their loans, and they became credit card slaves. This issue turned into bigger societal problems, many debtors and their families committed suicide due to huge loans, some became homeless, some debtors

S. K. Vishwakarma (✉) · A. Rasool · G. Hajela
Department of Computer Science and Engineering, MANIT, Bhopal, India
e-mail: sunil.manitb@gmail.com

A. Rasool
e-mail: akki262@gmail.com

G. Hajela
e-mail: contactgauravhajela@gmail.com

© The Editor(s) (if applicable) and The Author(s), under exclusive license
to Springer Nature Singapore Pte Ltd. 2021
S. Shakya et al. (eds.), *Proceedings of International Conference on Sustainable Expert
Systems*, Lecture Notes in Networks and Systems 176,
https://doi.org/10.1007/978-981-33-4355-9_12

141

involved in crime, and others could not afford to pay for their basic needs. According to the Taiwanese Department of health report, the suicide rate in the Taiwan is the second highest in the world, and debt is one main reason behind it. So to stop these kind of incident, it is important to having a relevant system to classify the appropriate customers for lending the credit or cash cards, which can save the both companies and customers and their relationship. The right acquisition of software solutions can help the banking-based companies to deal with these kind of situations [2]. A statistics Table 1 of 2005 from April to September has been attached which can be seen clearly that how much credit cards are issued and what are the delinquency rate with their revolving money [3]. In the field of transactional systems, the companies issuing credit cards face a debt crisis due to unexpected defaults. Different banks and companies offer loans and sometimes over issued the credit cards and cash to non-eligible customers [4] to raise market shares. But the problem happens when most of the cardholders overuse and accumulate huge cash and credit debt due to overconsumption. This uncertainty raises the question of the customer–finance relationship and their confidence which is a major challenge for companies and customers. So there is a need for a well-developed financial crisis management system where risk prediction is upstream. The Internet of things can also be used for better management of communication of these type of companies [5]. The prediction is possible only if available data of financial customers are used like transaction statements of customers, repayment information, other financial records, etc. [6]. It is important to predict every client's credit risk, so that the business performance can be increased to grow the company, and damage and uncertainty are reduced. For the sake of this, many methods have been studied such as Bayes classifiers, k-nearest neighbor, and some boosting methods like AdaBoost, CatBoost for risk prediction. Eight data mining techniques are reviewed (Random forest classifiers, Gaussian Naive Bayes, K-nearest neighbors, MLP classifier, AdaBoost, CatBoost, XGBoost, LightGBM classifiers), and their applications on credit scoring and the classification accuracy among them can be compared and found the boosting algorithm that performs well having the highest accuracy of the LightGBM algorithm.

Table 1 Credit card and financial information from April 2005 to September 2005

Month	Card in force	Revolving balance (in NT$ 1000)	Delinquency ratio (3–6 months)	Delinquency ratio (over 6 months)
April	44,924,431	473,665,343	2.18	0.56
May	45,147,399	470,077,082	2.24	0.52
June	45,385,369	473,539,271	2.26	0.50
July	45,472,639	480,421,836	2.34	0.47
August	45,656,778	477,656,247	2.20	0.34
September	45,606,672	488,331,243	2.33	0.34

2 Literature Review

Nowadays, a huge amount of information is collected in the form of text, image, audio, or video data by the companies. The mining of collected data is important to retrieve the useful knowledge from it for the growth and development of the company. In transactional system companies, like in Taiwan banking system, the forecast about the clients is very important for risk assessment [7]. So eight data mining techniques are studied and applied them on the dataset and measured the performance with the area under curve (AUC) with ROC chart [8]. AUC measures the area underneath the receiving operating charts (ROC) curve from (0,0) to (1,1). It is based on the confusion matrix. Using the confusion matrix, ROC curve is plotted as true positive rate (TPR) vs false positive rate (FPR). TPR is the probability that an actual positive will test positive. FPR is the probability that actual negative will test positive that is when false alarm is raised. The curves of different models can be compared directly, and area under the curve is used as a summary of the model skill. AUC is scale-invariant, and it is also classification-threshold-invariant which means it gives a quality of algorithms predictions irrespective of what the classification threshold is given. The scale range for AUC is from 0 to 1. AUC infers that how well the models can discriminate the class. If the AUC is 1, it means that the model can discriminate the classes perfectly and AUC 0.5 means, model cannot discriminate. In between 0.5 and 1, the discrimination power of models can be relatively compared as greater the value of AUC gives better model performance (prediction). AUC is 0 if the prediction of the model is 100% false, and it is 1 if the prediction of models 100% accurate. The mining technique used to predict the accuracy is random forest classifier, Gaussian Naive Bayes classifier, MLP classifier—a class of artificial neural network, K-nearest neighbors (KNN) classifier, AdaBoost classifier, CatBoost classifier, XGBoost classifier, and LightGBM classifier.

2.1 Random Forest Classifier

It is an ensemble tree-based learning algorithm used for classification and regression both. It combines the result of different subtrees to decide the final class of the test or new data object. Random forest [9] classifier runs smoothly on big datasets. It is also effective when there is a wide portion of data which is missing and maintains accuracy. The main cons of this classifier are sometimes overfitted for some datasets, and it also shows biased behavior in favor of those attributes which have more levels if they are categorical in nature.

2.2 Gaussian Naive Bayes Classifier

Naive Bayes classifier [10, 11] is based on Bayes theorem, and it is a probabilistic method to classify the data. The fundamental assumption of this classifier is that it assumes that all the features are independent and every attribute equally contributes to the outcomes. It is based on conditional probability as it gives the happening events with condition that the other event has already occurred. Let A and B are two events, and the probability of happening of B is not 0, then the mathematical equation is given as $P(A/B) = P(B/A)P(A)/P(B)$. In this method, the values of each feature are supposed to be Gaussian distribution or normal distribution. The main drawback of Naive Bayes is the performance accuracy of the model which is strongly related to the presumption made so for.

2.3 Multi-layer Perceptron (MLP) Classifier

MLP [12] is used mostly on labeled training data, so it is a type of supervised learning technique, and it is a type of feed-forward ANN. MLP uses the backpropagation method for teaching and training the model. It solves the problem stochastically, so it is widely used in research. The main pros of MLP classifiers are that it can learn nonlinear models, and it also learns various models using partial fit in real time too. The drawback of MLP classifier is that it is sensitive to feature scaling, and it requires to set different hyperparameters. Sometimes, it gives different accuracy during validation due to random weights initialization, for MLPs having hidden layers with non-convex loss functions.

2.4 K-Nearest Neighbors Classifier

KNN [10, 13] is nonparametric (means no assumption are made for data distribution) and lazy learning (does not require any training data points, all training data used in testing phase) algorithms. In KNN, K is the number of neighbors, which is the main deciding factor. In this algorithm, each point finds the closest similar point, which is measured in terms of distance such as Euclidean distance and Hamming distance. KNN is easy to use and does not require to establish a model before classification. The cons of this model are that its accuracy depends on the quality of data, a measure of distance, and cardinality k of the neighborhood.

2.5 AdaBoost Classifier

There are various ensemble boosting classifiers, and adaptive boosting (AdaBoost) [14] is one of them. It is an iterative ensemble method that combines multiple classifiers to build a stronger classifier with increased accuracy. The pros of this classifier are that it repetitively corrects the fault of the week classifier which increases the accuracy of the resulting classifier and is also easy to perform. It is not prone to overfitting. Since this model tries to fit every point accurately, so it is affected by outliers and is subtle to noisy data.

2.6 CatBoost Classifier

CatBoost [15] is the acronym for the category and boosting. This algorithm works well with multiple categorical data. It is based on a gradient boosting machine learning algorithm, which is a powerful ML algorithm widely used in business challenges like fraud detection. The main advantage of this algorithm is that it yields a good score without ample training compared to other ML algorithms and also gives extra support for descriptive data formats. This algorithm is robust in nature as there is less chance of overfitting and needs fewer hyperparameters setup.

2.7 XGBoost Classifier

XGBoost [16] is an acronym for extreme gradient boosting, and it is a more advanced version of the gradient boosting algorithm. The main focus of XGBoost is the speed and efficiency of the model, and for this reason, it has additional features. It works on parallelization by creating decision trees, instead of sequential modeling in computing algorithms. It is a more popular algorithm because it outperforms other algorithms and has a wide variety of tuning parameters like cross-validation and regularization.

2.8 LightGBM Classifier

LightGBM [17] is short for light gradient-boosted machine. It is also a type of gradient boosting method which uses a tree-based learning algorithm. It is different than other methods using tree learning as its tree grows vertically (means trees grow leaf by leaf), whereas other algorithm grows tree horizontally (grows level by level). It can handle large datasets and takes lesser memory to execute the algorithm smoothly. It focuses on accuracy and supports GPU learning that is why it is so popular nowadays.

But for small data, it can easily overfit as it is subtle to overfitting and also required tuning of parameters.

3 Experimental Setup and Results

3.1 Dataset Description

The dataset is taken from the UCI repository [18]. It contains the basic information of credit card clients' payments, credit amount, payment history, and their bill amount in Taiwan from April 2005 to September 2005. It consists of 30,000 instances and 25 features/variables. The description of features is as:

ID: Customer's unique ID, LIMIT BAL: Actual credit given in NT dollars (which consists customer's self and other supplementary credit like family), SEX: customer's Gender(1 = male and 2 = female), Education: Education level of clients (1 for school graduate, 2 for university level, 3 for high school level, 4 for other levels, and 5 for unknown education), Marital status: It tells whether the client is married = 1, unmarried = 2, and 3 for others, AGE: Customer's age (yr), Repay0: September month repayment done by the client (−1 stand for duly payment, 1 for delay of payment by one month, 2 for delay of payment by two months, …0.8 for delay of payment by eight months, 9 stands for delay of payment by nine months and above), Repay2: August month repayment done by the client (scale same as above), Repay3: July month repayment done by the client, Repay4: June month repayment done by the client, Repay5: May month repayment done by the client, Repay6: April month repayment done by the client, Bill1: September bill statement of the client (all the amount is in NT dollar), Bill2: August bill statement of the client, Bill3: July bill statement of the client, Bill4: June bill statement of the client, Bill5: May bill statement of the client, Bill6: April bill statement of the client, Payment1: Amount paid by the client (September)(NT dollar), Payment2: Amount paid by the client (August), Payment3: Amount paid by the client (July), Payment4: Amount paid by the client (June), Payment5: Amount paid by the client (May), Payment6: Amount paid by the client (April), Default: Customer's chance of default payment in the next month (1 for yes, 0 for no).

3.2 Data Visualization

As the dataset is explored, got all clients are distinct, and the average of the credit card limit of the dataset is 167,484 NT$ having a large standard deviation, where the maximum value is around one million. The education status of the clients is mostly university level and school graduates. Customers are either married or unmarried having an average age of 35.5 years, and the std. deviation is 9.2. Here also got that

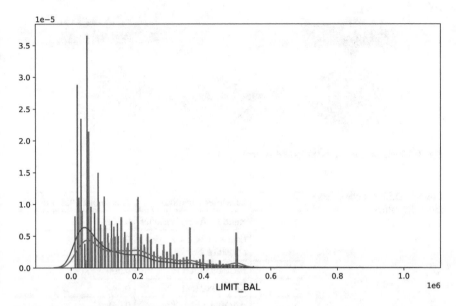

Fig. 1 Default amount of credit limit—grouped by payment next month

there are 22.1% of customers who will default next month which is around 6630 (defaults) clients in 30,000.

There is no missing data in the whole dataset, and data is balanced with respect to the target variable. The largest group of the amount of credit limit is observed apparently for the amount of 50,000 as there are 3365 clients with credit limit 50k followed by 1976 clients with a credit limit 20k, 1610 clients with 30k, 1567 clients with 80k, and 1528 clients with a credit limit 200k. From Fig. 1, for a credit limit up to 100k, they have the larger density for defaults, and larger default numbers are for the amounts of 50k, 20k, and 30k. The credit limit is equally spread between sexes, which is quite balanced. Married males have a mean age above married women.

3.3   Results Using Tables and Graphs

After the visualization, found that the dataset is already preprocessed, as there are no null values, the dataset is balanced, and there is no noise in the dataset. Then, different classifiers are applied on the dataset to classify the default customers. In order to classify, confusion matrix and ROC curve are drawn. The block diagram of the experimental process is shown in Fig. 2. From the confusion matrix and receiver operating characteristic curve of different classifiers, the area under the curve (AUC) is calculated to compare the performance of classifiers. The higher value of AUC gives a better classifier. From result Table 2 we can observe that For MLP classifier, the AUC value is 0.500, which is least among all the classifiers used here. There

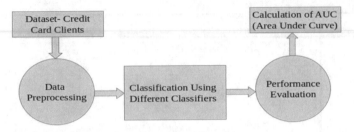

Fig. 2 Block diagram of the proposed system

Table 2 Different classifiers and AUC values

S. No.	Classifiers/methods	Area under ROC curve
1	Random forest classifier	0.662
2	Gaussian Naive Bayes classifier	0.552
3	MLP classifier	0.500
4	KNN classifier	0.542
5	AdaBoost classifier	0.658
6	CatBoost classifier	0.662
7	XGBoost classifier	0.782
8	LightGBM classifier	0.785

is little difference in the AUC values of GNB and KNN classifiers. AUC value for GNB is 0.552, and KNN has a value 0.542. AdaBoost has a value of 0.658. RFC and CatBoost have the same performance, both give AUC values 0.662. XGBoost with AUC value 0.782 and LightGBM with AUC value 0.785 perform well.

k-fold, cross-validation is done [19] to use data in a better way and to verify the performance given by the classifiers taking $k = 5$.

4   Conclusion and Further Extensions

In this paper, the eight most popular classification algorithms are examined to predict the default clients as per the historical data of the company. The boosting algorithm found to perform better than other algorithms, and among the boosting algorithm, XGBoost and LightGBM have little difference in their AUC values. LightGBM has a maximum AUC value for the ROC curve. The results strongly suggest that the LightGBM classifier gives the best performance accuracy among all classifiers, so it can be used for the prediction of the default probability of new clients. This study shows that there is a possibility of other algorithms of boosting family which may be stronger than LightGBM and can predict more accurately. The further visualization and exploration of data can give some more interesting insights. Some other methods and techniques can be used for the prediction of the probability of default clients, and it may result in more accurate performance. Some cascade learning systems, k-level classifier ensembles along with some more preprocessing steps can also be used for risk assessment. The real probability of default is difficult to predict, so for the perspective of risk control and clients confidentiality, it needs more research and study.

References

1. Taiwan's Credit Card Crisis: https://sevenpillarsinstitute.org/case-studies/taiwans-credit-card-crisis
2. Palanisamy, R., Verville, J., Taskin, N.: The critical success factors (CSFs) for enterprise software contract negotiations: an empirical analysis. J. Enterprise Inform. Manage. (JEIM) 28(1), 34–59 (2015)
3. Financial Supervisory Commision Republic of China (Taiwan): https://www.fsc.gov.tw/en/home.jsp?id=5&parentpath=0
4. Sharma, S., Mehra, V.: Default payment analysis of credit card clients (2018)
5. Wang, H.: Trust management of communication architectures of internet of things. J. Trends Comput. Sci. Smart Technol. (TCSST) 1(02), 121–130 (2019)
6. Anil Kumar, D., Ravi, V.: Predicting credit card customer churn in banks using data mining. Int. J. Data Anal. Tech. Strategies 1, 4–28 (2008)
7. Ajay, AV, Jacob, S.G.: Prediction of credit card defaulters: a comparative study on performance of classifiers. Int. J. Comput. Appl. 145, 36–41 (2016)
8. Bradley, A.E.: The use of the area under the ROC curve in the evaluation of machine learning algorithms. Patt. Recogn. 30, 1145–1159 (1997)
9. Kulkarni, V.Y., Sinha, P.K.: Effective learning and classification using random forest algorithm. Int. J. Eng. Innov. Technol. 3, 267–273 (2014)
10. Yeh, I.-C., Lien, C.: The comparisons of data mining techniques for the predictive accuracy of probability of default of credit card clients. Expert Syst. Appl. 36, 2473–2480 (2009)
11. Rish, I.: An empirical study of the Nave Bayes classifier (2001)
12. Murtagh, F.: Multilayer perceptrons for classification and regression. Neurocomputing 2, 183–197 (1991)
13. Cunningham, P., Delany, S.J.: K-nearest neighbour classifiers
14. Tharwat, A.: AdaBoost classifier: an overview (2018)
15. Dorogush, A.V., Ershov, V., Gulin, A.: CatBoost: gradient boosting with categorical features support (2018)
16. Chen, T., Guestrin, C.: XGBoost: A scalable tree boosting system (2016)
17. Ke, G., Meng, Q., Finley, T., Wang, T., Chen, W., Ma, W., Ye, Q., Liu, T.-Y.: LightGBM: a highly efficient gradient boosting decision tree (2017)
18. Yeh, I.-C.: UCI Repository, default of credit card clients Data Set. Available: https://archive.ics.uci.edu/ml/datasets/default+of+credit+card+clients
19. Refaeilzadeh, P., Tang, L., Liu, H.: Cross-validation (2009)

Investigation of Gait and Biomechanical Motion for Developing Energy Harvesting System

Nazmus Sakib Ribhu, M. K. A. Ahamed Khan, Manickam Ramasamy, Chun Kit Ang, Lim Wei Hong, Duc Chung Tran, Sridevi, and Deisy

Abstract Finding a means of clean and free source of energy has become vital in this age and era. With rapidly shifting technological advancements, the necessity for harvesting energy is a top priority. Moreover, the time has come to utilize the core of humanity, the human body itself, for greater purposes. In this report, thorough and proper research has been carried out to learn extensively about human gait and biomechanical movement and the means of extracting clean energy from that. Proper research along with practical experiments including that on different human test subjects of different body mass index, weight, and gender has been carried out in order to find the forces acting on the plantar region considering different test subjects of varying weight, age, gender, and body mass index to find out plantar pressure distribution and the precise distribution of forces acting. A working prototype has then been modelled and devised based on the findings from the research. The prototype has been made from 3D printing, a combination of piezoelectric materials is then tested and the results are tabulated which are provided herewith.

Keywords Biomechanical motion · Human gait · Clean energy

1 Introduction

Numerous researches have been carried out for decades to come up with energy extracting devices that can extract or harvest a significant amount of energy without compromising the environment or the stability of the factors involved [1]. A clean

N. S. Ribhu · M. K. A. Ahamed Khan (✉) · M. Ramasamy · C. K. Ang · L. W. Hong
Faculty of Engineering, UCSI University, 56000 Kuala Lumpur, Malaysia
e-mail: Mohamedkhan@ucsiuiversity.edu.my

D. C. Tran
FPT University, Ho Chi Minh City, Vietnam

Sridevi · Deisy
Thiagarajar College of Engineering, Madurai, India

S. Shakya et al. (eds.), *Proceedings of International Conference on Sustainable Expert Systems*, Lecture Notes in Networks and Systems 176,
https://doi.org/10.1007/978-981-33-4355-9_13

and safe method of extraction of energy, which can be used to power various useful objects, can be the pioneer to the sheer amount of possibilities that can lead to a noteworthy advancement in future technologies and feasibility of usable energy. As it has been divulged in the project title, this research is carried out to investigate the general gait of the human body and the biomechanical motion of it, to devise and develop an energy harvesting system that uses human gait and motion for harvesting the energy. The research is then followed by implementing a design along with the practical build of a prototype that demonstrates a functional way to generate energy with the help of biomechanical activities of the human body. The mainspring leading to the concept is to provide a means of yielding clean energy that could be of use to numerous situations and circumstances that might not always be favourable or in general conditions. There are many human-worn devices that are commercially available at the time being, different devices bear different characteristics and functions, while some may be available at an affordable price and vast quantity, some may be so expensive that it will be out of reach for general people and the availability might be limited.

At the same time, while some are suitable for robust use, some are better off as prototypes only and not quite practical for outdoor usage. While the general objective or outcome of these bio-mechanical motion utilizing devices is to extract or harvest energy, the method of extraction varies quite widely among manufacturers and their devices. Energy can be extracted from various parts of the human body [2]. In general, these devices are built to extract energy by using either the upper limbs or the lower limbs of the human body, where the upper limbs consist of the arms, elbows, hands, and shoulders and the lower limbs consist of the hip joints, knee joints, ankle joints, or heels of the feet.

During this work, first, an extensive investigation of the complete human gait and motion will be done to familiarize with a different types of movements of the human body, as well as the energy dissipation and distribution of the human body during these movements, and the suitable limb or part of the body from which this energy can be extracted. For practical usage and safety, proper analysis on the plantar pressure distribution, biological structure of the nerves, muscles, and bones of the selected area from where the energy is expected to be extracted is carried out. Once familiar with the biological structure of the human body, especially the part this research is concerned on, the author will be looking into the medical ethics and laws pertaining to the usability of the devices on a human body, since this work almost solely relies on human participation.

Finally, to illustrate the design of the energy harvesting system from the biomechanical motion of human anatomy, a prototype is to be made and the methods of building it and outcomes will be discussed and demonstrated in the suitable sections.

The main objectives of this study are to analyse and comprehend the overall human gait cycle, which also includes study about the distribution of pressure in the plantar region, the biological structure of the human body, especially the area this project will be related to. To understand and identify the appropriate sensors and suitable electro-mechanical systems to harvest or extract the energy also to design and develop a complete energy harvesting system considering the power/force distribution along the particular area of human body in terms of formulating data obtained from different human subjects.

Using the human body as a provenance of energy can be useful as it has been found that [3] just 0.2 kg of body fat can produce energy equating to about 800–2500mAh double-A batteries.

2   Material and Methods

2.1   Human Gait Analysis

There are eight phases in the gait cycle, and the different phases of the human gait cycle are shown in Fig. 1. The terminologies are to be discussed below:

1. Initial contact (IC): This is the foremost phase of the gait cycle (at 0%) where the hip is at 20° flexion, the knee is at 0 to 5° flexion, and the ankle joint is at 0° flexions. The heel is in contact with the ground.
2. Loading response (LR): The second phase of the gait cycle ranges from 0 to 12% where the hip is flexion towards the 20°, and the flexion near the knee is at 20°, and the flexion of the ankle joint is at the range of 5°–10°. Knee and talocrural joints are susceptible to absorption of shock, the hip is susceptible to load transmission and stability, forward motion by heel rocker.
3. Mid-stance (MST): Third phase of the gait cycle (at 12–31%) where the hip is at 0° flexion, the knee is at 0° flexion and the ankle joint is at 5° dorsal flexion. Tibia has a controlled motion forward, and ankle rocker shifts the centre of gravity to the front.
4. Terminal stance (TST): Fourth phase of the gait cycle (at 31–50%) where the hip is at −20° hyperextension, the knee is at 0°–5° flexion and the ankle joint is at

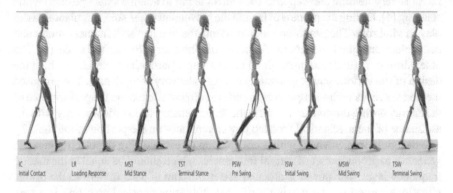

Fig. 1 Human gait phases [4]

10° dorsal flexion. The ankle joint performs controlled dorsal extension, while the heel is lifted on from the ground.

5. Pre-swing (PSW): Fifth phase of the gait cycle (at 50–62%) where the hip is at − 10° hyperextension, the knee is at 40° flexion and the ankle joint is at 15° plantar flexion, where the passive knee joint undergoes a 40° flexion, and the ankle joint is at plantar flexion.

6. Initial swing (ISW): Sixth phase of the gait cycle (at 62–75%) where the hip is at 15° flexion, the knee is at 60°–70° flexion and the ankle joint is at 5° plantar flexion. The knee bears a minimum of 55° of flexion for ground clearance.

7. Mid-swing (MSW): Seventh phase of the gait cycle (at 75–87%) where the hip is at 25° flexion, the knee is at 25° flexion and the ankle joint is at 0°. Ankle joint extends dorsally to neutral-zero-position.

8. Terminal swing (TSW): Eighth and final phase of the gait cycle (at 87–100%) where the hip is at 20° flexion, the knee is at 0°–5° flexion and the ankle joint is at 0°. Knee joint extension to neutral flexion and the next stance phase preparation initiates.

2.2 *Plantar Pressure Distribution*

As this work concentrates mainly on the heel strike/foot strike of human gait to build a device that extracts energy from the motion, it is essential to know the distribution of pressure on the inferior aspect or bottom of the foot, that is, the sole. A number of researches have been done on the topic; the data that are to be provided here are excerpted from few of the researches that have been previously done by researchers on the distribution of pressure on the plantar region of the feet.

A strong solid fibrous membrane that lies underneath the skin and surface layer of fat and binds is called the plantar fascia [5], which assists between the foot and calcaneus. The calcaneus helps to connect the talus and cuboid bones. The plantar fascia is very flexible and supports the human being to attain a stable position while standing [4]. During the process of having the movement of the sole, this plantar fascia plays a vital role. The proposed method adopts the two methods, namely rotary and linear electromagnetic generation for accumulating energy by the heel strike [6]. The other similar methods are highlighted in selecting a particular material in which the design of the various power generators directs to be very complicated. The proposed method succeeds with a simple design and cost effective to extend the performance of the energy during the conversion from the kinetic energy to electrical energy with the assistance of the mechanical footstep power generator on the portion of the foot [7]. The design is implemented with a critical step of establishing the electromagnetic generator to produce a set of natural frequencies which should be equal to the natural frequency level. The power is attained at a maximum of 12.5 mW from the bridge vibrations caused by the passing traffic [8]. Vibration energy harvesting is a new technique that mobilizes the unwanted energy that occurred inside the connections.

A huge quantity of vibration happens due to the movement of vehicles on the bridges and different machinery in industries and buildings [9, 10].

Data acquisition using Qualisys motion capture analysis: Human gait and complete bio-mechanical motion study is carried out by Qualisys Motion Capturing system which has precise motion capturing technology and 3D-positioning systems for engineering, bio-mechanics, virtual reality and movement sciences [11].

The system consists of a quadrant set-up of multiple high-performance motion capture cameras that can capture intricate details of motion and movement with the help of sensors attached to the subject. Along with that, a 2-tile pressure plate, which acted as the primary instrument in this research, helps to formulate the forces, moments and centre of pressure acting on a subject's body on the X-plane, Y-plane and Z-plane. All these hardware components are connected to a computer where the data are sent through and can be accessed by Qualisys's software Qualisys Track Manager (QTM) [12, 13].

In Fig. 2, the numbers 1–6 represent the main six motion capturing camera set-up, number 7 represents the 2-tile pressure plate with pressure sensors mounted underneath and number 8 represents the workstation with QTM software where the hardware is connected to. The experiments to find the forces acting, moment and centre of pressure were carried out in this exact set-up and three test subjects were used to find an average dissipated force, moment and the centre of pressure.

Figure 3 shows one of the six major motion capturing cameras at the Qualisys set-up. Special features of these cameras include: high-speed sensor mode—a subsampling mode with an optimum frame rate without interfering with the field of view. It captures data at 1740 frames per second with a resolution of 640 × 512 and 5 megapixels.

It also has lower latency settings for real-time applications, active and passive marker support, daisy-chaining and Wi-Fi connectivity. These cameras provide real-time data to the software which then can be stored for the analysis stage. All of

Fig. 2 Qualisys motion capture cameras, pressure plates and workstation

Fig. 3 Qualisys motion
capture camera

the six cameras are linked to each other by the process known as the daisy chain
method. In that manner, the cameras can visualize a three-dimensional plane and
using inputs from the sensors such as pressure sensors and the camera, the QTM
software establishes a proper rendering of the subject and the forces acting on the
body.

The slabs that are shown in Fig. 4 are the set of 2-tile pressure plate that is
interconnected with the Qualisys cameras and software system. It is activated by a
subject standing on it and moving from one tile to another in a walking or stepping
motion.

Real-time information such as the amount of forces acting in the X-plane, Y-plane
and Z-plane, the amount of moment acting at any particular stance and the centre of
pressure acting towards the Z-plane are shown and recorded in the QTM software.
The process is carried out by the numerous pressure sensing data that are underneath
the slab, as well as the interconnected cameras that help with obtaining the intricate
details of the outcome.

The pictures in Fig. 5 demonstrate how the pressure plates can be used. In these
photographs, the test subject can be seen in standing stance gait phase and in initial
swing (ISW) phase, multiple tests of 30 s each and different gait phases were done

Fig. 4 2-tile pressure plate
by Qualisys

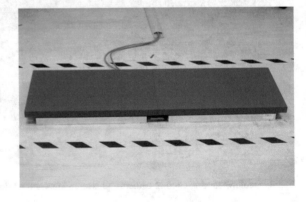

Fig. 5 Test subject using the
2-tile pressure plate

to four subjects of different gender, body mass index (BMI) and age, and the outputs
were recorded (Fig. 6) [14, 15].

Qualisys Software Interface

Testing Procedure

During this experiment, four test subjects were presented; each of the test subjects
had differing qualities than the other in terms of body mass index, weight, gender
and feet size. Multiple tests were done on each of the subjects and the results were
recorded. In this report, the results without outliers are presented to make it more
efficient and straight forward. Table 1 represents the details of the test subjects. QTM
software records the complete data till 30 seconds when the test subjects start walking
on the test tile.

Fig. 6 Graphical user
interface (GUI) of the
Qualisys Track Manager
application

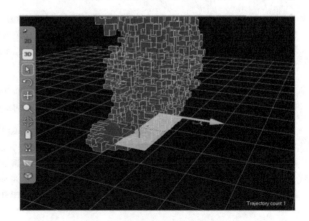

Table 1 Test Subject Information Qualisys

Test subject	Gender	Age (years)	Weight (kg)	BMI	Feet size (in.)	Number of tests done
1	Male	26	100	29.9	10.5	5
2	Female	24	96	29.5	10	5
3	Male	25	104	34.2	8	4
4	Male	25	98	30.8	9	4

Table 2 Test subject information DIERS

Test subject	Gender	Age (years)	Weight (kg)	BMI	Feet size (in.)	Number of tests done
1	Male	22	89	26.4	9	1
2	Female	20	65	21.0	7	1
3	Female	24	68	21.8	7	1

Plantar Pressure distribution data acquisition using DIERS insole

In this investigation, the EEG data set is reached from the resting state along with the eye closed (EC) position. Participants relax themselves with the EC condition while the other participants were employed in a multimedia learning task. This learning process was replicated three times to evaluate the cognitive capacity of each participant. These three states were implied as L1, L2 and L3.

Testing Procedure on DIERS

Table 2 contains a list of information related to the test subjects.

It is to be noted that the design process of the insole vastly relied on the outcomes of this particular set of experiments, as the pressure points were required to design the holes in the insoles.

Prototype Formation and Hardware Selection

As this research is based on the plantar region of humans, it leaves very less room for the flow of ideas to be implemented regarding making a device suitable to harness energy from. After thorough research, only two feasible ways were observed, in which some energy can be extracted from this region:

(i)   A shoe with a dynamo mechanism to manually harvest energy.
(ii) An insole with electrical connections to assist in energy harvesting.

To build an electricity-based generator, studies were done to find suitable hardware, in the end, piezoelectric transducers were selected as the main generator of energy. The effects of piezoelectric materials were first observed in the 1880s [12]. Piezoelectric materials bear something known as Weiss domain, and these are local regions that are polarized with steady moments that are magnetic (Fig. 7).

Fig. 7 Piezoelectric
transducers that are used in
the project

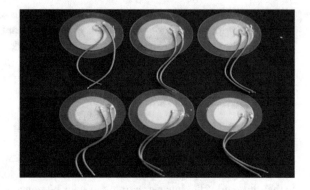

The general idea is to incorporate the piezoelectric tiles onto a shoe insole and demonstrate the current produced by stepping on them each time the wearer walks. However, setting it up is not as straightforward as it looks. The short bursts of current piezoelectric transducers produce are in an alternate form (AC current). To harvest the current and use it, it needs to be in direct form (DC current). And to do so, a certain filtration method needs to be used. After going through thorough research on which filtering system is the best for this project, it was decided that passing the AC through bridge rectifiers is the best solution to get a filtered DC as output that can be used to store in a battery and/or operate a string of LED to demonstrate harvest of energy. To make a bridge rectifier, 1N4007 diodes are used. Four diodes are connected to form one bridge rectifier. Figure 8 shows the diodes that were used in the project. For the piezoelectric transducers to work, they need to bend when mechanical stress is applied to them.

In general, polylactic acid material (PLA) is used for cheap and conventional 3D printing [13]. But in this case, the material did not deem suitable as PLA is very stiff and brittle, making it unsafe for its use as an insole. Chlorinated polyethylene (CPA) is another comparatively cheaper option for 3D printing; however, despite being less stiffer and brittle than PLA, it was not used since it is malleable, making it unsuitable for the piezoelectric discs to settle in. After more test and trials, it was found that the material acrylonitrile butadiene styrene (ABS) is the perfect choice for developing the insole as it is durable, water-resistant, heat resistant and has a very good stiffness

Fig. 8 Diodes that are used
in the project

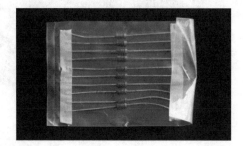

which is perfectly suitable for the piezoelectric discs to set on. Therefore, ABS was used as the printing element while printing the insole out.

What follows by is designing the insole that needs to be 3D printed, for that, specific software for 3D modelling has been referred to and design has been done (Fig. 9).

Insole Designing and Printing

The insole that is to be used in this project has been designed by using SolidWorks CAD drawing software. An initial sketch of the insole had been done on hand and then transferred on to the SolidWorks software and drawn there.

Advantage of using computer-aided design (CAD) software such as SolidWorks is that the work 2D planar drawing can later be converted into a 3D model, which this project needs. Figure 10 shows the CAD model of the sole.

Fig. 9 Bridge rectifier using four diodes

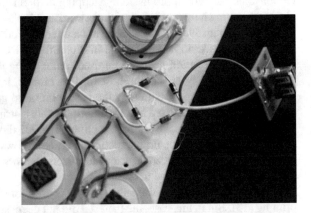

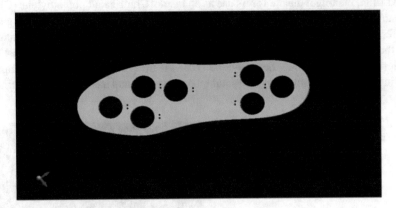

Fig. 10 CAD model of the insole (top view)

Assembling and connecting the prototype

Before connecting the piezoelectric transducers and the bridge rectifier, it was necessary to perform a set of tests to obtain the most efficient way of connecting them. From the plantar pressure distribution results obtained, the location where the piezoelectric sensors shall be installed was pinpointed and it was observed that a total of seven piezoelectric transducers can be installed in the selected regions.

Therefore, seven piezoelectric transducers were taken and varying circuit formations were formed and tested to find the most efficient set of connections. The insole was placed on a flat surface and piezoelectric transducers were glued according to the holes. The wiring was then soldered according to the circuit diagram. Figure 11 shows the finalized product. A rectifier was added to the circuit where seven piezoelectric transducers are connected in parallel to each other. A USB connector (Fig. 12) was then added to the rectifier to help obtain the output from the rectifiers. Rubber pads were cut according to a specific size and were placed on top of each piezoelectric transducer's ceramic disc. It was done to maximize the pressure point efficiency when being stepped on. Figure 13 shows the finalized product after foam pads were installed on the ceramic discs.

To visually demonstrate the output of the prototype, two mutually independent methods have been pertained to:

Fig. 11 Finalized top view of the insole

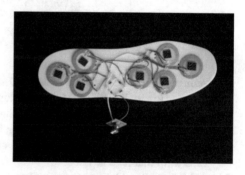

Fig. 12 USB connection

Fig. 13 Foam pads on the ceramic discs

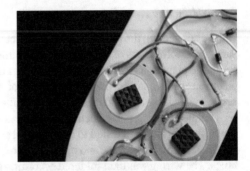

Fig. 14 LED strip connected to USB

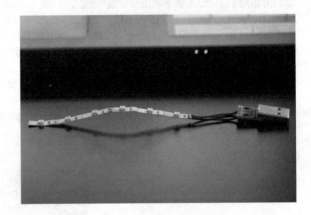

(a) A direct connection method to an LED strip contains six LED bulbs, which are attached to a male USB port to have a plug and play installation option to demonstrate instantaneous outputs whenever the insole generator is stepped upon. This method can be useful in demonstrating the energy harvesting capability of the prototype in any instance. Figure 14 shows an image of the aforementioned LED strip that is to be used.

(b) A 3.7 V, 2000mAh lithium-ion battery has been set up in a battery compartment and connected to a circuit to have two ends of a USB connecting port. The idea of this has been adapted from the design of a power bank module. Using a charging cable, this storage system can be connected to the prototype and every instance the piezoelectric set-up is to be pressed while walking. The battery will store the generated energy in voltage form.

3 Results and Analysis

Qualisys Analysis

Figures 15, 16, 17 and 18 contain force data, moment data and centre of pressure data for different individuals varying through weight. Each of the tests had been recorded

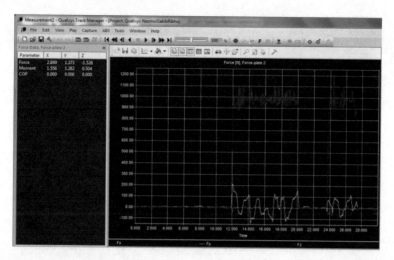

Fig. 15 Subject 1 force data

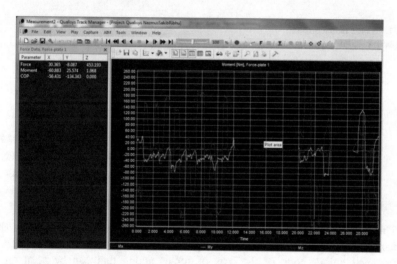

Fig. 16 Subject 1 moment data

for 30 s as the subjects walked across the two pressure plates. Four subjects took part in this experiment and the data that were obtained are discussed in this section .

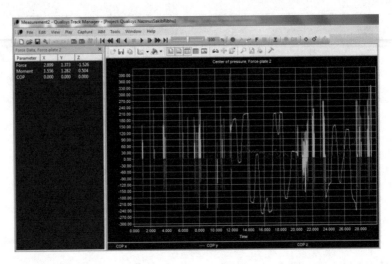

Fig. 17 Subject 1 COP data

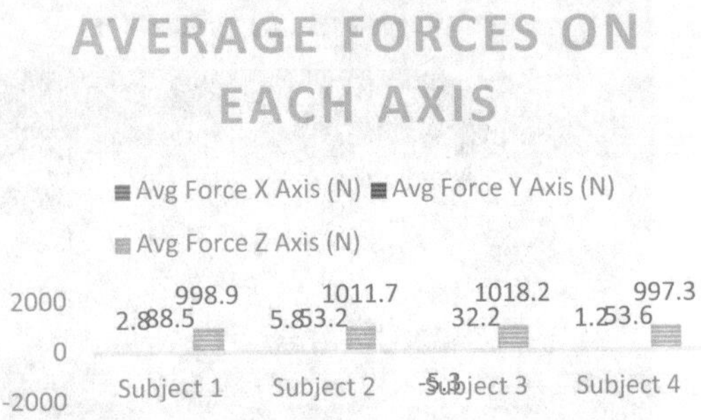

Fig. 18 Force distribution in
the plantar region

The results of the experiments are shown in the following images, and a short explanation is given for some of the images. Table 3 shows the individual denomination and colour code of the axis of forces acting that can be seen on the images.

SUBJECT 1

Subject 1's force data in Fig. 15 shows the force distribution in Newton's (N) for 30 s while the subject was moving from one tile to another. Data are absent for the time period 20.15–23.50 s and again on 28.00 s onward, it happened because the test subject was not on the pressure plate during that time period.

Figure 16 shows the moment distribution of the test subject while walking on the pressure plates. It shows the subject to have a varying amount of moment acting on each step, where each step is generating a considerable amount of moment on the x-axis and is subdued once the feet are moved across.

Table 3 Denomination and colour code of Qualisys output graphs

Parameter	Force (N) Denoted by	Moment (Nm) Denoted by	Centre of Pressure Denoted by	Colour of graph
X- Axis	Fx	Mx	COPx	Red
Y - Axis	Fy	My	COPy	Green
Z - Axis	Fz	Mz	COPz	Blue

Figure 17 shows the subject's centre of pressure data while performing the motion. It can be seen that with each step the COP data peaks at x-axis and y-axis and readily disintegrates.

Calculations

Power Analysis of the Prototype:

Each human being has varying types of force that they exert while walking, as it was seen from the Qualisys gait analysis part. Hence, rather than calculating the exact amount of power to be generated from a specific human being, the total force exerted by the four test subjects is taken and then divided by four to get an average estimation of force acting on the z-axis while an average human being is walking, which provides:

$$F = 1006.5 \text{ N}$$

Before setting up the circuit, for pin calculation, an average force of 1006.5 N produced a voltage of 12.5 V and a current of 18 μA This gives the circuit a power input of

$$
\begin{aligned}
P_{\text{in}} &= VI \\
&= 12.5 \text{ V} \times 18 \text{ μA} \\
&= 0.225 \text{ mW}
\end{aligned}
\tag{1}
$$

Therefore, 0.225 mW or mJ/s of power is given to the circuit per step.

The voltage and current output measured across the load (after constructing the circuit) were found to be 9.85 V and 15.24 μA, respectively. This gives a power output (per second) of

$$P_{out} = VI$$
$$= 9.85 \text{ V} \times 15.24 \text{ } \mu A$$
$$= 0.15 \text{ mW} = 0.15 \text{ mJ/s} \tag{2}$$

generated per step.

Therefore, the efficiency of the constructed circuit is

$$\eta = \left(\frac{P_{out}}{P_{in}} \right) \times 100$$
$$= \frac{0.15}{0.225} \times 100$$
$$= 66.7\% \text{ efficient} \tag{3}$$

Considering the output of the circuit, and also taking consideration of the fact, an average person walks 8000 steps a day [1], the average amount of energy that this prototype (worn in two shoes) can generate is:

$$0.15 \text{ mW} \times 8000 = 1.2 \text{ J}$$

4   Conclusion

This experiment was carried out with a specific goal in mind, to implement the results and see if it is fit for regular practice. Many reports and findings have been published where it states that it can be possible to extract energy from human biomechanical motion, and there have been other theories as well that tried to study the approach in harvesting energy from human motion; however, in very few cases, it was observed to see through the end of it and build a prototype suggesting the claims. That is exactly what has been done in this research, a prototype was created and it was tested to find out how much it supports the theory.

As the practical experiments, calculations, and tabulations have demonstrated, the amount of energy extracted by the insole may not be very suitable for practical use at the very stage. However, many factors can be related to this situation to help in perceiving the underwhelming achievement, as well as open up new ways to improve the outcome. Mainly, the piezoelectric transducers used in this project are not meant for robust use or applications such as using in a proper power generating circuit that can harvest enough energy to run a modern device. The piezoelectric transducers that are locally available are not viable for mass energy production. Hence, one way to maximize efficiency and generate a considerable amount of power is to use high-rated piezoelectric transducers. The overall cost of the devices must be analysed for better usage. However, without the core energy harvesting unit such as the piezoelectric transducer, it is improbable to generate positive results. So it is highly recommended to change the type of piezoelectric transducers to gain a substantial amount of energy. One other factor pertaining to the efficiency of the circuit is the wiring. Wiring must be done by good quality wires which have a very good conducting capability as well as less electrical wastage. Besides that, the connections have to be properly soldered and embedded to help lessen inefficiency.

References

1. Mittal, R.: Impact of population explosion on environment. 1(1). ISBN 978-1-62840-737-2 (2013)
2. Caineng, Z., Zhao, Q., Zhang, G., Xiong, B.: Energy revolution: from a fossil energy era to a new energy era. Nat. Gas Ind. B 3, 1–11 (2016)
3. Riemer, R., Shapino, A.: Biomechanical energy harvesting from human locomotion: theory, state of the art, design guidelines and future directions.J. Neuro Eng. Rehabil. 27 (2011)
4. Edwards, J., Davis, J.: Anatomy and biomechanics of the foot and ankle (2011)
5. Tudor-Locke, C, Craig, C.L., Aoyagi, Y., et al.: How many steps/day are enough? For older adults and special populations. Int. J. Behav. Nutr. Phys. Act. 8, 80 (2011). https://doi.org/10.1186/1479-5868-8-80
6. Gurusamy, N., Elamvazuthi, I., Yahya, N., Parasuraman, S., Ahamed Khan, M.K.A.: Biomechanical energy harvesting from human lower extremity gait: a comparative analysis. In: IEEE ROMA2017, 19–21 Sept 2017.IEEE Explorer, Scopus indexed.https://doi.org/10.1109/ROMA.2017.8231741
7. Houng, H.O.C., Sarah, S., Parasuraman, S., Ahamed Khan, M.K.A., Elamvazhuthi, I.: Energy harvesting from human locomotion: gait analysis, design and state of art. Procedia Comput. Sci. 42, 327–335 (2014)
8. Houng, H.O.C., Parasuraman, S., Ahamed khan,M.K.A.: Energy Harvesting from Human Locomotion, pp. 1–6 (2013). https://doi.org/10.1109/INDCON.2013.6726020
9. Ang, C.K., Al-Talib, A.A., Tai, S.M., Lim, W.H.: Development of a footstep power generator in converting kinetic energy to electricity. In: E3S Web of Conferences (2019)
10. Davidson, J., Mo, C.: Recent advances in energy harvesting technologies for structural health monitoring applications.2014, ID 410316. https://doi.org/10.1155/2014/410316
11. Bajrang, C., Suganthi, G.V., Tamilselvi, R., Parisabeham, M., Nagaraj, A.A.: Systematic review of energy harvesting from biomechanical factors. Biomed. Pharmacol. J. 12(4) (2019)
12. Loreto Mateu Sáez, M., Echeto, F.M.: Energy harvesting from passive human power. Ph.D. thesis (2004)
13. Mason, W.P.: Piezoelectricity, its history and applications. J. Acoust. Soc. Am. (1981)
14. Proto, A., Penhaker, M. et al.: Measurements of generated energy/electrical quantities from locomotion activities using piezoelectric wearable sensors for body motion energy harvesting. Sensors (2016)
15. Ryokai, K., Su, P., et al.: EnergyBugs: energy harvesting wearables for children. CHI (2014)

Hybrid Genetic Algorithm and Machine Learning Method for COVID-19 Cases Prediction

Miodrag Zivkovic⑩, Venkatachalam K⑩, Nebojsa Bacanin⑩, Aleksandar Djordjevic, Milos Antonijevic⑩, Ivana Strumberger⑩, and Tarik A. Rashid⑩

Abstract A novel type of coronavirus, now known under the acronym COVID-19, was initially discovered in the city of Wuhan, China. Since then, it has spread across the globe and now it is affecting over 210 countries worldwide. The number of confirmed cases is rapidly increasing and has recently reached over 14 million on July 18, 2020, with over 600,000 confirmed deaths. In the research presented within this paper, a new forecasting model to predict the number of confirmed cases of COVID-19 disease is proposed. The model proposed in this paper is a hybrid between machine learning adaptive neuro-fuzzy inference system and enhanced genetic algorithm metaheuristics. The enhanced genetic algorithm is applied to determine the parameters of the adaptive neuro-fuzzy inference system and to enhance the overall quality and performances of the prediction model. Proposed hybrid method was tested by using realistic official dataset on the COVID-19 outbreak in the state of China. In this paper, proposed approach was compared against multiple existing state-of-the-art techniques that were tested in the same environment, on the same

M. Zivkovic (✉) · N. Bacanin · A. Djordjevic · M. Antonijevic · I. Strumberger
Singidunum University, Danijelova 32, 11000 Belgrade, Serbia
e-mail: mzivkovic@singidunum.ac.rs

N. Bacanin
e-mail: nbacanin@singidunum.ac.rs

A. Djordjevic
e-mail: adjordjevic@singidunum.ac.rs

M. Antonijevic
e-mail: mantonijevic@singidunum.ac.rs

I. Strumberger
e-mail: istrumberger@singidunum.ac.rs

V. K
School of computer science and Engineering, VIT Bhopal University, Bhopal, India
e-mail: venkatachalam.k@vitbhopal.ac.in

T. A. Rashid
Computer Science and Engineering Department, University of Kurdistan Hewler, Erbil, KRG, Iraq
e-mail: tarik.ahmed@ukh.edu.krd

169

S. Shakya et al. (eds.), *Proceedings of International Conference on Sustainable Expert Systems*, Lecture Notes in Networks and Systems 176,
https://doi.org/10.1007/978-981-33-4355-9_14

datasets. Based on the simulation results and conducted comparative analysis, it is observed that the proposed hybrid approach has outperformed other sophisticated approaches and that it can be used as a tool for other time-series prediction.

Keywords COVID-19 · Genetic algorithm · Adaptive neuro-fuzzy inference system · Machine learning · Optimization · Prediction

1 Introduction

COVID-19 (now officially known as SARS-CoV-2) [1] is a novel respiratory virus, which was recently discovered in Wuhan, China [2, 3]. Since its discovery in late December 2019, the COVID-19 has spread worldwide, on all continents. It is now affecting more than 210 countries, with the number of confirmed cases of infection rising to over 14 million, and the number of deaths rising to 600,000, as of July 18, 2020. Since it is a newly discovered type of virus, scientists, virologists, and epidemiologists still have a lot to learn about its characteristics.

According to the initial estimations from the World Health Organization (WHO), the COVID-19 is extremely contagious and dangerous [4]. In just three months from the start of pandemic, the virus had reached all continents and practically every country worldwide. Many countries were forced to declare the state of emergency and enforce serious measures such as curfew, gathering restrictions and social distancing, in order to try to slow down the virus spread and minimize the number of deaths. In order to decide which measures to apply, the authorities had utilized numerous epidemiological models to try to predict the virus spread, to estimate the peak of the epidemic, and to predict the number of deaths.

The recent literature overview that deals with the prediction of the virus outbreak shows a great interest of the research community about this hot topic. The majority of the recent papers focus on the prediction of the number of infections, serious cases, and deaths. The machine learning approach in outbreak prediction was discussed in [5]. It surveys several machine learning models and suggests that two models that have shown very promising results (MLP or multi-layered perceptron, and ANFIS—adaptive network-based fuzzy inference system). The same approach was conducted on the case of outbreak in Hungary [6], with a goal to show the potential of utilizing the machine learning technique in this domain. Machine learning was also used in [7] in order to estimate the number of reported cases in individual provinces of South Korea, by utilizing a combination of XGBoost and MultiOutputRegressor as a machine learning model.

The most important goal of the research proposed within this paper was set toward enhancements in time-series predictions by using hybrid approaches between machine learning and nature-inspired algorithms. To complete this goal, an improved version of the genetic algorithm was implemented, which is then utilized to determine the parameters of the adaptive neuro-fuzzy inference system (ANFIS) model. In the research presented in this paper, antecedent and conclusion ANFIS parameters were taken into consideration. However, the types of membership functions were not considered in the process of optimization.

Therefore, as a part of research conducted in this paper, first, the basic GA has been enhanced. Next, the ANFIS model trained by the improved GA has been utilized to create a hybrid prediction model for the virus outbreak estimation. The proposed approach has been tested on the COVID-19 dataset in China.

The rest of the paper is organized as follows: Sect. 2 gives an overview of the ANFIS and swarm intelligence metaheuristics application in solving various NP-hard problems, Sect. 3 provides insights of ANFIS method. In Sect. 4, details of the original and improved GA algorithm, as well as the proposed ANFIS framework implementation are deployed. Section 5 exhibits simulation results and discussion, while Sect. 6 provides a conclusion and final remarks of this research along with the future work.

2 Background and Related Work

The ANFIS is a popular and effective artificial intelligence technique, which merges two approaches together: artificial neural networks and fuzzy inference systems, respectively. Neuro-fuzzy systems have been used in the past to solve many real-world problems. ANFIS model was initially introduced by Jang in 1993 [8], and since then, it became one of the most popular neuro-fuzzy systems. ANFIS has been applied in wide range of fields, including traffic control, economic data, image processing, feature extraction, prediction, etc [9]. The ANFIS was already used for disease spread forecasting. It was applied in forecasting Measles cases in Ethiopia [10]. Other researches include Hepatitis C [11], tuberculosis [12], and finally, COVID-19 outbreak prediction [5, 6, 13].

The most important challenge with machine learning algorithms is to find the appropriate values of the parameters specific to a given problem. Finding the optimal or near-optimal (but still good enough) values of these parameters is considered to be an NP-hard problem, and it is possible to use metaheuristics approach to solve it. NP-hard problems are the group of problems which cannot be solved within the polynomial time by using traditional (deterministic) methods. Practical NP-hard problems can be found in a wide spectrum of domains, including machine learning, cloud computing, and wireless sensor networks, to name the few. To solve NP-hard problems within the reasonable amount of time, it is necessary to use the stochastic approach.

Metaheuristics are one group of algorithms that belongs to the stochastic approaches, with a goal to find an approximate solution which not necessarily optimal but good enough within the reasonable time [14–16]. The most famous family of metaheuristics is nature-inspired algorithms. Generally speaking, nature-inspired metaheuristics can be separated into two distinctive groups: evolutionary algorithms (EA) and swarm intelligence algorithms, respectively.

The EA tries to mimic the process of natural selection, which is defined as the survival of the fittest individuals, which are selected for breeding and producing the offspring for the next generation. The most important example of the evolutionary algorithms is a genetic algorithm (GA) [17]. The GA was used to solve numerous NP-hard real-life problems in the past, including task scheduling and load balancing in the cloud computing [18], designing convolutional neural networks [19], feature selection for machine learning [20], etc.

The second large group of nature-inspired algorithms, swarm intelligence, was inspired by the behavior exhibited by the group of otherwise primitive individuals: ants, bees, fireflies, dragonflies, bats, elephants, wolves, fish, etc. One of the first swarm intelligence algorithms is the particle swarm optimization algorithm (PSO) [21]. Another important representative of this group of algorithms is artificial bee colony (ABC), which has been applied to various practical NP-hard problems, as stated in [22, 23]. Some representatives of other important examples of the swarm intelligence algorithms with numerous practical applications either in original or modified forms are the bat algorithm (BA) [24, 25], cuckoo search (CS) [26, 27], whale optimization algorithm (WOA) [28], firefly algorithm (FA) [14, 29], and monarch butterfly optimization (MBO) [30, 31].

Literature overview shows that, in [32], ABC algorithm was applied for the optimization process of ANFIS, with a goal to estimate the number of tourists coming to Turkey. The same authors also suggested training ANFIS model with a hybrid ABC algorithm, as stated in [33]. Another recent research paper [34] proposes ANFIS paired with PSO to estimate gas density based on the gas parameters (temperature, volume, etc.). The suggested ANFIS-PSO model was more precise when compared to other gas prediction models.

3 Overview of Adaptive Neuro-Fuzzy Inference System

Neuro-fuzzy systems are nowadays utilized to model a wide spectrum of real-life problems. The fuzzy logic part is responsible of the learning abilities, while the artificial neural network is responsible for the feature interpretation which comes from the fuzzy logic. Combining two approaches allows elimination of the drawbacks of individual components, and resulting neuro-fuzzy systems have superior features and performances.

The error of the model is defined as the difference between the output of the system during the training process with the actual output of the observed system. According to the error status, the ANFIS parameters are repeatedly updated in order to achieve the optimum structure. The neural network architecture in ANFIS consists of five fixed layers: fuzzification (layer one), fuzzy inference system (layer two and layer three), defuzzification (layer four), and aggregation (layer five).

The first layer contains adaptive nodes with one parametric activation function. The calculated membership values fall inside the [0, 1] range. Parameters a_i, b_i, c_i are used in ANFIS training, with a role to set the shape of the applied membership function. These parameters are also known under the name antecedent parameters. The output is the membership degree of inputs which satisfy the membership functions. As an example, the generalized bell membership function is given with Eqs. (1) and (2).

$$\mu_{A_i} = gbellmf(x; a, b, c) = \frac{1}{1 + \left| \frac{x-c}{a} \right|^{2b}} \tag{1}$$

$$O_i^1 = \mu_{A_i}(x) \tag{2}$$

On the second layer, every node is fixed and calculates the product of the inputs. In most cases, it applies the fuzzy operation AND. Firing strengths w_i are obtained by using the membership values which are the output of the first layer. Values w_i are calculated as a product of the membership values, as given in Eq. 3:

$$O_i^2 = w_i = \mu_{A_i}(x) \times \mu_{B_i}(y), \quad i = 1, 2 \tag{3}$$

Third layer consists of fixed nodes. Every node calculates the normalized firing strengths for each rule by taking into the account the firing strengths that are the output of the second level. Normalized firing strength for the rule i is calculated as given in Eq. 4:

$$O_i^3 = \overline{w_i} = \frac{w_i}{w_1 + w_2}, \quad i = 1, 2 \tag{4}$$

Fourth layer is a defuzzification layer, where every node is adaptive. Output for each rule is computed by multiplying the normalized firing strength from the third layer and a first-order polynomial. The set of polynomial's parameters $\{p_i, q_i, r_i\}$ is called the conclusion parameters, and these parameters are used in the training of the ANFIS model. The output of each rule is calculated by using the Eq. 5:

$$O_i^4 = \overline{w_i} f_i = \overline{w_i}(p_i x + q_i y + r_i) \tag{5}$$

Fifth layer consists of fixed nodes. Every node adds all input values. As the result, the final output of ANFIS is calculated as a sum of outputs of every rule from the fourth layer, as given by the Eq. 6:

$$O_i^5 = \sum_i \overline{w_i} f_i = \frac{\sum_i w_i f_i}{\sum_i w_i} \tag{6}$$

The training process of ANFIS represents the process of the optimization of the ANFIS parameters, including the number of inputs, the number of the membership functions that are used within the model and their types, and a total number of rules required by the model. The total set of parameters for optimization also includes the antecedent and conclusion parameters.

In the research presented in this paper, an enhanced GA was used to perform the optimization process. Only optimization of the antecedent and conclusion parameters was considered, while for membership function, the generalized bell membership function was selected and utilized.

4 Proposed Hybrid Machine Learning Method

GA is the most famous representative of the evolutionary algorithms family, and it was proposed by John Holland in 1992 [35]. The main characteristics of GA are that the potential solution to the given problem is encoded by binary strings of fixed length, cross-operators are executed over the pair of individuals, and the mutation operator is used to randomly modify the solutions. Potential solutions within the GA context are called chromosomes, and their parts are called genes.

GA belongs to the group of stochastic methods of global search which simulate the process of biological evolution. The initial population is typically consisting of 10 to 200 individuals. To solve a particular problem with GA, it is required to encode the potential solutions (individuals). The greatest challenge of the GA is that there is no universal coding scheme, but it rather depends on the particular problem. Each individual is given a certain quality, calculated by the fitness function. During the search phase, GA repeatedly enhances both the absolute fitness of every individual in the population and the average fitness of the complete population, therefore converging to the optimum.

Convergence is achieved by repeatedly applying the genetic operators: crossover, mutation, and selection. The selection favors the individuals with above-average fitness, and their parts (genes) have greater chance to survive and take role as a parent in the forming of the next generation. Individuals which are less adapted have smaller chance for reproduction and will eventually die out. The basic implication of this process is that the new individuals will be more adapted to the environment than their parents, as it is the case in the biological systems as well.

The simple GA is characterized with the crossover with the single breaking point, simple mutation, and proportional, roulette selection.

4.1 Improved GA

It is known that the GA also exhibits some disadvantages. The search process of the GA is very susceptible to modifications of the control parameters, such as the population size, employed selection criteria, as well as crossover and mutation probabilities [36]. If the combination of parameters is not chosen carefully, the search process may easily converge to some of suboptimal domain of the search space.

By conducting simulations with the traditional GA on standard unconstrained benchmark functions, it frequently happens that some solutions do not improve in consecutive iterations and this is good indicator that the search process has trapped in local optimum. To overcome this, exploration process from the ABC metaheuristics was adopted [37]. Each solution (chromosome) is encoded with the additional attribute *counter*, and each time when a solution is not improved and is being propagated into next generation, the *counter* is incremented. When a *counter* for a particular solution reaches a predetermined threshold value (*tvalue*), this solution is replaced with the random solution and the *counter* of newly created individual is initialized to 0. In approach proposed in this paper the same expression for generating random solution was used, as it is used in the initialization phase:

$$c_{i,j} = lb_j + \text{gauss} * (u_j - l_j), \tag{7}$$

where $c_{i,j}$ represents j-th component of the i-th chromosome, l_j and u_j are lower and upper bounds of j-th component, respectively and *gauss* denotes pseudo-random number drawn from Gaussian distribution with mean 0 ($\mu = 0$) and standard deviation of 1 ($\sigma^2 = 1$).

Moreover, proposed approach utilizes stochastic universal sampling (SUS) selection mechanism and uniform crossover operator. According to performed empirical simulations, it was concluded that combination of SUS selection and uniform crossover operators yields best results. More about GA recombination and selection methods can be read in [38].

Since ABC's exploration was incorporated in the GA method, the proposed approach was named GA ABC exploration (GAAE). The GAAE pseudo-code is presented in Algorithm 1.

Algorithm 1 Pseudo-code of the GAAE

Initialization: generate initial population of n chromosomes and adjust control parameters (p_c, p_m and $tvalue$)
for $gen = 1$ to $maxGen$ **do**
   Replace all solutions for which the condition $counter == tvalue$ is satisfied with the random solution by using Eq. (7)
   Calculate fitness $f(x)$ for each chromosome in the population
   while ! n offspring created **do**
      Select the pair of parent chromosomes from the current population by using SUS method
      Apply the uniform crossover operator to the selected parents with the probability p_c and create two offspring
      Apply mutation operator to the generated offspring with probability p_m
   end while
   Rank all solutions according to fitness and choose n best chromosomes for next generation
end for

4.2 Hybrid GAAE-ANFIS Method

Devised GAAE metaheuristics was incorporated into the ANFIS training process. Generally speaking, the training process of ANFIS is optimization of its structure and parameters for a specific problem and represents one of the greatest issues in this domain. In optimization process, antecedent and conclusion ANFIS parameters were taken into the account.

Each GAAE individual represents potential ANFIS structure with the length that is equal of the sum of ANFIS's antecedent and conclusion parameters. Membership function was not taken in optimization process, and the generalized bell membership function was used. By incorporating GAAE into the ANFIS, we devised GAAE-ANFIS method for forecasting time-series. Employed solutions encoding scheme is shown in Fig. 1.

Proposed method has similar structure as approach presented in [13]. It employs standard ANFIS model with five layers, where inputs are given in the Layer 1, while the Layer 5 generates forecasted outputs. The weights between Layer 4 and 5 are obtained by the GAAE metaheuristics.

Inputs to GAAE-ANFIS are first formatted in a form of time-series by employing autocorrelation function (ACF and variables with the ACF value grater than 0.2 were considered. The fuzzy c-mean (FCM) method was used for constructing ANFIS model. The 25% of dataset was utilized for testing and remaining 75% for training. The best generated solution (ANFIS structure) by the GAAE is returned, and this solution was used in the testing phase.

Metric mean square error (MSE) was taken as objective function:

$$\text{MSE}_x = \frac{1}{N} \sum_{i=1}^{N} (\hat{y}_i - y_i)^2 \tag{8}$$

where $\hat{y}_i$ and y_i represent predicted and real data for each observation, respectively, and the total number of observations is denoted as N.

Flowchart diagram of proposed GAAE-ANFIS is given in Fig. 2.

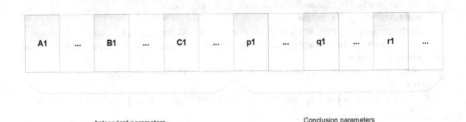

Antecedent parameters   Conclusion parameters

Fig. 1 GAAE-ANFIS solutions encoding

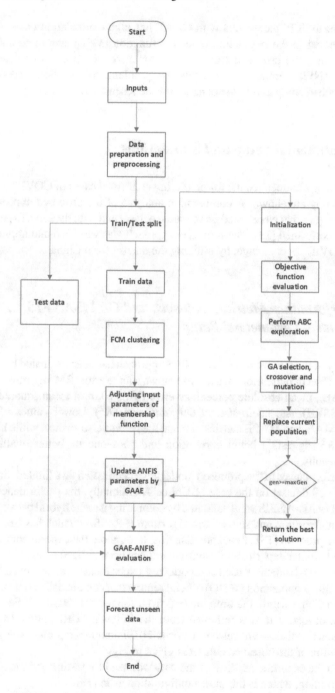

Fig. 2 GAAE-ANFIS flowchart diagram

Updating ANFIS parameters with GAAE has high computation costs, since for each individual in the population objective function (MSE) should be calculated. However, since the proposed framework automatizes the process of determining satisfying ANFIS parameters for a concrete problem, computational costs can be neglected when compared to the benefits for the researcher.

5   Experimental Setup and Simulations

In this section, obtained results for predicting confirmed cases of COVID-19 on one practical study are shown. A comparative analysis of the proposed approach with other techniques that were tested on the same dataset and with the same experimental setup was performed [13]. The proposed approach was validated and applied to the current COVID-19 challenge, by utilizing the dataset from China.

5.1   Performance Metrics, Datasets, and GAAE-ANFIS Control Parameters' Setup

The quality of the proposed GAAE-ANFIS approach has been evaluated by applying the standard set of regression metrics: root mean square error (RMSE), mean absolute error (MAE), mean absolute percentage error (MAPE), root mean squared relative error (RMSRE), and coefficient of determination (R^2). Lower values of RMSE, RMSRE, MAE, and MAPE metrics indicate the better performance, while the higher value of R^2 suggests a better correlation and consequently better quality of the obtained results.

The performances of the proposed model have been tested on a limited time period of COVID-19 dataset on the case of China. Additionally, the performances of the proposed GAAE-ANFIS were validated by comparing it to the hybrid between flower pollination algorithm and salp swarm algorithm (FPASSA) which has been applied to the same problem [13]. To accomplish this, throughout the experiments, the same dataset and similar experimental environment has been utilized as in [13].

The COVID-19 dataset used in conducted experiments was retrieved from the World Health Organization (WHO) official reports that contain daily confirmed cases reported in China within the time interval from January 21, 2020, till February 18, 2020. The official data was retrieved from the following URL: https://www.who. int/emergencies/diseases/novel-coronavirus-2019/situation-reports/. The graphical representation of the observed dataset is given in Fig. 3.

Also, in the experiments, 25% of the data was used for testing and the remaining 75% for training, which is the same configuration as in [13].

Additionally, as we wanted to provide a better analysis of GA-ANFIS model performances, we have performed additional simulations by utilizing one more dataset

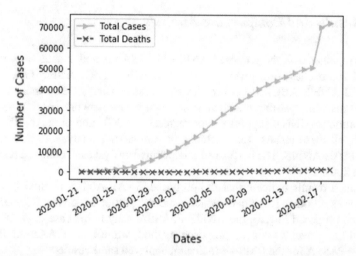

Fig. 3 Graphical representation of the number of reported cases of COVID-19 in China in the interval January 21, 2020 to February 18, 2020, obtained from WHO reports

containing confirmed influenza cases, as in [13]. The dataset (influenza dataset—IDS) was obtained from the Center for Disease Control and Prevention (CDS). The IDS contains weekly reported cases for time period between fourteenth week in 2015 and sixth week in 2020 (URL: https://www.cdc.gov/flu/weekly/). Global GAAE-ANFIS control parameters were adjusted in a similar fashion as in [13]: The size of population was fixed to 25, while the number of generations was fixed to 100. Algorithms that were proposed in [13] were tested by using 25 solutions in the population. Specific GAAE-ANFIS parameters were adjusted as: $tvalue$ was set to 4, while p_c and p_m were established to 0.1 and 0.005, respectively. All implemented algorithms were executed in 30 independent runs, and we have taken the best values and noted them for the comparative analysis.

In both set of experiments (COVID-19 and influenza cases), the past time-series were considered as independent variables, while the prediction of new cases of COVID-19 and influenza was observed as dependent variables. The dataset with time-series was prepared as follows: Time-series data were categorized into four inputs for the last four consequently odd days of confirmed cases which were then used for predicting x_t, as the number of confirmed cases for the following day.

The GAAE-ANFIS framework was implemented in Python programming language. The ANFIS 0.3.1 module for Python was utilized along with the following Python libraries for data preprocessing and visualization: Scipy, Pandas, Pyplot, and Seaborn. The platform that was used for testing the algorithms was Intel i7 CPU, with 32GB of RAM and Windows 10 × 64 operating system. Additionally, CUDA with one GPU NVIDIA 1080 with 8GB RAM was utilized throughout the experiments.

5.2 Results and Comparative Analysis

In [13], together with the proposed ANFIS-FPASSA method, results for other bio-inspired approaches were given: ANFIS-GA, ANFIS-PSO, ANFIS-GA, ANFIS-FPA, and ANFIS-ABC. Moreover, the comparative analysis also included other standard machine learning techniques: artificial neural network (ANN), K-nearest neighborhoods (KNN), support vector regression (SVR) and pure ANFIS. In this research, all these results were included in comparative analysis to evaluate proposed GAAE-ANFIS. The performed comparative analysis and obtained results are given in Table 1.

Obtained results indicate that the proposed GAAE-ANFIS method in average outperforms all other techniques which were the part of the comparative analysis. Approach proposed in [13], the ANFIS-FPASSA, only in the case of MAE metric obtained better result than the proposed method, while both, GAAE-ANFIS and ANFIS-FPASSA for the RMSRE indicator, achieved same results.

It is important to compare results of proposed hybrid GAAE with the basic GA and ABC metaheuristics. From the Table 1, it can be plainly concluded that the GAAE-ANFIS obtained much better results than the ANFIS-GA and ANFIS-ABC and that the proposed approach efficiently combines advantages of basic GA and ABC. In detailed analysis of each run, it can be observed that the GAAE overcomes premature convergence of the original GA, while it also performs more efficient exploitation than the basic ABC metaheuristics.

Table 1 Comparative analysis between GAAE-ANFIS and other techniques for COVID-19 outbreak prediction results, based on the observed China dataset obtained from the official WHO reports—results for methods included in comparative analysis were retrieved from [13]

Method	RMSE	MAE	MAPE	RMSRE	R^2	Time
ANN	8750	5413	13.09	0.204	0.8991	–
KNN	12100	7671	8.32	0.130	0.7710	–
SVR	7822	5354	8.40	0.080	0.8910	–
ANFIS	7375	5523	5.32	0.09	0.9032	–
ANFIS-PSO	6842	4559	5.12	0.08	0.9492	24.18
ANFIS-GA	7194	4963	5.26	0.08	0.9575	27.02
ANFIS-ABC	8327	6066	6.86	0.10	0.7906	46.80
ANFIS-FPA	6059	4379	5.04	0.07	0.9439	23.41
ANFIS-FPASSA	5779	**4271**	4.79	0.07	0.9645	23.30
GAAE-ANFIS	**5293**	4305	**4.62**	0.07	**0.9721**	16.34

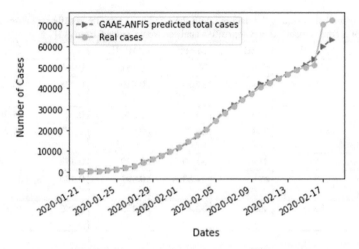

Fig. 4 Predicted cases of COVID-19 in China by GAAE-ANFIS

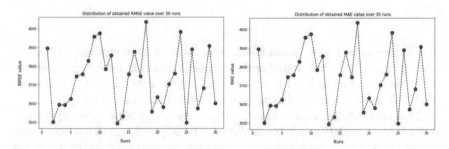

Fig. 5 Distribution of RMSE (left) and MAE (right) indicators over 30 runs

Visual representation of predicted COVID-19 for proposed GAAE-ANFIS is shown in Fig. 4, while the distribution of RMSE and MAE values obtained by the GAAE-ANFIS over 30 runs is depicted in Fig. 5.

Comparative analysis with influenza dataset (IDS) is presented in Table 2.

It can be seen that the results presented in Table 2 are very similar to those shown in Table 1 (COVID-19 dataset vs. influenza dataset). The second best approach, ANFIS-FPASSA, for MAE performance metrics only obtained better result than the proposed GAAE-ANFIS, while for the R^2 indicator both methods showed the same performance. Moreover, in simulations with influenza dataset, GAAE-ANFIS significantly outscored original ABC (ANFIS-ABC) and GA (ANFIS-GA) adaptations.

Table 2 Simulation results for datasets with the number of confirmed influenza cases (IDS)—results for methods included in comparative analysis were retrieved from [13]

Method	RMSE	MAE	MAPE	RMSRE	R^2	Time
ANFIS	952	570	37.61	0.551	0.969	–
ANFIS-PSO	798	494	34.13	0.510	0.978	25.43
ANFIS-GA	766	480	35.44	0.530	0.98	28.70
ANFIS-ABC	878	564	39.79	0.593	0.972	49.27
ANFIS-FPA	618	411	37.69	0.570	0.979	24.58
ANFIS-FPASSA	609	**391**	32.58	0.497	0.986	24.55
GAAE-ANFIS	**605**	403	**32.49**	**0.495**	0.986	19.64

6 Conclusion

In this manuscript, a new approach has been proposed to predict new COVID-19 cases by employing hybridized algorithm between machine learning adaptive neuro-fuzzy inference system (ANFIS) and enhanced genetic algorithm (GA) metaheuristics which has been employed in parameters' optimization and adjustments.

GA algorithm has been incorporated for updating ANFIS parameters and tested in on practical COVID-19 new cases prediction. The proposed method has been tested on the COVID-19 case study because it is currently the most important challenge the entire humanity faces. However, the method can be generalized and applied to predict any time-series. The proposed GAAE-ANFIS was compared against several existing successful techniques that were tested in the same experimental environment, with the same datasets. It was showed that proposed methods can be successfully applied in this domain.

Our plan is to continue research from this domain and to devise other hybridized methods between GA and swarm intelligence. Moreover, we plan to adapt other swarm intelligence methods for machine learning approaches.

References

1. C. S. G. of the International et al.: The species severe acute respiratory syndrome-related coronavirus: classifying 2019-ncov and naming it sars-cov-2. Nat. Microbiol. **5**(4), 536 (2020)
2. Yadav, T., Saxena, S.K.: Transmission Cycle of SARS-CoV and SARS-CoV-2, pp. 33–42. Springer, Singapore (2020)
3. Chan, J.F.-W., Yuan, S., Kok, K.-H., To, K.K.-W., Chu, H., Yang, J., Xing, F., Liu, J., Yip, C.C.-Y., Poon, R.W.-S., et al.: A familial cluster of pneumonia associated with the 2019 novel coronavirus indicating person-to-person transmission: a study of a family cluster. Lancet **395**(10223), 514–523 (2020)

4. W. H. Organization et al.: Coronavirus disease 2019 (covid-19): situation report, 72 (2020)
5. Ardabili, S.F., Mosavi, A., Ghamisi, P., Ferdinand, F., Varkonyi-Koczy, A.R., Reuter, U., Rabczuk, T., Atkinson, P.M.: Covid-19 outbreak prediction with machine learning. Available at SSRN 3580188 (2020)
6. Pinter, G., Felde, I., Mosavi, A., Ghamisi, P., Gloaguen, R.: Covid-19 pandemic prediction for hungary; a hybrid machine learning approach. In: A Hybrid Machine Learning Approach (2020)
7. Suzuki, Y., Suzuki, A.: Machine learning model estimating number of covid-19 infection cases over coming 24 days in every province of South Korea (xgboost and multioutputregressor)
8. Jang, J.-S.: Anfis: adaptive-network-based fuzzy inference system. IEEE Trans. Syst. Man Cybern. **23**(3), 665–685 (1993)
9. Karaboga, D., Kaya, E.: Adaptive network based fuzzy inference system (anfis) training approaches: a comprehensive survey. Artif. Intell. Rev. **52**(4), 2263–2293 (2019)
10. Uyar, K., Ilhan, U., Iseri,E.I., Ilhan, A.: Forecasting measles cases in Ethiopia using neuro-fuzzy systems. In: 2019 3rd International Symposium on Multidisciplinary Studies and Innovative Technologies (ISMSIT), pp. 1–5 (2019)
11. Khodaei-mehr, J., Tangestanizadeh, S., Vatankhah, R., Sharifi, M.: Anfis-based optimal control of hepatitis c virus epidemic. IFAC-PapersOnLine **51**(15), 539–544 (2018)
12. Uçar, T., Karahoca, A., Karahoca, D.: Tuberculosis disease diagnosis by using adaptive neuro fuzzy inference system and rough sets. Neural Comput. Appl. **23**(2), 471–483 (2013)
13. Al-qaness, M.A.A., Ewees, A.A., Fan, H., Aziz, M.A.E.: Optimization method for forecasting confirmed cases of COVID-19 in china. J. Clin. Med. **9**(3), 674–674 (2020)
14. Strumberger, I., Bacanin, N., Tuba, M.: Enhanced firefly algorithm for constrained numerical optimization, ieee congress on evolutionary computation. In: Proceedings of the IEEE International Congress on Evolutionary Computation (CEC 2017), pp. 2120–2127 (2017)
15. Tuba, E., Strumberger, I., Bacanin, N., Tuba, M.: Bare bones fireworks algorithm for capacitated p-median problem. In: Tan, Y., Shi, Y., Tang, Q. (eds.) Advances in Swarm Intelligence. pp. 283–291. Springer International Publishing Berlin (2018)
16. Strumberger, I., Minovic, M., Tuba, M., Bacanin, N.: Performance of elephant herding optimization and tree growth algorithm adapted for node localization in wireless sensor networks. Sensors **19**(11), 2515 (2019)
17. Goldberg, D.E. : Genetic Algorithms in Search, Optimization and Machine Learning, 1st ed. Addison-Wesley Longman Publishing Co., Inc., Boston (1989)
18. Wang, T., Liu, Z ., Chen, Y., Xu, Y., X. Dai, "Load balancing task scheduling based on genetic algorithm in cloud computing. In: 2014 IEEE 12th International Conference on Dependable, Autonomic and Secure Computing, pp. 146–152 (2014)
19. Suganuma, M., Shirakawa, S., Nagao, T.: A genetic programming approach to designing convolutional neural network architectures. In: Proceedings of the Genetic and Evolutionary Computation Conference (GECCO '17), New York, NY, USA, pp. 497–504. ACM (2017)
20. Xue, B., Zhang, M., Browne, W.N., Yao, X.: A survey on evolutionary computation approaches to feature selection. IEEE Trans. Evolution. Comput. **20**(4), 606–626 (2015)
21. Kennedy, J., Eberhart, R.: Particle swarm optimization. In: Proceedings of the IEEE International Conference on Neural Networks (ICNN '95), vol. 4, pp. 1942–1948 (1995)
22. Tuba, M., Bacanin, N.: Artificial bee colony algorithm hybridized with firefly metaheuristic for cardinality constrained mean-variance portfolio problem. Appl. Math. Inform. Sci. **8**, 2831–2844 (2014)
23. Bacanin, N., Tuba, M., Strumberger, I.: Rfid network planning by abc algorithm hybridized with heuristic for initial number and locations of readers. In: 2015 17th UKSim-AMSS International Conference on Modelling and Simulation (UKSim), pp. 39–44. IEEE (2015)
24. Yang, X.-S. : A New Metaheuristic Bat-Inspired Algorithm, pp, 65–74. Springer, Berlin (2010)
25. Tuba, M., Bacanin, N.: Hybridized bat algorithm for multi-objective radio frequency identification (rfid) network planning. In: 2015 IEEE Congress on Evolutionary Computation (CEC), pp. 499–506 (2015)

26. Tuba, M., Alihodzic, A., Bacanin, N.: Cuckoo search and bat algorithm applied to training feed-forward neural Networks, pp. 139–162. Springer International Publishing, Cham (2015)
27. Bacanin, N.: Implementation and performance of an object-oriented software system for cuckoo search algorithm. Int. J. Math. Comput. Simulation **6**, 185–193 (2010)
28. Strumberger, I., Bacanin, N., Tuba, M., Tuba, E.: Resource scheduling in cloud computing based on a hybridized whale optimization algorithm. App. Sci. **9**(22), 4893 (2019)
29. Bacanin , N., Tuba, M.: Firefly algorithm for cardinality constrained mean-variance portfolio optimization problem with entropy diversity constraint. In: Sci. World J. **2014**, 16. Article ID 721521 (2014)(special issue Computational Intelligence and Metaheuristic Algorithms with Applications)
30. Strumberger, I., Tuba, E., Bacanin,N., Beko, M., Tuba, M.: Monarch butterfly optimization algorithm for localization in wireless sensor networks. In: 2018 28th International Conference Radioelektronika (RADIOELEKTRONIKA), pp. 1–6 (2018)
31. Bacanin, N., Bezdan, T., Tuba, E., Strumberger, I., Tuba, M.: Monarch butterfly optimization based convolutional neural network design. Mathematics **8**(6), 936 (2020)
32. Karaboga, D., Kaya, E.: Estimation of number of foreign visitors with anfis by using abc algorithm. Soft Comput. 1–13 (2019)
33. Karaboga, D., Kaya, E.: Training anfis by using an adaptive and hybrid artificial bee colony algorithm (aabc) for the identification of nonlinear static systems. Arab. J. Sci. Eng. **44**(4), 3531–3547 (2019)
34. Mir, M., Kamyab, M., Lariche, M.J., Bemani, A., Baghban, A.: Applying anfis-pso algorithm as a novel accurate approach for prediction of gas density. Petrol. Sci. Technol. **36**(12), 820–826 (2018)
35. Holland, J.H., et al.: Adaptation in Natural and Artificial Systems: An Introductory Analysis with Applications to Biology, Control, and Artificial Intelligence. MIT press, Cambridge (1992)
36. "Nature-Inspired optimization algorithms. In: Yang, X.-S., Nature-Inspired Optimization Algorithms, p. i. Elsevier, Oxford (2014)
37. Karaboga, D., Akay, B.: A modified artificial bee colony (ABC) algorithm for constrained optimization problems. Appl. Soft Comput. **11**(3), 3021–3031 (2011)
38. Mitchell, M.: An Introduction to Genetic Algorithms. MIT Press, Cambridge (1998)

Bidirectional Battery Charger for Electric Vehicle

Nooriya Shahul and Siddharth Shelly

Abstract Dynamic pricing is expected to become the main pricing scheme in case of smart grids. A comprehensive prototype of battery charger is proposed here for the electric vehicle. The proposed prototype can attain energy saving by deploying additional features in the design. Exchange of energy transfer between consumer and grid reveals the energy management system to become more convenient and safe. For the switching purpose, a PWM algorithm is used. In addition, power electronic components and DSP (TMS320f28335) processor make the system to be more efficient and faster with a frequency of 150 MHz. In this paper, a suitable bidirectional battery charger design is presented. The implemented bidirectional battery charger is established by H bridge converter (combination of AC-to-DC and DC-to-DC converter) which is having the same DC-based capacitor. The first stage (AC-to-DC conversion) persists between the power grid and the DC-based capacitor. The proposed system is constructed with a parallel combination of power converters which control the current through the grid and voltage across the DC-based capacitor. The chopper (DC-to-DC conversion) subsists between the DC-based capacitor and the battery. Control of phase amplitude and frequency is done by the processor. The proposed system enhances different applications such as charging of battery, act as inverter during load shedding period and vehicle-to-vehicle operation in an emergency situation. The experimental validation was performed on MATLAB/Simulink and the operation of grid-to-vehicle (G2V) and vehicle-to-grid (V2G) are exhibited in the simulation results.

Keywords Bidirectional charger · Grid to vehicle · Vehicle to grid · H Bridge · DSP processor

N. Shahul (✉) · S. Shelly
Department of Electronics & Communication Engineering, Mar Athanasius College
of Engineering,Kothamangalam, Kerala, India
e-mail: nooriya.shahul777@gmail.com

S. Shelly
e-mail: sidhushelly@mace.ac.in

S. Shakya et al. (eds.), *Proceedings of International Conference on Sustainable Expert
Systems*, Lecture Notes in Networks and Systems 176,
https://doi.org/10.1007/978-981-33-4355-9_15

1 Introduction

The term "Electricity" has gained a huge change in our technology, especially in case of transport sector. Moreover, the expeditious growth of industrialization and population has fallen to tremendous utilization of conventional energy sources. The availability of fossil fuels will be mitigated in the coming years. Nowadays, most of the consumers are using internal combustion engine (ICE) vehicle which is powered by fossil fuels like petrol, diesel, coal, etc., and emit large amount of greenhouse gases into the environment that are detrimental to the lives. To overcome the disadvantage of ICE vehicle, a paradigm is introduced, "Electric Vehicle" (EV) which drives with electricity. People began to switch from ICE vehicle to electric vehicle by realizing the supremacy of the new paradigm. So that this idea enhances the sustainability and efficiency of transportation side. However, charging of electric vehicle will become a burden to the current energy resources. Charging is the crucial part while considering the idea of electric vehicle. It is required to defeat such scarcity of energy by implementing a proper Energy Management System (EMS). The growing evolution of present grid into smart grid furnishes an elegant energy management system. The concept of smart grid reveals the improvement in reliability, security, and efficiency of the electric grid. Nevertheless, it correlates the smart grid and electric vehicle to build a new renewable energy source in the form of a battery charger.

There are mainly two types of battery chargers available in market, namely on-board charger and off-board charger. On-board charger is located inside the vehicle, which is charged through plug in a socket while the vehicle is at parking area. The main concerns about this charger topology are cost, minimal size, distance covered per charging, high energy density and light weight. Besides this topology widely enacted with low power converters. Off-board charger is charged by removing the battery from the vehicle and to be recharged via fast charger. They do not consider the weight and cost, but small strategy about this charger topology is to lower the price of architecture for public charging. In addition, on-board charger can be charged in everywhere rather than the off-board charger [1]. This paper presents a comprehensive prototype of battery charger for electric vehicle, which is bidirectional and controllable with the help of power converters. It operates in two modes which are grid to vehicle and vehicle to grid. First mode is widely used when the electric vehicle battery is not having required charge; second mode can be used when there is load shedding and electric scarcity. In V2G mode, considerable amount of charge is stored in the battery that can be exploited effectively thereby stabilizing the power grid. It can conserve energy through the charging and discharging of battery. The proposed system has many applications. The various functions of proposed system are shown in Fig. 1. The charging of battery is done through plug-in socket and is used to charge the battery. In grid-to-vehicle, the main application is to load the battery to drive the vehicle. However, the main utilization of our system is in vehicle-to-grid mode which can be used for many applications. In that scenario, energy from the vehicle can be utilized for any home applications even in the sudden load shedding times (like an inverter); furthermore, another application of the proposed system is to

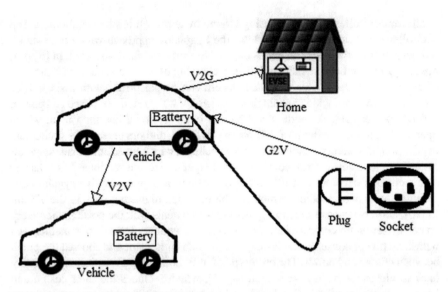

Fig. 1 Applications of the proposed system

transfer energy from one electric vehicle to another electric vehicle in an emergency situations. People are switching from conventional vehicle system to electric vehicle by realizing the advantages and possibilities of electric vehicle. In that scenario, where most of the people will be using electric vehicle, if a running vehicle is turned off suddenly due to the shortage of charge, the neighboring vehicle can be utilized to give energy from its battery, which makes the application of the proposed system more wider.

The first section describes the difficulties in the conventional system, motivation, and how to solve the bug. The second section reveals the ideas that have been done in different methods and topologies and also depicted the previously done works. Next two sections discuss the methodology, topology, operation modes, and simulation results of the proposed system. The fifth section deals the experimental setup and results. Last section gives the overall conclusion about this project.

2 Literature Review

To attenuate the obstructive impact of electric vehicle on the energy system, various researchers have suggested smart charging approaches. A lot of research works on vehicular networks, electric vehicles and hybrid vehicles are going on today [2]. A model predictive control for off-board plug-in electric vehicle (PEV) and its algorithm have been proposed in [3, 4]. It is implemented with photovoltaic integration using two-level four-leg-inverter topology. In [5], numerical simulation results of a nonlinear controller is designed using Lyapunov approach has been exhibited. The controller is nonlinear and also follows the Lyapunov approach which is based on ordinary differential equation and is originated from Russia. Sousa et al. in [6] propose a system for both traction and battery charging of electric vehicle with the help of universal interface. This proposed system can be performed with mainly three types of power grid, which are single-phase AC power grid, three-phase AC power, and DC-power grid. A single-phase battery charger model has innovated, which operates in all four quadrant of P-Q plane [7]. In [8], authors present the design and installation of a battery charger for electric vehicle, which uses Landsman converter instead of diode converter in conventional charger. A SiC bidirectional LLC charger architecture is developed in [9] and also obtained a digital adaptive synchronous rectification driving method, which can be provided high immunity to the circuit. In [10], authors propose a charging method which deals with the power supervisory characterization on constant current and constant voltage. Indeed, this method can withstand power consumption inside the available aggregator and showed the graph between charge and power. The prototype of 3kW bidirectional converter for battery bank in electric vehicle is presented in [11], which includes the filter design and impedance measurement. The distributed system of mobile ad hoc network is very difficult, and each node operates individually with their requirements. Moreover, this system operates independently and gives fluctuations in random node. Harikumar et al. in [12] propose a architecture which is used to maintain the mobility problem using the three structures which are (1) mesh tree, (2) mesh cluster and (3) mesh backbone. A modular charge equalizer circuit with bidirectional battery charger has been implemented in [13], and the circuit operates in two modes, namely grid-to-vehicle and vehicle-to-grid. In addition, the implemented system has active and reactive capabilities while operating in the respective modes. The paper [14] reveals a charging strategy to minimize charge loss in Li-ion battery, and it uses with required current status based on the changes of internal resistance of battery. A 3.6kW bidirectional charger prototype and valid experimental results have been discussed in [15] ,where five modes of operations are illustrated. The operation modes described are vehicle-to-grid (V2G), grid-to-vehicle (G2V), vehicle-to-home (V2H), home-to-vehicle (H2V), and vehicle-for-grid (V4G). The aforementioned papers exhibit the idea of a controlled electric vehicle charging contours to alleviate energy price, eliminate operative limits of power system, and reduce the system loss. It controls the phase, amplitude, and frequency of the generated output. Also, unlike previous papers, a new technology has been implemented for controlling these characteristics, that is a hardware PLL DSP controller is used for controlling the phase, amplitude, and frequency. The full control of phase, amplitude, and frequency in both vehicle-to-grid and grid-to-vehicle scenario and the control of these parameters using TMS320f28335 DSP Processor with PLL method in a hardware arrangement are the uniqueness of our project.

In this proposed context, a controllable and bidirectional battery charger is proposed to enhance the grid capacity and efficiency of charging. The proposed charger topology for EV can be integrated with smart grid. The implemented bidirectional battery charger is established by H bridge converter which is having the same DC-based capacitor. The proposed system has mainly two stages. The first stage (AC-to-DC conversion) persists between the power grid and the DC-based capacitor. The proposed system is developed with a parallel combination of power converters which control the current through the grid and voltage across the DC-based capacitor. The second stage (DC-to-DC conversion) subsists between the DC-based capacitor and the battery. It can operate in two modes which are grid-to-vehicle (G2V) and vehicle-to-grid (V2G). Charging of battery is happening in G2V mode, and discharging process is happening in V2G mode, so that modes are also called as charging mode and discharging mode, respectively. Bidirectional property is achieved through H-bridge converter to superintend the overall system by a DSP processor named TMS320f28335. Moreover, the processor controls the phase, amplitude and frequency of the signal. It is implemented with power electronics components and uses PWM algorithm for the effective switching of these components. The developed prototype can attain energy saving in addition, to be more convenient and safer to the energy management system.

3 Proposed System

3.1 Methodology

The block diagram of proposed system is shown in Fig. 2. This system consists of bidirectional converter, battery, and controller. The bidirectional converter has mainly two power stage conversions. The first stage is AC-DC converter, and second stage is DC-DC converter. It can operate in two modes: grid-to-vehicle and vehicle-to-grid. The bidirectional converter can act as both converter and inverter in G2V and V2G modes, respectively. When the charger is in charging mode, rectification is performed by AC-DC converter which is followed by DC-based capacitor. The chopper (DC-DC) operates as buck converter and the energy is stored in the battery. When the vehicle is in rest condition, the battery is integrated with grid; moreover, AC-DC converter acts as an inverter. During this time, DC-AC conversion is initiated by switching of power converters in H-bridge circuit. Entire system is under the control of digital signal processor (TMS320f28335). The detailed circuit topology and used tools are illustrated below.

3.2 Topology

The circuit topology of proposed bidirectional battery charger for electric vehicle is depicted in Fig. 3. It is based on a bidirectional converter using the same DC-based capacitor. The circuit has mainly two sections which are AC-DC converter and DC-DC converter. The power converters are implemented by insulated gate bipolar

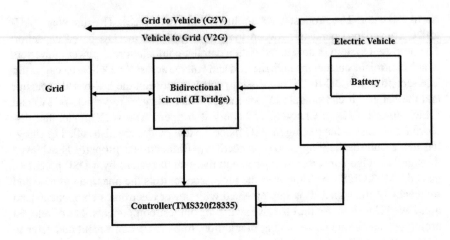

Fig. 2 Block diagram of bidirectional battery charger

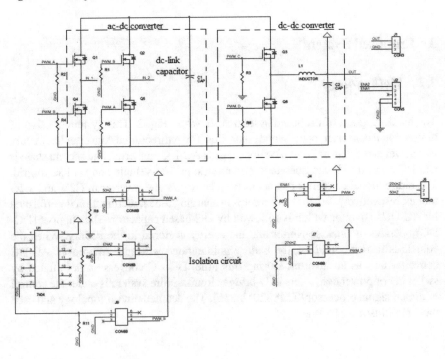

Fig. 3 Circuit diagram of bidirectional battery charger

junction transistors (IGBT). Furthermore, an isolation circuit for the protection of controller is fabricated with opto-couplers. The DC-link capacitor is used for the filtering process. This same circuit can work in two modes so that the proposed system can ensure the bidirectional property. The PWM signal which is generated from the controller that will switch the corresponding IGBTs for the respective periods. Different modes and operations of developed system are detected as follows.

3.3 Grid to Vehicle (G2V)

The electric vehicle battery is connected to grid when the battery is not having sufficient charge; thus, this mode is also called as charging mode. For this mode, 230V is taken from grid which is stepped down to required AC value; moreover, only the body diodes of AC-DC converter is turned on and performs rectification (AC-to-DC conversion). It does not require PWM signal for the switching of IGBTs during this mode. The converted DC signal may have some ripples, in order to filter such ripple it is passed through the DC-based capacitor. The voltage across the capacitor will be high, DC-DC converter will act as buck converter and steps down to the needed voltage. Finally, the voltage will be stored in battery. If the voltage that appears across the capacitor is of a low value, the DC-DC converter will act as boost converter at that time. So it will operate according to the voltage across the DC-based capacitor. Moreover, the DC-DC converter needs PWM signal for the proper working buck/boost action. The controller will generate the PWM signal with required duty cycle and is fed to the gate terminal of IGBT through an opto coupler.

3.4 Vehicle to Grid (V2G)

This mode contributes the idea of integrating grid and battery for an efficient energy management system. This time, energy flows from vehicle to grid after some conditions to access the energy from the battery. The condition is that whether the battery is having minimal voltage and does the owner ready for taking energy from the battery. If the aforementioned conditions are satisfied, then user can perform this mode of operation. The reverse action of G2V mode will be happening in this mode. First, the DC-DC converter will be initiated for the boosting up the voltage from battery which is followed by a DC-link capacitor. The boosted voltage is given to the AC-DC converter, it will act as an inverter and convert the AC to DC by turning on the IGBTs in complementary manner. The generated PWM signals is given to a NOT gate, input pin and output pin of the NOT gate is connected to each opto-couplers. The controller operates in 5V, if the controller and hardware set up connected directly, the controller will be damaged when an excess voltage passes to the controller from the hardware. To avoid such burden an isolated circuit is included.

4 Experimental Results

The circuit topology of proposed system is simulated on MATLAB/Simulink. Two modes of operations are generated and verified. Energy flows from grid to vehicle in the forward mode and vice versa (vehicle to grid) occurs in the reverse mode. The simulated circuits and results of respective modes are exhibited below.

4.1 Charging Mode (Forward Mode)

The simulation circuit of grid-to-vehicle mode is illustrated in Fig. 4. H bridge has mainly two power stage conversions. In G2V, the first stage acts as converter, and second stage acts as a buck converter. In this mode, 230V AC supply is taken from the power grid which is reduced by a step-down transformer. The step-down voltage is given to the first stage. The first stage consists of four IGBTs, in which the body diodes of IGBTs get ON, resulting in AC-DC conversion. So that, this mode does not require PWM signals for turn ON the IGBTs. The rectified voltage is having some ripples in order to filter such ripples by passing through a DC link capacitor and which is followed by a DC-DC converter. The voltage across the capacitor is a high value, to reduce that value by passing to the DC-DC converter. For that the DC-DC converter is enabled for buck operation. The final output will be approximately 12V. So the 230V AC value is converted to 12V DC value. The simulated result of grid-to-vehicle mode is depicted in Fig. 5.

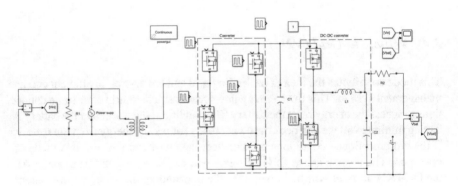

Fig. 4 Simulation circuit for G2V

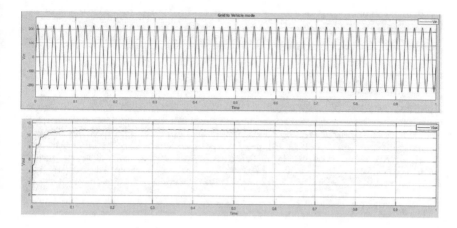

Fig. 5 Obtained waveform for G2V

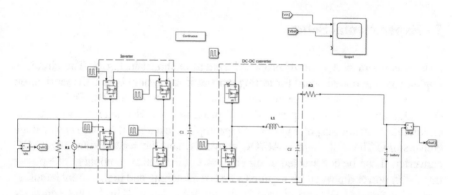

Fig. 6 Simulation circuit for V2G

4.2 Discharging Mode (Reverse Mode)

The simulation circuit of discharging mode is shown in Fig. 6. In V2G, the first stage acts as inverter and second stage acts as a boost converter. The operation of this mode is started from the right side. In this mode, 15V DC voltage is boosted by the DC-DC converter. For that DC-DC converter is enabled as boost operation by setting the lower IGBT is triggered PWM signal. The output is passed through the body diode of the upper IGBT of the DC-DC converter. The boosted volatge is given to inverter circuit, where all IGBTs are triggered by the PWM signal. During the positive half cycle, two opposite IGBTs are turned ON by PWM signal without phase shift. During the negative half cycle, other two IGBTs are turned ON by PWM signal with phase shift. Finally, the 15V DC value is converted to 200V AC approximately. The obtained output of discharging mode is represented in Fig. 7.

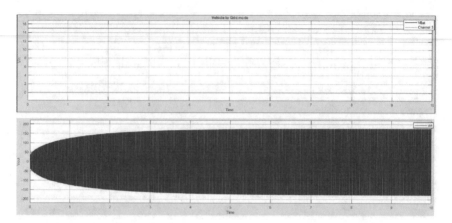

Fig. 7 Obtained waveform for V2G

5 Experimental Setup

Overall experimental setup of proposed system is shown in Fig. 8. The circuit is implemented in dotted board for testing our proposed system before the fabrication of PCB. The experimental setup consists of AC/DC converter, DC-DC converter (H bridge circuit), opto-coupler and DSP processor. An H bridge is an electronic circuit that switches the polarity of a voltage applied to a load. It has two power stage conversions. The first stage is AC/DC converter, and the second stage is DC-DC converter. It can be constructed using any power electronics component like power transistor, power diode, etc. It requires a fastest switching and high current handling capability. So that IGBTs are chosen for the implementation. H bridge has a property of bidirectionality. So it can act as both converter and inverter in G2V and V2G mode, respectively. An opto-coupler is a protection component which is used to isolate the processor and implemented circuit. Five opto-couplers are used for the protection of processor. If there is any voltage higher 5 V comes the coupler will not pass it to the processor and protects the processor. So the PWM signal from the processor is given to the H bridge through the opto-coupler. TMS320f28335 is used as a control unit. It has many features over other processors. It is required to monitor the system in real time. It runs with a clock frequency 150 MHz and also contains 6 internal PWM channels. The main functions of control unit are in controlling the overall system; PWM signals are generated for switching the IGBT, provides authorization, and controls the phase, amplitude and frequency. Transformer is used to step down the voltage, which is taken from the grid in the forward mode. A 15V battery will be taken as the DC source.

Initially, the generated PWM signals from the processor is given to a NOT gate for getting phase-shifted signal of the generated PWM for triggering the IGBTs in the corresponding modes. The 230 V AC signal is taken from grid fed to H-bridge circuit for conversion of AC to DC. Moreover, PWM signal is not required at the beginning

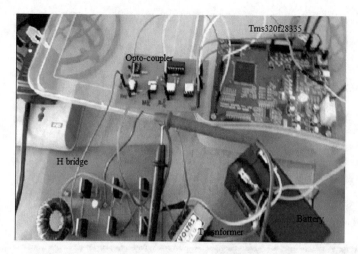

Fig. 8 Experimental setup of proposed system

Fig. 9 Output for G2V
mode

of G2V mode. Hence, the input of NOT gate will be zero and output will be high. The input and output of the NOT gate are connected to opto-couplers for the safety purpose. The input of the NOT gate will turn OFF a pair of IGBTs in the H-bridge circuit, and the output of the NOT gate will turn ON the other pair of IGBTs. Both pair of IGBTs should be turned OFF at this mode. In order to OFF the pair of IGBTs that are turned ON, an extra opto-coupler will be used. Then, the body diode of each IGBTs get activated and performs rectification. The rectified signal is filtered by using a DC-link capacitor, which is followed by buck/boost converter and is stored in the battery. The obtained value is displayed on the multimeter 19.08 V, is shown in Fig. 9 which is enough for the charging 15 V battery. There should 16.6–20 V range voltage for the charging 15 V battery. In case, if there is any voltage higher 20 V the buck operation is initiated and voltage is reduced to the required voltage.

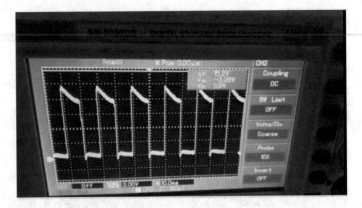

Fig. 10 Output for V2G mode

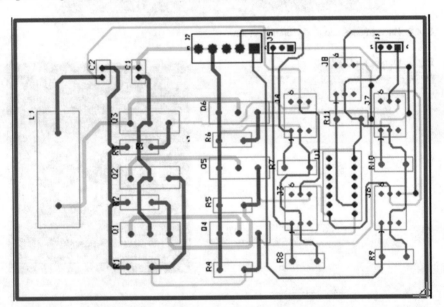

Fig. 11 Layout of proposed system

In V2G, voltage taken from the battery is boosted up by buck/boost converter and is fed to H bridge circuit. The generated PWM signal from the processor is given to gate terminals of IGBTs via an opto-couplers. The input of the NOT gate is PWM signal without phase shift and output of NOT gate is its phase shifted signal. The PWM signals will turn ON the pair of IGBTs in the respective cycles. The obtained waveform of AC waveform of 15.5 V is shown in Fig. 10 which can be boosted to our required level by a step up transformer for other applications. Altium designer software is used to draw the circuit diagram and layout of proposed system which is shown in Fig. 11. It is fabricated as a two-layer board which is represented as colored lines in the layout.

6 Conclusion

This paper proposes a controllable bidirectional charger for electric vehicle which enhances the power grid capabilities. It is fabricated in simple and affordable cost with power converter circuits. The proposed system ensures the energy saving in the energy management system. This paper deals with two modes, namely grid-to-vehicle and vehicle-to-grid. A single circuit can be operated in two modes; moreover, it can provide the property of bi-directionality. The proposed system is very useful for the charging of vehicle, different home appliance requirements like an inverter (when sudden load shedding is happened), and charge can share to another vehicle in an urgent situations. The circuits for the corresponding modes are evaluated and also simulated on MATLAB/Simulink. The board layout is drawn on Altium Designer tool. The efficiency of the system is bit less due to conversions of AC to DC and vice versa, and cost of DSP processor is high. The recognition of proposed system has been done in a small-scale prototype. The proposed system is compatible that delivers power from grid to battery and battery to grid with required power factor.

References

1. Patil, D., Agarwal, V.: Compact onboard single-phase EV battery charger with novel low-frequency ripple compensator and optimum filter design. IEEE Trans. Veh. Technol. **65**(4), 1948–1956 (2015)
2. Shelly, S., Babu, A.V.: A probabilistic model for link duration in vehicular ad Hoc networks under Rayleigh Fading Channel Conditions. In: 2015 Fifth International Conference on Advances in Computing and Communications (ICACC), Kochi, pp. 177–182 (2015). https://doi.org/10.1109/ICACC.2015.16
3. Tan, A.S., Theam, D.I., Mohd-Mokhtar, R.: Sze Sing Lee, and Nik Rumzi Nik Idris, "Predictive control of plug-in electric vehicle chargers with photovoltaic integration". Journal of Modern Power Systems and Clean Energy **6**(6), 1264–1276 (2018)
4. Pedrosa, D., Gomes, R., Monteiro, V., Afonso, J.A., Afonso, J.L.: Model Predictive Current Control of a Slow Battery Charger for Electric Mobility Applications, pp. 643-653. Springer, Cham (2017)
5. Rachid, A., Fadil, H.E., Belhaj, F.Z., Gaouzi, K., Giri, F.: Lyapunov-based control of single-phase AC–DC power converter for BEV charger. In: Recent Advances in Electrical and Information Technologies for Sustainable Development, pp.115–121. Springer, Cham (2019)
6. Sousa, T.J.C., Monteiro, V., Afonso, J.L.: Integrated system for traction and battery charging of electric vehicles with universal interface to the power grid. In: Doctoral Conference on Computing, Electrical and Industrial Systems, pp. 355–366. Springer, Cham (2019)
7. Restrepo, M., Morris, J., Kazerani, M., Canizares, C.A.: Modeling and testing of a bidirectional smart charger for distribution system EV integration. IEEE Trans. Smart Grid **9**(1), 152–162 (2016)
8. Kushwaha, R., Singh, B.: Power factor improvement in modified bridgeless landsman converter fed EV battery charger. IEEE Trans. Veh. Technol. **68**(4), 3325–3336 (2019)
9. Li, H., Zhang, Z., Wang, S., Tang, J., Ren, X., Chen, Q.: A 300-kHz 6.6-kW SiC bidirectional LLC onboard charger. IEEE Trans. Ind. Electron. **67**(2), 1435–1445 (2019)
10. Pai, F.-S., Tseng, P.-S.: A Control approach integrating electric vehicle charger and dynamic response of demand-side resource aggregators. J. Electr. Eng. Technol. 1–10 (2020)

11. Dam, S.K., John, V.: Battery impedance spectroscopy using bidirectional grid connected converter. Sadhana **42**(8), 1343–1354 (2017)
12. Harikumar, R., Raj, J.S.: Ad hoc node connectivity improvement analysis-Why not through mesh clients? Comput. Electr. Eng. **40**(2), 473–483 (2014)
13. Tashakor, N., Farjah, E., Ghanbari, T.: A bidirectional battery charger with modular integrated charge equalization circuit. IEEE Trans. Power Electron. **32**(3), 2133–2145 (2016)
14. Ahn, J.-Ho., Lee, B.K.: High-efficiency adaptive-current charging strategy for electric vehicles considering variation of internal resistance of lithium-ion battery. IEEE Trans. Power Electron. **34**(4), 3041–3052 (2018)
15. Monteiro, V., Pinto, J.G., Afonso, J.L.: Operation modes for the electric vehicle in smart grids and smart homes: present and proposed modes. IEEE Trans. Veh. Technol. **65**(3), 1007–1020 (2015)

Correlative Study of Image Magnification Techniques

Sangeetha Muthiah and A. Senthilrajan

Abstract A digital image is often required to be rescaled. When the pixels in the image are mapped to a larger grid, new coordinate positions will be formed. The interpolation procedures are used to approximate the value for the unknown sample points set between the initial sample points. The concept is to replace the missing values using the known values of the initial sample points. The interpolation techniques are intended to effectively preserve the characteristic features of an image. There is an indeed greater number of interpolation algorithms available which have their strengths and challenges. In this study, interpolation techniques based on the nearest neighbour, bilinear interpolation, and high-quality magnification (hq3x) scaling were examined. The qualitative and quantitative properties of these three interpolation techniques were analysed. The parameters considered are the SSIM and SNR measure. In the experiment, the performance of the three methods was compared and measured from objective and visual assessments. The aim of this study is to analyse the visual quality of the output to determine the suitable method. From the analysis, it is observed that hqx scaling performs better than nearest neighbour interpolation and bilinear approach performs fairly closer to hqx scaling.

Keywords Image interpolation · Magnification · Nearest neighbour · Bilinear · Hqx scaling · SSIM

1 Introduction

Interpolation is a method of estimating the values at any given position between the range of distinct sets of points [1]. For several image processing applications, this is one among the basic procedures [2]. In digital imaging, scaling refers to

S. Muthiah (✉) · A. Senthilrajan
Department of Computational Logistics, Alagappa University, Karaikudi, Tamil Nadu, India
e-mail: sangeethsaro@yahoo.com

A. Senthilrajan
e-mail: agni_senthil@yahoo.com

© The Editor(s) (if applicable) and The Author(s), under exclusive license to Springer Nature Singapore Pte Ltd. 2021
S. Shakya et al. (eds.), *Proceedings of International Conference on Sustainable Expert Systems*, Lecture Notes in Networks and Systems 176,
https://doi.org/10.1007/978-981-33-4355-9_16

expanding or reducing the number of pixels in the image. When the image dimension is extended, an additional number of sample points is created. Most commonly, magnifying the image involves interpolation procedures to approximate the numerical values of unknown sample points. A technique called image panning is related to image magnification in a way how the images are viewed in the display. In the context of image display, panning allows viewing of the image horizontally or vertically in the display system, whereas image magnification provides a deeper insight into the image or a wider perspective of the viewed object. The issue with inspecting the zoomed image, however, will only reveal a portion of the image. Panning makes it possible to look at the different parts of the zoomed image by shifting the focus inside the image to the desired area. Similarly, contrast enhancement is done to maximize the intensity values present in the image to give a better visual perception. This process differentiates objects and the background in a more distinguishable way [3, 4]. Image rescaling is often necessary for many image operations such as zooming or shrinking images for digital display devices, geometric transformations and subpixel registration, and decompression [5–7]. Interpolation algorithms are structured to achieve optimal results with a trade-off between maximum performance, edge smoothness and sharpness [8]. Interpolation, however, is meant to retain the image the details as the image is extended, adding new details would never make the original image. Image interpolation is the problem of approximating the value for the new sample point with the known points located around the unknown point. There are a number of practical methods for upscaling images which obtain the desirable result, but it is valuable to find the appropriate method of significantly lower distortion. In this study, to evaluate an efficient approach to achieving optimum performance, the efficiency of the image interpolation methods is tested from objective criteria along with visual criteria.

This paper is structured as follows. The second section provides a review of relevant literature on methods of image interpolation. The third section describes the three interpolation techniques. The fourth section presents the experimental analysis and performance comparisons and the conclusion is drawn in the final section.

2   Related Work

The classical interpolation approaches do not consider the local details of the image and the unknown pixel values are interpolated in the same way across the whole image. Dianyuan Han presents a comparison of widely used interpolation methods for image upscaling. The results show the bicubic interpolated image scores better both in SNR measures as well as the visual quality and take less computation time compared to cubic B spline method. The nearest neighbour interpolation is fast however it produces uneven edges. According to the author, the bilinear interpolation method makes an effective option with respect to the processing speed and visual quality [9]. There has been many study addresses interpolation based on the differences in local intensity features to preserve sharp edges. An edge-directed interpolation employed by Allebach et al. proposes expanding the image with the high-resolution edge map generated based on subpixel estimation of the original image and the edge map is used as a direction to interpolate the smooth regions with bilinear interpolation and the visual quality is achieved through iteration [10]. Another approach by Li et al. proposed a model to preserve edges using local covariance measures to estimate the covariance pixels of the high-resolution image. This estimation of covariance is done based on the geometric duality of the pixels of the low-resolution and high-resolution image. To minimize the complexity, a hybrid method is adopted to interpolate exactly the edge region with covariance-based interpolation and to interpolate smooth regions with bilinear interpolation [11]. Zhang presents a nonlinear interpolation method to retain the edge structure using directional filter and data fusion. The directional filter estimates two values along with the orthogonal directions of a missing sample from a local window and the directional estimated values are adaptively merged together to calculate a more robust value with linear MMSE [12]. To reduce the computational complexity and to achieve high-resolution image, the method focuses on isolating edge pixels from non-edge pixels. To do so, it measures the local variance of the unknown pixel's nearest pixels. With bilinear interpolation, the non-edge pixels are interpolated and the edge pixels are adaptively interpolated.

3 Image Interpolation

Image interpolation is the process by which the unknown points are estimated using the known points. It generates high-resolution image from the low-resolution image without affecting the visual information present in the original image. Unknown sample points are created to extend the original image grid according to the interpolation ratio. Accuracy of the image being constructed depends on the number of identified samples. In interpolation, it is presumed that the closest known points have a greater impact when measuring value at the unknown point than those farther away [13].

The brightness values of points P1, P2, P3, P4, for example, represent the pixels in the original image as shown on the left in Fig. 1, and its original dimension is magnified three times as seen in Fig. 1 on the right. The number of unknown sample points generated is determined by the desired magnification ratio. The values of original points P1, P2, P3, P4 are mapped to the new position in the enlarged grid, respectively. Notice the number of unknown points such as X, between these four known sample points. Now the values of those unknown samples are to be estimated by interpolation procedure using the brightness values of original points P1, P2, P3, P4 [9].

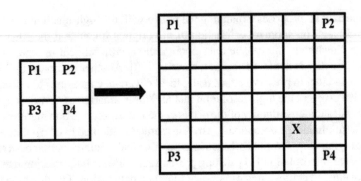

Fig. 1 Image interpolation

3.1 *Nearest Neighbour Interpolation*

The nearest neighbour interpolation is the simplest approach which assigns each sample with the value of a sample closest to the point. It is a piecewise polynomial. It is a single point approach; therefore, the function g(x) at any point is interpolated based on just one nearest sample value. The nearest neighbour approximation kernel is described by a box filter kernel

$$s(t) = \begin{cases} 1 \text{ if } |t| < 0.5 \\ 0 \text{ otherwise} \end{cases} \tag{1}$$

This filter kernel is the first-order family of B-splines. The samples are interpolated by convolving the sampled signal with the box filter in the spatial domain. This is equivalent to multiplication of the transformed function with a sinc function in the frequency domain [5]. Since sinc function has endless side lobes on both sides of the main lobe, it is not an optimal filter for interpolation. If the sinc function is approximated with the box window function defined in Eq. (1), the inverse transform of this filter displays ringing effect as shown in Fig. 2. This is due to the poor absorption of the high-frequency elements which often contributes to the discontinuity of the intensity surface of the image. This method is preferred, however, because of its simple operations which make implementation easier with little computing time.

As shown in Fig. 3, the pixel $P(x,y)$ is at non-integer location and is interpolated using the closest single known neighbour, if, for example, the input coordinates of the original sample points $P(x_1,y_1)$, $P(x_1,y_2)$, $P(x_2,y_1)$, $P(x_2,y_2)$ are the four nearest neighbours to the new sample point. Now to find the value for the new coordinate point $P(x.y)$, the distance between $P(x,y)$ and the four neighbouring points $P(x_1, y_1)$, $P(x_1,y_2)$, $P(x_2,y_1)$, $P(x_2,y_2)$ are computed. The closest point is determined and its intensity value is assigned to the new sample point $P(x,y)$ [9].

The nearest neighbour method, if scaled to an integer coordinate, will reproduce the original image. This approach can induce distortion if it is mapped partially within

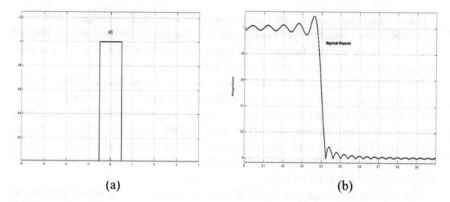

| | (a) | | (b) |

Fig. 2 **a** Nearest neighbour interpolation box kernel, **b** frequency response

Fig. 3 Illustration of nearest
neighbour interpolation

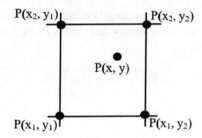

the original location. However, the arrangement of pixels which form the image
becomes visible as the scale factor is increased. Olivier et al. proposed a new nearest
neighbour value interpolation to overcome this problem. According to the proposed
approach, determining the nearest value is guided by bilinear interpolation. The
value for the missing pixel is estimated by taking the minimum difference between
the values of four nearest cells and the value obtained by bilinear interpolation. The
value which is almost equal to the nearest cells is set as the value for the missing
pixel [14].

3.2 Bilinear Interpolation

The box filter in Eq. (1) is convolved with itself produces a triangular function and
the triangular filter kernel is defined as [15]

$$s(t) = \begin{cases} 1 - |t| & |t| < 1 \\ 0 & \text{otherwise} \end{cases} \tag{2}$$

It is a linear piecewise polynomial where straight lines are used to connect between the points. The sampled image is convolved with the triangular filter and the Fourier transform will be sinc function multiply with a sinc function. This filter is a low-pass filter and gives smoother frequency response as seen in Fig. 4.

The bilinear interpolation takes into account two closest sample points in each direction located near to the undefined sample point. The linear interpolation along with the horizontal directions and on the vertical direction is computed. The final value for the undefined sample point is the weighted average value of the closest neighbours. The weights are calculated based on the distance from the undefined sample point to the four nearest points and the points located close to an undefined point are assigned greater weights.

Figure 5 illustrates the process of bilinear interpolation. The sampling point at $P(x,y)$ is to be interpolated. This point is surrounded by two nearest neighbour on the bottom row denoted by $P(1,1)$ and $P(2,1)$ and two nearest neighbour on the top row denoted by $P(1,2), P(2,2)$. The value for sample point Q is determined, by performing two linear interpolations: one in the x-direction of the bottom row to estimate the value for intermediate point $P(x,y_1)$ and top row to estimate the point.

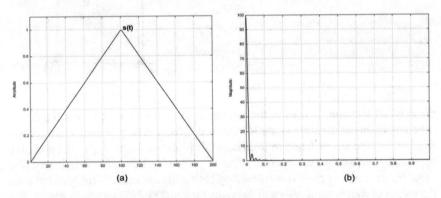

(a)   (b)

Fig. 4 Bilinear interpolation **a** triangle kernel, **b** frequency response

Fig. 5 Illustration of bilinear interpolation

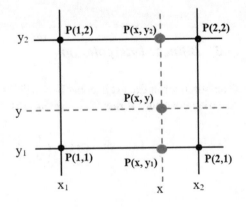

$P(x,y_2)$ and one in the y-direction using the previously estimated points $P(x,y_1)$, $P(x,y_2)$ to interpolate $P(x,y)$.

To calculate the value for the sample point $P(x,y_1)$ and $P(x,y_2)$, the bottom row is linearly interpolated followed by top row in the x-direction

$$P(x, y_1) = P(x_1, y_1) + (P(x_2, y_1) - P(x_1, y_1)) * \frac{x - x_1}{x_2 - x_1} \qquad (3.1)$$

$$P(x, y_2) = P(x_1, y_2) + (P(x_2, y_2) - P(x_1, y_2)) * \frac{x - x_1}{x_2 - x_1} \qquad (3.2)$$

and one linear interpolation on the y-direction to estimate the point)(x,y).

$$P(x, y) = P(x, y_1) + (P(x, y_2) - P(x, y_1)) * \frac{y - y_1}{y_2 - y_1} \qquad (4)$$

3.3 High Quality Magnification (Hqx) Interpolation

HQX stands for high quality magnification [16]. It was created by Maxim Stepin. It is created for pixel art scaling algorithm. This interpolation approach checks for lines in the image and gives smooth interpolation. The hqx scaling algorithm compares the color of the 8 immediate neighbours to the color of each sample point. The difference in colour between the source and each neighbour is calculated based on the YUV threshold and further, the Y component has a greater weight assigned compared to U and V components. The pixels compared is either classified as close or distant which gives 256 possible combinations. Further, If the colour difference in any one of the three planes is above the threshold, that bit is 1 otherwise 0. The magnification of each input sample points to the corresponding 9 output sample points.

The interpolation pattern for each 3×3 is searched in the preconstructed lookup table. This table directs which pixel color is assigned to a particular pattern. For each close or distant combination, there is an entry in the lookup table. Each entry states how to blend the colors of the source pixels and its neighbours to interpolate nine output pixels. The hqx works well when the input is not anti-aliased. This algorithm produces an image of better visual quality and retains the edges without much blurriness.

4 Experiment and Discussion

The widely used algorithm for image interpolation is the nearest neighbour and bilinear interpolation. The high-quality magnification algorithm considered in pixel art scaling is compared with these approaches. The consistency of each of these methods is analysed. To determine the accuracy of the magnified images, a test image taken is reduced to a third of its original size and using different interpolation algorithms the image is extended to its original arrangement. The original size of the test image is compared with the reconstructed image for similarity measure.

4.1 Objective Assessment

In this study, the performance of the selected image magnification algorithms is compared. To evaluate the fidelity of the interpolated images, signal-to-noise ratio (SNR) and structure similarity index (SSIM) of those scaled images are measured. The SSIM measures the structure content variation between the original image and the magnified image and the SNR computes the difference between two images. These methods are tested in 20 images and the results of the different interpolation algorithms are shown in Table 1.

The SNR measure between two images produces larger value, represents better quality of an image and the mean SSIM value should be close to one. The results of SSIM and SNR values of different interpolation algorithms from the table show the bilinear interpolation and hqx scaling performed better than nearest neighbour interpolation. Importantly, if the image is not antialiased the hqx scaling provides better image quality. The SSIM and SNR values of bilinear interpolated images are greater than the nearest neighbour approach but fairly close to hqx scaling algorithm.

Table 1 Evaluation of SSIM and SNR of different interpolated method

Metrics	SSIM			SNR		
Interpolation algorithm	Nearest neighbour	Bilinear	HQX	Nearest neighbour	Bilinear	HQX
Baboon	0.7922	0.8774	0.9101	20.43559	23.05296	23.08568
Leaf 1	0.6357	0.7215	0.7495	13.96609	17.17062	17.38480
Pepper	0.7502	0.8235	0.8402	18.40475	21.39917	22.29207
Lena	0.6342	0.7615	0.7621	14.32169	18.42498	18.45049
Leaf 2	0.6502	0.8374	0.8322	16.75714	21.97753	21.84567
Leaf 3	0.7689	0.8199	0.8593	18.07040	20.15241	20.65962
Leaf 4	0.7844	0.8603	0.8592	19.99927	23.05181	23.49497
Leaf 5	0.8003	0.8533	0.9077	20.84055	22.90923	23.36608

4.2 Visual Quality Assessment

Generally, the quality of the magnified images by various interpolation methods can be determined by performing a visual examination of the results. For visual evaluation, the test images are enlarged to a scale factor of 3 to illustrate the visual contrast between nearest neighbour, bilinear and hqx scaling methods. An example of a magnified image for visual comparison is shown in Fig. 6. The nearest neighbour interpolation displays the presence of serration near the borders, curves and lines which can be realized from Fig. 6, since this filter suppresses the stopband frequencies which results in unwanted deviations in the frequency response. It also generates significant blockiness over the entire surface of the image. Hence the accuracy of the image is reduced.

On the other hand, the bilinear interpolation eliminates the serrated edges which are more apparent in the nearest neighbour approach. Due to the improved stopband attenuation performance, this filter gives a smoother frequency response. But this is achieved by blurring the high-frequency data present in the original image. This produces extensive smoothing on the image surface which makes the sharp edges and other features appear smudged. The hqx approach is a pattern matching algorithm which interpolates pixels from the pregenerated table. It begins to search for edges in the image and smoothly interpolate those edges. However, it has slightly visible

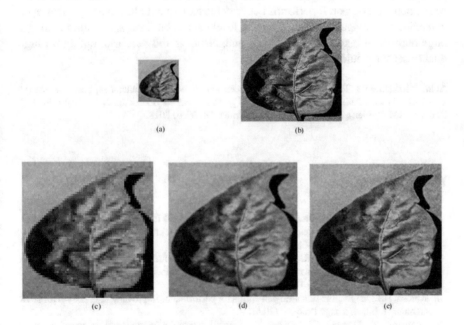

(a)   (b)

(c)   (d)   (e)

Fig. 6 Interpolated images for visual contrast **a** shrunken image, **b** original image, **c** nearest neighbour, **d** bilinear, **e** HQX

serrated edges in the image. It retains much of the sharp features contained in the original image.

5 Conclusion

The nearest neighbour interpolation can be used for its speed and ease of implementation. It replicates the closest sample point intensity. This possibly brings a stair step pattern across the image surface, specifically around the edges. This method, however, does not add any new information to the image. The bilinear interpolation which works in the opposite way to the latter approach in computing time as well as many sample points involved for an interpolating point. It requires additional calculation as it stretches around interpolating point to four nearest pixels. It is a low-pass filter with a decent frequency response that softens the high-frequency features such as image borders which looks indistinct. Unlike the nearest neighbour method, the bilinear interpolation provides an image of continuous intensity. As a consequence, the quality of the image is more satisfactory for viewing. The hqx algorithm can get reasonably better image quality with fast computing time except the difficulty involved in constructing the custom lookup table for interpolation. The hqx produces an image of very good in quality. All the edges and lines contained in the image appears sharper. It performs better in the case of antialiased image. However, extending the image size past a certain level cannot be done since interpolating a large number of pixels would become unrealistic. In this scenario, opting for high quality sensor would be useful.

Acknowledgements This research work has been written with the financial support of Rashtriya Uchchatar Shiksha Abhiyan (RUSA Phase 2.0) grant sanctioned vide Letter No. F.24-51/2014-U, Policy (TNMulti-Gen), Dept. of Edn. Govt. of India, Dt. 09.10.2018.

References

1. Diana Earshia, M.V.: A Comprehensive study of 1D and 2D Image Interpolation Techniques. In: International Conference on Communications and cyber physical Engineering (2018)
2. Miklos, P.: Image interpolation techniques
3. Agarwal, M., Mahajan, R.: Medical image contrast enhancement using range limited weighted histogram equalization. In: 6th International Conference on Smart Computing and Communications (ICSCC 2017), Kurukshetra (2018)
4. Koresh, H.J.D.: Quantization with perception for performance improvement in HEVC for HDR Content. J. Innov. Image Process. (2020)
5. Lehmann, T., Gonner, C., Spitzer, K.: Survey: interpolation methods in medical image processing. IEEE Trans. Med. Imagıng **18**, 1049–1075 (1999)
6. Rajan, A. S.: Image reduction using edge based region of interest. In: IOP Conference Series: Materials Science and Engineering (ICMAEM-2017) (2017)

7. Smys, S., Taj, J.S., Krishna Raj, N.: Virtual reality simulation as therapy for posttraumatic. J. Electron. Inform. **01**, 24–34 (2019)
8. Asghar, Z., Naeem, R., Jarral, N.K.: Correlative analysis of different image scaling algorithms in graphics. Int. J. Adv. Res. Comput. Sci. Electron. Eng. **6**(1), 6–10 (2017)
9. Han, D.: Comparison of commonly used image interpolation methods. In: International Conference on Computer Science and Electronics Engineering, Paris, France (2013)
10. Allebach, J., Wong, P.: Edge-directed interpolation. In: IEEE, pp. 707–710 (1996)
11. Li X, Orchard, M.T.: New edge-directed interpolation. IEEE Trans. Image Process. **10**, 1521–1527 (2001)
12. Zhang, L., Wu, X.: An edge-guided image interpolation algorithm via directional filtering and data fusion. IEEE Trans. Image Process. **15**, 2226–2238 (2006)
13. Chang, K.-T.: Introduction to Geographic Information Systems. McGraw Hill, Idaho (2018)
14. Rukundo, O., Cao, H.: Nearest neighbor value interpolation. Int. J. Adv. Comput. Sci. Appl. **3**, 1–6 (2012)
15. Dodgson, N.A.: Image resampling. University of Cambridge, Computer Laboratory, Cambridge (1992)
16. Stepin, M.: HQX. https://web.archive.org/web/20070717064839/www.hiend3d.com/hq4x.html (2003)

Healthy Sri Lankan Meal Planner with Evolutionary Computing Approach

M. I. M. Jayarathna and W. A. C. Weerakoon

Abstract Due to the busy lifestyle and the negligence of health, human beings are becoming the victims of non-communicable diseases (NCD). Further, the hereditary and lifestyle factors are major causes of NCD. My Smart Diet is a menu planning Android application with the genetic algorithmic (GA) approach. It suggests healthy meals with Sri Lankan foods and beverage. Moreover, My Smart Diet app uses calorie requirements and nutrition needs per day of a person according to age, height, weight, and daily activities. These parameters are taken from the user. Foods and actual daily calorie and nutrients that suggested by nutritionist and other sources have been stored in MySQL database. In this application, evolutionary computing has been applied on two levels. In the first level, the chromosomes have been designed using a dietary nutrient with 21 genes, where the nutrients and quantities were used as the genes. In the second level, the chromosomes have been designed using dietary foods with a variable length. The numbers of genes rely on user inputs. There, the Sri Lankan food items and quantities have represented the genes. Further, in the first level, the fitness function has been designed using user inputs and sample personal data based on nutritional data and chromosomes, while in the second level, the fitness function has been designed using foods that insert by the user, nutrition recommendation that given by the first genetic algorithm and chromosomes. Moreover, in these two levels, the individual with the lowest fitness value has been selected as the output. In addition to that, application testing was conducted by colleagues for its functionality and user satisfaction. Finally, the app has been evaluated using the preference given by its users compared with the preference received by the same group of users who used the Fitness Meal Planner app (an existing western food meal planner). After that, the preferences have been evaluated by applying the sign test. Overall, the Smart Diet app ranks higher than Fitness Meal Planner. Ultimately, Sri Lankan users can live a healthy life using My Smart Diet app.

Keywords Genetic algorithm · Android application · My smart diet · PHP restful API

M. I. M. Jayarathna (✉) · W. A. C. Weerakoon
Department of Statistics and Computer Science, University of Kelaniya, Kelaniya, Sri Lanka
e-mail: mihiranijayarathna@gmail.com

© The Editor(s) (if applicable) and The Author(s), under exclusive license
to Springer Nature Singapore Pte Ltd. 2021
S. Shakya et al. (eds.), *Proceedings of International Conference on Sustainable Expert Systems*, Lecture Notes in Networks and Systems 176,
https://doi.org/10.1007/978-981-33-4355-9_17

1 Introduction

Due to the busy lifestyle and the negligence of health, human beings are becoming the victims of non-communicable diseases (NCD). In fact, the dietary and lifestyle factors such as inadequate physical activities and unhealthy diets are major causes for NCD. As a trend, people those who have concerns on this tend to surf the Internet to find healthy diets or exercise plans. A healthy diet plan helps to maintain or improve overall health with various foods. In app stores, it could find plenty of diet planner apps. Several such food planning applications exploit GA. It could find a menu planning application for diabetic mellitus patients [1] and a diet schedule for kidney patients using GA with a fuzzy expert system [2] In addition to those, in 2016, a personal guidance application was developed with the GA approach [3]. Further, an automated food menu planning application was developed in 2012 with evolutionary algorithms [4]. Another such attempt was a Web-based weekly menu planning application [5]. However, most of these planners were not suitable for the Sri Lankans as the foods and the others environmental factor such as a climate could be varied. Furthermore, the daily calorie consumption rate of a Sri Lankan could be varied. As such, it was difficult to find a diet app with Sri Lankan foods that matches to the environment in Sri Lanka. My Smart Diet app [6] is the solution for above-mentioned problem which was developed for the Sri Lankan context with an Android and genetic algorithmic approach. It suggests healthy diet with Sri Lankan foods. Moreover, My Smart Diet app uses calorie requirements and nutrition needs per day of a Sri Lankan according to age, height, weight, and daily activities. The values for these parameters are taken from the user. Foods and actual daily calorie and nutrients that suggested by nutritionist and other sources was stored in a MySQL database. These data were collected through interviews and questionnaires. A multiobjective genetic algorithm was used to suggest appropriate daily meals based on personal inputs. Daily calorie requirement is calculated using the Harris Benedict equation. The genetic algorithm approach has the initial population, chromosome coding, fitness score calculating, crossover, and mutation operators. In this application, five meals were suggested for every day according to user's preference. Those are three main meals and two snacks. Accordingly, each chromosome has not in a fixed length. Each gene includes nutrients and the quantity of it per day. Nutrients include carbohydrate, protein, calcium, fats, salt, sugar, vitamin B12, water, fiber, iron, and folic acid. This application implement in two parts. First genetic algorithm implement with PHP Restful Application Programming Interface (API). This API passes the result of the genetic algorithm to an Android Studio. My Smart Diet app [6] was developed with the Android Studio to get user inputs and to display the meals. The final result of this project is to suggest daily meals according to personal inputs. Further, application testing was conducted by the colleagues to check its functionality compared the user preferences with respect to another existing meal planner app. The user preferences were statistically analyzed using the sign test. Finally, it could prove that My Smart Diet app is a user-friendly meal planner for Sri Lankans.

2 Background Related Works

Many recent studies on meal planners have focused on menu planning with a genetic algorithm. In 2017, M. F. Syahputra, V. Felicia, R. F. Rahmat, and R. Budiarto researched Schedule Diet for Diabetes Mellitus (DM) Patients using genetic algorithm. This has produced a menu plan for DM patients for a period of a week as per their calorie needs. This consists of two data types, such as DM patients' details and details of food nutrition. Calorie needs can be calculated using the Harris Benedict equation. The initial population was randomly generated through foodstuffs, and the chromosome was composed of 15 genes. Then, a gene represents the total number of calories required from an item. The fitness score was calculated using the total caloric requirements and the total number of calories in the foods. This rank-based fitness assignment method was used for the selection, hence selecting the individual with the lowest fitness value. Further, the two-point crossover method and random mutation are used in the genetic algorithm. Here, the test was carried out for twelve times with different individuals and generations. It recommends that genetic algorithms perform well in menu designing for diabetes mellitus patients [1].

In 2012, a project on automatic dietary menu planning [4] based on an evolutionary algorithm was conducted by Kołodziejczyk and Przybyłek. This article presents an advisory system for dietitians to arrange a set of divers daily meals that satisfying personalized nutrition standard. The initial population was randomly generated through foodstuffs, and the chromosome consisted of 20 genes. It suggested five meals per day, and each meal had four products, where a gene represents a food product. Fitness scores are calculated by measuring the nutrient requirements depending on the food references of the diet. Further, the tournament selection was used to select the best individual with uniform crossover method and standard mutation. Here, the test was conducted using 10 menus with a fitness value with more than 0.75. The advantage of this method was it generates a variety of meals. However, it did not suggest the type of cooking, which was a disadvantage of the system [4].

Next, a Dietary Menu Planning application [5] using an evolutionary method was introduced by Barbara Koroušić Seljak in 2007. This paper has proposed a computer-aided Web-based application for weekly menu planning considering diet-planning principles and the aesthetic standards. Here, GA was used in a multilevel way. The initial population was constant for all levels. There were three levels of chromosomes. At the first level, seven genes were there and a gene in the chromosome has represented one daily meal. Then, the middle-level chromosome had five genes, where a gene has represented one meal. Further, the end-stage chromosome does not have a fixed length, and a gene has represented food item and quantity. In fact, the aesthetic standards considered in calculating the fitness scores were cost, color, and nutrients. In addition to that, the binary tournament selection method was used with two-point crossover and a linear descent mutation, where the individual with the largest crowding distance and the non-dominant solution was selected as the best individual. Finally, the testing was conducted using the dietary requirements of a local hospital and the experimental results were obtained by running GA 25 times.

Next two kinds of research used a genetic algorithm with fuzzy system. The first research was conducted for patients suffering from kidney and urinary tract (UT) diseases [2] by Sri Hartati in 2013. In this work, an initial population of 100 individuals were randomly selected from food indices and the chromosome was composed of 10 genes. Then, a gene represents one food index. Further, they have identified five types of kidney disease and have used different fitness functions for each type of kidney disease. Dietary supplements were used for the fitness function. Genetic algorithms use crossover probability and mutation probability given by Mamdani's fuzzy inferential system. Here, the testing was conducted using real test cases, and it has recommended the genetic algorithm perform well in designing menus for patients with kidney and UT diseases.

Next research was personalized dietary guidance conducted by Petri Heinonen and Esko K. Juuso in 2016. As an initial population, 100 individuals are randomly selected from the diet and there is no fixed length for the chromosome. Use this tournament selection method as a selection and select the individual with the lowest fitness rating. Genetic algorithms use arithmetic crossover and mutation probability given by Mamdani's fuzzy inferential system. Except for these operators, elitism was used here. Here, the validity of the system was done with specialist expertise, comparison of nutritional status and monitoring of key aspects of the work of the Gas [3].

3   Genetic Algorithm

Genetic algorithm is a heuristic search algorithm [7]. It is based on the idea of natural selection [7] and genetics. It uses to select fittest individuals for reproduction to produce offspring of the next generation. They are commonly used for genetic high-quality solutions for optimization problems and search problems. It is especially efficient with an optimization problem. There are five phases [7] in GA. The first phase is the initial population. The initial population is a subset of all possible solutions to a given problem. The chromosome is one such solution to the given problem. Gene is one element position of chromosomes. Fitness function is the first step of creating a genetic algorithm. It is a function which gets solutions as an input and makes a suitable solution as output. Mainly, there are three types of genetic operators in GA. These are selection, crossover, and mutation. The parent selection is the process of selecting parents which mate and recombine to create offsprings [7] for the next generation. Crossover is combining two individuals to create new individuals for possible inclusion in the next generation. Most popular crossover operators are uniform crossover [4], two-point crossover [5], and arithmetic crossover [3]. The mutation may be defined as a small random tweak in the chromosome, to get a new solution. "Apply random changes to individual parents from children" [8]. It is used to maintain and introduce diversity in the genetic population. It is usually applied in low probability. Most popular mutation operators are standard mutation [4], linear descending mutation [5], and random resetting scramble mutation [1].

4 Methodology

My Smart Diet application was developed as an Android app. Figure 1 shows the design of My Smart Diet. Implementation of My Smart Diet was done covering the phases illustrated in Fig. 2. First, the database was created using the MySQL server. There were nine tables, namely the personal table, the recommended nutrition table, cereals, vegetables, leafy vegetables, fruits, juices, meat and fish and dairy product. Group of 100 people has participated in the personal data collection process. After that, the data collected were stored in the personal data table. Age, gender, BMI, activity level, and pregnancy have been used as personal data. In addition to that, according to the personal data table, 21 essential nutritional data have been inserted into the nutrition data table. These are energy (kcal), fat (g), carbohydrate (g), protein (g), calcium (mg), thiamin (mcg), riboflavin (mg), vitamin C (mg), iron (mg), sodium (mg), cholesterol (mg), fiber (mg), moisture (g), potassium (mg), zinc (mg), magnesium (mg), phosphorus (mg), niacin (mg) carotenoid (mcg), and vitamin A (mcg).

Subsequently, seven tables have been added to the database under the Sri Lankan food category. They are whole grains, vegetables, green leafy vegetables, fruits, juices, meat and fish and dairy products. Sample personnel data were collected through questionnaires. Recommendations on nutrients and quantity for personal data samples have been collected using interviews with a nutritionist. The nutrition information on Sri Lankan food and beverages was collected using the Web site of biodiversity for food and nutrition project. According to the information gathered on food nutrition, energy, protein, calcium, moisture and carotenoids are found in abundance in vegetables, green vegetables, fruits, juice, and cereal. As shown in Fig. 3,

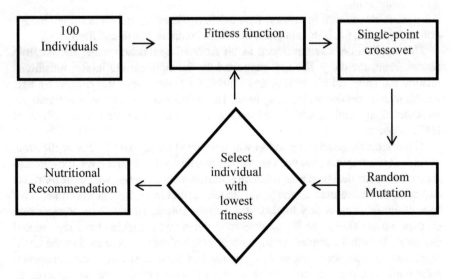

Fig. 1 Design

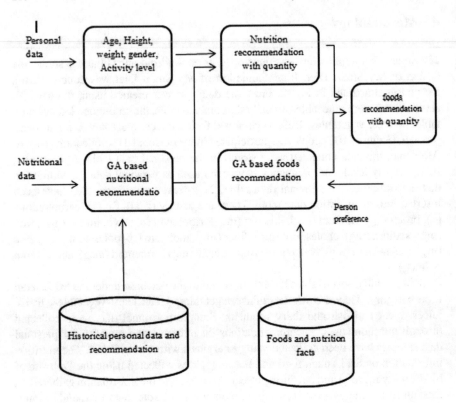

Fig. 2 Implementation

dairy products are rich in energy, fats, calcium, vitamins, and cholesterol, energy, protein, calcium, fat, iron, and phosphorus are common in fish and meat.

This application was developed as an Android application using the Android Studio. Here, the daily diet was suggested by the daily caloric intake, nutritional factors, the daily caloric intake, and nutritional factors were determined by age, height, weight, gender, and activity level. These data were inserted as user inputs via an Android application, and those inputs were sent to the server via the API as an HTTP request.

The respective genetic algorithm was developed using PHP. In this application, evolutionary computing has been applied on two levels. In the first level, the chromosomes have been designed using a dietary nutrient with 21 genes, where the nutrients and quantities were used as the genes. In the second level, the chromosomes have been designed using dietary foods with a variable length. The numbers of genes rely on user inputs. There, the Sri Lankan food items and quantities have represented the genes. In both the implementation stage, 100 individuals were used as the initial population, single-point crossover operation has been used and selects crossover point randomly, and random mutation operator has been used in the mutation phase. In the first level, the fitness functionality has been designed using user inputs and

(a)

Food's Name	Quantity	Calorie	Fats(g)	Carbohydrate(g)	Protein(g
rice	100g				
Imbul kiribath	200g	425	13.2	75.6	5
Milk rice & lunumiris	135 15	293	10.9	6.6	
Kee roti &kiripeni	35g	173	6.8	27.7	41
Kolakend	200ml	232	5.3	44.1	4

(b)

Foodname	Quanti	Energy	Fats(g)	Carboh	Protein	Cal
thumba Karawila	100g	0	0	10.6	2.1	
Bandakka	100g	35	0.2	6.4	1.9	
Del	100g	113	0.4	26	1.5	
Kos	100g	51	0.3	9.4	2.6	
puhul	100g	10	0.1	1.9	0.4	

(c)

Foodname	Quar	Energ	Fats(	Carb	Prote	Ca
cow milk-with suger(skim)	100g	67	3.2	4.4	3.2	
Buffalo milk	100g	117	4.3	5	4.3	
Goat milk	100ml	71	4.27	4.59	3.67	
Butter-salted	100g	717	81.11	0.06	0.85	
Butter	100g	102	11.52	0.01	0.12	

(d)

Foodname	Quant	Energ	Fats(g	Carbo	Protei	Calsiu	Thiar
Andha	100g	184	12.4	0	16.8	118	
paraw	100g						
Katuwalla	100g	126	4.1	0	21.5	0	
Katta	100g	114	3.2	0	20	12	
Magura	100g	85	1	4.2	15	410	
Saleya	100g	273	19.4	0	21.8	180	
Kunissa	100g	349	8.5	0	68.1	0	

(e)

Foodname	Quantity	Energy(l	Fats(g)	Carbohy	Protein(	Calsi
Bilin	100g	19	0.3	3.5	0.5	
Cashew	100g	51	0.1	12.3	0.2	
Delum	100g	65	0.1	14.3	1.6	
Durian	100g	183	3.9	34.1	2.8	
Madan	100g	62	0.3	14	0.7	
Ambrella	100g	46	0.1	12.4	0.2	

Fig. 3 Foods **a** cereal, **b** vegetables, **c** diary products, **d** fish and meats, **e** fruit

Fig. 4 Fitness function $F_{Cj} = \sum_{k=1}^{n} | O_{ik} - C_{jk} | + \sum_{x=1}^{5} | P_{ix} - I_x |$

Fig. 5 Fitness function $F = | \sum_{i=0}^{21} (Xi - \sum_{j=0}^{n} (Qj * Ni)) |$

sample personal data based on nutritional data and chromosomes as shown in Fig. 4. The individual with the lowest fitness value has been selected as the output.

In the fitness equation, the output recommendation (O_i) represents records on the nutrition recommendation table, for example, O_1 is the first record of the nutrition recommendation table and the chromosome is represented by C_J. As an example, C_1 is the first chromosome. Here, the initial population is equal to 100. So, j is changed from one to one hundred. n denotes the number of genes in the chromosome, where a gene represents the required intake of the nutrition and nutrition such as carbohydrates, fats, calcium, and sodium. Further, chromosome length is not fixed. Therefore, the parameter k is changed from 1 to the length of the particular chromosome. For example, the chromosome contains 21 nutrients, such as carbohydrate, energy, fat, calcium, sodium, and then k has been varied 1–21 and n equal to 21.

Further, personal data records (P_i) represent the records on the personal data table. For example, P_1 is the first record of the personal data table. The input personal data (I_x) variable represents the data entered by the user. Age, BMI, gender, activity level, and pregnancy are represented by x. For example, I_1 is the age of the user. The personal data table has five columns, so x is changed from 1 to 5.

In the second level, the fitness functionality has been designed using foods that input by the user, nutrition recommendation that given by the first genetic algorithm, and chromosomes as shown in Fig. 5. The individual with the lowest fitness value has been selected as the output.

In the fitness equation, the nutrition recommendation (X_i) represents nutrition given by the first genetic algorithm. Quantity of food item (Q_j) represents a gene on chromosome, n represents the number of genes in chromosome, and n depends on user preference. Nutrition (N_i) represents the nutrition of each food item.

After the foods and quantity suggested by the genetic algorithm, the Android application has been referred to as a JSON object which has been displayed to the user through the Android app. This suggests the quantity of foods that the user chooses using genetic algorithms.

5 Discussion

The system was tested with a group of 75 people which includes university students, school students, and members of the village death donation society, collecting their personal information. Their responses were collected by distributing a questionnaire among them. The questionnaire covered the questions to check the satisfiability of the non-functional requirements of the system. Under the non-functional requirements, whether the system's interfaces match the resolution of the user's phones, appropriateness of colors, texts, and font sizes, whether the users can understand the application's request data, and the attributes in the interface such as buttons, cursors, and dropdown buttons separately. Analyzing data under these requirements ensures system usability. Functional requirements have been verified using the feedback provided by the user in the questionnaires. Under the functional requirements checking, it was required to check the functionality of the system. For examples, are the submission forms functioning properly, are the inserted data passing through the API, is the fitness function operated according to the included data calculating the fitness value and producing the desired output, which is a constructive food recommendation. Analyzing the data under this criteria, it could ensure the system is functioning desirably. The app has been evaluated using the preference given by its users compared with the preference received by the same group of users who used the Fitness Meal Planner app (an existing western food meal planner). After that, the preferences have been evaluated by applying the sign test. Further, a descriptive analysis was conducted over the preferences. For example, 83% of the participants like the food variations produced by the application, 85% understood the inquired data in the application, 89% satisfied with the response time, and 94% agreed that the cost incurred due to the meals suggested by the application was lower than the other application. Overall, the Smart Diet app ranks higher than Fitness Meal Planner to the food preferences and respective questions. However, only 23% indicated that the interfaces of these applications were attractive.

Moreover, this application can be enhanced by combining food recipes with meal recommendations in the future. In addition to that, this app can also be developed as a Web-based application.

Ultimately, Sri Lankan users can live a healthy life using My Smart Diet app.

References

1. Syahputra, M.F., Felicia, V., Rahmat, R.F., Budiarto, R.: scheduling diet for diabetes mellitus patients using genetic algorithm. J. Phys. Conf. Ser. **801**, 012033 (2017)
2. Hartati, S., Uyun, S.: Computation of diet composition for patients suffering from kidney and urinary tract diseases with the fuzzy genetic system, June 2013
3. Heinonen, P., Juuso, E.K.: Development of a genetic algorithms optimization algorithm for a nutritional guidance application. Presented at the Proceedings of The 9th EUROSIM Congress on Modelling and Simulation (EUROSIM 2016,) The 57th SIMS Conference on Simulation and Modelling (SIMS 2016) (2018), pp 755–761.
4. Kołodziejczyk, J., Przybyłek, Ł.: Automatic dietary menu planning based on evolutionary algorithm (2012)
5. Seljak, B.K.: Dietary menu planning using an evolutionary method. In: 2006 International Conference on Intelligent Engineering Systems, pp. 108–113. (2006)
6. Weerakoon, W. A. C., Jayarathna, M. I. M.: My smart diet: genetic algorithm approach for healthy Sri Lankan Meal Planner. In: Proceedings of international research symposium on pure and applied sciences, Faculty of Science, University of Kelaniya, Sri Lanka, Nov 2019

7. Mallawaarachchi, V.: Introduction to genetic algorithms—including example code. Towards Data Sci. 08 Jul 2017. [Online]. Available: https://towardsdatascience.com/introduction-to-genetic-algorithms-including-example-code-e396e98d8bf3. Accessed: 11 Jun 2019
8. What Is the Genetic Algorithm? MATLAB & Simulink. [Online]. Available: https://www.mathworks.com/help/gads/what-is-the-genetic-algorithm.html. Accessed: 11 Jun 2019

Fuzzy Logic-Based Approach for Back Analysis of VISA Granting Process

Susara S. Thenuwara and R. A. R. C. Gopura

Abstract In this paper, a fuzzy logic-based approach is proposed for the back analysis of the VISA granting process in common practice. VISA granting process is varying with countries, regions as well as other factors. Different types are granted and they have different criteria that apply to VISA a particular country or a region. The final results of VISA decisions affect the passport ranking index in a particular country. The passport ranking index is the most common factor for the measurement of the wealthiness and healthiness of a country. Hence, it is very important to study the procedure of modifying, updating, and maintaining the rules and regulations of the VISA granting process in common practice. The proposed fuzzy logic-based approach is developed by considering vague factors that impact the final decision making for VISA. After applying the fuzzy logic-based approach, VISA granting process can be modified according to the particular country as well as maintaining the quality of the decision.

Keywords Fuzzy logic · VISA grating process · Passport ranking index

1 Introduction

In general, the VISA is granting permission to enter a country. The Latin meaning of VISA is "a paper that has been seen" [1]. However, it is a document or seal issued by a country to a person to enter, leave, or stay in a region for a specific period. The most common VISA types are tourist, work, student, and transit, but there are many subcategories and different types as well. Depending on the specific region of travel, it can be valid for single or multiple entries. In common practice, approval of VISA depends on several factors: validity and availability of primary documents,

S. S. Thenuwara (✉)
Department of Computational Mathematics, University of Moratuwa, Moratuwa, Sri Lanka
e-mail: 179418b@uom.lk

R. A. R. C. Gopura
Department of Mechanical Engineering, University of Moratuwa, Moratuwa, Sri Lanka
e-mail: gopurar@uom.lk

S. Shakya et al. (eds.), *Proceedings of International Conference on Sustainable Expert Systems*, Lecture Notes in Networks and Systems 176,
https://doi.org/10.1007/978-981-33-4355-9_18

221

proof of income (livelihood), sponsorship and relatives, illegal, criminal and overstay activities, travel history, health condition (health index), and other straight forward factors.

The results of VISA decisions influence the passport ranking index in a particular country. The passport ranking index is the most common factor for the measurement of the wealthiness and healthiness of a country. Hence, it is very important to study the procedure of modifying, updating, and maintaining the rules and regulations of the VISA granting process in common practice.

After analyzing the previous VISA granting data, issues have been identified in the common practice of the VISA granting process. Major issues identified are VISA guidelines are not up to date and passport ranking index is not even slightly changed in a very long period. To study the procedure of modifying, updating, and maintaining the rules and regulation of the granting process, a proper back analysis of the process is required. Therefore, in this paper, a fuzzy logic-based approach is proposed for the back analysis of the VISA granting process [2].

The research paper is structured as follows. A brief literature review of the current complex decisions is available in the next section. Section 3 explains the solution suggested. This involves designing and implementing the program proposed. Section 4 presents results and discussions. The final section concludes the paper by outlining the research conclusions, limitations, and future directions.

2 Literature Review

Many types of research have been conducted on fuzzy-based decision-making systems. They have used popular fuzzy inference systems like Mamdani and Sugeno. The fuzzy logic algorithm helps to solve many vague decision-making problems that humans involved in. In this section, the authors investigate the use and how their involvement and weakness influence final decisions of fuzzy logic systems.

The VISA process is a kind of decision-making process. Therefore, it is potentially important to study decision-based fuzzy applications. Zaher et al. discussed an artificial intelligence approach for decision making in investment [3]. The proposed fuzzy inference system is Mamdani and the major objective of the research was to advise their clients to allocate the portion of their investments. The paper discussed different membership functions (MF) and comparison with other methods as well.

Sajfert et al. introduced a fuzzy logic framework into decision-making processes concerning managers selection [4]. The manager's qualifications considered in this research were wealth, expertness, leadership, and status. The purpose of this paper was to use dynamic logic to construct a list and to achieve an advantageous solution.

Li et al. [5] proposed an extended Takagi–Sugeno–Kang inference system (TSK+) with fuzzy interpolation and rule base generation. This paper introduced an entirely distinctive, fuzzy interpolation response to the TSK statement. It also suggested a database method for producing the expanded TSK thinking framework. The proposed system enhances the standard TSK thought in two ways. The experimental result is unquestionable because the program with compact rules and competitive efficiency is very significant.

Vermonden and Gay presented the Migration from Oaxaca, Mexico fuzzy modeling [6]. This study shows the fluid model of logic based on migration factors such as higher primary employment, high unemployment rates, and a high marginalization index. The model shows the tendency of primary-sector workers to migrate and a growing effect on migration patterns of soil degradation. The proposed approach uses the Sugeno model to evaluate the two selected input variables.

The next literature based on VISA granting-based decision-making problems that can give solutions based on fuzzy inference systems. When the first papers by Hamedi and Jafari are considered and discussed on fuzzy logic decision making in the e-tourism industry: This is a case report on e-tourism in Shiraz region [7]. This study is based on fuzzy knowledge and inference for the city of Shiraz, as a case study and more related to our preliminary analysis. Membership functions are seen as triangular and models are input variables, with various limiting values: lodging, distances, and facilities. The authors conclude the result with a comparison of fuzzy decision making and Euclidean distance method.

Fuzzy-based approach is done after analyzing the previous VISA granting data, and issues have been identified in the common practice of the VISA granting process. Furthermore, Major issues identified are VISA guidelines are not up to date and passport ranking index is not even slightly change in a very long period, Therefore, risk assessment using fuzzy systems was guided using the following paper. Sabokbar et al. discussed risk assessment in the tourism system explored using an unusually rough and dominant set [8]. The purpose of this research is to identify the risk of Iran tourism destination. The models are considered as inputs of political, economic, social, cultural, technological, environmental, functional, and security factors. The model proposed for creating a comprehensive, rational expert system was to allow both organizations and planners to meet their needs and conditions.

It is important to study a fuzzy-based system in a critical situation like an emergency case solution. Sundharakumar et al. discussed a cloud-based fuzzy healthcare system [9]. The system is very familiar with the healthcare sector to organize their work. Within this paper, a modern, cloud-based healthcare system is built where the wireless body space networks merge knowledge with a non-public repository like STORM, the period measurement method, and the flouted logical thinking method. In the case of harmful conditions, the system shall promptly supply important patient health information to physicians, careers, and hospital management. The research shows that real-time cloud analytics help boosts the system's performance and quality of life with timely medical support.

Mammadli presented a fuzzy logic-based loan evaluation system [10]. This paper proposes a fuzzy logic model for the analysis of retail loans. There are five input variables in the fuzzy model: "income," "financial history," "job," "character," "collateral condition," and a single financial-standing output variable. The model evaluated and tested with leading banks in Azerbaijan. For practical outcomes, a wide range of linguistic variables and additional advanced "IF...Then..." guidelines is suggested.

After a comprehensive literature review and a study, the authors proposed a fuzzy logic-based approach to VISA decision making. Furthermore, it is significantly important for the evaluation of the VISA decisions and improves the whole process with priority to passport ranking index.

3 Proposed Approach

In common practice, the VISA grating process depends on many factors. Among all the factors, livelihood and health index are the vaguest factors. Those factors directly influenced in the final decision making as well. Other factors are fuzzy singletons. For example, if a person applies with forge documents final decision will exactly be VISA rejection. These types of variables are fuzzy singletons and they are neglected in the proposed approach. Hence, health and livelihood are considered as inputs of the proposed fuzzy-based approach. The output of the approach is the decision. It can have three categories such as granted, conditional, and rejected. Two input variables are fuzzy and output is not fuzzy. Therefore, Takagi–Sugeno–Kang (TSK) method is used for fuzzy inferencing [11]. After analyzing previous VISA cases, fuzzy membership values for health factor are selected as "Poor," "Normal," and "Good" and for livelihood factor as "Bad," "Good," "High." Decisions of the proposed approach have been compared with the actual decision to evaluate the model. Figure 1 shows TSK inferencing using the fuzzy logic toolbox of MATLAB.

The first input variable is a health condition and after analyzing the past data (250 VISA cases), sigmoid (1) and Gaussian (2) functions have been selected as fuzzy membership functions [12]. A Gaussian MF is determined completely by c and σ; c represents the MFs center and σ determines the MFs width.

$$\text{Gaussian}(x; c, \sigma) = e^{\frac{1}{2}\left(\frac{x-c}{\sigma}\right)^2} \tag{1}$$

$$\text{Sigmoid}(x; a, c) = \frac{1}{1 + \exp[-a(x - c)]} \tag{2}$$

Table 1 lists the membership values of health factor and their parameters.

Normally, parameters of health condition membership functions are defined based on the prescriptions of physicians. Figure 2 shows the MF of the health fuzzy variable using the fuzzy logic toolbox.

The second input variable is the livelihood and after studying past data the Gaussian (1) and sigmoid (2) functions have been selected as the membership functions. Table 2 lists the membership values of livelihood.

Figure 3 shows the MF of livelihood fuzzy variable in the fuzzy logic toolbox of MATLAB. The output of the system is the VISA decision of the VISA granting process. Most common VISA decisions are "Rejected," "Conditional," and "Granted." VISA decisions are not fuzzy and it makes the following function.

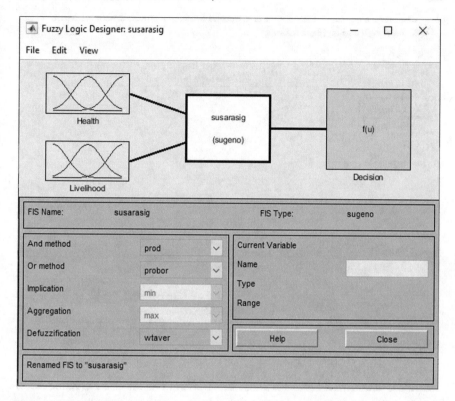

Fig. 1 TSK fuzzy inference

Table 1 Membership values of health condition fuzzy variable

Membership values	Parameters
Poor	<0.2
Normal	>0.4 and <0.6
Good	>0.8

$$F(x) = \text{VISA decision} \begin{cases} \text{Rejected} & x = 1 \\ \text{Conditional} & x = 0.5 \\ \text{Granted} & x = 0 \end{cases} \tag{3}$$

VISA decisions are taken by a panel after receiving the necessary supporting documents from the applicant. There is a point system based on the documents they apply and it will take as an input of the actual system. Figure 4 shows MF after inserting the values of VISA decisions to the fuzzy logic toolbox.

After a comprehensive study of previous VISA decisions, nine fuzzy rules have been defined. Figure 5 represents the generated fuzzy rules in the fuzzy logic toolbox. Figure 6 shows the surface view of the fuzzy rules.

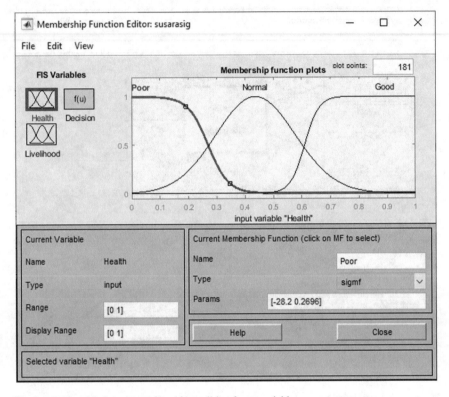

Fig. 2 Membership functions of health condition fuzzy variable

Table 2 Membership values of livelihood fuzzy variable

Categories	Values
Bad	<0.15
Good	>0.4 and <0.55
High	>0.8

1. If (Health is Poor) and (Livelihood is Bad) then (Decision is Rejected)
2. If (Health is Poor) and (Livelihood is Good) then (Decision is Conditional)
3. If (Health is Poor) and (Livelihood is Good) then (Decision is Conditional)
4. If (Health is Normal) and (Livelihood is Bad) then (Decision is Rejected)
5. If (Health is Normal) and (Livelihood is Good) then (Decision is Conditional)
6. If (Health is Normal) and (Livelihood is High) then (Decision is Granted)
7. If (Health is Good) and (Livelihood is High) then (Decision is Granted)
8. If (Health is Good) and (Livelihood is Bad) then (Decision is Rejected)
9. If (Health is Good) and (Livelihood is Good) then (Decision is Granted).

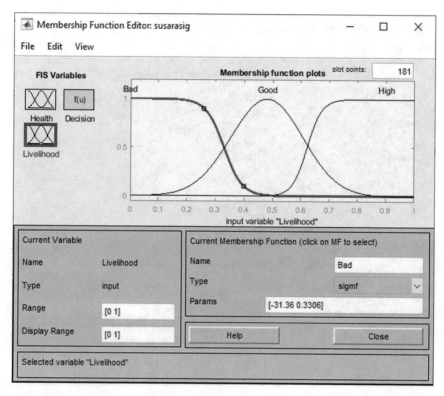

Fig. 3 Membership function of livelihood fuzzy variable

4 Results and Discussion

500 sets of previous VISA cases have been selected and 250 cases are randomly chosen to identify the membership functions and build the fuzzy-based approach using MATLAB. The following twelve cases were presented to show the performance of the proposed approach.

VISA Case 1:

Health condition = 0.353.

Livelihood = 0.653.

The actual VISA decision is rejected.

Figure 7 shows the related fuzzy rules for case 1. The VISA decision from the proposed approach is conditionally granted, but the actual grant decision was VISA rejection.

VISA Case 2:

Health factor = 0.28.

Livelihood = 0.314.

The actual VISA decision is rejected.

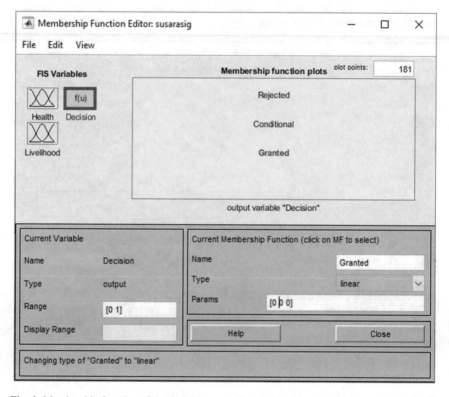

Fig. 4 Membership function of the decision

Figure 8 shows the related fuzzy rules for case 2. Since health conditions and livelihood are low, VISA decision is rejected and the actual VISA decision is to reject as well. Table 3 shows a comparison of all twelve VISA cases. From a total of twelve cases, two cases are not matched with actual decisions due to reasons such as human biases and issues of definition of fuzzy rules and membership functions.

5 Conclusion

The proposed fuzzy logic-based approach for back analysis of the VISA granting process can be used to discuss the human subjective judgment by linguistic terms. A fuzzy model has been built based on previous information of VISA cases and input variables have been selected on prior knowledge. Fuzzy input variables are health condition and livelihood and respective membership values are "Poor," "Normal," "Good" and "Bad," "Good," "High." The output of the fuzzy system is the VISA decision which is not a fuzzy variable. "Granted," "Conditional," and "Rejected" are the three VISA decisions. This system has been evaluated with previous VISA cases

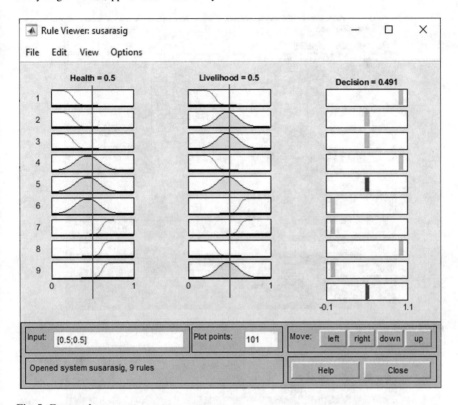

Fig. 5 Fuzzy rules

with permission for the research purpose. The proposed fuzzy approach can be used to improve the VISA decisions as well as analysis of the previous data. It will directly affect the world passport rank index.

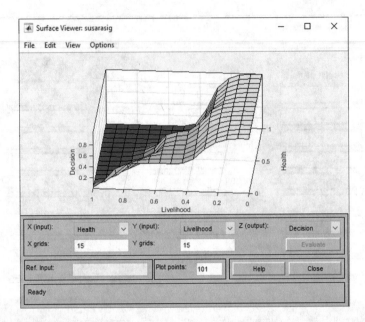

Fig. 6 Surface view

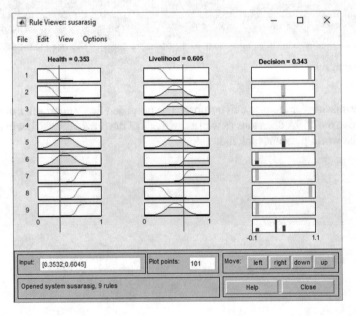

Fig. 7 Fuzzy rules for case1

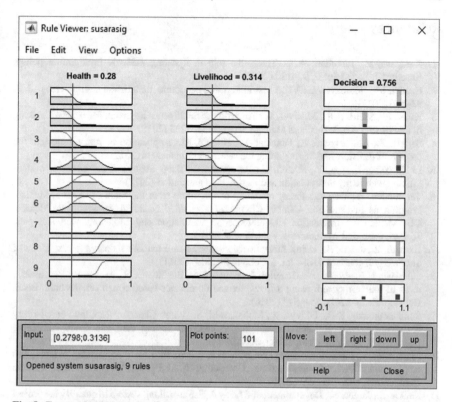

Fig. 8 Fuzzy rules for case 2

Table 3 Comparison of VISA decisions

Case	Health condition	Livelihood	Actual decision	Proposed decision
1	0.353	0.653	Rejected	Conditional
2	0.280	0.314	Rejected	Rejected
3	0.363	0.258	Granted	Granted
4	0.322	0.333	Granted	Granted
5	0.253	0.552	Granted	Granted
6	0.453	0.650	Granted	Granted
7	0.300	0.373	Conditional	Rejected
8	0.153	0.253	Rejected	Rejected
9	0.103	0.159	Rejected	Rejected
10	0.243	0.553	Rejected	Rejected
11	0.153	0.650	Rejected	Rejected
12	0.053	0.553	Rejected	Rejected

References

1. Kampana, P., Tanielian, A.R.: Thailand's role in updating ASEAN immigration policy. Kasetsart J. Social Sci. **2**, 92–126 (2017)
2. Czaika, M., Neumayer, E.: VISA restrictions and economic globalization. Appl. Geogr. J. **3**, 92–526 (2017)
3. Zaher, H., Saeid, N.R., Moshref, W.: An artificial intelligence approach for decision making in investment. Afr. J. Business Manage. **6** (9), 3221–3233 (2018)
4. Sajfert, Z., Atanasković, P., Pamučar, D., Nikolić, M.: Application of fuzzy logic into the process of decision-making regarding the selection of managers (2012)
5. Li, J., Yang, L., Qu, Y., Sexton, G.: An extended Takagi–Sugeno–Kang inference system (TSK+) with fuzzy interpolation and its rule base generation (2017)
6. Vermonden, A., Gay, C.: Fuzzy modeling of migration from the State of Oaxaca, Mexico. Program de Investigation Cambio Climatic, Universidad Nacional Autonomy de México, CU, Mexico 4th International Conference on Simulation and Modeling Methodologies, Technologies, and Applications, pp. 845–851 (2014)
7. Hamedi, Z., Jafari, S.: Using fuzzy decision-making in e-tourism industry: a case study of Shiraz city E-tourism. IJCSI Int. J. Compu. Sci. **8**(3) (2011)
8. Sabokbar, H.F., Ayashi, A., Hosseini, A., Banaitis, A., Banaitienė, N., Ayashi, R.: Risk assessment in tourism system using a fuzzy set and dominance-based rough set. Technol. Econ. Develop. Econ. **22**(4), 554–573 (2016)
9. Sundharakumar, K.B., Dhivya, S., Mohanavalli, S., Vinob Chander, R.: Cloud-based fuzzy healthcare system. In: 2nd international symposium on big data and cloud computing (ISBCC'15). Procedia Comput. Sci. **50**, 143–148
10. Mammadli, S.: Fuzzy logic based loan evaluation system. In: 12th International Conference on Application of Fuzzy Systems and Soft Computing (ICAFS 2016), Vienna, Austria. Procedia Comput. Sci. **102**, 495–499 (2016)
11. Shukla, S., James, N.: Development of a fuzzy decision-making system to quantify the project management efficiency using interval type-2 fuzzy logic. IRACST—Int. J. Commerce Bus. Manage. (IJCBM) **3**(6) (2014). ISSN: 2319–2828
12. Teixeira, M.S., Maran, V., Lima, J.C.D., Augustin, I., Machado, A.: Fuzzy-based model to detect patient's health decline in ambient assisted living. In: 19th International Conference on Enterprise Information Systems (ICEIS 2017), vol. 1, pp. 659–666 (2017)

A Lucrative Model for Identifying Potential Adverse Effects from Biomedical Texts by Augmenting BERT and ELMo

Jarashanth Selvarajah and Ruwan D. Nawarathna

Abstract This study copes with extracting adverse effects (AEs) from biomedical texts. An adverse effect is a noxious, unintended, and undesired effect caused by the administration of an external entity such as medication, dietary supplement, radiotherapy, and others. A binary classifier is proposed to filter out irrelevant texts from AE assertive texts and a sequence labeling model for extracting the AE mentions. Both models are built by consolidating the cutting-edge deep learning technologies: Bidirectional Encoder Representations from Transformers (BERT), Embeddings from Language Models (ELMo), and Bidirectional Gated Recurrent Units. The performances of our models are evaluated on an Adverse Drug Effects dataset constructed by sampling from Medline case studies. Both models perform significantly better than previously published models with an F1 score of 0.906 for binary classification and an approximate match $F1$ score of 0.925 for text labeling. The proposed models can be adapted to any tasks with similar interests.

Keywords Adverse effects · Biomedical text · Deep learning · Binary classifier · Sequence labeling

1 Introduction

Public health is one of the major factors that influence a nation's economy greatly. A recent survey carried out by the World Health Organization (WHO) on global health expenditure reveals that the annual spending on health care is growing on an average of 4% in high-income countries and 6% in low- and middle-income countries [1]. Sri Lanka's Health Expenditure as a percent of gross domestic product accounted for 1.480 in December 2017 [2]. Although the healthcare system has been

J. Selvarajah
Postgraduate Institute of Science, University of Peradeniya, Peradeniya, Sri Lanka

R. D. Nawarathna (✉)
Department of Statistics and Computer Science, Faculty of Science, University of Peradeniya, Peradeniya, Sri Lanka
e-mail: ruwand@pdn.ac.lk

© The Editor(s) (if applicable) and The Author(s), under exclusive license to Springer Nature Singapore Pte Ltd. 2021
S. Shakya et al. (eds.), *Proceedings of International Conference on Sustainable Expert Systems*, Lecture Notes in Networks and Systems 176,
https://doi.org/10.1007/978-981-33-4355-9_19

revolutionized to its peak, the efficacy of the treatments is still questionable. This is because the medications/treatments are tested and perfected almost exclusively on a small group of men in a narrowed geographic region, and hence they may not be effective on (pregnant) women, elderly people, patients with comorbidities or patients from another geographic region [3]. As an example, the Food and Drug Administration (FDA), USA, recommended cutting the dosage of a popular sleep aid, known as Ambien, to half on women, after 20 years of its release as they realized that women metabolize the drug in a slower rate than men [3]. This brings to a legitimate concern that the healthcare system should do more research on personalized health care for everyone, also known as precision medicine. To achieve this goal, the manual processing of health-related records, such as patient's clinical reports, globally would not even scratch the surface. A machine learning model is needed to be created that can robustly identify and extract health-related anomalies from colossal digital data such as electronic health records, clinical narratives, case reports, scientific articles, product reviews, health forums' user discussions, and health-related social media data on a global scale.

During the infant stages of natural language processing (NLP) and machine learning for automatic processing of texts, researchers used clinical narratives as data sources for studying health-related adversities [4]. However, the accesses for the clinical narratives were limited only to the researchers affiliated with the medical centers. Since there was a dearth of health-related information for data mining, researchers brought their attention to case studies published in bibliographic databases of life sciences and biomedical information (e.g., Medline), and social media data, that is overwhelming and representing real-time data on a global scale.

The future of precision medicine highly relies on health-related experiences of individuals. The widespread use of social media brings disclosing personal information, including health-related experiences, which enlightens the healthcare researchers to use them as a substantial yet powerful data source. However, the irregularities of social media data, such as vernacular language, idiomatic expressions, descriptive medical terminologies, use of abbreviations, and ambiguities, introduce additional challenges to the NLP researchers to process them for automatic data mining.

While extensive health-related research is going on over the past few years on social media data, some of the nourishing low-hanging fruits such as biomedical texts are still unnoticed or kept aside for future consumption. Though a significant number of studies have fostered biomedical data mining for health care, their explorations on primary areas such as identifying adverse effects caused by dietary supplements, food products, radiation therapies, and prescribed medications under regular dosages, also known as adverse drug reactions (ADRs), are underutilized.

The objective of this study is to build deep learning models to identify texts that mention health-related adverse effects and extract them. Moreover, the existing models for health-related texts mining are obsolete, and hence models are built using advanced deep learning technologies. One important expertise in modern deep learning technology that language modeling-based transfer learning is adapted where a language model is trained on a massive raw text and it is fine-tuned for a downstream task using an annotated dataset. The advantage of this technology is that the model can perform exceptionally well even with the small amount of annotated dataset, thus minimizes the annotation cost considerably.

In this study, two deep learning models are developed by merging and adjusting the salient features of the cutting-edge deep learning technologies, namely Bidirectional Encoder Representations from Transformers (BERT) [5], Embeddings from Language Models (ELMo) [6], and Bidirectional Gated Recurrent Units (BiGRUs) [7]. The first model is a binary classifier which discards the irrelevant topics from the intended health topic that is to be further analyzed for insights, whereas the other model extracts the knowledge. To be more specific, if the intended health topic is to identify ADR, the input to the first model is a drug-related sentence, and the output is a label which claims whether an ADR is present or not. The second model is fed with ADR positive sentences, where the ADR assertive tokens are extracted. The performance of the models is tested using drug-related topics, particularly identifying ADRs. However, these models can be adapted to any similar domains, including, but not limited to identifying adverse effects caused by dietary supplements, processed foods, and radiotherapy. The outcomes of these types of studies may bring several hypotheses that would be validated by healthcare professionals through a series of real-world experiments. Since our models identify potential AEs, medical scientists can take precautions to avoid any dangers and thus become a lucrative model in the sense of healthcare domain.

Several studies have been carried out on automated health-related text mining over the past decade under three major themes: disease surveillance, post-marketing drug surveillance, also known as pharmacovigilance, and behavioral medicine. PubMed abstracts, generic- and health-related social media data have been extensively analyzed for identifying potential health-related hazards. Some of the most important studies are listed down below.

The pioneering work from Laeman et al. used user posts from a health discussion forum and identified ADRs using context matching [8]. Nikfarjam and Gonzalez addressed several limitations of this study by utilizing association rule mining for identifying lexical patterns [9]. Sarker et al. applied a traditional supervised-based machine learning algorithm with linguistic and domain-specific features, to classify ADR assertive texts from a Twitter dataset and a dataset created from PubMed abstracts of case reports also referred to as Adverse Drug Effects (ADE) corpus [10]. Their results were outperformed by a deep learning model designed by Huynh et al. that integrates recurrent neural networks with convolutional neural networks [11]. The current state-of-the-art model on ADR classification for Twitter dataset combines the BERT output with features from knowledge bases [12].

A prominent study by Nikfarjam et al. extracted ADRs from Twitter and Daily Strength, a health forum, datasets by designing a concept extraction system using conditional random fields (CRF) [13]. Peng et al. developed a nifty model that generates character-level embedding and integrates it with word vectors via an embedding-level attention mechanism, and the output is passed to a bidirectional long short-term memory networks (BiLSTM) for ADR extraction. They test the performance of the model on a Twitter dataset and a dataset prepared from PubMed abstracts [14].

The ability to buy medicines over the counters has opened rooms for non-therapeutic or abusive uses. A substantial number of studies have mined social media data for monitoring abusive consumption of drugs [15].

Moreover, social media data has been widely monitored for detecting outbreaks of diseases such as influenza, dengue fever, and others [16]. Another area of research uses social media data for learning health behaviors such as food consumption patterns, smoking cessation patterns, and substance abuse [16]. In addition to drug-related ADRs, Sullivan et al. identified potentially dangerous dietary supplements, based on the product reviews from Amazon.com [17].

Most of the previously proposed models employ traditional machine learning algorithms with feature engineering and deep neural networks. The downside of the machine learning algorithms is that they highly rely on task-specific features generation. Though the earlier deep neural technologies, such as convolutional neural networks and BiLSTMs, are task-independent models, their performances are tied with the number of training samples. Moreover, the inputs to those deep networks solely depend on word embedding, which is keen enough to neither distinguish polysemous words nor handle out of vocabulary words. Both ELMo and BERT have been pretrained on large corpora so that they have learned the structure of the trained language very well. These models can be fine-tuned with even smaller task-specific datasets for exceptional results (i.e., transfer learning). Moreover, ELMo gives word embedding based on its context rather than fixed representation.

The rest of this paper is organized as follows. Section 2 gives a brief overview of the technologies used in this study including word embedding, BiGRU, ELMo, transformer networks, and BERT. Section 3 dissects our proposed models for adverse effects classification and labeling. Section 4 presents experiments that are conducted to interpret the results of our models against various other models and state-of-the-art models. Furthermore, this section discusses their performances as well. Finally, a conclusion is given in Sect. 5.

2 Background

Our study leverages the modern deep learning innovations that excelled in NLP research such as ELMo, transformer networks, and BERT. Each method has its associated pros and cons. Our objective is to devise a new model for adverse health effects classification and labeling by integrating the above-mentioned models to subside the cons of a model by the pros of another model. A brief overview of these technologies is given below.

2.1 Word Embedding

The advent of word embedding techniques (e.g., Word2Vec [18] and GloVe [19]) has brought the NLP research to a whole new level of success. These embeddings encode the semantic properties of a word into a low-dimensional vector where words with similar meanings tend to have similar representations. A drawback of word embedding is that they give fixed vector representation for each word without accounting the context in which the word is used. For example, word embeddings are the same for the word "playing" in both of the sentences: "A song is playing on a radio" and "Children are playing cricket". These words that have multiple meanings are called "polysemy". Another downside is the inability to provide a meaningful vector representation for the words that are not present in the embedding's lookup table, also identified as out-of-vocabulary (OOV) words. These OOV words are usually given a random representation that would impact the performance of a model considerably.

2.2 Bidirectional Gated Recurrent Units (BiGRUs)

Bidirectional Gated Recurrent Units (BiGRUs) [7] are an enhanced version of vanilla recurrent neural networks (RNNs). Gated Recurrent Units (GRUs) address the adversities involved in RNN, namely the vanishing- and exploding- gradient problems to keep the long-term dependencies among the tokens in a sequence. Since GRU only remembers the tokens that it has seen in the past and the meaning of a word often depends on its surrounding context, BiGRU was introduced by combining forward and backward GRUs. Hence, BiGRU generates a vector representation for each word by utilizing both past and future contexts. GRU resembles the architecture of long short-term memory (LSTM) networks, but in GRU, some parameters of the LSTM are merged, and hence the training time is minimized.

2.3 Embeddings from Language Models (ELMo)

Embedding from Language Models (ELMo) provide an embedding for each word based on its context, and hence the embedding for a word is dynamic in different contexts [6]. The core components of an ELMo model are a character-level CNN layer and two BiLSTM layers. The character-level CNN learns the interconnections between morphemes, whereas the BiLSTM layers learn the contextual information. The ELMo language model has been trained on a large text corpus and the pretrained model with its weights is available to download in TensorFlow Hub. The vectors for each word are calculated as a function of the entire contextual words within a sentence. The input to the model can be passed as a whole sentence or a set of tokens. The model calculates the weighted sum of all three layers. Either an embedding for each token in a sentence or an average vector representation for the whole sentence can be obtained. The use of character-level CNN gives a rectified representation for each word, including the OOV tokens. The model parameters can be fine-tuned for our needs.

2.4 Transformer Networks

In the earlier days, BiLSTM or BiGRU was used for learning context-dependent vector representations for each word for sequential data processing. However, they account each context word equally to represent the meaning of a word. In reality, the meaning of a word may not be equally determined by all the surrounding words. For example, in a sentence like "the movie was good but it was too long", the meaning of the word "it" highly relies on "movie" rather than the other words. This limitation was resolved by incorporating a mechanism, called "attention" along with BiGRU. The "attention" assigns higher weights to the context words that contribute significantly to the meaning of a word while keeping the weights lower to the rest. In later stages, the researchers from the "Google Brain" contended that the recurrent relationships between sequential inputs are redundant in a paper titled "Attention Is All You Need", and hence the era of transfer learning has begun [20]. The notion behind this approach is that since the "attention" mechanism can access all the context words to learn the meaning of a word, the long-term dependencies may not be needed to be stored. A rational question that would arise is how a sequential order is kept. To do that, the authors use "positional encodings" to keep the words intact.

The transformer network is designed as an encoder–decoder architecture where the encoding component is a stack of six encoders, whereas the decoding component is a stack of decoders of the same number. All the encoders are identical in structure. The decoders are also identical in structure but slightly differ from the encoders. Each encoder has a self-attention layer and a feed-forward network. Each input token is fed to the first encoder and the output of the first encoder is forwarded to the next encoder, and so on. Each decoder has a self-attention layer, encoder–decoder attention layer, and a feed-forward network. The output of the last encoder is sent to the first decoder. The transformer network has been trained on a large corpus as a unidirectional language model, and the model is publicly available. The pretrained model can be fine-tuned for various downstream tasks (i.e., transfer learning in NLP). More importantly, since the transformer network does not need to process the data sequentially, the model accommodates parallelization. It has been proved that transformers deal with long-term dependencies better than LSTMs/GRUs [21].

In this section, two variants of the transformer architecture are presented: Google's Universal Sentence Encoder [22] and OpenAI transformer [23]. The universal sentence encoder uses the encoding part of the transformer to encode texts into high-dimensional vectors. The model outputs a 512-dimensional vector for each input text (e.g., sentences or phrases). The OpenAI transformer uses the decoder part of the transformer to generate vectors for the inputs. The model incorporates twelve decoder layers, and the encoder–decoder sublayer from the transformer decoder is discarded. These models are trained to predict the next word in a sentence using a large book corpus so that its layers are tuned to reasonably handle language. The pretrained model can be fine-tuned for downstream tasks.

2.5 Bidirectional Encoder Representations from Transformers (BERT)

As mentioned in Sect. 2.4, the OpenAI transformer gives fine-tunable pretrained model based on the transformer. A downside of the model is that the OpenAI transformer only trains a forward language model. The BERT trains a language model exploiting both forward and backward contexts [5]. The training is executed by a technique known as "masked language model" where a word is masked, and the masked word is predicted using past and future contexts. Unlike OpenAI transformers, BERT uses the encoder part of the transformer. BERT can also be fine-tuned for downstream tasks. BERT claimed their model outperformed the state-of-the-art models' performances for eleven NLP tasks at the time of its publication [5].

3 Methodology

In this section, our models that are applied to confront the objective of our study is defined. Two robust models are used for two different purposes: Binary classifier for identifying health-related adverse effects (AE) assertive sentences from biomedical texts and text labeler for extracting those AEs. For systematic demonstrations of our models, text labeler is explained first, followed by the binary classifier.

3.1 Text Labeler for Extracting Adverse Effects from Biomedical Texts

A hybrid model is built(see Fig. 1) that leverages the potentials of BERT and ELMo. ELMo handles out-of-vocabulary (OOV) tokens effectively using character-level CNN whereas BERT uses WordPiece tokenization to achieve the same. ELMo and BERT learn contextual information via BiLSTM networks and bidirectional transformers, respectively. However, several studies have shown that the multiple self-attention layers within the BERT store more information than ELMo. These two models are combined with a hypothesis in mind that the output probabilities are

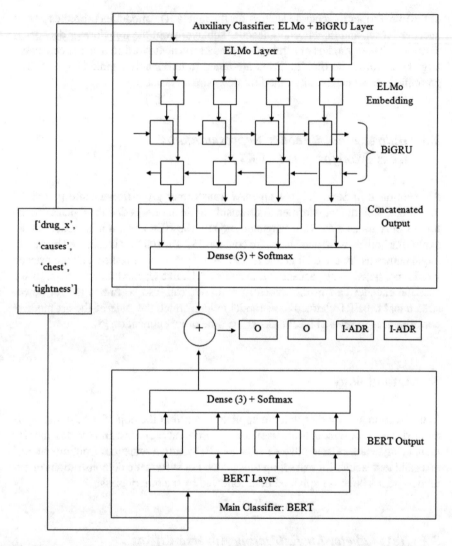

Fig. 1 Proposed hybrid text labeler for adverse effect labeling using BERT, ELMo, and BiGRU

more deterministic. Since ELMo only provides deep contextualized vector representations, a BiGRU is incorporated to learn more information about the problem in
hand. An overview of our model is as follows:

The input to the model is a set of tokens where each token holds a label complying
IOB scheme, a standard method for text labeling. An IOB scheme is adopted, where
"-PAD-", "I-AE", and "O" denotes "padding token", "part of an AE", and "outside of any AE", respectively [24]. The entire corpus is tokenized and the length of
the sentence that has the maximum number of tokens is identified (i.e., maximum
sequence length). The rest of the sentences are padded to the maximum sequence

Table 1 An example of ADR labeling

Tokens	drug_x	causes	chest	tightness	PAD	PAD
Labels	O	O	I-AE	I-AE	PAD	PAD

Table 2 Mapping original tokens to BERT specific tokens

Original tokens		drug_x	causes	chest	tightness		PAD	PAD	
BERT tokens	[CLS]	drug_x	causes	chest	tight	#ness	PAD	PAD	[SEP]
Labels	PAD	O	O	I-AE		I-AE	PAD	PAD	PAD
Original to BERT token map	0	1	2	3		5	6	7	8

length to keep all the sentences of the same length. For example, Table 1 illustrates how a sentence is labeled with a maximum sequence length of 6. Here, "chest tightness" is an AE entity.

This input tokens are fed with their corresponding labels to a BERT model. The BERT further analyzes each token and divides the tokens into subtokens if necessary, using an inbuilt tokenizer named "WordPiece". In addition to that, the BERT adds BERT specific tokens "[CLS]" and "[SEP]" at the beginning and end of the sentence, respectively. The BERT-specific tokenization for the example mentioned above is given in Table 2.

According to Table 2, only the token "tightness" has been further divided into subtokens "tight" and "##ness". The BERT gives 768-dimensional vector for each subtoken (or token if no subtoken is present). To maintain consistent labeling, a vector representation is chosen among "tight" or "ness" to represent the token "tightness". Our preliminary studies showed that keeping the last piece of the tokenized term (i.e., using the BERT vector for the last subtoken to represent an original token) gives better results, and hence our model utilizes last-level subtoken (i.e., using the vector representation of "##ness" to represent "tightness" rather than using "tight" to represent "tightness"). As shown in Fig. 1, the sequential output of the BERT is forwarded to a 3-neuron dense layer. As an auxiliary classifier, the input tokens are fed to an ELMo layer as well. ELMo provides a 1024-dimensional vector for each token. A BiGRU analyzes these inputs and forwards the outputs to a three neurons dense layer. The output probabilities of the BERT and ELMo are summed up to provide a final value.

3.2 Binary Classifier for Identifying Adverse Effects in Biomedical Texts

As can be seen from Fig. 2, the architecture of the binary classifier is almost identical to the text labeler, but there are slight changes. Here, the output of the BERT layer

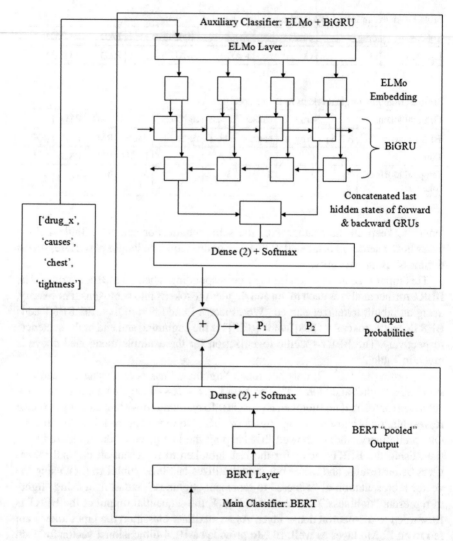

Fig. 2 Proposed binary classifier for identifying health-related adverse effects assertive texts

returns a "pooled output" rather than an individual embedding for each token. The "pooled output" gives the hidden state vector representation of the [CLS] token (i.e., the first token). Because BERT is bidirectional, the [CLS] token encodes the contextual information of the entire tokens, and hence it gives a meaningful representation of the whole sentence. The output of the classifier returns two probability values for the whole sentence: positive and negative class probabilities.

4 Results and Discussion

In this section, a set of experiments is explained that is conducted to evaluate the performances of our models. These experiments are limited specifically for adverse effects that are caused by prescribed medications under regular doses, also known as adverse drug reactions (ADRs). Our experiments are carried out on a manually annotated Adverse Drug Effects (ADE) dataset, which includes sentences collected from case studies that are available on PubMed [25]. The authors of this dataset label each sentence as an "ADR instance" or a "Non-ADR instance". Moreover, they annotate the ADRs in each ADR instance. A summary of the datasets is given in Table 3. Some basic preprocessing steps such as lowercasing and punctuations removal are performed, and some of the text preparations carried out by Peng et al. on PubMed dataset for ADR labeling are also adopted [14].

4.1 Performance of the Binary Classifier

The first set of experiments use ELMo embeddings. The input to the model could be either an untokenized sentence or a set of tokens. The output could be received as a fixed mean pooling of all contextualized word representations or a sequence of contextual vectors for each token. Model C1 and Model C2 use a feed-forward network (FFN) with ELMo embedding, however Model C1 and Model C2 use "mean pooled" output, and "sequential output" of the ELMo embedding, respectively. Model C3 applies BiGRU for "sequential output" of the ELMo embedding.

In the next set of experiments, the pretrained BERT model is tweaked to learn how well a better performance can be achieved. Like ELMo, input to the BERT model could be either a whole sentence or a set of tokens. There is an option to get the output as a "pooled output" or a "sequence output". The "pooled output" returns the hidden state of the BERT-specific first token [CLS] as a single vector representation for the entire sentence. Since BERT is bidirectional, the [CLS] token can be a meaningful representation of the whole sentence. In Model C4, 'sequence-output' is obtained and it is passed to a 256-neuron dense layer which is flatten before feeding into a classification layer. In Model C5, the "pooled output" is passed to a 2-layer FFN. Model C6 is our proposed model for this task (please refer to Sect. 3.2 for more details).

Table 3 Summary of the datasets

Dataset	Total number of instances
PubMed ADE (binary classification)	23,516 ADR Positive: 6821 ADR Negative: 16,695
PubMed ADE (text labeling)	6821

Table 4 Performance matrices for binary classifications

Model name	Model components	Precision	Recall	F1 score
Model C1	ELMo (sentence-level embedding) + FFN	0.854	0.749	0.798
Model C2	ELMo + FFN	0.875	0.793	0.832
Model C3	ELMo + BiGRU	0.879	0.888	0.883
Model C4	BERT ("sequence_output") + FFN	0.903	0.895	0.899
Model C5	BERT ("pooled_output") + FFN	0.904	0.902	0.903
Model C6	BERT ("pooled_output") + (ELMo + BiGRU)	0.902	0.911	**0.906**
State-of-the-art model [11]	Convolutional neural network	0.850	0.890	**0.870**

The 5-fold cross-validated results of the standard performance matrices are reported: precision, recall, and $F1$ score. Table 4 compares the outcomes of the model variants.

According to Table 4, Model C2 performs much better than Model C1. Both models use ELMo with an FFN; however, Model C1 leverages the mean pooled value of the entire contextual vectors, whereas Model C2 concatenates the contextual vectors of each token. This performance disparity brings to an inference that representing the entire context of a sentence by a single vector causes loss of information. It is to be noted that the rest of the models use ELMo with token-level embedding. There is a significant performance difference between Model C2 and Model C3. Though ELMo gives contextualized meaning for each token, it does not retain the positions of them. As the primary contribution of ELMo is distinguishing polysemous words and handling OOV tokens, along with ELMo a classifier is needed to understand the meaning of the entire sentence. Unlike FFN that does not account positional embedding, BiGRU effectively learns contextual information and carries long-term dependencies in both forward and backward directions. Thus, ELMo with BiGRU achieves better results than ELMo with an FFN. There is no substantial difference in performance between Model C4 and Model C5. Since the "pooled output" carries a lesser number of parameters, the Model C5 trains faster than Model C4 and hence "pooled output" is utilized in Model C6. Although it is evident that the BERT performs better than ELMo, our results indicate that incorporating the ELMo model as an auxiliary classifier and combining its output with the BERT's output improves the overall performance (Model C6). This method could help the BERT model to resolve any perplexities in determining a label for a sentence. As indicated in Table 4 (in bold), our proposed model (Model C6) provides the best F1 score of 0.906 whereas the current state-of-the-art method performs with an F1 score of 0.870. That means, our proposed model (Model C6) outperforms the current state-of-the-art model for ADR classification on the same benchmark dataset by 3.6% [11].

Table 5 Performance matrices for text labeling

Model name	Model components	Precision	Recall	F1 score
Model L1	ELMo + FFN	0.802	0.965	0.876
Model L2	ELMo + BiGRU	0.876	0.941	0.907
Model L3	BERT + FFN	0.911	0.920	0.916
Model L4	BERT + (ELMo + BiGRU)	0.904	0.947	**0.925**
State-of-the-art model [14]	Attentive word- and character-level embedding + BiGRU	0.867	0.948	**0.906**

4.2 Performance on the Text Labeler

The models assessed for text labeling are almost similar, except the classification label provides a sequential output indicating a label for each token rather than a fixed output for the entire sentence. A standard evaluation matrix is adopted for this task, named "approximate matching" from Cocos et al. [24]. Table 5 presents the findings of the model variants for ADR labeling. Since the models are nearly identical to the models that are discussed for classifications, the reasons behind the discrepancies between the results are easily deduced. Since it is a labeling task, sentence-level embedding is not suitable. As given in Table 5 (in bold), the best F1 score value of 0.925 is obtained for our proposed model (Model L4) while the state-of-the-art model provides an F1 score of 0.906. Therefore, our model for ADR labeling (Model L4) has overcome the current state-of-the-art model on the same task by 1.9% [14].

Because ELMo and BERT models are pretrained using large Wikipedia and book corpus and the datasets used are written in formal English, these models perform competently. However, the BERT and ELMo models may not achieve the same results on informal datasets such as social media texts. To cope with colloquial texts, these models should be pretrained on large datasets of similar domains.

5 Conclusion

Although healthcare has extended its frontiers to a whole new level, many diseases can only be controlled rather than curing them completely. Treatments available for disease do not account the anatomy and physiology of people with different background, and hence a treatment that works for one group of people may not work for another or cause serious adverse effects. To monitor the safety of medication after its release into the market, healthcare professionals and drug manufacturers use clinical reports, case studies from health databases, and more importantly, health-related personal experiences on social media. Since manual processing of these data is infeasible, the focus has been directed toward machine learning with natural language processing.

Our research identifies sentences from health-related case reports from a biomedical corpus that mention adverse effects (AEs) and extract them. Two novel deep learning models are introduced that exploit the potentials of BERT and ELMo models: a binary classifier for distinguishing AE assertive texts from others and a text labeler for extracting the AEs. Both models are almost identical in nature. An input sentence is fed to an ELMo and a BERT independently. The BERT output is directly sent to a dense layer for classification, whereas the ELMo embeddings are further analyzed by a BiGRU before processed by a classification layer. The output probabilities of both BERT and ELMo are summed to get the final value. The performance of our models is determined specifically for drug-related AEs known as adverse drug reactions (ADRs) on a benchmark dataset known as ADE corpus. The binary classifier obtains an $F1$ score of 0.906 which is 3.6% better than the current state-of-the-art model. The text labeler yields and $F1$ score of 0.925 excel its current state-of-the-art model by 1.9%.

In this study, biomedical texts are mainly focused on which are usually written in formal English and are well formulated. Even though our models are tested for ADRs, these models can be accommodated for any research area with similar domains. In future, our research will be expanded on health-related user-generated contexts such as social media data. Since the derivatives of the transformer networks are generally trained on Wikipedia and book corpus, they will be pretrained on large social media corpus and used for our motives.

References

1. W. H. Organization: Countries are spending more on health, but people are still paying too much out of their own pockets (2019). [Online]. Available: https://www.who.int/news-room/detail/20-02-2019-countries-are-spending-more-on-health-but-people-are-still-paying-too-much-out-of-their-own-pockets. Accessed: 13 Sep 2019
2. C. Data: Sri Lanka Health Expenditure as % of GDP (2017). [Online]. Available: https://www.ceicdata.com/en/sri-lanka/health-statistics/health-expenditure-as--of-gdp. Accessed: 13 Sep 2019
3. McGregor, A.: Why medications often have dangerous side effects for women? TEDx (2014). [Online]. Available: https://www.ted.com/talks/alyson_mcgregor_why_medicine_often_has_dangerous_side_effects_for_women. Accessed: 13 Sep 2019
4. Sarker, A., et al.: Utilizing social media data for pharmacovigilance: a review. J. Biomed. Inform. **54**, 202–212 (2015)
5. Devlin, J., Chang, M.-W., Lee, K., Toutanova, K.: BERT: pre-training of deep bidirectional transformers for language understanding. no. Mlm, Oct 2018
6. Peters, M. E., et al.: Deep contextualized word representations, Feb 2018
7. Cho, K.: et al.: Learning phrase representations using RNN encoder-decoder for statistical machine translation, June 2014
8. Leaman, R., Wojtulewicz, L.: Towards internet-age pharmacovigilance: extracting adverse drug reactions from user posts to health-related social networks, pp. 117–125, July 2010
9. Nikfarjam, A., Gonzalez, G.H.: Pattern mining for extraction of mentions of adverse drug reactions from user comments. AMIA Annual Symposium Proceedings Archive, vol. 2011, pp. 1019–1026, Jan 2011

10. Sarker, A., Gonzalez, G.: Portable automatic text classification for adverse drug reaction detection via multi-corpus training. J. Biomed. Inform. **53**, 196–207 (2015)
11. Huynh, T., He, Y., Willis, A., Stefan, R.: Adverse drug reaction classification with deep neural networks. In: Proceedings of 26th International Conference on Computational Linguistics, pp. 877–887
12. Chen, S., Huang, Y., Huang, X., Qin, H., Yan, J., Tang, B.: HITSZ-ICRC : a report for SMM4H shared task 2019-automatic classification and extraction of adverse drug reactions in tweets, pp 47–51 (2019)
13. Nikfarjam, A., Sarker, A., O'Connor, K., Ginn, R., Gonzalez, G.: Pharmacovigilance from social media: mining adverse drug reaction mentions using sequence labeling with word embedding cluster features. J. Am. Med. Inform. Assoc., pp. 671–681, Mar 2015
14. Ding, P., Zhou, X., Zhang, X., Wang, J., Lei, Z.: An Attentive neural sequence labeling model for adverse drug reactions mentions extraction. IEEE Access **6**, 73305–73315 (2018)
15. Sarker, A., et al.: Social media mining for toxicovigilance: automatic monitoring of prescription medication abuse from Twitter. Drug Saf. **39**(3), 231–240 (2016)
16. Paul, M.J., et al.: Social media mining for public health monitoring and surveillance. Biocomputing **2016**, 468–479 (2016)
17. Sullivan, R., Sarker, A., O'Connor, K., Goodin, A., Karlsrud, M., Gonzalez, G.: Finding potentially unsafe nutritional supplements from user reviews with topic modeling. Pac. Symp. Biocomput. **21**, 528–539 (2016)
18. Mikolov, T., Sutskever, I., Chen, K., Corrado, G., Dean, J.: Distributed representations ofwords and phrases and their compositionality. Adv. Neural Inf. Process. Syst. 1–9 (2013)
19. Hanson, E.R.: Musicassette interchangeability. the facts behind the facts. AES J. Audio Eng. Soc. **19**(5), 417–425 (1971)
20. Vaswani, A., et al.: Attention is all you need. Adv. Neural Inf. Process. Syst.**2017**, 5999–6009
21. Lakew, S.M., Kessler, F.B., Kessler, F.B., Federico, M., Srl, M.M. T., Kessler, F.B.: A comparison of transformer and recurrent neural networks on multilingual neural machine translation, pp. 641–652 (2018)
22. Cer, D., Yang, Y., Kong, S., Hua, N., Limtiaco, N.: Universal Sentence Encoder
23. Radford, A., Salimans, T.: Improving language understanding by generative pre-training. OpenAI, 1–12 (2018)
24. Cocos, A., Fiks, A.G., Masino, A.J.: Deep learning for pharmacovigilance: recurrent neural network architectures for labeling adverse drug reactions in Twitter posts. J. Am. Med. Informatics Assoc. **24**(4), 813–821 (2017)
25. Gurulingappa, H., Rajput, A.M., Roberts, A., Fluck, J., Hofmann-Apitius, M., Toldo, L.: Development of a benchmark corpus to support the automatic extraction of drug-related adverse effects from medical case reports. J. Biomed. Inform. **45**(5), 885–892 (2012)

An Intelligent Framework for Online Product Recommendation Using Collaborative Filtering

Ganesh Chandrasekaran and D. Jude Hemanth

Abstract Recommendation systems have become a vital area of research in recent times. These recommendation systems are very much needed for e-commerce applications to identify the products liked by a customer which helps the companies to promote product sales and improve their product quality. It also helps the users to arrive at the purchasing decision without reading the online reviews about the product. The key idea behind the proposed work is to analyze the user preference for the products from the online data by employing the collaborative filtering-based recommendation framework. The concept of collaborative filtering is best suited for recommendation systems involving a large set of product users. It generates a user-item matrix and finds the list of products liked by the individual users. It gives prediction regarding the product that a user could buy in the future and also recommends the products which are liked by the customers who have similar interests. It gives a comparative analysis in terms of performance metrics and accuracy of different collaborative filtering techniques.

Keywords Recommendation systems · Sentiment analysis · Collaborative filtering · Machine learning

1 Introduction

The extraordinary growth of the e-commerce industry has increased the number of products that are available online. The users get large collections of products, and they find it difficult to choose the correct product which satisfies their needs. Recommender systems are developed to make use of the online data and help users to choose the right products. It aims to predict the items or products that a user is interested in [1]. The companies use different strategies to promote their product, and identifying the user interest is very important to increase sales. The performance of recommendation systems depends on the quality of reviews that it uses to give the

G. Chandrasekaran · D. J. Hemanth (✉)
Department of ECE, Karunya Institute of Technology and Sciences, Coimbatore, India
e-mail: judehemanth@karunya.edu

© The Editor(s) (if applicable) and The Author(s), under exclusive license
to Springer Nature Singapore Pte Ltd. 2021
S. Shakya et al. (eds.), *Proceedings of International Conference on Sustainable Expert Systems*, Lecture Notes in Networks and Systems 176,
https://doi.org/10.1007/978-981-33-4355-9_20

249

predictions. Nowadays, the customers read the online reviews of the product and get to know what the existing customers of the same product feel before buying it. These online reviews have a huge impact on the sales of the product and the company's growth. [2, 3]. If the user's opinions on a product are not studied, it is difficult to design a product that will be liked by many customers. Many online sites use a recommendation system to understand user opinion like: Amazon (Books review), Netflix (Movie review), Flipkart (Products rating), Trip Advisor (Hotel review), etc. Due to the abundant availability of online sentiment data [4], it is required to filter data using some filtering method. The collaborative filtering is one such approach which uses the idea that people get the finest recommendations from other users who have similar preferences.

The collaborative filtering methods will improve the way in which the recommendation systems work. If a customer A likes most of the products liked by customer B, then it is not good to recommend a product that is not liked by customer B to A and vice-versa. This is because they (customer A and B) have similar taste, i.e., they like the features of the product which make them interested in it. This paper deals with the design of a recommendation system that will take the product purchase data from the customer and will tell whether the user likes the product or not. The system also predicts the products which the customer will be buying in the future.

2 Related Works

The concept of sentiment analysis has been applied in many areas like e-commerce and social media applications. The wordnet was used to classify the text with the assumption that words with similar polarity have the same orientation [5]. This lexicon approach was used in identifying sentiments from microblogs and user reviews [6]. They have drawbacks when applied to big datasets. To determine the polarity (positive, negative, and neutral) of movie reviews on imdb.com movie dataset, the researchers [7] have used NB, SVM, and other machine learning techniques. The SVM classifier performed well when compared to others and its classification accuracy was about 86.5%. A hybrid algorithm with the combination of Naïve Bayes (NB) and maximum entropy (ME) was used for the analysis of the same dataset which gave good results with high accuracy [8]. The SVM approach was used successfully in perform opinion mining [9], and it was used along with some weighting procedure for sentiment classification [10]. Collaborative filtering is one of the finest techniques for recommendation systems. It is also preferred when the number of users is higher than the number of products/items for which the recommendation has to be given. The memory-based collaborative filtering technique is successfully applied to the movie lens and Netflix datasets to study the sentiment of movie reviews by the researchers [11]. Several works used model-based collaborative filtering, and one such work is opinion mining with model-based collaborative filtering in the online newspaper [12].

3 Background

3.1 Phases of Recommendation System

The recommendation system follows certain steps which are termed as phases. and it is an iterative process. They are: data acquisition, learning, and recommendation phase.

3.1.1 Data Acquisition and Learning Phases

It involves the collections of data to provide the product recommendation. It is essential to get the attributes from these collected data regarding the various products and build the recommendation system according to that. Before the start of the recommendation phase, one has to get an idea on the data to give a good quality recommendation. The learning involves the application of filtering techniques and algorithms to the dataset.

3.1.2 Recommendation Phase

It gives the recommendation of products to the user by analyzing the data acquired and employs an appropriate learning algorithm. This gives an output whether the customer likes the product or not and also recommends a set of the product that would be bought by the same customer in the future by comparing his likes with the other users.

3.2 Types of Recommender Systems

Recommender systems are based on two entities: user and the product. The users buy the products based on their preferences, and understanding their preferences from the data is important for any application. It is possible to recommend a product to the user based on the similarity among the users of the product. Recommendation systems can be classified [13] into three categories as shown in Fig. 1.

3.2.1 Content Based

The concept of content-based filtering deals with the evaluation of attributes of the products to give predictions. It takes the purchase history of the users and their interests [14]. The users give their preferences for a product in the form of ratings or

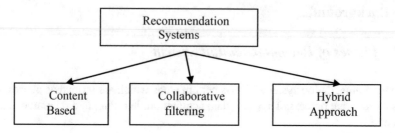

Fig. 1 Categories of recommendation systems

by purchasing a particular item frequently. This technique is used mostly in the case of news and Web pages which involve textual information [15].

3.2.2 Collaborative Filtering

This filtering technique is also termed as social filtering which filters information by making use of recommendations given by the other users of the product [16]. Collaborative filtering (CF) techniques can be classified into two categories: Memory based and model based [17]. Memory-based CF uses historical data of the user to give the prediction and provide recommendations for him. They are classified into user and item-based approaches [18]. Model-based CF uses machine learning techniques to give the prediction based on the known data (Fig. 2).

(A) **User-based and Item-based Collaborative filtering**

    User-based CF: It is based on the calculation of similarity of the user under consideration and the existing users. The neighbors of the users are selected such that the other users have high similarity with the current user. The neighbor's ratings can be used to predict the rating of the current user. The similarity measures like cosine-based similarity, Pearson similarity, and adjusted cosine similarity can be used. This approach has limitations like sparsity, scalability, and cold-start issues.

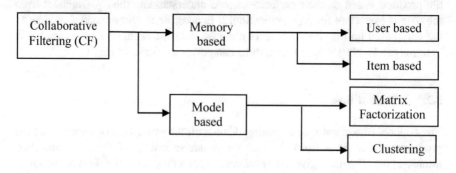

Fig. 2 Collaborative filtering techniques

Item-based CF: It makes use of similarity among items rather than similarity among users like user-based CF. The similarity among items are calculated using the similarity measures, and prediction is done by a weighted average method.

(B) Matrix factorization and Clustering

Matrix factorization is the unsupervised learning which reduces the dimensionality. It decomposes the user-item matrix into two lower dimensionality matrices. One of the key advantages of this approach is that it eliminates the data scarcity problem [19]. It was effectively used in Netflix prize challenge and thus became popular. Clustering is one of the data mining methods which reduce the computing resources and time for the recommendation task. Input data is divided into several partitions based on similarities.

3.2.3 Hybrid Approach

This technique combines the previous two techniques discussed, i.e., content-based filtering and collaborative filtering to increase the overall performance of the recommender system. This technique can be further classified as: weighted hybridization, mixed hybridization, switched hybridization, etc., based on their operation.

4 Proposed Methodology

See Fig. 3.

4.1 Amazon Products Review—Dataset

The data used for the experiment is the Amazon product dataset which consists of customer reviews on various products. The dataset of five categories of products like Automotive (8,523 reviews), Health (6523 reviews), Beauty (7582 reviews), Office products (8671 reviews), and Baby products (6784 reviews) have been used. Five reviews per user have also been collected for a specific product. The dataset also contains the ratings given by the customer for each product along with the review.

Fig. 3 Block diagram of the proposed methodology

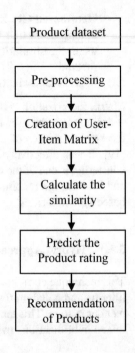

4.2 Data Preprocessing

Raw data regarding the user reviews and ratings cannot be directly used for the analysis. The user reviews about a product is tokenized, i.e., splitting the sentences into tokens. Then, redundant words and start and stop words are removed.

4.3 User-Based Collaborative Filtering

It uses a user-item matrix which contains the list of items and user rating for the items for calculating the similarity among the users. The users who are in the neighborhood of the current user, i.e., who are similar to the current user, are calculated using the Pearson correlation. The rating is predicted from these neighbors, and top items are recommended to the user under consideration.

Pearson Correlation

$$\text{Sim}(cu, v) = \frac{\sum_{u=1}^{k}(r_{cu,i} - \hat{r}_{cu})(r_{v,i} - \hat{r}_{cu})}{\sqrt{(r_{cu,i} - \hat{r}_{cu})^2}\sqrt{(r_{v,i} - \hat{r}_v)^2}} \tag{1}$$

where 'cu' is the current user.

'v' is another user.

'i' is the item.

'r' is the rating.

After determining the similarity measure between each user and the current user, 'K' users who share the similarity with the current user 'cu' are determined. The rating for the items/ products purchased by neighborhood of users and not purchased by the user 'cu' is found by,

$$P_{cu,i} = \hat{r}_a + \frac{\sum_{u=1}^{k} \left(r_{cu,i} - \hat{r}_{cu}\right) * \text{sim}(cu, v)}{\sum_{u=1}^{k} \text{sim}(cu, v)} \tag{2}$$

The items with top N ratings are recommended to the user 'cu.'

4.4 Item-Based Collaborative Filtering

It is based on the similarity between item preferences and not user preferences. The top N items are recommended based on the similarity value calculated for the items/products. The past purchase history of the current user 'cu' is collected. With this information, an item-item matrix is constructed. Items 'i' which are most similar to the items 'j' rated by the user in the past are taken, and the Pearson correlation computes the similarity between the items.

$$\text{Sim}(i, j) = \frac{\sum_{u=1}^{k} \left(r_{cu,i} - \hat{r}_i\right)\left(r_{cu,j} - \hat{r}_j\right)}{\sqrt{\left(r_{cu,i} - \hat{r}_i\right)^2}\sqrt{\left(r_{cu,j} - \hat{r}_j\right)^2}} \tag{3}$$

After computing the similarity, the top N items are recommended to the user based on the highest similarity value calculated.

5 Results and Discussion

Recommendations of products to a customer are generated using their ratings and sentiment scores. The evaluation of the recommended products for the users is done by dividing the product dataset into training and testing samples. The training data consists of 80% of the samples, and the test data consists of the remaining 20% of samples. The user-based collaborative filtering was applied to the dataset during the training. The generated predictions were evaluated with the actual value using the following performance metrics:

Mean Absolute Error (MAE)

It calculates the average value of the absolute difference between the predicted value and the true or the rated value

$$\text{MAE} = \frac{\sum_{i,j} |p_{i,j} - r_{i,j}|}{n} \tag{4}$$

where $p_{i,j}$ is the predicted rating for the user i on item j and $r_{i,j}$ is the actual rating.

Normalized Mean Absolute Error (NMAE)

It estimates the overall deviations between the predicted and measured values, and it expresses error as a percentage of full scale.

$$\text{NMAE} = \frac{\text{MAE}}{r_{\max} - r_{\min}} \tag{5}$$

where $r_{\max}$ and $r_{\min}$ are the maximum and minimum values of user ratings.

Root Mean Square Error (RMSE)

It is a quadratic scoring rule which computes the average of squared differences between predicted observation and the actual observation (Table 1).

$$\text{RMSE} = \sqrt{\frac{1}{n} \sum_{j=1}^{n} (p_{i,j} - r_{i,j})^2} \tag{6}$$

Accuracy

It is a measure of correctness of the classifier, and it is defined as the ratio between the number of correct predictions and the total number of input samples (Fig. 4).

$$\text{Accuracy} = \frac{\text{Number of correct predictions}}{\text{Total number of predictions}}$$

Table 1 Performance metrics calculation for User Based – Collaborative Filtering (UB-CF)

S. No	Product category	MAE	NMAE	RMSE	Accuracy (%)
1	Automotive products	0.65	0.62	0.68	84.3
2	Health products	0.58	0.55	0.52	90.3
3	Beauty products	0.52	0.51	0.58	90.6
4	Office products	0.65	0.62	0.68	86.7
5	Baby products	0.53	0.51	0.57	91.4

Table 2 Performance metrics calculation – Hybrid Approach

S. No	Product category	MAE	NMAE	RMSE	Accuracy (%)
1	Automotive products	0.52	0.48	0.50	83.7
2	Health products	0.42	0.45	0.42	92.8
3	Beauty products	0.38	0.36	0.32	94.6
4	Office products	0.50	0.42	0.48	90.3
5	Baby products	0.42	0.43	0.47	95.2

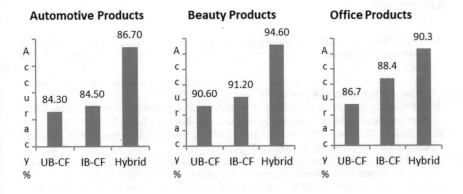

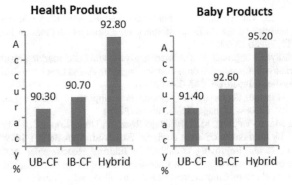

Fig. 4 Accuracy (in %) comparison of user-based (UB-CF), item-based (IB-CF), and hybrid techniques for various products recommendation

6 Conclusion

In this paper, the collaborative filtering-based product recommendation system is presented. The purpose of the work is to help the users in choosing the right product and assist them in taking the buying decision. It will also help the product manufacturers to increase their sales and plan accordingly. The proposed technique takes product reviews in the form of text and gives prediction using collaborative filtering techniques. Comparison listed in Table 2 justifies that the hybrid technique yields the best result for this application. The product dataset is first preprocessed, and sentiment scores are calculated for each review.

Acknowledgements The proposed work has achieved good accuracy for different product categories, and the future work will be to handle the sarcastic reviews and ratings by the user which would affect the accuracy.

References

1. Adomavicius, G., Tuzhilin, A.: Toward the next generation of recommender systems: A survey of the state-of-the-art and possible extensions. IEEE Trans. Knowl. Data Eng. **17**(6), 734–749 (2005)
2. Senecal, S., Nantel, J.: The influence of online product recommendations on consumers' online choices. J. Retail. **80**, 159–169 (2004)
3. Chevalier, J., Mayzlin, D.: The effect of word of mouth on sales: online book reviews, NBER Working Paper Series, National Bureau of Economic Research, USA (2003)
4. Dellarocas, C.: The digitization of Word-of-Mouth: promise and challenges of online feedback mechanisms. Manage. Sci. **49**(10), 1407–1424 (2003)
5. Agarwal, A., Xie, B.,Vovsha, I., Rambow, O., Passonneau, R.: Sentiment analysis of twitter data. In: Annual International Conferences. Columbia University, New York (2012)
6. Turney, P.D., Litman, M.L.: Measuring praise and criticism. ACM Trans. Inf. Syst. **21**(4), 315–416 (2002)
7. Pang, B., Lee, L., Vaithyanathan, S.: Sentiment classification using machine learning techniques. Proceedings of the ACL- Conference on Empirical Methods in Natural Language Processing **10**, 79–86 (2002)
8. Appel, O.: A hybrid approach to sentiment analysis with benchmarking results. In: Proceedings of International Conference on Industrial Engineering and Other Applications of Applied Intelligent Systems, Japan, pp. 242–254 (2016)
9. Severyn, A., Moschitti, A., Uryupina, O., Plank, B., Filippova, K.: Multi lingual opinion mining on youtube. Inf. Process Manage. **52**(1), 46–60 (2016)
10. Saleh, M.R., Martín-Valdivia, M.T., Montejo-Ráez, A., Ureña-López, L.A.: Experiments with SVM to classify opinions in different domains. Exp. Syst. Appl. **38**(12), 14799–14804 (2011)
11. Bobadilla, J., et al.: Improving collaborative filtering recommender system results and performance using genetic algorithms. Knowledge Based Systems. **24**(8), 1310–1316 (2011)
12. Claypool, M., et al.: Combining content-based and collaborative filters in an online newspaper. In: Proceedings of ACM SIGIR Workshop on Recommender Systems. Citeseer (1999)
13. Isinkaye, F.O., Folajimi, Y.O., Ojokoh, B.A.: Recommendation systems: principles, methods and evaluation. Egypt. Inf. J. **16**, 261–273 (2015)
14. Das, D., Sahoo, L., Datta, S.: A survey on recommendation system. Int. J. Comput. Appl. **160**(7), 6–10 (2017)
15. Pereira, N., Varma, S.: Survey on content based recommendation system. Int. J. Comput. Sci. Inf. Technol. (IJCSIT) **7**(1), 281–284 (2016)
16. Aggarwal, C.C.: An introduction to recommendation system. In: Recommender Systems: The Textbook. Springer, Switzerland (2016)
17. Tang, G., Zhang, Z., Kang, G., Liu, J., Yang, Y., Zhang, T.: Collaborative web service quality prediction via exploiting matrix factorization and network map. IEEE. Trans. Netw. Serv. Mange. **13**, 126–137 (2016)

18. Adomavicius, G., Tuzhilin, A.: Toward the next generation of recommender systems: A survey of the state of the art and possible extensions. IEEE Trans. Knowl. Data. Eng **17**, 734–749 (2005)
19. Bokde, D., Girase, S., Mukhopadhayay, D.: Role of matrix factorization model in collaborative filtering algorithm: a survey (Online) (2014)

Novel Method to Analyze and Forecast Social Impact on Macro- and Micro-Economies Using Social Media Data

Shakthi Weerasinghe, Aeshana Udadeniya, Nisal Waduge, Randilu de Zoysa, and Upeksha Ganegoda

Abstract Stock markets are often seen to be the backbone of an economy, the variations in which could be indicative of the prevailing condition of the economy. Although the high vulnerability to external factors, it is often a natural challenge to quantify the impact. The research aims to identify and isolate the true performance of the listed companies by adjusting the impact from the external events—which is quantified using a novel approach considering social media as a platform. The resultant approach is highly generalizable and could be applicable for any industry which is highly significant compared to previous work. Hence, this could be seen used in the perspective of stakeholders such as potential investors, clients, and governments as well.

Keywords Stock market prediction · Social media analysis · Natural language processing · Performance isolation · Macro-economic forces

1 Introduction

The economy of a country often dictates the purchasing power of its citizens and hence the operational capacity of its industries and capability to serve the global market which enables a healthy flow of finances. Stock markets are often seen as the

S. Weerasinghe (✉) · A. Udadeniya · N. Waduge · R. de Zoysa · U. Ganegoda (✉)
Faculty of Information Technology, University of Moratuwa, Katubedda, Sri Lanka
e-mail: shakthi.14@itfac.mrt.ac.lk

U. Ganegoda
e-mail: upekshag@uom.lk

A. Udadeniya
e-mail: aeshana.14@itfac.mrt.ac.lk

N. Waduge
e-mail: nisal.14@itfac.mrt.ac.lk

R. de Zoysa
e-mail: randilu.14@itfac.mrt.ac.lk

S. Shakya et al. (eds.), *Proceedings of International Conference on Sustainable Expert Systems*, Lecture Notes in Networks and Systems 176,
https://doi.org/10.1007/978-981-33-4355-9_21

central point in which these factors are visible at both the micro- and macro-levels of the economy considering many case studies used by economists to elaborate on the cause and effect of internal and external factors to cooperates—largely known as the market forces. Hence, Silva et al. [1] indicate a 0.905 correlation between the share price index (SPI) and the gross domestic product (GDP), while it also indicates that 81.9% of the economic growth is determined by the stock market performance considering the Colombo Stock Exchange (CSE) in the study. However, these forces are largely unpredictable and vary that the stock brokering is famously known to be one of the most knowledge-intensive trades, which requires manual man-hours in disposing to establish a considerably accurate prediction to the stock market, upon which the investors are likely to make capital investments on positively performing companies.

However, it is highly unlikely to access statistics which indicates *pure performances* of companies due to the effects of market forces in the resulting available data—which stipulates the need to isolate the actual performance of companies less of external factors to obtain and verify the actual health of the organization in a stock market perspective. This is in contrast to usual prediction strategies such as the technical analysis [2] through which it is attempted to forecast stock performance solely on the historical data—a common error observable in many kinds of research as it violates the fundamental principles of economics, ideally described in the random walk theory [3] and efficient market hypothesis [4]. These facts are indicated in the previous study of this work [5], which underlines the deficiencies of existing systems which are addressed by the approach of isolating the impact of market forces, followed in this research work.

Among the topmost contributing industries to GDP, the tea plantation industry was selected as a sample for the research considering its 15% share in the Sri Lankan economy [6] and pure dependency on demand–supply forces by which the prices are settled compared to price arbitration policies. This approach adopts a systematic procedure which establishes a novel framework to predict organizational performance based on indications of probable events in the future which could be considered a major contribution to the field. Hence, the framework has been evaluated to have produced highly accurate forecasts based on which a highly lucrative and effective methodology to automating stock trading is suggested in further work.

The rest of the paper is organized as follows. Section 2 reviews the related work in the area. Section 3 provides an insight to the design and analysis of the work, while Sect. 4 describes the methodology in which the research was executed followed by the details of the results obtained by execution of the experiments, critical evaluation of the results, and conclusion and further work in Sects. 5, 6, and 7, respectively.

2   Related Work

An extensive review of previous work had been conducted in [5], in relation to the work described in this paper. This outlined the approaches in which typical stock predictions are made based on the historical variations of stocks and that as a binding and definitive factor, which was argued contrarily against common economical models and theories. Evidence can be further justified in [7] where the researchers have indicated the deficiencies of adopting linear regression models against the experimental results using auto-regressive moving average model (ARIMA) through which the hypothesis of statistical independence of stock variation from external factors could be nullified.

However, as it is indicated in [8], it could be argued based upon the results that predictions can be improved by adjusting for the effects of subtle sentimental variations in society atop an ARIMA model. Hence, it has been established that an increase in search volumes of keywords related to some topics tend to precede falls in stock prices—a similar baseline assumption considered in many works. Although, in relation to [9], it could be seen that a common pit hole in the area of study is the assumption of a generic time series variation for stocks—in which the residuals are attributed to the event specifically not accounting for the exponential decay of hype, which implicitly amounts for a common time-boxed trend, differentiated among events by the news value and sentiment—which are commonly relevant to the consensus of the society in that particular point of time. In this respect, the application of Kaplan Meier estimator to quantify the level of persistence of lagged correlation of search trends and stock market index in [10] could be considered remedial.

This is justifying evidence that multivariate time series models common in research work are vulnerable to residual events not captured in the model, especially considering the long-term variations in the state of each external factor. This is indicative even in the ARIMA (0, 1, 0) model given in (1), which indicates a random walk with drift—the ideal case for describing stock variations.

$$X_t = c + X_{(t-1)} + \varepsilon_t, \tag{1}$$

However, it is also established in [10] that economic or stock variations do not solely vary on social consensus as would be portrayed in proportional variations in Google Trends or search volumes of relevant keywords, which essentially prove the above further, as the current state is determined by the immediate past and the residual adjustment. This is further evident in the results and the conclusions were drawn by the researchers in [11, 12] which considers tweet for analysis of social consciences in which a domain-based multivariate correlation analysis had been performed. An important aspect to consider in this respect, however, is the language complexity in expression—where grammatical patterns, such as passive constructs, are likely to affect traditional sentiment results such as a bag of words (BOW) indicating impacts contrary to a general consensus. Hence, in a highly unbiased tweet set, it is highly likely the aggregate effect would be neutral—a misleading figure toward a skewed social opinion in response to a particular incident.

It should also be noted, [13, 14] discusses machine learning approaches for stock predictions with special consideration for polynomial regression techniques such as radial basis function (RBF), sigmoid, and the LSboost random forest (LS-RF) regression, however failing to derive conclusions on an optimized solution, although noting the complexity of accurate predictions which is a significant remark in the area of study as explained above. This could also be the reason that the researchers in the latter work have excluded the concept of random walk—a strategy that could be seen to have adopted to for the sole comparison purposes of regression models, which therefore had derived only considerable results.

However, [12] contrary has been theoretically successful in adopting the macroeconomic effects in accordance with generally accepted theories for economic variations although in highly infrastructural respect of a big data platform.

Therefore, in relation to the above work, the research work described in the forthcoming sections was designed to overcome major issues in previous work to develop a generalized model, which a common shortcoming of the above work as specific domains or scenarios had been considered.

3   Stock Markets and Social Media

Considering the research problem defined in Sect. 01, the methodology was designed as a combination of four unique modules. The rationale of the design is to programmatically realize the common events to which a particular company is sensitive to in the form of textual analysis on historical financial reports, through which a model is developed to isolate and apply the effect of such a residual identified through social media on the pure performance of a company at an extrapolated point of time. Therefore, the tightly coupled architecture in this approach is particularly adopted to generalize the technique to allow the approach to be useful at national or strategic-level planning for industries or governments as well, although it is external to the scope of this research. Figure 1 summarizes the overall process of the system and the modules could be described as follows.

3.1   Keyword Extraction

The objective of the module is to extract the unique keywords which reflect the overall portfolio of a company to provide the historical financial reports of the company. This would provide the platform upon which the model is built, specifically for a company given the industrial and micro-level independence of organizational performance which would further be explained in Sect. 06. Therefore, it was designed to reduce the content of a financial statement to several significant keywords using an NLP approach which considered a novel method of extracting named entities (NE) using semantic similarity filters using WordNet [15] and corpus statistics, coupled with a dependency parser to optimize the results. The solution design for the module is indicated in Fig. 2.

The core functionality designed for the text extractor and preprocessor submodule was to convert PDF documents to editable text documents from which the keywords

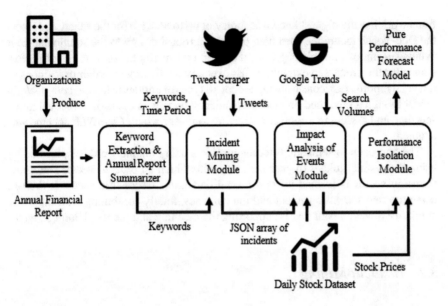

Fig. 1 Summary of the designed process of the solution

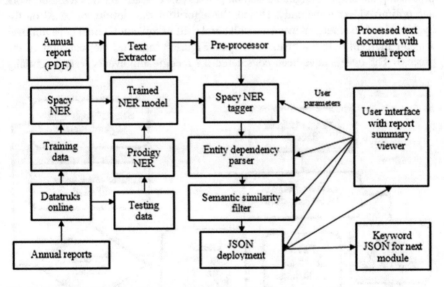

Fig. 2 Overall architecture of the keyword extraction module

are extracted. *Pdfminer.six* was adopted in this matter due to its robustness in programmatically rendering and converting recognized characters to ASCII or Unicode strings. *Spacy* was adopted for named entity filtering process based on the key advantages of using a convolutional neural network approach [16], provided the framework has the highest known accuracy of up to 86.81% for the function. Hence, the Doc-Vocab technique used here provides a robust design to the solution since it avoids storing of multiple copies of data—an optimizing factor in terms of performance for the process in extracting NEs over a typically large number of lexicons in financial reports. Lexicons were tagged for 18 types of entities using *en_core_web_lg* model though a pipelined process. Dependency parsing is the task of recognizing a sentence and assigning a syntactic structure to it –for which *ClearNLP* scheme was adopted.

The approach to semantic filtering considers using unique word sets, formed by using the distinct tokens in each sentence with which an ordered vector and a raw semantic vector is formed using a lexical database. A semantic vector is derived using the raw semantic vectors and the corpuses, finally combining with the order similarity to get a result that is supporting the meaning and the word order as well.

3.2   Incident Mining

Incident mining based on keywords from the module above adopted a pseudo-online processing technique designed based on [17–19] specifically for this research work and optimized experimentally. Hence, the algorithm was developed based on the hypothesis of an emergent event—a collateral shift of *topic* in Internet behavior over a major incident within a given time span. Figure 3 indicates the workflow of the process. The tweets have been considered for recognizing events were essentially

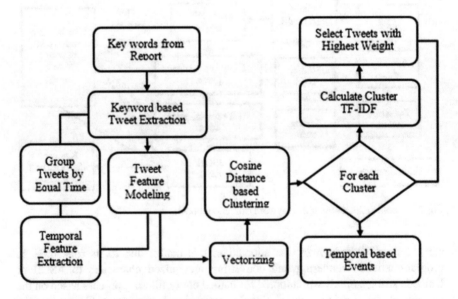

Fig. 3 Process of pseudo-online clustering

historical, because of which the algorithm was implemented to execute similar to online processing based on temporal features of the tweets.

The algorithm was designed for two phases of execution, each of which independently executes preprocessing steps to support the core functions of each phase—NE recognition and feature vectorizing, respectively. It should be noted that the research adopted Stanford NER tagger [20] for NE tagging and *senti-strength* [21] for sentiment scoring while the clustering was performed using the cosine distance of vectorized features. However, it should be noted, contrary to previous work hashtag decomposition was experimented as an optimizing feature in the research.

3.3 Impact Analysis

As it is indicated in Fig. 4, trends for keywords are fetched from Google Trends, where the non-trending keywords are removed based on rules prior to preprocessing for developing the feature vector which especially included the normalizing steps for removing non-functioning days.

Raw stock data are processed using Facebook Prophet [22] to develop a nonlinear saturating growth time series model of the data upon which a change point analysis was performed which could be given as follows in (2).

$$g(t) = C/\left(1 + e^{(-k(t-m))}\right), \tag{2}$$

where C is the carrying capacity or the maximum value of the curve. k is the growth rate which is a measure of the steepness of the curve, and m is an offset parameter.

This was selected prematurely to the research based on experimentations considering flexibility and suitability to the task. It should be noted that this approach underlined the use of *L-BFGS* as an algorithm having used Facebook Prophet [22], optimizing for the number of change points. These change points mapped on to events with a degree of temporal error were designed to produce a component to the

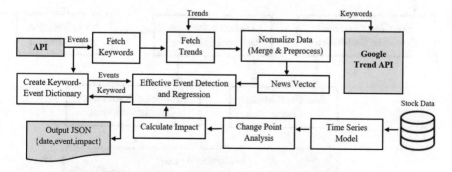

Fig. 4 Process of the impact analysis module

impact score based on the degree of variance from the calculated moving average. Therefore, an aggregate score was produced considering all trend, baseline variation at change point and sentiment.

3.4 Performance Isolation and Prediction

The final module of the solution was designed as two tools as indicated in Fig. 5, intended to provide services given by the primary objective of the research—forecasting and technical analysis tool, respectively. The tools were designed to deploy in AWS cloud for the benefit of serving users as SaaS application. In the case of forecasting, the module would perform pure performance isolation in stock data through adopting a noise filtering strategy, using Facebook Prophet [22] as a tool.

The research considered the hypothesis of stock market data being distorted with noise (macro-economic effects) which therefore results in difficulty to identify a proper regression within the raw set of stock price data. Although the previous research had adopted moving average and/or exponential moving average techniques to remove this noise from the stock market data, these two approaches the main problem was that the changes (ups and downs) followed the exact trend by after consuming some time. In other words, those two approaches were lacking in

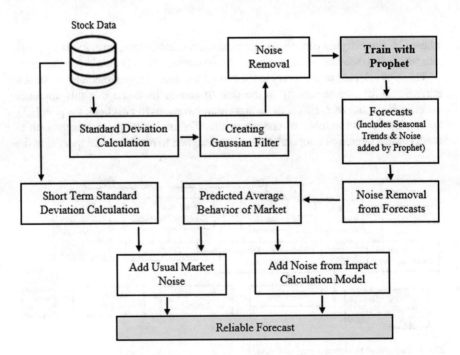

Fig. 5 Process of the forecasting tool

responding to market changes in quick time. Therefore, as a countermeasure, the research used optimized Gaussian filters, the results from which is modeled as a time series upon which the impact scores of events are applied to produce the forecast in response to an event. Here, a technical analysis tool was also developed, to provide a statistical comparison of against the results from the research approach.

4 Research Methodology

The research considered the financial reports of Kelani Valley (KV), Bogawanthalawa and Namunukula plantations for analysis. For the keyword extraction module, over 10,000 entries of data for named entities were manually annotated in compliance with the Gold Standards [23], developing a unique dataset relevant for the local context, using which the NER model was trained adopting the active learning method to avoid catastrophic forgetting issue when training such models. WordNet and Brown Corpus were used as databases for calculating purposes of the semantic similarities of words and sentences executed a sequence of experiments to optimize the process in the following respects. The word similarity was calculated by the combination of the long distance of synset pairs as well as the hierarchy distance of synset pairs as they appear in WordNet synset trees.

The incident mining module adopted several third-party datasets such as the English slang [24] and UCSC Sinhala Corpus [25]—which was translated to English using Google Translator (the results were manually verified and rectified for any errors). Historical tweets were scraped based on the reporting periods stated in the financial reports considered for the research therefore varying in volume for each case. But, three samples of macro-level datasets were extracted using "Sri Lanka" as a keyword for each consecutive year from 2016 for evaluation of effects of granularity to analysis.

The research focused on experimenting the approach of incident mining against testing hashtag decomposition as an optimizing feature in the process, for feature selection and threshold optimization for clustering, temporal lexicon size and incident mining accuracy. This was carried out for both macro- and micro-levels based on context granularity of keywords from which the tweets were mined. It should be noted that days particularly with tweet volume of equal or less than 2 were ignored based on the distribution of volume against the dates.

The work also focused on optimizing the change point analysis experimentally. Here, a change point range—the total portion of the time series where the change points should be extracted was specified. These also considered *change point prior index* to define the strength of the sparse that is used to filter out the most effective change points out of the extracted ones. These two measures allow fine-tune the model so that higher accuracy levels could be achieved while extracting the change points. By running the model, change points ordered with respect to the value of the deltas was extracted, which was then classified through utility code to negative and positive change points with respect to the delta values so to perform manual inspection in removing ignorable change points. A prediction window for short run—30 to 90 days—was defined to evaluate performance of the model with regard to identification of change points for different predefined ranges and prior indices. Hence, based on the results, 0.8 was considered optimal for the model.

After the combination of the sets of resulting impact point dates and the news vectors, these points were mapped to the relevant event. The news vectors were regressed using *ScikitLearn* along with the impact score to evaluate the accuracy of the results and to attempt performing a probable prediction if a substantial number of events exist.

The prediction platform was developed to execute in accordance with the process depicted in Fig. 5. Gaussian filter—1D filter—was adopted in particular from the discipline of image processing considering its noise filtering properties. The following (3) was used in creating the kernel for removing noise.

$$g_\sigma(x) = 1/\sqrt{2\pi}\sigma \, \exp(-x^2/(2\sigma^2)), \tag{3}$$

Hence, Fig. 6 indicates the comparison between the methods against the moving average techniques, where it is clear that Gaussian filter outperformed the rest, therefore it being the method of choice for the solution. The noise removed data were fed to Facebook Prophet [22] for development of the model and hence perform future predictions. A novel approach is considered in this respect here while feeding the noise removed data to the model. However, since the prediction from the model should produce the average future behavior of the stock market, intuitive market noise in the predictions induced by the model has been removed using Gaussian filtering again to accommodate adding residual noise from prospective events from impact calculation module in the latter stages of the process.

Residual noise was introduced to obtain the predictions based on the size of the projection, i.e., if the prediction is performed for 30 days, the standard deviation of past 30 days is calculated using actual stock market data. Then that standard

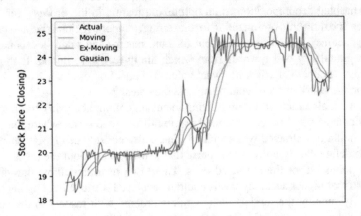

Fig. 6 Process of the forecasting tool

deviation and predicted average market behavior were fed to white noise creation function. The noise created will be added to the predicted average market behavior. Here, the noise was not an arbitrary value but calculated using the actual stock prices. The results from the predictions are discussed in Sect. 5. But, it should be noted that it was more likely actual market behavior is similar to the predictions compared against the statistical and technical analysis methods such as on-balance volume, accumulation/distribution line, and moving average analyses.

5 Results and Experiments

The research adopted *Prodigy* as a tool to test the NER models for three approaches using an 80–20 hold-out approach, where each of the models was trained over 20 iterations using the same batch size. The semantic filter was evaluated based on heuristic methods of randomly choosing a set of sentences and comparing them with the similarity scores produced by a set of participants. The active learning approach for NER produced the best performance in terms of accuracy at 78.5% for NER and was identified that similarity of more than 0.6 between the keywords could be considered optimal for semantic similarity, producing 0.82 strong correlation.

The event recognition algorithm has experimented among the combinations of hashtag decomposition and hashtags in its actual form as features. Accordingly, hashtag decomposition approach was seen to improve the event recognition capability with the highest precision of 91% because it was seen to normalize the distribution of vocabulary to a limited scope that improves the cohesiveness among content in adjoining time frames that allows them to cluster better if they refer the same event.

Three numerical values—20, 50, 83—were selected for optimizing the lexical size of the temporal dictionary corresponding to least common, average, and most common lexicons. However, it was seen that adopting 20 as the lexical size would optimize the cause as a larger size would generalize the scope reducing the accuracy of event recognition, given the maximum precision achieved of 0.6585 at $L = 20$.

Event recognition algorithm was optimized considering combinations of the four features as in [20] against the threshold value for distinguishing clusters. Accordingly, a threshold value of $T = 0.4$ was selected based on performance. However, as it is indicated in Table 1, micro-level analysis taking Kelani Valley Plantation as a sample indicated a better performance using all features. But, for generalizability, it was decided to consider only the temporal features for analysis as it is in line with the

Table 1 Best event recognition performance with respect to context-based granularity levels

Granularity	Feature set	Precision	Recall
Macro	Temporal only	0.8571	0.4524
Micro	Temporal only	0.7222	0.6191
Micro	All features	0.8333	0.6250

concept of emergent events where it is natural to cluster based on shifts in vocabulary over time captured by the temporal vocabularies built for each day.

Changepoint prior scale (CPS) was used to control the generalizability of the model as higher CPS would lead to overfitting and lower—an under fitted one. Therefore, an appropriate CPS was evaluated visually and quantitatively as it is indicated in Fig. 7 and Table 2, considering 30–60 days predictions for different CPSs. The optimum CPS for this data was chosen as 0.05 after evaluating these results.

The prediction tools were comparatively studied among the techniques mentioned above in Sect. 04. The research considered the error measures for the evaluation against the variable of *time*, for which short (20–30 days), medium (40 days), and long (100 or more days) runs were considered. Figure 8 indicates the comparative analysis of the techniques while Tables 3 and 4 indicate the numerical values of evaluation for each time frames for Watawala Plantations.

It could be observed that the Gaussian filtered approach is more likely to produce less error than forecasting with raw data. It is also depicted that the forecasting period has a proportional relation to the error for which reason it could be concluded that

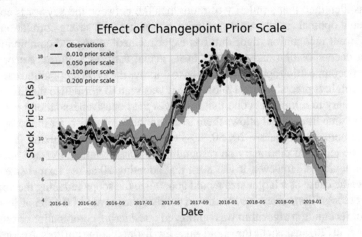

Fig. 7 Effect of CPS on predictions for Bogawantalawa Tea Estates PLC

Table 2 Comparison of predictive performance for different CPS values

CPS	Window (days)	MMSE	RMSE	MAE	MAPE
0.01	30	5.9481	2.4389	1.7046	0.134
0.01	60	10.7859	3.2842	2.1491	0.1784
0.05	30	0.7077	0.8779	0.6633	0.0515
0.05	60	3.8053	1.9507	1.2845	0.0972
0.1	30	0.6537	0.8085	0.623	0.0509
0.1	60	4.3026	2.0743	1.5039	0.1197

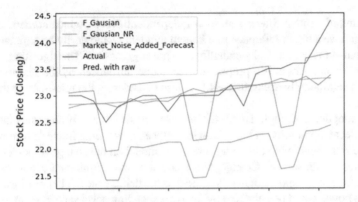

Fig. 8 Distribution of the values of daily news vector sums

Table 3 Evaluation of prediction for the short run

Approach	RMSE	MAE	MAPE
Gaussian filtered stock data	0.78872	0.58152	2.50840
Gaussian filtered stock data (noise removed)	**0.60261**	**0.54438**	**2.34242**
Allowing market noise	0.72451	0.61119	2.63487
Raw stock data	2.09104	2.04613	8.83917

Table 4 Comparison of prediction for a medium and long run using the best approach

Approach	RMSE	MAE	MAPE
Medium run	2.42326	2.38252	10.33180
Long run	4.4554	3.52112	15.82404

the tool is more suitable for short-term forecasts. Although it is justifiable as it is unlikely to predict forthcoming event at least heuristically for a long period of time provided the market variability. Hence, it is highly relevant to use the system in the event where significant market indicators are available, typically occurring within very small-time spans.

6 Discussion

The novel technique for analyzing and forecasting stocks based on social impacts overcame several critical issues in previous work in developing a robust technique and was successful in bringing about several novelties that would be explained below.

The incident mining algorithm has been capable of recognizing abstract events—a set of collective independent incidents or actions which could be recognized to have caused by a common reason. Contemporary research has neglected such events by the means of content filtering as such independent events, which sometimes may not even have a direct reference to a known incident would easily be annotated as a non-event as the content independently may not suggest such. The algorithm used for hashtag decomposition is a proprietary technique introduced by the research work which is significantly accurate at 84.8%, also considering the relevance to the local context.

The adoption of Google Trends for impact analysis intuitively considers the exponential decay popularity and growth within the analyzed time frame in resemblance with the sentiment, which provides a realistic model on the change of perception or consensus within society. Changepoint analysis to map significant events could be considered a novel approach in this context as the deltas (rate of change of gradients) of those points could be validated by the corresponding news vectors from the news event analysis. Therefore, the technique provides a hybrid model for news analysis and time series impact point analysis.

Although for 80% of the tweets had been annotated to be garbage by 9 independent annotators over a sample of 8952 tweets it was observed that proportionately large-sized clusters of tweets had been automatically removed from being recognized as events by the algorithm. The problem of having such large proportion of garbage could also affect the event identification performance of the system as well since tweets constructed using different language structure or usage could easily be garbage collected as they may not represent the token space of the particular day.

Tweets collected using keywords relevant only to the company were produced with events that were significant only for the company. However, the datasets collected at a coarse grain such as those with the national interest recognized events that had national importance. Therefore, the algorithm could be seen to be highly adaptive to size and context. However, the solution could be noted to be prone to contextual granularity as it has been mentioned before. It should also be noted that, in any case, the geotagging of the data was not of high relevance at least in recognizing incidents at the national level, which is in contrary to some of the approaches of several research where the researchers have adopted geotagging as a feature [26, 27].

With respect to evaluating impact analysis for news events, conventional linear regression model considers linear relationships between the variables although it is not to commonly exist. Hence, such would only analyze the historical stock prices ignoring the macro-economic effects which leads to lower accuracy. A more reliable prediction was observed to be obtained by adopting a nonlinear saturating growth model. Comparatively, creating a regression model by taking the macro-economic factors into account via sentiment analysis of related news and events during a given period of time could be considered a viable option for research in the area. As the news vector is the cumulative news value for each day, it performs well in highly sensitive markets such as the CSE where the market indices are highly dependent on external news and events.

Another approach is to calculate the Granger causality of selected keywords to determine whether the signals from the Google trend data are capable of unleashing useful information of the market as a whole. This method is suitable to predict bi-directional movements in the S&P 500 index as it allows forecasts with over

60% accuracy, therefore most suitable for predictive techniques targeted at the stock market level. In developing a stock forecasting tool, it was established through the research that macro-economic effects on the organization should be captured in the process or the architecture in relevance to theories such as random walk theory and efficient market hypothesis—the baseline reason to invalidate fundamental analysis, therefore.

7 Conclusion and Further Work

The research would like to restate the fact that the stock market is affected by many forces, much of which is internal to the organization's financial performance and the internal stock market's forces such as changes in policy and market parameters. Therefore, the probability of an external incident effecting the stock market performance at a substantial level is low, but not significantly low as it would commonly be referred. The research work has established—especially through the results obtained through the performance isolation module that the considerable error between the models with and without market noises have indicated the hypostasis. Therefore, the model could be used as a significant tool for decision-makers, especially investors as a company's pure performance can be isolated and tested upon for realistic phenomenon for a foreseeable future.

To reach a higher accuracy in this area, further work has to be aimed at overcoming several drawbacks of the solution such as reducing the effect of seasonality produced in the resultant time series models by using Facebook Prophet [22]. Hence, it should be noted that it is recommended to execute Gaussian filtering prior to feeding data to ANNs used in this solution, which is expected to produce a better performance as it is observed that the randomness of stock variations could affect the performance of the model outputs.

An important further work in the area in relation to this work is the possibility to automate stock trade. This could be performed by adopting the event identification framework for real-time social media streams and by application of the related known impact on the current market state at a certain point of time, from which the divesting and investing opportunities could be quantified across the entire stock market providing a versatile platform which could be optimized to yield the maximum for an investor. Therefore, a tool such as Apache Spark could be a viable option to adopt the solution in a distributed environment, providing the robustness and scalability to operate on a large scale—ambitiously across stock markets as well. Hence, the models developed through the methodology could be improved to be *evolutionary* by the means of introducing a similarity metric against known impact factors allowing consistency and improved reliability on a *timeless scale*, adapting with the changes in the market. Therefore, the event-impact dictionary been constructed in this work could be adopted as *seeds* to such an evolutionary approach, thereby ensuring scalability and uniform applicability across organizations as well.

References

1. Silva, N.L., Perera, P.R., Silva, N.C.: Relationship between stock market performance and economic growth: empirical evidence from Sri Lanka. J. Acc. Res. Edu. (JARE) **1**, (2018)
2. Attigeri, G.V., MM, M.P., Pai, R.M., Nayak, A.: November. Stock market prediction: A big data approach. In: TENCON 2015–2015 IEEE Region 10 Conference, pp. 1–5. IEEE
3. Mathivannan, S., Selvakumar, M.: Test of random walk theory in the national stock exchange. Asian J. Manage. Sci. **4**, 21–25 (2015)
4. Țiṭan, A.G.: The efficient market hypothesis: review of specialized literature and empirical research. Proc. Econ. Fin. **32**, 442–449 (2015)
5. Waduge, N., Ganegoda, U.: Stock market prediction: forecasting stock price of a company considering macroeconomic effect from news events. In: 2018 3rd International Conference on Information Technology Research (ICITR), pp. 1–5 (2018)
6. Industry Capability Report.: Sri Lanka Exports Development Board (EDB) (2016)
7. Ariyo, A.A., Adewumi, A.O., Ayo, C.K.: Stock price prediction using the ARIMA model. In: 2014 UKSim-AMSS 16th International Conference on Computer Modelling and Simulation, pp. 106–112 (2014)
8. Curme, C., Preis, T., Stanley, H.E., Moat, H.S.: Quantifying the semantics of search behavior before stock market moves. Proc. Natl. Acad. Sci. **111**(32), 11600–11605 (2014)
9. Kaushal, A., Chaudhary, P.: News and events aware stock price forecasting (BID), pp. 8–13. Pune, India (2017)
10. Huang, M.Y., Rojas, R.R., Convery, P.D.: Forecasting stock market movements using Google Trend searches. Empir Econ, pp. 1–19 (2019)
11. Bollen, J., Mao, H., Zeng, X.-J.: Twitter mood predicts the stock market. J. Comput. Sci. **2**(1), 1–8 (2011)
12. Lee, C., Paik, I.: Stock market analysis from Twitter and news based on streaming big data infrastructure. In: 2017 IEEE 8th International Conference on Awareness Science and Technology (iCAST), Taichung, pp. 312–317 (2017)
13. Sharma, A., Bhuriya, D., Singh, U.: Survey of stock market prediction using machine learning approach. In: 2017 International conference of Electronics, Communication and Aerospace Technology (ICECA), Coimbatore, pp. 506—509. (2017).
14. Sharma, N., Juneja, A.: Combining of random forest estimates using LSboost for stock market index prediction. In: 2017 2nd International Conference for Convergence in Technology (I2CT), Mumbai, pp. 1199–1202 (2017)
15. Miller, G.A., Beckwith, R., Fellbaum, C., Gross, D., Miller, K.J.: Introduction to WordNet: An on-line lexical database. Int. J. Lexicogr. **3**(4), 235–244 (1990)
16. Kiperwasser, E., Goldberg, Y.: Simple and accurate dependency parsing using bidirectional LSTM feature representations. Trans. Assoc. Comput. Linguist. **4**, 313–327 (2016)
17. Alsaedi, N., Burnap, P., Rana, O.: Identifying disruptive events from social media to enhance situational awareness. In: 2015 IEEE/ACM International Conference on Advances in Social Networks Analysis and Mining 2015—ASONAM '15, Paris, France, pp. 934–941 (2015)
18. Alsaedi, N., Burnap, P.: Feature extraction and analysis for identifying disruptive events from social media. In: 2015 IEEE/ACM International Conference on Advances in Social Networks Analysis and Mining 2015, pp. 1495–1502 (2015)
19. Alsaedi, N., Burnap, P., Rana, O.: Sensing real-world events using social media data and a classification-clustering framework. In: International Conference on Web Intelligence (WI), 2016 IEEE/WIC/ACM, pp. 216–223 (2016)
20. Manning, C., Surdeanu, M., Bauer, J., Finkel, J., Bethard, S., McClosky, D.: The stanford CoreNLP natural language processing toolkit. In: 52nd Annual Meeting of the Association for Computational Linguistics: System Demonstrations, pp. 55–60 (2014)
21. Thelwall, M., Buckley, K., Paltoglou, G., Cai, D., Kappas, A.: Sentiment strength detection in short informal text. J. Am. Soc. Inf. Sci. Technol. **62**(2), 419 (2011)
22. Taylor, S.J., Letham, B.: Forecasting at scale. Peer J Preprints 5:e3190v2 (2017). https://doi.org/https://doi.org/10.7287/peerj.preprints.3190v2

23. SpaCy API Documentation, https://spacy.io/api/#nn-models
24. Infochimps. https://www.infochimps.com
25. Welgama, V., Herath, D.L., Liyanage, C., Udalamatta, N., Weerasinghe, R., Jayawardana, T.: Towards a sinhala wordnet. In: Conference on Human Language Technology for Development. (2011)
26. Salas, A., Georgakis, P., Petalas, Y.: Incident detection using data from social media. In: Intelligent Transportation Systems (ITSC), 2017 IEEE 20th International Conference, pp. 751–755 (2017)
27. Abdelhaq, H., Sengstock, C., Gertz, M.: Eventweet: Online localized event detection from twitter. VLDB Endowment **6**, 1326–1329 (2013)

Prediction of Malocclusion Pattern of the Orthodontic Patients Using a Classification Model

A. M. Imanthi C. K. Jayathilake, Lakshika S. Nawarathna, and P. N. P. S. Nagarathne

Abstract Malocclusion is concerned, when the teeth of one arch are misplaced relative to teeth of other arches in the anteroposterior transverse or vertical planes of space, and they are referred to as malocclued teeth. There are two factors, general and local, contributing to developing the malalignment and malocclusion. The skeletal factor is one of the general and very important factors to identify malocclusion patterns. The dentist prescribes treatment according to the malocclusion pattern of the patients before starting treatment. The analysis was performed on patients who attend the health center in the Orthodontist Division, Faculty of Dental Sciences, University of Peradeniya, and the sample included 60 subjects, the patient's age ranged from 11 to 15 years. The study was carried out using the cervical vertebral maturation phase, which was evaluated on cephalometric radiographs. Although it is a very convenient method to identify the malocclusion pattern of the patients, it is time-consuming and expensive. The key feature of the research was to propose a classification model to predict the malocclusion patterns of orthodontic patients. Multinomial logistic regression model, k-NN algorithm, random forest model, and Naïve Bayes model were used to predict the malocclusion pattern. Accuracy of the multinomial logistic regression model, k-NN algorithm, random forest, and Naïve Bayes classification of malocclusion patterns were 88.89%, 83.33%, 88.89%, and 55.56%, respectively. Further, areas under the curve (AUC) of the multinomial logistic regression model, k-NN algorithm, random forest model, and Naïve Bayes classification were 0.9889, 0.6176, 0.5389, and 0.5333, respectively. The results of the research recommend a method based on a classification technique to anticipate malocclusion patterns. In

A. M. I. C. K. Jayathilake · L. S. Nawarathna (✉)
Department of Statistics and Computer Science, Faculty of Science, University of Peradeniya, Peradeniya, Sri Lanka
e-mail: lakshikas@pdn.ac.lk

A. M. I. C. K. Jayathilake
e-mail: imanthi.kghs@gmail.com

P. N. P. S. Nagarathne
Division of Orthodontics, Faculty of Dental Sciences, University of Peradeniya, Peradeniya, Sri Lanka
e-mail: spnpnagarathne@gmail.com

279

S. Shakya et al. (eds.), *Proceedings of International Conference on Sustainable Expert Systems*, Lecture Notes in Networks and Systems 176,
https://doi.org/10.1007/978-981-33-4355-9_22

addition, the multinomial logistic regression model is well succeeded for all accuracy and area under the curve values relative to other classification models.

Keywords *k*-NN algorithm · Malocclusion pattern · Multinomial logistic regression model · Naïve bayes classification · Random forest model

1 Introduction

Malocclusion is the relationship between the two dental arches. If the teeth of one arch are misplaced relative to teeth of other arches in the anteroposterior transverse or vertical planes of space, they are referred to as malocclued teeth [1]. The continuing growth of the face may affect the developing occlusion resulting in increasing the severity of the malocclusion. Usually, malocclusion grows with growing children and it is possible to affect developing occlusion. There are four main factors causing malocclusion: genetic syndromes, defects of embryological development, trauma, and anomalous postnatal development [2]. The etiology of malocclusion is not discussed under the main factors here, but under two factors, general and local which help to develop malalignment and malocclusion of teeth. The skeletal factor is one of the general factors used for the study. There are three types of skeletal pattern, skeletal pattern I, skeletal pattern II, and skeletal pattern III [3]. In skeletal pattern I, both jaws are in an ideal anteroposterior relationship to each other, while the lower jaw is positioned further back than in skeletal pattern I and the lower jaw is positioned farther forward than in skeletal pattern I for skeletal pattern II and skeletal pattern III, respectively. Figure 1 shows the skeletal patterns of the human. Besides,

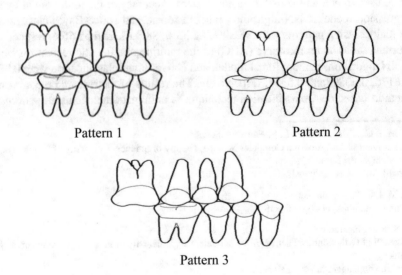

Pattern 1   Pattern 2

Pattern 3

Fig. 1 Types of skeletal pattern of the patients

Angle's classification is used to classify malocclusion for orthodontic diagnosis. Edward Angle introduced Angle's classification method in 1899. Angle's classification classified malocclusion into three main categories, malocclusion I, II, and III [4, 5].

Al-Jundi (2015) has conducted a study under the topic "pattern of malocclusion in a sample of orthodontic patients from a hospital in the Kingdom of Saudi Arabia" [6]. The main aim of the research was to identify the skeletal and dental malocclusion of Saudi Arabians. The selected 510 patients from September 2013 to May 2015 were analyzed, and different cephalometric values were used to evaluate the type of malocclusion. They found class II malocclusion to be frequent with class III subdivision the least frequent. They also observed a statistical difference in normal overjet, overbite, crossbite, slight midline deviation, and ANB according to gender. Ghiz (2005) was done the study, "cephalometric variables to predict the future success of early orthopedic class III treatment" [7]. The sample contained sixty-four patients, and the sample was divided into successful and unsuccessful clusters in the opinion of overjet and molar relationship. Five angular measurements and eleven linear measurements are considered for the study. A logistic regression model was fitted to distinguish the dentoskeletal variables which are an influence for the projection of the two clusters. The logistic regression model gave the results, an accurate rate for the successful group is 95.5% and 70% for the unsuccessful group. Sandeep (2012) has carried out a study to identify occlusal issues, their prevalence, and the necessity for treatment that collaborate to decide the convenient perception plans, preventive and interceptive treatment, and manpower needed. The main focus of this research was to evaluate and come up with accurate details on dental malocclusion patients in Rwanda [8]. They have checked several measurements of the patients' records of 243 selected patients with dental malocclusion who visited the Dental Department of King Faisal Hospital, Rwanda, between 2009 and 2012. Angle's class I malocclusion (60.9%) is the most common malocclusion in Rwanda comparative to class II (28.8%) and class III malocclusions (10.3%).

None of the previous studies indicates a statistical method to identify the malocclusion patterns of patients in Sri Lanka. Therefore, this pioneering research focuses on predicting the malocclusion patterns of Sri Lankan children between the ages of 11 and 15 according to angular measurements.

The article is arranged as follows. Section 2 describes the dataset used for the study and the classification model used to predict the malocclusion pattern along with its model validation techniques. Section 3 expands on the statistical analysis performed with the assist of R statistical software and the main outcomes obtained. In Sect. 4, the conclusion declares the most important results of the study.

2   Material and Methods

2.1   Data

The dataset contains data about sixty patients collected from patients who presented for treatment at the orthodontic clinic, Faculty of Dental Sciences at the University of Peradeniya. Pretreatment and post-treatment radiographs were obtained for each case, and patients were selected during their pubertal growth spurt.

Sella–Nasion–A point (SNA), Sella–Nasion–B point (SNB), A point–Nasion–B point angles (ANB) were the cephalometric variables considered in this study [9]. The skeletal pattern was usually determined by using maxilla and mandible with the cranial base using SNA and SNB angles or the difference between SNA and SNB angles [10]. The patients were classified into malocclusion patterns according to the ANB angle. There are three types of skeletal patterns, the skeletal pattern I, skeletal pattern II, and skeletal pattern III. Skeletal pattern 1 is the normal or standard skeletal pattern while skeletal pattern II and skeletal pattern III are deviations of pattern 1. The groups were identified according to the following criteria: if $0 < ANB < 4$ then the skeletal pattern is 1, if $ANB > 4$ then the skeletal pattern is 2, and if $0 > ANB$ then the skeletal pattern is 3.

The error of the method [11] was determined replicating the evaluation and measurements of 12 lateral cephalometric radiographs. This random selection of data was sampled to ensure acceptable intra-examiner reliability. A paired sample t-test was carried out to determine intra-examiner measurement error between before and after SNA and SNB measurements.

The paired sample t-test was performed to determine the intra-examiner measurement error between before and after angular measurements. Twenty-four patients were selected from the sample to check angular reliability. The sample was selected systematically, and a paired t-test was applied to contrast two populations' means. Moreover, the test can be used to compare before and after observations on the same subjects and to compare two different methods of measurements or two different treatments.

2.2   Classification Models

Classification models, namely multinomial logistic regression, random forest model, k-NN algorithm, and Naïve Bayes classification models were used to predict the skeletal patterns of the patients. The multinomial logistic regression model is used when the dependent variable had more than two classes [12, 13]. In this model, first, one needs to find the relationship between malocclusion I and malocclusion II and malocclusion I and malocclusion III. These relationships are expressed as $g_1(x)$ and as $g_2(x)$ follows,

$$g_1(x) = \ln\left(\frac{\Pr(y = II|x)}{\Pr(y = I|x)}\right) = \beta_{01} + \beta_{11}x_{11} + \cdots + \beta_{n1}x_{n1} \tag{1}$$

$$g_2(x) = \ln\left(\frac{\Pr(y = III|x)}{\Pr(y = I|x)}\right) = \beta_{02} + \beta_{12}x_{12} + \cdots + \beta_{n2}x_{n2} \tag{2}$$

where $\Pr(y = i|x)$ is the probability of having ith skeletal pattern; $i = I, II, III$, and x_{ij} is the ith variable in the jth equation and β_{ij} is constant for $i = 0, j = 1,2$.

The natural log of $\left(\frac{\Pr(y=i|x)}{\Pr(y=i|x)}\right)$ is the same as the logit in standard logistic regression. The probability of being in each given category can be found by using Eqs. (1) and (2). The logistic transformation was used to find the probability according to the logistic cumulative distribution. The probability of being in the category for logistic regression with one predictor is given by,

$$\Pr(y|x) = \frac{1}{1 + e^{\beta_0 + \beta_1 x_1 + \cdots + \beta_n x_n}} \tag{3}$$

The extended form of Eq. (3) was used to find the probability of being in each category: exponent functions of the intercepts and the coefficients for each of the comparisons to the reference category. Following equations were modified by using Eq. (3),

$$\Pr(y = I|x) = \frac{1}{(1 + \exp(g_1(x)) + \exp(g_2(x)))} \tag{4}$$

$$\Pr(y = II|x) = \frac{\exp(g_1(x))}{(1 + \exp(g_1(x)) + \exp(g_2(x)))} \tag{5}$$

$$\Pr(y = III|x) = \frac{\exp(g_2(x))}{(1 + \exp(g_1(x)) + \exp(g_2(x)))} \tag{6}$$

The k-NN algorithm is easy to implement, effective, and simple. In the case of classification, new data points get classified in a particular class, and in the case of regression, new data gets labeled based on the average value of kth nearest neighbor. There are some shortcomings of the k-NN algorithm which affect the precision of the classification [14]. Besides, all the features should be on the same scale, attributes have been normalized to obtain zero mean and unit variance. The similarity or distance among training and test set is measured using Euclidean distance. [15, 16]. The standard Euclidean distance $d(x_i, x_j)$ can be expressed as,

$$d(x_i, x_j) = \sum\left(a_r(x_i) - a_r(x_j)\right)^2 \tag{7}$$

Random forest or random decision forest is a classification model that functions by developing multiple decision trees during the training time [17]. Random forest is a type of machine learning algorithm and supervised learning. Multiple trees in the random forest algorithm reduce the risk of overfitting, shorter training times, high accuracy, and estimation of missing data [18].

The Bayesian classification algorithm is a statistical algorithm which predicts the class membership probabilities as the probability of a given tuple belonging to a specific class [19]. The Naïve Bayes classification operates as follows: S is a training set of tuples and their related class labels and there are n classes, $G_1, G_2,.., G_i,.., G_n$. The classifier will predict the class of a given class Y which has the highest posterior probability, condition on Y [20]. The Naïve Bayesian classifier predicts the class of tuple Y, G_i if and only if

$$P(G_i|Y) > P(G_j|Y), \quad \text{for } 1 \leq j \leq n, j \neq i$$

$P(C_i|Y)$ should be maximized and by Bayes' theorem,

$$P(G_i|Y) = \frac{P(Y|G_i)P(G_i)}{P(Y)} \tag{8}$$

$P(Y)$ is an unchanging factor for all the groups and only $P(G_i|Y)$ require to be maximized. If class probabilities are unknown, then assume all the classes are equally like. Therefore, $P(Y|G_i)$ should maximize otherwise $P(Y|G_i)P(G_i)$ maximizes. It is very difficult to calculate $P(Y|G_i)$ when the dataset has many attributes. Hence, the class conditional independence is used and that assumes each class is conditionally independent of one another. Thus,

$$P(Y|G_i) = \prod_{k=1}^{m} P(y_k|G_i)$$
$$= P(y_1|G_i) \times P(y_2|G_i) \times \cdots \times P(y_m|G_i) \tag{9}$$

The train tuples are used to estimate probabilities, then $P(Y|G_i)P(G)$ is evaluated for each class G_i to project a category label of Y. Moreover, the Naïve Bayes predicts the class of tuple Y if and only if,

$$P(Y|G_i)P(G_i) > P(Y|G_j)P(G), \quad \text{for } 1 \leq j \leq n, j \neq i. \tag{10}$$

2.3 Model Validation

The confusion matrix is a kind of contingency table that describes the performance of the model used for predictions [21]. The confusion matrix is easy to understand and easy to interpret. Actual and prediction are the two dimensions. Table 1 gives the confusion matrix for the binary case below.

Table 1 Confusion matrix for classification models

		Predicted	
		Yes	No
Actual	Yes	True positive	False negative
	No	False positive	True negative

Accuracy indicates how often a classifier is correct, and it can be found by dividing the sum of the diagonal of the confusion matrix by the total number of observations. The misclassification rate can be found by dividing off-diagonal elements by the total number of observations, which indicates how often a classification is wrong [22].

$$\text{Accuracy} = \frac{\text{Sum of the diagonal elements}}{\text{total number of observations}} \tag{11}$$

$$\text{Misclassification rate} = \frac{\text{Sum of the off} - \text{diagonal elements}}{\text{total number of observations}} \tag{12}$$

The area under the curve (AUC) is used to measure or interpret the quality of the classification models [23]. The AUC value ranges from zero to one and most of the classification models have values from 0.5 to 1. The AUC value is equal to 1 for the best classification model. Moreover, the best classification method among the other classification methods can be determined considering AUC values, and the best model has the highest AUC value.

3 Results and Discussion

Sixty cases were recorded at the Clinic at the Faculty of Dental Sciences, University of Peradeniya. Out of these, thirty patients were female and thirty patients were male, and Class I malocclusion was found in four patients, which illustrated 6.67%. Class II malocclusion was discovered in 53 patients representing 83.33% of the sample. Finally, three patients have diagnosed with class III malocclusion, which represented 5.00% of the sample.

Table 2 gives the outcome of the paired sample t-test which is used to determine intra-examiner measurement error between first and second measures. The p-value $= 0.0693 > 0.05$, indicates there is no significant difference between first and second

Table 2 Summary results of the t-test of angular measurements

	Mean difference	Test statistics	DF	p-value	95% confidence interval	
					Lower	Upper
Angular measurements	0.5	1.9057	23	0.0693	− 0.0428	1.0428

Table 3 Accuracy, misclassification rate, and AUC for fitted classification models

Model	Accuracy (%)	Misclassification rate (%)	AUC
Multinomial logistic regression	88.89	11.11	0.9889
k-NN algorithm	83.33	16.67	0.6176
Random forest classification	88.89	11.11	0.5389
Naïve Bayes classifier	55.66	44.44	0.533

measurements. Moreover, the 95% confidence interval determines a range of values that can be 95% certain contains the population mean. In other words, the true mean of the population is in the range between -0.0428 and 1.0428 with 95% confidence.

Multinomial logistic regression, k-NN algorithm, random forest, and Naïve Bayes classifier models were used to determine the malocclusion patterns of the patients. Table 3 gives the accuracy, misclassification rate, and AUC value of fitted classification models.

According to Table 3, the accuracy of the multinomial logistic regression model, k-NN algorithm, random forest, and Naïve Bayes classification of malocclusion patterns are 88.89%, 83.33%, 88.89%, and 55.56%, respectively, and misclassification errors for classification models are 11.11%, 16.67%, 11.11%, and 44.44%, respectively. The area under the curve (AUC) of the multinomial logistic regression model, k-NN algorithm, random forest model, and Naïve Bayes classification are 0.9889, 0.6176, 0.5389, and 0.5333, respectively.

4 Conclusion

The error of the method was computed by performing a paired sample t-test and the p-value for the test indicated that both the samples have the same mean, and there is no significant difference between before and after treatments. In this paper, classification models were used to predict the malocclusion patterns of the patients and the multinomial logistic regression model is suggested to predict the malocclusion patterns of patients considering cephalometric measures. The outcome of the classification model shows that it effects with an accuracy of 88.89% and 0.9889 AUC value, providing that this model can be applied to predict the malocclusion pattern of the patients utilizing cephalometric measurements.

References

1. Nagarathna, N.: Orthodontic Diagnosis. Prof. Nandani Nagarathne, Peradeniya (2017)

2. Zhou, Z., Liu, F., Shen, S., Shang L, Shang, L., Wang, X.: Prevalence of and factors affecting malocclusion in primary dentition among children in Xi'an, China. BMC Oral Health. **91** (2016)
3. Proffit, W.R., Fields, H.W., Moray, L.J.: Prevalence of malocclusion and orthodontic treatment need in the United States; estimates from the NHANES-III survey. Int. J. Adult Orthod. Orthogn. Surg. **13**(2), 97–106 (1998)
4. Gong, A., Li, J., Wang, Z., Li, Y., Hu, F., Li, Q., Miao, D., Wang, L.: Cranial base characteristics in anteroposterior malocclusions: a meta-analysis. Angle Orthod. **86**(4), 668–680 (2016)
5. Ackerman, J.L., Proffit, W.R.: The characteristics of malocclusion: a modern approach to classification and diagnosis. Am. J. Orthod. **56**(5), 443–454 (1969)
6. Al-Jundi, A., Riba, H.: Pattern of malocclusion in a sample of orthodontic patients from a hospital in the kingdom of Saudi Arabia. Savant J. **1**(1), 14–21 (2015)
7. Ghiz, M.A., Ngan, P., Gunel, E.: Cephalometric variables to predict future success of early orthopedic class III treatment. Am. J. Orthod. Dentofacial Orthop. **127**(3), 301–306 (2005)
8. Sandeep, G., Sonia, G.: Pattern of dental malocclussion in orthodontic patients in Rwanda: a retrospective hospital based study. Rwanda Med. J. **69**(4), 13–18 (2012)
9. Rochi, P., Cinquini, V., Ambrosli, A.M., Caprioglio, A.: Maxillomandibular advancement in obstructive sleep apnea syndrome patients: a restrospective study on the sagittal cephalometric variables. J. Oral. Maxillofac. Res. **4**(2), 1–9 (2013)
10. Dhopatkar, A., Bhatia, S., Rock, P.: An Investigation into the relationship the cranial base angle and malocclusion. Angle Orthod. **72**(5), 456–463 (2002)
11. Cançado, R.H., Lauris, J.R.P.: Error of the method: what is it for? Dental Press J. Orthod. **19**(2), 25–26 (2014)
12. El-Habil, A.M.: An application on multinomial logistic regression model. Pak. J. Stat. Oper. Res. **8**(2), 271–291 (2016)
13. Fox, J., Hong, J.: Effect displays in R for multinomial and proportional odds logit models: extensions to the effects package. J. Stat. Softw. **32**(1), 1–24 (2009)
14. Baobao, W., Jinsheng, M., Minru, S.: An Enhancement of k Nearest Neighbor Algorithm using Information Gain and Eextension relativity. In: Proceedings of International Conference on Condition Monitoring and Diagnosis, pp. 1314–1317 (2008)
15. Sun, S., Huan, R.: An adaptive k-Nearest Neighbor Algorithm. In: Proceedings of 7th International Conference on Fuzzy Systems and Knowledge Discovery, pp. 91–94. IEEE Press, China (2010)
16. Taneja, S., Gupta, C., Goyal, K., Gureja, D.: An enhanced K-nearest neighbor algorithm using information gain and clustering. In: 4th International Conference on Advanced Computing & Communication Technologies, pp. 326–328. IEEE Press, India (2014)
17. Gislason, P.O., Benediktsson, J.A., Sveinsson, J.R.: Random forest for land cover classification. Pattern Recogn. Lett. **27**(4), 294–300 (2006)
18. Vladimir, S., Aand, L., Christopher, C.J., Sheridan, R.P., Feuston, B.P.: Random forest: a classification and regression tool for compound classification and QSAR modeling. J. Chem. Inf. Comput. Sci. **43**(6), 1947–1958 (2003)
19. Farid, D.M., Zhang, L., Rahman, C.M., Hossain, M.A., Strachan, R.: Hybrid decision tree and Naïve Bayes classifiers for multi-class classification tasks. Expert Syst. Appl. **41**(4), 1937–1946 (2014)
20. Yager, R.R.: An extension of the Naïve Bayesian classifier. Inf. Sci. **176**(5), 577–588 (2006)
21. Jain, A., Zongker, D.: Feature selection: evaluation, application, and small sample performance. In: IEEE Transaction on Pattern Analysis and Machine Intelligence, pp. 153–158. IEEE Press (1997)
22. Kohavi, R., John, G.H.: Wrappers for features subset selection. Artif. Intell. **97**(1–2), 273–324 (1997)
23. Hajian-Tilaki, K.: Receiver operating characteristic (ROC) curve analysis for medical diagnostic test evaluation. Caspian. J. Int. Med. **4**(2), 627–635 (2013)

An Ensemble Learning Approach for Automatic Emotion Classification of Sri Lankan Folk Music

Joseph Charles and Sugeeswari Lekamge

Abstract Music experience is closely associated with our moods and emotions. Even though data mining techniques have been widely adopted in computational analysis of music-emotion, traditional music including Sri Lankan folk music is less explored computationally. Therefore, considering a Sri Lankan folk music dataset, performed the classification using Support Vector Machines, Naive Bayes, Random Forest (RF), k-Nearest Neighbor (k-NN), and Logistic Regression (LR), employing dynamics, rhythm, timbre, pitch, and tonality features. k-NN achieved the maximum accuracy (78.44%) while RF and LR achieved accuracies of 76.19% and 73.42%, respectively. Combining the above three classifiers, an ensemble model was developed. Max-voting was applied, and the results were further enhanced using ensemble boosting. With optimized features, AdaBoost (RF as base estimator) achieved the highest accuracy (95.23%) while reducing the training time significantly. Expanding the dataset in terms of the number of music stimuli and emotion categories looked progressive.

Keywords Music-emotion classification · Ensemble learning · Max-voting · Sri lankan folk music · Computational musicology

1 Introduction

Music is a ubiquitous phenomenon present in our daily lives. It has become a highly influential factor in entertaining and healing humans by evoking various emotions.

J. Charles (✉)
Department of Physical Sciences and Technology, Faculty of Applied Sciences, Sabaragamuwa University of Sri Lanka, Belihuloya, Sri Lanka
e-mail: jcharles@appsc.sab.ac.lk

S. Lekamge
Department of Computing and Information Systems, Faculty of Applied Sciences, Sabaragamuwa University of Sri Lanka, Belihuloya, Sri Lanka
e-mail: slekamge@appsc.sab.ac.lk

S. Shakya et al. (eds.), *Proceedings of International Conference on Sustainable Expert Systems*, Lecture Notes in Networks and Systems 176, https://doi.org/10.1007/978-981-33-4355-9_23

The tremendous growth in the music industry within the recent past has motivated the music information retrieval (MIR) research community worldwide to identify the significant relationships between music and emotion [1]. With the ever-increasing amount of musical content available in digital libraries, keeping track of them and discovering relationships of music with other key aspects including emotion has become a challenging task. Therefore, in this regard, data mining has been extensively utilized which is a broad area of research comprising feature extraction, co-occurrence analysis, similarity measures, and classification and clustering [1, 2].

However, due to the subjectivity connected with emotion perception, discovering the emotional correlates of music has become a complex task warranting further research; different emotions may be elicited among different listeners for the same music stimulus. As well as, the same listener may express another form of emotional expression at different times for the same stimulus [3]. Automatic recognition of music-emotion is a major area of research that utilizes the ability of a computer for sensing and recognizing emotions expressed in music. Usually, musical information is retrieved from music depending on the name of the composer or the title of the composition. The techniques that are used to organize and retrieve music information need to be further improved to cater to the growing demand for effective information access and retrieval. As a result, organizing and retrieving music by emotion has drawn the attention of researchers in the field of computational musicology. Making digital platforms that are capable of recognizing the emotions expressed in music further helps enhance human–computer interaction.

Sri Lankan folk melodies which usually emerged to express the true emotions and feelings of the community have an immense potential in emotional expression. However, these melodies warrant further research on their emotional expression, harnessing the potential of data mining and machine learning, whereas the literature provides evidence for similar investigations very frequently on Western or Western classical music [4]. Further, a study preceding to the current study investigating the generalizability of existing emotion classifiers for classifying the above folk melodies revealed that the generalizability varies depending on the type of emotion [5]. Moreover, the need for developing new models for such cultural-specific melodies also has been emphasized. Therefore, the study aimed to develop an emotion classification model for Sri Lankan folk melodies aiming to achieve an enhanced classification performance through an ensemble approach with different learning classifiers.

2 Related Works

Music-emotion recognition (MER) is one of the major fields in developing music recommender systems. To the best of our knowledge, the very first MER paper was published in 2003, by Feng et al. [6]. The study was carried out using 200 audio clips considering two major attributes namely tempo and articulation. Four predominant emotion categories (happiness, sadness, anger, and fear) were considered in the classification process. By using neural technology, the classifier is reported to have achieved precision and recall of 67% and 66%, respectively.

The performance of any prediction model relies on the number and type of features used to train the model and as revealed by the review of literature, limitations reported in several previous studies [7, 8] are attributed to the feature selection. The majority of the studies in the past have proposed comparatively similar classification approaches utilizing different sets of musical features such as sharpness, timbre, width, tonal dissonance, multiplicity, intensity, and rhythm [7, 8].

Mokhsin, et al. [9] have used Artificial Neural Networks (ANN) for the emotion classification of vocal and instrumental sound timbres of Malay popular music. Timbral features namely spectral centroid, zero-cross, and spectral roll-off were considered in the study in identifying the emotions: anger, happiness, calmness, and sadness. The final trained system was able to detect the emotions with an accuracy of 75%.

Introducing a computational framework, Lu et al. [8] attempted to detect and track mood based on acoustic features related to intensity, timbre, and rhythm. A dataset of 800 classical music pieces were used in the study with an accuracy of 86.3%.

As opposed to using musical acoustic features for emotion classification, song lyrics are also employed. On a dataset comprising metadata on mood about songs from the blog LiveJournal [10], Dang and Shiraj [11] used Naive Bayes, Support Vector Machines (SVM), and graph-based methods. Mood categories as defined in Music Information Retrieval Evaluation eXchange (MIREX 2007) were used in the study. Poor classification performance in this study was reported to have been due to the inclusion of extremely subjective metadata and song lyrics which contain numerous metaphors that are only human-understandable.

As evident by a recent review of literature on music-emotion classification [12], some of which were discussed above, the studies vary depending on several factors including the datasets, acoustic features, and feature extraction tools, listener populations in the creation of ground-truth, and the classification models used. Therefore, the discrepancies in the classifier performance could be partially attributed to the above differences while they could also be affected by the specific limitations within each study.

3   Methodology

3.1   Tools and Resources

The study was conducted using MATLAB MIRToolbox version 1.7.2, for the feature extraction. Among other feature extraction tools including Marsyas [4, 7, 14, 15] and Psysound [7, 8] which are reported to have been frequently used in the related literature, MIRToolbox which is free and open-source has been widely employed [8, 9, 13, 16] by previous researchers as well. For scientific programming, Spyder IDE was used with the Python programming language to train and test the classifiers. Further, the study was carried out through specific hardware configuration (2.3 GHz Intel Core i5, 8 GB 2133 MHz LPDDR3 memory, macOS High Sierra Version 10.13.6 operating system).

3.2 Dataset

The dataset used in this study consisted of 206 Sri Lankan folk music stimuli. Each stimulus was of 30 s duration which was intentionally composed (Disura Institute of Music, Ruwanwella, Sri Lanka) and recorded (Audio Visual Unit, Sabaragamuwa University of Sri Lanka) for emotion-based research purposes. The stimuli were annotated with the predominant emotion labels: happiness, sadness, and fear. The above three emotions were considered since they are easily distinguishable from all other emotions. The annotation task was performed by a panel of three experts in the field of musicology. The music stimuli contained vocal and instrumental content where there were no lyrical contents.

3.3 Audio Preprocessing and Feature Extraction

All music files must be preprocessed before any computational simulation for improved mining and to obtain correct output values [9]. All the music files were resampled into a common standard format of 44100 Hz; Stereo; 32-bit;.wav PCM, and compressed to the same volume scale that is most applicable for processing using MIRToolbox. To train the machine classifier via supervised learning, 22 acoustic features were extracted using MATLAB MIRToolbox employing various dimensional feature methods defined in the toolbox (Table 1). These 22 features belong to five selected perceptual dimensions of music namely: dynamics, rhythm, timbre, pitch, and tonality.

3.4 Classifier Training

Emotions perceived by listening to music are a result of different musical-feature statistical values. Therefore, musical features that are highly significant for assessing emotions should be considered when developing a music-emotion classification (MEC) system. As shown in Fig. 1, the MEC process consists of three stages. Stage 1 refers to audio preprocessing which was described in 3.3.

In Stage 2, the dataset was split into 80% and 20%, respectively, for training and testing purposes. Five basic classifiers identified by the literature review as often used classifiers for the classification of music emotions: Support Vector Machine [4, 7,

Table 1 Dimensional feature methods used in the study

Dimension	Used method (MIRToolbox)
Dynamics	*mirrms()*
	mirlowenergy()
Rhythm	*mirfluctuation()*
	mireventdensity()
	mirtempo()
	mirmetroid()
	mirmetroid()
	mirpulseclarity()
Timbre	*mirzerocross()*
	mirrolloff()
	mirbrightness()
	mircentroid()
	mirspread()
	mirskewness()
	mirkurtosis()
	mirentropy()
	mirroughness()
	mirroughness()
Pitch	*mirpitch()*
	mirinharmonicit()
Tonality	*mirkey()*
	mirtonalcentrid()

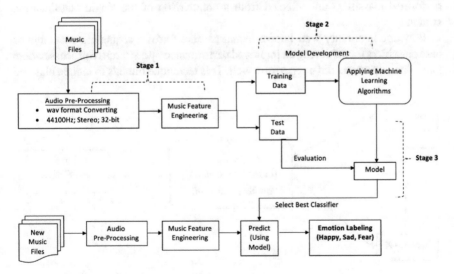

Fig. 1 Music-emotion classification process

13–15], k-Nearest Neighbor [7, 15], Naïve Bayse [4, 11], Logistic Regression, and Random Forest were then trained.

In Stage 3, the trained classifiers were evaluated using the testing dataset, and the classifiers with the best performance as indicated through precision, recall, and F-measure values were identified. The use of such common evaluation measures enables us to compare the performance of our approach with those of the previous approaches and thereby to analyze the impact of each approach. In addition, all the classifier models were tested by k-fold cross-validation ($k = 5$) as it ensures that every observation from the original dataset has the chance of performing in the training set and testing set. Based on a comparison among the performances of the classifiers, the best classifiers were identified.

3.5 Ensemble Learning

With the aim of creating an improved classification model, the study further applied an ensemble learning employing multiple learning algorithms. Each individual algorithm predicts a discrete value according to the output. To obtain a better predictive performance, different weak classifiers are integrated to generate a strong ensemble of classifiers. This technique directs to achieving an enhanced accuracy by reducing the individual error rate [16]. Each classifier applied in the proposed machine learning approach has its individual contribution in the final predictive model performance. Here, an ensemble technique is used for voting and boosting. Voting is an advanced method of single classifier performance, which returns individual classifier vote and final prediction label that performs majority voting as shown in Fig. 2. The best-performed classifiers are selected from a comparison of the single classification system.

Boosting is another well-known technique that forms a sequential ensemble of base classifiers for achieving an improved performance by an iterative process, where individual classifiers did not perform well. This technique intends to reduce bias and

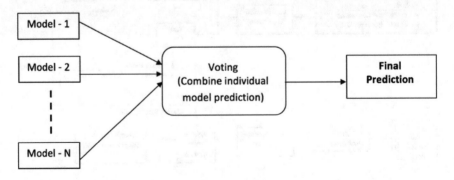

Fig. 2 Ensemble learning approach with max-voting

increase the variance [17]. Here, a new training dataset is formed according to its sample distribution. Adaptive boosting (AdaBoost), gradient boosting, and extreme gradient boosting (XGBoost) algorithms were used for ensemble boosting.

3.6 Feature Selection

Including more features in the model during training makes the model more complex and overfitting with the dataset. Proper feature selection significantly reduces the overfitting problem and makes the model more generalizable. Therefore, using two different techniques namely feature importance and correlation matrix, a highly optimized set of features were obtained.

4 Results and Discussion

The individual classifier performances obtained through the experiments are discussed first under this section. Then, the three-classifier ensemble results are discussed. Finally, a performance comparison between the individual classifier models and the ensemble model is presented.

4.1 Performance of Individual Classifiers

Table 2 depicts the performance of individual classifiers. Accordingly, the highest accuracy (78.57%) was achieved by k-NN.

Therefore, k-NN was identified as the best classifier, while Random Forest and Logistic Regression also performed comparatively well, achieving accuracies of 76.19% and 73.42%, respectively.

Further, the experiment was continued to evaluate the performance of classifiers by using fivefold cross-validation. Apart from the conventional training and testing approach, cross-validation randomly partitions the dataset into equal-sized subsamples. Thereafter, each fold splits the dataset in training and testing with randomly selected data. After the training, the test data is experimented and evaluated. Likewise, the training and testing process is repeated until each fold serves as the test data. In each iteration, the accuracy is returned for k-folds, and finally, the average of all k accuracies is calculated. Table 3 provides a comparison of the performance of individual classifiers resulting from fivefold cross-validation. During this step also, k-NN was identified as the best classifier among others, achieving a mean accuracy of 78.44%. The Logistic Regression and Random Forest achieved accuracies of 73.12% and 69.79%, respectively, as given in Table 3.

Table 2 Performance of individual classifiers (SVM—Support Vector Machine, NB—Naïve Bayes, k-NN—K-Nearest Neighbor, LR—Logistic Regression, RF—Random Forest, H—Happiness, S—Sadness, F—Fear)

Algorithm	Confusion matrix				Evaluation metrics			Accuracy (%)
					Precision (%)	Recall (%)	f1-score (%)'	
SVM		H	S	F	68	70	76	71.19
	H	8	2	0				
	S	5	13	1				
	F	1	1	11				
NB		H	S	F	69	68	67	66.66
	H	6	2	6				
	S	8	11	0				
	F	2	0	11				
k-NN		H	S	F	94	79	86	78.57
	H	10	0	0				
	S	4	15	0				
	F	4	1	8				
LR		H	S	F	70	74	70	73.42
	H	6	2	2				
	S	3	14	2				
	F	1	2	10				
RF		H	S	F	81	71	74	76.19
	H	5	4	1				
	S	1	18	0				
	F	0	4	9				

Table 3 Performance of Individual Classifiers (after fivefold cross-validation) (SVM—Support Vector Machine, NB—Naïve Bayes, k-NN—K-Nearest Neighbor, LR—Logistic Regression, RF—Random Forest)

	SVM (%)	RF (%)	KNN (%)	NB (%)	LR (%)
1st fold	76.47	73.53	88.24	52.49	76.47
2nd fold	70.59	73.53	79.41	47.05	73.53
3rd fold	66.67	90.91	81.82	79.74	75.76
4th fold	62.50	59.38	75.00	34.67	65.63
5th fold	48.39	51.61	67.74	37.13	74.19
Mean	64.92	69.79	78.44	50.22	73.12

Table 4 Ensemble learner results (RF—random forest, LR—logistic regression, k-NN—k-nearest neighbor)

Algorithm's evaluation performance	Max-voting classifier (RF, LR, k-NN)	AdaBoost		Gradient boosting	XGBoost
		RF	LR		
Accuracy (%)	81.42	88.28	74.57	54.76	83.8
Precision	0.76	0.72	0.61	0.59	0.86
Recall	0.81	0.64	0.72	0.55	0.74
F-measure	0.71	0.63	0.72	0.56	0.72
Model training time / seconds	0.1555	0.0227	0.1046	0.4373	0.1155

4.2 Performance of the Classifier Ensemble

Based on the classification accuracies of individual classifiers (fivefold cross-validation applied), k-NN, LR, and RF were selected for the ensemble model, and the max-voting approach was used. Each classifier was used as a base estimator and was evaluated with the same training dataset. Using *VotingClassifier* class from *sklearn.ensemble*, the models were pitted against each other. Accordingly, the max-voting ensemble experiment achieved an accuracy of 81.42%, which is far better than the performance of the individual classifiers.

To obtain a more reliable and accurate prediction result, boosting which is another technique of ensemble approach was applied to the dataset. As given in Table 4, the AdaBoost technique was found to be promising, which yielded the highest accuracy (88.28%) for Random Forest base estimator, whereas the lowest accuracy (54.76%) was for Logistic Regression. Further, XGBoost and Gradient Boosting ensemble boosting algorithms yielded accuracies of 83.8% and 74.57%, respectively. Based on the above results and comparisons concerning the selected performance measures, AdaBoost and XGBoost algorithms were identified as the well-performing.

4.3 Classifier Performance After Applying Feature Selection Techniques

Through applying the feature importance technique for feature selection, the score for each feature of the dataset was obtained. In this regard, extra tree classifier was used which comes with an inbuilt class of tree-based classifiers which extracted the ten most important features as shown in Fig. 3. These ten features included the pulse clarity (M), fluctuation (M), pitch (M), entropy of spectrum (M), tempo (M), tonal centroid (M), spectral flatness (M), spectral kurtosis (M), roughness (SD), and spectral skewness (M).

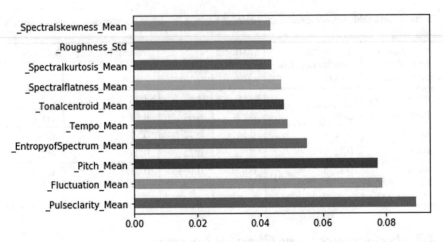

Fig. 3 Top-ten features from the feature importance technique

Figure 4 shows the correlation matrix heatmap with positively correlated and negatively correlated values between features. This makes it easy to identify the most related features of the target variable.

The results of the ensemble classifier model showed considerable improvement with the feature selection techniques. With the feature importance technique, the best results for emotion classification surpassed those with the correlated features technique. Accordingly, the highest classification performance was achieved with AdaBoost (RF) and XGBoost with accuracies of 95.23% and 93.23%, respectively as shown in Fig. 5.

Therefore, the AdaBoost algorithm with RF as the base classifier was finally identified for the emotion classification model. Not only in respect of the accuracy, but also in respect of the training time, the selected model outperformed the other models as shown in Fig. 6, which is efficient, effective, and economically beneficial for emotion classification. These are among the key concerns when integrating the developed model into any prospective commercial application for music recommendation and retrieval.

5 Conclusion

Performance of the five supervised learning classification algorithms which were identified through the literature review was evaluated using a collection of 206 Sri Lankan folk music stimuli that were intentionally composed to express happiness, sadness, and fear as the principal emotions. Among them, k-NN, Logistic Regression, and Random Forest which yielded comparatively higher accuracies were selected for the ensemble model. Voting and boosting techniques were applied which finally revealed AdaBoost (RF as the base estimator) as the most suitable algorithm which

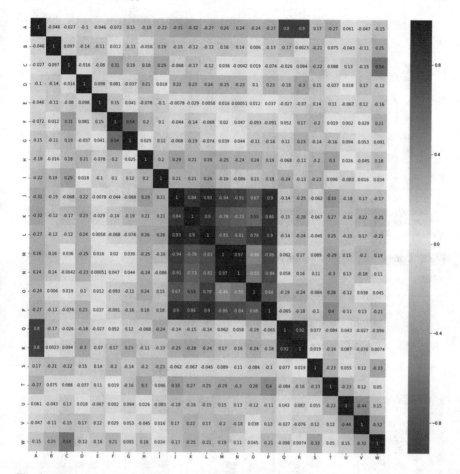

Fig. 4 Correlation matrix heatmap

yielded an accuracy of 95.23%. The model consisted of ten musical acoustic features identified through feature selection techniques. The selected model outperformed the other classifiers not only in terms of the accuracy but also in terms of the classifier training time. Taken together, the study makes a significant contribution to the domain of music data mining through comprehensively applying machine learning techniques to Sri Lankan folk music which is new and yet to be explored dataset. However, several limitations could be identified in the present study which need to be overcome in future studies, e.g., the classifiers were trained on a comparatively small dataset which may lead to model overfitting and thereby produce inaccurate results. Moreover, as an initiating step, the study considered only three key emotions, whereas for achieving an enhanced validity and a wider acceptance of the model, it is required to expand the dataset in terms of the number of music stimuli while introducing various other emotions representing the entire emotional spectrum.

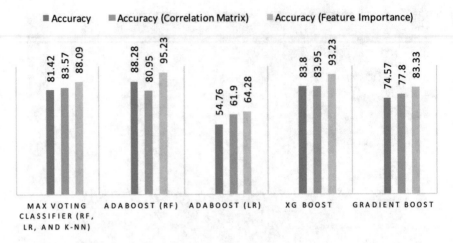

Fig. 5 Comparison of classification accuracy

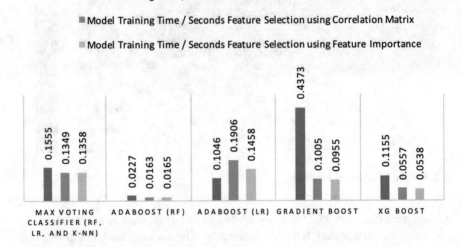

Fig. 6 Comparison of classifier training time

References

1. Liu, Y., Liu, Y., Zhao, Y., Hua, K.A.: What strikes the strings of your heart?—feature mining for music emotion analysis. IEEE Trans. Affect. Comput. **6**, 247–260 (2015)
2. Pickens, J., Crawford, T.: Harmonic models for polyphonic music retrieval. In: Proceedings of the 11th International Conference on Information and Knowledge Management—CIKM 02 (2002). doi: https://doi.org/10.1145/584861.584863
3. Ren, J.-M., Chen, Z.-S., Jang, J.-S.R.: On the use of sequential patterns mining as temporal features for music genre classification. In: 2010 IEEE International Conference on Acoustics, Speech and Signal Processing (2010). doi: https://doi.org/10.1109/icassp.2010.5495955

4. Sun, X., Tang, Y.: Automatic music emotion classification using a new classification algorithm. In: 2009 Second International Symposium on Computational Intelligence and Design (2009). doi: https://doi.org/10.1109/iscid.2009.28
5. Charles, J., Lekamge, L.S.: Generalizability of music emotion classifiers: an evaluation of the applicability of *miremotion* in emotion classification of Sri Lankan folk melodies. In: Abs. International Research Symposium on Pure and Applied Sciences (IRSPAS) p. 153. (2019)
6. Feng, Y., Zhuang, Y., Pan, Y.: Popular music retrieval by detecting mood. In: Proceedings of the 26th Annual International ACM SIGIR Conference on Research and Development in Information Retrieval—SIGIR 03 (2003). doi: https://doi.org/10.1145/860500.860508
7. Liu, C.C., Yang, Y.H., Wu, P.H., Chen, H.: Detecting and classifying emotion in popular music. In: Proceedings of the 9th Joint Conference on Information Sciences (JCIS) (2006). doi: https://doi.org/10.2991/jcis.2006.325
8. Lu, L., Liu, D., Zhang, H.-J.: Automatic mood detection and tracking of music audio signals. IEEE Trans. Audio Speech Lang. Process. **14**, 5–18 (2006)
9. Mokhsin, M.B., Rosli, N.B., Zambri, S., Ahmad, N.D., Rahah, S.: Automatic music emotion classification using artificial neural network based on vocal and instrumental sound timbres. Journal of Computer Science **10**, 2584–2592 (2014)
10. Discover global communities of friends who share your unique passions and interests. (Online). Available: https://www.livejournal.com/. Accessed 03 Nov 2019
11. Dang, T.-T., Shirai, K.: Machine learning approaches for mood classification of songs toward music search engine. In: 2009 International Conference on Knowledge and Systems Engineering (2009). doi: https://doi.org/10.1109/kse.2009.10
12. Charles, J., Lekamge, L.S.: Machine learning approaches for emotion classification of music: A systematic literature review. In: IEEE International Conference on Advancements in Computing (ICAC 2019). https://doi.org/10.1109/ICAC49085.2019.9103378
13. Ardakani, M.S., Arbabi, E.: A categorical approach for recognizing emotional effects of music. arXiv (2018)
14. Trohidis K, Tsoumakas G, Kalliris G, Vlahavas I: Multi-label classification of music by emotion. In: EURASIP Journal on Audio, Speech, and Music Processing. doi: https://doi.org/10.1186/1687-4722-2011-426793. (2011)
15. Panda R and Paiva RP: Using Support Vector Machines for Automatic Mood Tracking in Audio Music. In: Proceedings of the 130th Audio Engineering Society Convention – AES 130, September. (2011)
16. Laurier C, Herrera P: Automatic Detection of Emotion in Music: Interaction with Emotionally Sensitive Machines. In: Machine Learning 1330–1354. (2012)
17. Greenberg, D.M.: RentfrowPJ: Music and big data: a new frontier. Current Opinion in Behavioral Sciences **18**, 50–56 (2017)
18. Lekamge, S., Marasinghe, A., Kalansooriya, P., Nomura, S.: A Visual Interface for Emotion based Music Navigation using Subjective and Objective Measures of Emotion Perception. International Journal of Affective Engineering **15**, 205–211 (2016)
19. Jr. CNS, Koerich AL, Kaestner CAA: A machine learning approach to automatic music genre classification. In: Journal of the Brazilian Computer Society 14:7–18. (2008)
20. Lekamge S, Marasinghe A.: Identifying the associations between music and emotion in developing an emotion based music recommendation system. In: Transactions on GIGAKU 2:2 (2014)
21. Chen, Y.-L., Chang, C.-L., Yeh, C.-S.: Emotion classification of YouTube videos. Decis. Support Syst. **101**, 40–50 (2017)
22. Iloga, S., Romain, O., Tchuenté, M.: A sequential pattern mining approach to design taxonomies for hierarchical music genre recognition. Pattern Anal. Appl. **21**, 363–380 (2016)
23. Zhang, J., Huang, X., Yang, L., Nie, L.: Bridge the semantic gap between pop music acoustic feature and emotion. Build an Interpretable Model. Neurocomputing **208**, 333–341 (2016)
24. Babu, K.B.: Joonwhoan, L: Rough Set-Based Approach for Automatic Emotion Classification of Music. J. Inf. Process. Syst. (2017). https://doi.org/10.3745/jips.04.0032

25. Bergstra, J., Casagrande, N., Erhan, D., Eck, D., Kégl, B.: Aggregate features and ADABOOST for music classification. Machine Learning **65**, 473–484 (2006)
26. Tzanetakis, G., Cook, P.: Musical genre classification of audio signals. In: IEEE Transactions on Audio and Speech Processing (2002)

Mobile Application to Identify Fish Species Using YOLO and Convolutional Neural Networks

K. Priyankan and T. G. I. Fernando

Abstract Object detection is one of the sub-components of computer vision. With recent development in deep neural networks, many day-to-day problems can be solved. One of the practical problems faced by shoppers is the difficulties in identifying the fish species correctly. Even though there are few studies to solve this problem, those implemented solutions are not easily accessible. The goal of this research is to implement a mobile application based on deep learning that can detect the fish species and provide information on vitamins, minerals, prices, and recipes. For this study, top-selling 16 Sri Lankan fish species are used. This study proves that it is possible to build a model using a YOLO-based convolutional neural network. Mobile application takes 3–20 s to detect the fish species based on Internet speed.

Keywords Fish detection · Convolutional neural network · YOLO · Detection and classification

1 Introduction

People visit the fishing market very often since fish products have vital vitamins and minerals which are required for a healthy life. Since the fish products fulfill 53% of animal protein per capita consumption, the country has achieved 44.6 g per day [1]. The supermarket industry offers a wide variety of food and household items where customers have many advantages compared to traditional grocery shops. Most of the supermarkets have tags specifying the product details and price details for the convenience of the customer. One of the sections in a supermarket is the fresh fish/meat section.

The current approach followed in buying fish in most of the supermarkets and fish markets is, sliced pieces/whole fish are displayed where customers must ask the shopkeeper and get to know the fish type and price details and buy according

K. Priyankan (✉) · T. G. I. Fernando
Department of Computer Science, University of Sri Jayewardenepura, Gangodawila 10250, Nugegoda, Sri Lanka
e-mail: priyankankiru@gmail.com

© The Editor(s) (if applicable) and The Author(s), under exclusive license to Springer Nature Singapore Pte Ltd. 2021
S. Shakya et al. (eds.), *Proceedings of International Conference on Sustainable Expert Systems*, Lecture Notes in Networks and Systems 176,
https://doi.org/10.1007/978-981-33-4355-9_24

303

to the requirement. This consumes plenty of time, and sometimes can lead to false information. Customers can spend less time in the fish market if the fish species can easily be identified like the tags available on each product at the supermarkets.

Japan has a high life expectancy linked to diet. This is achieved due to proper fish consumption. Some fish products are used to cure many diseases and lead to a healthy lifestyle. People must be able to get information about vitamins and minerals in each fish species and what type of diseases it can cure. This information must be simple in such a way that the general public must understand and also easily available at any time [2].

Machine learning and deep learning play a major role in computer vision. Computer vision applications are widely used with the advancement in technology and resources. It is used in a simple scenario like character recognition to a complex scenario like self-driving vehicles.

Object detection is one of the sub-components in computer vision that solves many day-to-day problems. Object detection is the process of finding the exact location of objects in an image. Even though humans and animals can easily detect objects, it is difficult for a machine to detect objects in a scene. But with the use of computer vision, the objects can easily be detected. Although much research in the object detection field has been done, very few have tried in solving the fish detection problem [3]. Fish detection can be considered as one main application of computer vision. Object detection differs from object classification, i.e., object classification shows which object is depicted in the image, and object detection shows where the object is in the image. In this study, the main focus is to implement a simple deep learning solution for detecting the fish species accurately.

This study aims to implement a mobile application that can identify and detect fish species and then view the price details, nutrients, and recipes efficiently. The objectives of this research are studying the existing tools and technologies that are applied to classify fish types, studying the deep learning algorithms that can be applied to detect and identify objects and develop a novel method to detect and identify the fish type even if the image contains a sliced fish using deep learning algorithms and develop a mobile application which can detect and identify the fish type and then provide the vital information about the fish such as price, nutrients, and recipes to cook the fish.

2 Related Work

In the study conducted by Eiji et al. [4], a neural network was developed to identify the fish species in Japan by using reference points. By applying the truss protocol, getting the characteristic points from the edges of the fish in an image is called reference points. The ratio of specific truss lengths between the 'reference points' relative to the total body is used to compile the dataset, and it is used as the network input. The study says that the neural network developed provides higher accuracy in identifying the fish species. But this study was conducted only for three fish species.

Fish species are identified using the head section and color of the body section. After using the RGB values, the accuracy was enhanced proving that the color plays a major role in identifying the species of the fish.

Another study conducted by Michael Chatzidakis [5] has used a convolutional neural network (CNN) to identify fish species in the camera footage. For this study, the researcher has considered only eight species of fish. This study says that 97.4% accuracy was obtained before overfitting. The pretrained model has been used for this study with many augmentation techniques applied to the training dataset. Dropout, weight decay, and batch normalization have been effectively used to improve the convergence and decrease the overfitting. But this solution cannot be embedded in a mobile device since this need high-performance computing devices to classify fish species.

Li et al. [6] have researched on fast accurate fish detection and recognition of underwater images with fast R-CNN. In this study, they stated that fast R-CNN is more suitable for underwater fish detection, and it is comparatively faster than a CNN. For this study, 12 types of fish species are used which are found in the deep ocean. This design contains a RoI layer, two sibling layers (a fully connected layer and softmax layer over 12 fish classes plus background class and bounding-box regressors), two fully connected layers, and five convolutional layers. This system produces output on classes and bounding-box values by taking an RGB image and its 2,000 Region of Interests (ROIs) collected by the selective search. After the first and second convolutional layers, max-pooling layers and necessary normalization are applied. All the fully collected layers and convolutional layers are subjected to a rectified linear unit (ReLU) nonlinearity.

Another study "Automatic Nile Tilapia Fish Classification Approach Using Machine Learning Techniques" identifies one single fish using support vector machines [7]. This study uses the scale-invariant feature transform (SIFT) and speeded-up robust Features (SURF) algorithms to extract features. Furthermore, the study states that the experimental results obtained from the support vector machine algorithm outperformed other machine learning techniques, such as artificial neural networks (ANNs) and K-nearest neighbor (KNN) algorithms, in terms of the overall classification accuracy.

Hnin and Lynn [8] proposed a system to identify species based on taxonomic characters and specimens. The study provides a statistical approach for helping taxonomists correctly identify the species and comment wrongly identified species. This kind of system has a key factor which is feature selection that reduces the data dimension. This system created the training dataset using a combination theory. Attribute pairs generated from the system were tested using two classifiers. Most matching features are selected based on the accuracy of each classifier for each attribute pair. Effectiveness of the features verified by selected feature sub-set based on three supervised classifiers.

Salimi et al. [9] introduce a system based on the 'Otolith' contours to identify the fish species with high classification accuracy. They have identified 14 fish species. Short-time Fourier transformations (STFT) to extract features of the 'Otolith' contours, and then, discriminant analysis (DA) has been used to classify the fish species from the extracted features. This study states that they were able to get 90% accuracy for nearly all the 14 classes that they have used.

How fast a trained model takes to detect an image is important in a practical point of view. Most of the above studies are not capable enough to detect fish species quickly, so that they can be implemented in handheld devices. This research is unique and differs from others since there is no other model that is capable to detect Sri Lankan fish species instantly.

3 Methodology

According to the report [1] published by the Ministry of Fisheries and Aquatic Resources Development in Sri Lanka, fish production in Sri Lanka is mainly from the marine sector, coastal water, offshore seawater, agriculture, and shrimp farming. There are around 90 major species of fish that are consumed by the public. It is very difficult to collect a dataset for all the fish species with a limited time of this undergraduate research project. So, the plan is to reduce the number of fish species but also solve the problem faced by the general public in identifying fish species. Mainly images were collected from the sectors such as people who consume fish (general public), people who take part in fish production (fisherman), and also from people who sell fish products (supermarkets/fishmarket owners).

By considering all the information collected from all three sectors, choose the top 16 fish species that are consumed by the general public of Sri Lanka [1]. Figure 1 depicts the top 16 species selected for this study according to the data provided by three groups of people. The most important task in this research is to have a dataset that can be used to train the model. But there is no such dataset of fish species of Sri Lanka. So, collecting and preparing the dataset plays a major role in this study.

3.1 Data Collection and Preprocessing

Before starting to collect the dataset, the plan is to decide how to collect the fish images. Since this study is based on object classification, we need plenty of images for each fish species. We should have nearly 800–900 images of each species to get better accuracy. First, we found out the places that we can get the top 16 fish species that we selected. Accordingly, the following are the places that images were collected:

- John Keells Distribution Center (Wattala)
- Fisheries Cooperation (Nawina)
- Fisheries Cooperation (Moratuwa)
- Fish Market (Colombo 06)
- Fish Market (Colombo 04)

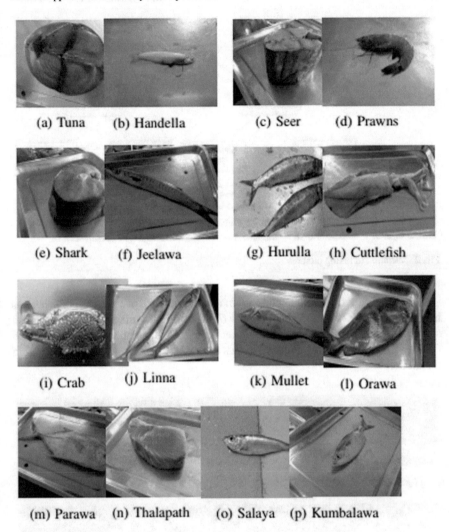

(a) Tuna   (b) Handella   (c) Seer   (d) Prawns

(e) Shark   (f) Jeelawa   (g) Hurulla   (h) Cuttlefish

(i) Crab   (j) Linna   (k) Mullet   (l) Orawa

(m) Parawa   (n) Thalapath   (o) Salaya   (p) Kumbalawa

Fig. 1 Top 16 fish species

- Keells Super Centre (Dehiwala)
- Internet images.

Using the plan given in Fig. 2, collect 120 pictures for each species from each shop and hence collected 1920 (=120*16) images. But this is not enough to train a deep neural network. Therefore, it is decided to use image augmentation techniques to increase the number of images without distorting the images. The techniques are followed to increase the number of pictures of fish in the dataset which are translation, rotation, and flipping.

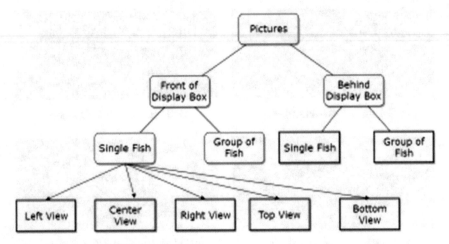

Fig. 2 Variation in taking pictures

After applying the augmentation methods, the number of images for each class is increased to 150 for each shop. Finally obtained 14,400 images for all top 16 fish species, i.e., 150 (original + images after applying augmentation techniques) * 6 (shops) * 16 (fish species). Figure 3 shows the variation in images taken for training the model.

(a) Group View,Behind Fish Display Box

(b) Single View,Behind Fish Display Box

Fig. 3 Variations in images taken according to the plan

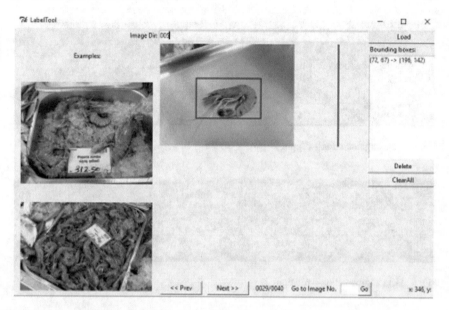

Fig. 4 Preprocessing using the BBox tool

This study is not only fish classification but also handles fish detection, where the model should identify where the fish is in the image. To train this kind of model, it is required to have training dataset with a bounding box specifying where the fish is in the image. This process takes a very long period since this must be done manually by the developer one by one specifying the fish species class and bounding box, respectively. An open-source Python program called the "BBox" tool [10] has been used to do this process. This tool has a graphical user interface where the bounding box can be drawn, and then, it generates the number of boxes drawn with, Xmin, Ymin, width, and height of the bounding boxes. Figure 4 shows the user interface of the tool.

Dataset is the key to all machine learning problems. The number of examples in the dataset affects the accuracy of the model [11]. As discussed earlier, only three augmentation techniques were used to increase the number of images in the dataset without distorting images. Only 20 images from each species were taken and translated 5 to 20 pixels up, down, left, and right. Similarly, those images were rotated 1 to 2 degrees to left and right. An image flipping technique was also used since flipping fish images does not affect the accuracy of the model. Figures 5 and 6 depict the images after applying the two augmentation techniques—flipping and rotation.

After applying these techniques, the final dataset was spitted into training and testing datasets as to the ratio 7:3, respectively.

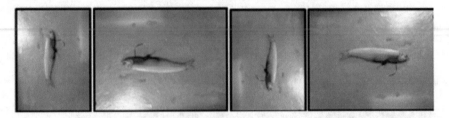

Fig. 5 Data augmentation—flipping

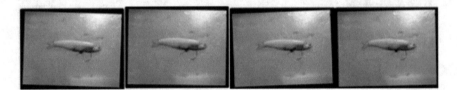

Fig. 6 Data augmentation—rotation

3.2 YOLO

YOLO [12] is a concept called "You Only Look Once." In this method, bounding boxes with the highest score are assumed as detection. This is achieved by applying the model to the image at multiple locations and multiple scales. Unlike other architecture, YOLO applies a single neural network for the entire picture. An image is split into multiple regions and probabilities and bounding boxes are predicted for every region. Each bounding box weights its corresponding output probabilities.

This YOLO divides the image into 5*5 grid cells. Each cell is responsible for predicting two bounding boxes, which are called the anchors. So, in the final layer, the tensor of size 5*5*(5*2 + 16) is obtained.

Since after producing two bounding boxes for each cell, the first ten values of the 1*26 are X coordinate of the bounding-box center inside the cell, Y coordinate of the bounding-box center inside the cell, width of the bounding box, height of the bounding box, and confidence of the class—Pr(Class—object) (Fig. 7).

The remaining 16 cells denote the conditional probability of the object belongs to the class i if an object is present in the box. Next, multiply all of these class scores with bounding boxes for each grid cell. So now possess a 5*5*2 = 50 bounding box of (1*16) tensor. Now utilize a non-maximum suppression algorithm [13] to set the score to zero for redundant boxes. After that, it is left with only 2 or 3 bounding boxes where others are set to zero. From these bounding boxes, select the boxes according to the class score.

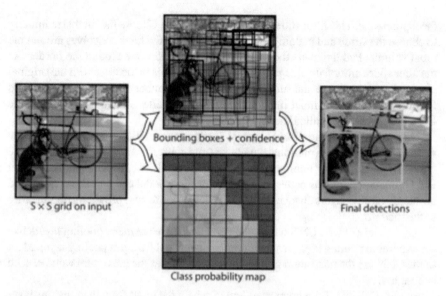

Fig. 7 YOLO architecture

3.3 Convolutional Neural Network (CNN)

A CNN [14–17] is a popular deep learning technique for visual recognition tasks because of its proven quality of performance in image classification with less image preprocessing. In machine learning, CNN is a type of feedforward artificial neural network (Fig. 8). They are widely used in the field of pattern recognition within images and videos. CNN consists of several layers. These layers are convolutional layers, ReLU layers, pooling layers, and fully connected layers. When these layers are stacked together, CNN architecture has been created. The neurons of the CNN layers are arranged in three dimensions (width, height, and no. of channels). The no. of channels of the input layer is three for color images, and it is one for gray-scaled images.

The convolutional layer [17] is the first layer on CNN. This layer has a filter/kernel which has weights and biases. The depth of the filter must be the same as the depth of

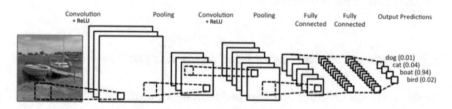

Fig. 8 Convolutional Neural Network (CNN)

the input image. This filter slides over the input image. Choose the filter size initially to choose the stride and padding. Stride controls how the filter convolves around the input volume. Padding pads the input volume with the value around the border. As the filter slides around the image, it multiplies the values in the filter with the original pixel value of the image and summed up as a single number. This process is repeated for every region of the input image. The final region after sliding is completed; the output of this layer is called the feature map.

The purpose of adding a ReLU layer [17] after a convolutional layer is to introduce nonlinearity since convolutional layers operate with linearity. Even though there are many functions for nonlinearity like sigmoid and tanh, researchers found that ReLU function performs better than those since the model can be trained faster and accurately. ReLU layers change the negative activation to zero providing nonlinearity to the model.

After a ReLU layer, CNN has a pooling layer. There are many pooling layers like the maximum pooling layer, average pooling layer, and L2 norm pooling layer. Most of the CNN use the maximum pooling layer which takes the maximum value in each sub-region.

Finally, CNN has fully connected layers, where the input from the previous layer is flattened and sent so as to transform the output into the number of classes as desired by the network.

In addition to that, dropout layers [16] are added between layers to prevent the overfitting problem. Dropout is a technique where randomly selected neurons are ignored during the training.

3.4 Implemented Architecture

YOLO has an advantage over the other CNN. YOLO applies a single CNN for both classification and localization of objects. YOLO can process images at about 40–90 FPS. Our network uses 24 convolutional layers followed by two fully connected layers in the first layer. To build this model, 90 layers of filters have been used in the second layer. This model takes the input image and resizes to 448*448 pixels. Then, the image passes through the first and second layers of the network above and outputs 7*7*30 tensor. This output provides the coordinates of the bounding box and the probability of the detected class.

3.5 Hardware and Software Used in the Experiment

For this study, the following hardware and software are used:

- A GPU computer was used with Intel Core i7 CPU, DDR3 16 GB of memory and NVIDIA GeForce GTX 960 processor (2 GB) and Tesla K40 GPU.

- Ubuntu 14.04 (64 bit)
- NVIDIA DIGITS 5
- MATLAB R2012
- BBox-Label

4 Results

The accuracy of this model is 77%. Even though there were many studies with better accuracy than this, they require high computing resources or a considerable amount of time to detect the fish. Those implemented solutions are not available for easy access to the general public, so that they can solve their problem in identifying the fish species correctly.

The average time taken for detecting the fish species depends on the Internet speed and the quality of the image. On average, this takes 3–20 s per image to detect.

Dataset greatly affects the accuracy and efficiency of a trained model. Table 1 and Table 2 show that the model needs to be more improved. The accuracy of the model can be further improved by introducing new images for some classes and retrain the model with these images.

A mobile application has been developed to implement the model which is displayed in Fig. 9. Since a mobile device has a limited capacity and speed, a script running on an AWS machine is called from the mobile device when a user inputs an image for the detection of a fish. Then, the output of the application which is

Table 1 Confusion matrix

n = 1607	Predicted yes	Predicted no
Actual yes	768	202
Actual no	166	471

Table 2 Confusion matrix measures

Measure	Value	Derivation
Sensitivity	0.8143	TPR = TP/(TP + FN)
Specificity	0.6999	SPC = TN/(FP + TN)
Precision	0.7828	PPV = TP/(TP + FP)
Negative predictive value	0.7394	NPV = TN/(TN + FN)
False positive rate	0.3001	FPR = FP/(FP + TN)
False discovery rate	0.2172	FDR = FP/(FP + TP)
False negative rate	0.1857	FNR = FN/(FN + TP)
Accuracy	0.7652	ACC = (Tp + TN)/(P + N)
F1 score	0.7982	F1 = 2TP/(2TP + FP + FN)

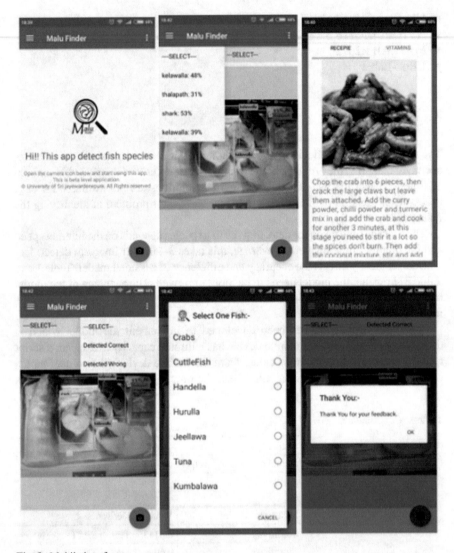

Fig. 9 Mobile interfaces

depicted in Fig. 10 is the detected fish with a bounding box and the accuracy of the model. Further, the user can select the detected fish from a dropdown menu to view the details of vitamins, minerals, and price of the fish, and its recipes. The application is also capable of detecting sliced fish and multiple fish types in a single image. The model has been trained in a way that the user can input an image taken in any orientation.

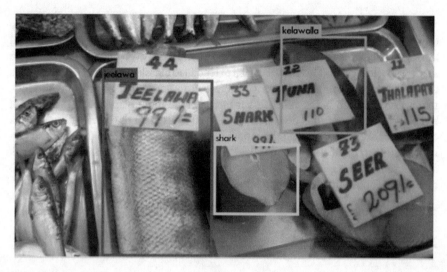

Fig. 10 Output of the application

Also, a user can send feedback to the system after it detects a fish correctly or incorrectly. These images and labels are saved in the server, so that these images can be used for future training of the model and fine-tune it further.

For more details of the mobile application, a demonstration is available at the YouTube link "https://www.youtube.com/watch?v=PuuBARG4S-0."

5 Conclusion

The main goal of this research is to identify the fish types correctly. The complex part of this study is to identify the sliced fish images correctly. The investigation is to implement a convolutional neural network using YOLO to identify fish species and locate where the fish are in the image. Sliced fish species and images with multiple fish species in a single image also can be classified and detected properly using this neural network.

To review the related work on our problem domain, this is fish species detection. Related studies proved that time consumption for identifying with low computational resources is high, but to implement a mobile application that uses this trained neural network successfully, testing is performed by the neural network with reasonable accuracy and low time latency.

CNN requires less image preprocessing compared with other approaches. To conclude, the CNN can be used to detect the fish species with acceptable accuracy.

To implement a successful mobile application that can detect the fish species, and customers can access this service from any place and at any time. But there are several enhancements that can be incorporated into this study.

It is feasible to add the application with the user and can send their feedback to the system, and this feedback can be used to fine-tune the model later. The developed model is tested only for 16 fish species found in Sri Lanka even though there are more than 90 species found. This model can be extended to classify and detect other species as well. And training dataset can be increased to get higher accuracy. Currently, the model is hosted in the AWS server in which the mobile application accesses it remotely. On average, this detection takes 3–20 s based on the Internet speed. This can be enhanced in a way that the model runs inside the mobile application, so that it is possible to reduce the time for detection. As another enhancement, the network architecture can be changed and tested in a different environment to minimize the time taken for training and also to increase the accuracy.

References

1. Welcome to the Department of Fisheries and Aquatic Resources. (Online). Available: https://www.fisheriesdept.gov.lk/. Accessed: 23 Mar 2019
2. Health and Families.: The Independent. (Online). Available: https://www.independent.co.uk/life-style/health-and-families. Accessed 23 Mar 2019
3. Wu, J., Peng, B., Huang, Z., Xie, J.: Research on computer vision-based object detection and classification. Comput. Comput. Technol. Agric. **6**, 183–188 (2013)
4. Morimoto, E., Taira, Y., Nakamura, M.: Identification of Fish Species using Neural Networks. J. Nat. Fish. Univ. **58**(1), 65 (2009)
5. Using Convolutional Neural Networks to Identify Fish Species in Camera Footage. Michael Chatzidakis. (Online). Available: https://www.mikechatzidakis.com/home/2017/7/30/using-convolutional-neural-networks-to-identify-fish-species-in-camera-footage. Accessed: 23 Mar 2019
6. Li, X., Shang, M., Qin, H., Chen, L.: Fast accurate fish detection and recognition of underwater images with Fast R-CNN. In: OCEANS 2015—MTS/IEEE Washington, pp. 1–5 (2015)
7. Fouad, M.M.M., Zawbaa, H.M., El-Bendary, N., Hassanien, A.E.: Automatic Nile Tilapia fish classification approach using machine learning techniques. In: 13th International Conference on Hybrid Intelligent Systems (HIS 2013), pp. 173–178 (2013)
8. Hnin, T.T., Lynn, K.T.: Fish classification based on robust features selection using machine learning techniques. In: Genetic and Evolutionary Computing, pp. 237–245. Springer, Cham (2016)
9. Salimi, N., Loh, K.H., Dhillon, S.K., Chong, V.C.: Fully-automated identification of fish species based on otolith contour: using short-time Fourier transform and discriminant analysis (STFT-DA). Peer J. **4**, e1664 (2016)
10. Agrawal, V.: bbox: 2D/3D bounding box library for Computer Vision
11. Huang, J., et al.: Speed/accuracy trade-offs for modern convolutional object detectors (2016). arXiv:1611.10012 [cs]
12. YOLO: Real-Time Object Detection. (Online). Available: https://pjreddie.com/darknet/yolo/. Accessed: 24 Mar 2019
13. Rothe, R., Guillaumin, M., Van Gool, L.: Non-maximum suppression for object detection by passing messages between windows. In: Computer Vision—ACCV 2014, pp. 290–306 (2015)
14. Hubel, D.H., Wiesel, T.N.: Receptive fields and functional architecture of monkey striate cortex. J. Physiol. (Lond.) **195**(1), 215–243 (1968)
15. Krizhevsky, A., Sutskever, I., Hinton, G.E.: ImageNet Classification with Deep Convolutional Neural Networks. In: Pereira, F., Burges, C.J.C., Bottou, L., Weinberger, K. Q. (eds.) Curran Associates, Inc., pp. 1097–1105 (2012)

16. Srivastava, N., Hinton, G., Krizhevsky, A., Sutskever, I., Salakhutdinov, R., Bengio, Y.: Dropout: a simple way to prevent neural networks from overfitting. J. Mach. Learn. Res. **15**(1), 1929–1958 (2014)

17. Aarshay, J.: Deep learning for computer vision—introduction to convolution neural networks. 04 Apr 2016. (Online). Available: https://www.analyticsvidhya.com/blog/2016/04/deep-learning-computer-vision-introduction-convolution-neural-networks/.

Pseudoscience Detection Using a Pre-trained Transformer Model with Intelligent ReLabeling

Thilina C. Rajapakse and Ruwan D. Nawarathna

Abstract Often dismissed as a harmless pastime for the gullible, pseudoscience nonetheless has devastating effects, including the loss of life. Its menace is amplified by the difficulty of differentiating pseudoscience from science, especially for the untrained eye. A novel method recognizes pseudoscience in the text by utilizing a fine-tuned Robustly Optimized Bidirectional Encoder Representation from Transformers Approach (RoBERTa) model. The dataset of 112,720 full-text articles used in this work is made publicly available to remedy the lack of datasets related to pseudoscience. A novel technique, Intelligent ReLabeling (IRL), is employed to minimize mislabeled data, enabling the rapid creation of high-quality textual datasets. IRL eliminates the need for expensive manual verification processes and minimizes domain expertise requirements in many applications. The final model trained with IRL achieves an F_1 score of 0.929 on a separate manually labeled test dataset.

Keywords Natural language processing · Classification · Data cleaning · Pseudoscience dataset

1 Introduction

Pseudoscience is a collection of beliefs or processes that are an imitation of science. Despite a superficial resemblance in the eyes of a casual observer, science and pseudoscience are radically different [1]. Science is both a body of knowledge and a process that seeks to understand the behavior of the physical and natural world through systematic observation and experimentation.

T. C. Rajapakse (✉)
Postgraduate Institute of Science, University of Peradeniya, Peradeniya 20400, Sri Lanka
e-mail: chaturangarajapakshe@gmail.com

R. D. Nawarathna
Department of Statistics and Computer Science, University of Peradeniya, Peradeniya 20400, Sri Lanka
e-mail: ruwand@pdn.ac.lk

© The Editor(s) (if applicable) and The Author(s), under exclusive license to Springer Nature Singapore Pte Ltd. 2021
S. Shakya et al. (eds.), *Proceedings of International Conference on Sustainable Expert Systems*, Lecture Notes in Networks and Systems 176,
https://doi.org/10.1007/978-981-33-4355-9_25

319

Although it is difficult to define exactly what does and does not fall under the umbrella of pseudoscience, an activity or teaching that satisfies the following criteria may be considered to be pseudoscientific [2] such as it is not scientific, and it is part of a doctrine whose major proponents try to create the impression that it represents the most reliable knowledge on its subject matter.

Examples of popular pseudoscientific beliefs include, and are unfortunately not limited to, faith healing, homeopathy, magnetic therapy, various "cleanses", aromatherapy, creationism, and astrology. Science denial, which includes climate change denial and the anti-vaccine movement, can also be considered to be a form of pseudoscience [3]. Pseudoscience is not merely a harmless delusion of the ignorant. It can, and does, impact people's welfare, sometimes on massive scales. Millions of people have died of AIDS because they (or their governments) ignored scientific findings and instead relied on folk remedies and "snake oil" therapies [4]. Numerous other cases of loss of life or health due to the rejection of modern medicine in favor of "alternative medicine" steeped in pseudoscience can be found easily. The true extent of repercussions of climate change denial is yet to be fully realized, but the threat faced is undeniable [5].

In light of these dangers, the ability to detect pseudoscience is imperative, particularly in online news articles and blog posts where it is most commonly found and spread. Unfortunately, most of the general public does not possess a good understanding of science and find it difficult to differentiate science from pseudoscience. Pseudoscience practitioners will often prey on this fact by using scientific terms like energy, vibrations, frequencies, and quantum effects to convey a false sense of scientific credibility. Other common tactics include claiming that their intervention is derived from the latest discoveries in science, and claiming that large institutions or companies (so-called Big Pharma) are withholding knowledge of cures and treatments for profit. Another important factor behind the popularity of pseudoscience is that it offers a version of reality that people desperately want to be true. They want to believe that there is a miraculous cure for a given disease, or that sickness can be avoided by eating purely organic food, or even that there is life after death. Combining these factors with cognitive biases such as confirmation bias and a paucity of critical thinking skills results in a population that is alarmingly susceptible to pseudoscience.

Organizations and Web sites such as Snopes, Skeptic Society, and RationalWiki provide information on pseudoscience and attempts to expose any falsehoods being spread to the public. However, it is impossible to manually fact check all of the vast volumes of content that is published online on any given day. As such, an automated tool that can quickly and accurately predict whether a given article of text is based on pseudoscience will be of immense help in stopping the spread of misinformation to the public.

1.1 Deep Learning for Pseudoscience Detection

The dearth of research into automated pseudoscience detection, despite the critical need for an efficient and accurate method of separating science from pseudoscience, may be attributed to two major factors. The first and perhaps the most obvious factor is the inherent difficulty of detecting pseudoscience even for (non-expert) humans. The process typically involves arduous fact-checking and tracing of sources to conclusively determine whether or not a given claim is pseudoscientific. In certain cases, it can be near impossible to verify the presence or absence of pseudoscience without extensive knowledge and expertise in the relevant domain. The second contributing factor is the need for large amounts of labeled data. Self-training language models such as Bidirectional Encoder Representations from Transformers (BERT), XLNet, XLM (Cross-lingual Language Models), and RoBERTa have advanced the field of natural language processing at dizzying speeds and have demonstrated significant improvements over more traditional methods such as recurrent neural networks (RNNs) and long short-term memory (LSTMs) in various NLP tasks. While these BERT-like models perform well on performance benchmarks (GLUE, SQUAD, etc.) used to evaluate NLP capabilities, their real-world applicability is hindered by the lack of sufficient amounts of labeled data. The Internet can serve as a virtually unlimited repository of textual data for nearly any conceivable subject, but it involves a troublesome caveat. Data on the Internet is relatively easy to obtain, but there is no obvious way to efficiently label this data without compromising on accuracy or without resorting to prohibitively time-consuming manual verification processes.

Considering the challenging nature of labeling textual data, it is unsurprising to find that no readily available dataset could be leveraged to train a deep learning model to detect pseudoscience. Torabi Asr and Taboada [6] observe a similar lack of datasets concerning "fake news" articles. Any existing datasets, related to any domain, are often too small to be used for training deep learning models. To fill this gap for the domain of pseudoscience, a new dataset containing 112,720 articles collected from 20 Web sites that regularly publish content related to science and/or pseudoscience was presented. The articles were labeled according to the source from which they were extracted,[1] i.e., articles from reputable Web sites were tagged as non-pseudoscientific and articles from dubious Web sites tagged as pseudoscientific.

This method of labeling, although commonly used [6], inevitably leads to misclassified data. For example, it is highly unlikely that even the most misguided of Web sites would publish solely pseudoscientific articles. Conversely, even the most reliable of news sources may not be immune to the occasional lapse of judgment. However, it is impractical to manually verify sufficiently large numbers of articles to effectively fine-tune a pre-trained model. A novel method, Intelligent ReLabeling (IRL), was proposed to automatically relabel the misclassified data. IRL relies on the fact that the accuracy of the labels of one class of the dataset can be asserted with reasonable confidence. The articles labeled as non-pseudoscientific were all obtained from credible sources, and as such, it is safe to assume that at least a substantial portion of these

---

[1] See Sect. 3.1 for details.

articles are correctly labeled, and thus this class can be "trusted." New datasets were formed, one for each dubious source, consisting of all the articles except those from the given dubious source. A separate RoBERTa model is trained on each of the new datasets, and each dubious source is relabeled using the corresponding RoBERTa model (which has not been trained on the given source). This relabeling technique was shown to significantly mitigate the effects of "noisy labeling" by comparing two models, one trained on data with IRL applied, and the other trained on data with the original labels.

In summary, the contributions of the proposed work are as follows; a new dataset of 112,720 articles related to science and/or pseudoscience news collected from 20 Web sites, a new technique, IRL, for automatically minimizing the number of misclassified samples in situations where manual labeling is prohibitively time-consuming and/or difficult without subject expertise, and also a fine-tuned RoBERTa model capable of accurately and efficiently detecting pseudoscience in text.

2 Background

2.1 Indicators of Pseudoscience

As previously noted, there is no single, agreed-upon definition of pseudoscience. However, pseudoscience, by most definitions, contains some common traits that can be used as identifiers. The following is a list of 10 such warning signs of pseudoscience [7]; lack of falsifiability, lack of self-correction, emphasis on the confirmation, evasion of peer review, over-reliance on testimonial and anecdotal evidence, absence of connectivity, extraordinary claims, ad antequitem fallacy (appealing to the antiquity of claims as a sign of their credibility), use of hypertechnical language, and absence of boundary conditions.

These traits, among others, can be used to evaluate the authenticity of claims in news media, on the Internet, and in peer-reviewed literature. A robust delineation of pseudoscience and science was not attempted, instead existing guidelines [2, 7] were followed to differentiate pseudoscience and science.

2.2 Detecting Deception

Research has shown that humans are not good at making correct lie–truth judgments. In a study comprising of 206 documents and 24,483 judges, an average of 54% correct lie–truth judgments was observed [8]. The participants correctly classified 47% of lies as deceptive and 61% of truths as non-deceptive. From the results of this study, it can be seen that people are particularly bad at identifying lies or deceptions, with an

Table 1 Table of cognitive biases and errors

Error	Description
Naive realism	The belief that the world is precisely as one sees it
Confirmation bias	Tendency to seek out evidence consistent with our beliefs, and deny, dismiss, or distort evidence that is not
Belief perseverance	Tendency to cling to beliefs despite repeated contradictory evidence
Illusory correlation	Tendency to perceive statistical associations that are objectively absent
Over-reliance on heuristics	Tendency to place too much weight on mental shortcuts and rules of thumb

accuracy score of 3% below that of random chance. This serves to further highlight the need for automated, accurate methods of detecting deception and misinformation.

A similar problem to the detection of pseudoscience exists in the issue of detecting fake news. Fake news is a category of misinformation that mainly deals with the political realm. There have been many attempts to harness the power of machine learning to detect fake news in online text. Shu et al. [9] present a comprehensive review of detecting fake news on social media. Although the two cases share certain common traits, it should be noted that they are fundamentally different in nature. Fake news attempts to deceive and sway public opinion on political issues whereas pseudoscience typically masquerades as genuine science with the goal of selling questionable products, services, and treatments.

The same challenges and difficulties arise when attempting to decipher pseudo-science from science, as pseudoscience is, at its core, a subset of deception and misinformation. It is, perhaps, an even greater challenge as pseudoscience cunningly preys on the cognitive biases present in all people [7] and tunes in on comfortable intuitive representations of the world, making it much harder to resist the allure of pseudoscience. Table 1 provides some examples of widespread cognitive biases and errors [7].

The methods currently used in detecting deception and misinformation can be broadly categorized into two major categories: linguistic approaches and network approaches [10]. Linguistic approaches analyze the content of the text and attempt to find language patterns that indicate deception. Network approaches use information such as article metadata and/or structured knowledge network queries to predict deception. This method can utilize existing knowledge networks or publicly available structured data, such as the Google Relation Extraction Corpus [10].

Linguistics-based methods rely on the fact that text containing misinformation and deception tend to be created for financial or political gains rather than to report objective truths. As such, they often use opinionated and inflammatory language, and "clickbait" headlines (headlines designed to entice the reader into clicking on a link to read the full article [9]). These features can be exploited to identify patterns in writing styles and language used that differ across articles reporting the truth and articles peddling misinformation. Commonly used linguistic features include [9] lexical features such as total words, characters per word, frequency of large words, and unique words and syntactic features such as frequency of function words and phrases (n-grams and bag-of-words) or punctuation and parts-of-speech tagging.

Automated detection of pseudoscientific publications was performed using automatic text analysis with fairly good results on a limited dataset [11]. The constraint of using only publications (scientific or pseudoscientific) simplifies the problem as there is a high degree of similarity in language and structure except in cases where pseudoscientific claims are made. Online articles, on the other hand, have much higher variance in both the language used and the structure of the text. Despite the current crisis of misinformation, the literature on automatic detection of pseudoscience in news articles, blog posts, and other online media remains scarce.

Although there have been several attempts to utilize machine learning and other algorithmic tools to automate the detection of misinformation, particularly fake news [12], success, and applicability have thus far been limited. Despite dramatic increases in the performance of deep learning models in natural language processing, it is yet to be tried substantially in the task of misinformation detection. Currently, no literature on using deep learning to detect pseudoscience was found.

2.3 Natural Language Processing

Until recently, recurrent neural networks, long short-term memory [13], and gated recurrent neural networks [14] have demonstrated the state-of-the-art performance in most natural language processing tasks. Recurrent models are an extension of the conventional feed-forward neural network, with a special ability to process variable-length input sequences. This is achieved by having a recurrent hidden state whose activation at a given time step is dependent on that of the previous time step. While this enables the handling of variable-length inputs, the sequential nature of the process prevents parallelization within training examples. This, in turn, limits the ability to train the RNN on batches of longer sequences due to memory constraints.

The Transformer model [15] offered a solution to this problem of parallelization by dropping recurrence and relying instead on attention mechanisms. Not only was the Transformer capable of significantly more parallelization, but it also surpassed the benchmarks in translation tasks previously set by models based on recurrent architectures.

2.4 BERT

Bidirectional Encoder Representations from Transformers (BERT) is a new language representation model that uses bidirectional pre-training to reduce the need for heavily engineered task-specific architectures [16]. BERT's model architecture is a multi-layer bidirectional transformer encoder, similar to the original Transformer model [15]. The input to BERT is a concatenation of two segments (token sequences), $x_1, \ldots, x_N$ and $y_1, \ldots, y_M$. A segment will typically contain more than one natural sentence. Formula (1) shows the form of a BERT input segment.

$$[CLS], x_1, \ldots x_N, [SEP], y_1, \ldots, y_M, [EOS] \tag{1}$$

Here, $N + M < T$ and T is the maximum sequence length.

The model is initially pre-trained on a large corpus of unlabeled text data and is then fine-tuned for a required task with labeled data.

The pre-training procedure for BERT was further improved in the model RoBERTa, producing state-of-the-art results on GLUE [17], RACE [18], and SQuAD [19] tasks. BERT was trained using two pre-training objectives: masked language modeling and next sentence prediction [16].

Masked Language Model (MLM) A random selection of tokens in the input sequence is replaced with a special token [MASK]. The MLM objective is to minimize the cross-entropy loss on predicting the masked tokens. BERT uniformly selects 15% of tokens as candidates for replacements. Of this selection, 80% are masked (replaced with [MASK]), 10% left unchanged, and the remaining 10% replaced by a randomly selected token from the vocabulary [20]. Masking is performed once in the beginning and kept for the duration of training.

Next Sentence Prediction (NSP) NSP is a binary classification loss for predicting whether two segments occur consecutively in the original text [20]. This was designed to improve performance on downstream tasks such as natural language inference.

2.5 RoBERTa

RoBERTa makes several modifications to the pre-training strategy used by BERT. Specifically, increased training steps (500 K) and bigger batch sizes (8 K sequences per batch) [20], removal of the next sentence prediction objective [20], trained on longer sequences [20], and use of dynamic masking patterns on training data [20].

RoBERTa has shown a state-of-the-art performance in many NLP tasks while requiring significantly less computational resources compared to other models such as XLNet.

One limitation of BERT, and by extension RoBERTa, is that it will only accept an input sequence that is less than or equal to 512 WordPiece [21] tokens in length. Accounting for special tokens used by RoBERTa, the actual maximum sequence length comes down to 509 tokens. According to the Pew Research Centre's State of the News Media, the average article length at the largest papers was about 1,200 words, about 800 words at mid-sized papers, and about 600 words at the smaller papers [22]. Our method uses a "sliding window" approach to enable training and inference on sequences longer than 512 WordPiece tokens.

3 Method

For this work, consider any text displaying the traits of pseudoscience and attempting to hijack the legitimacy of science to promote any non-scientific (not evidence-based or experimentally shown to be correct) product or idea as pseudoscientific *(class 1)* and text which does not display these traits as reliable *(class 0)*. The RoBERTa model is fine-tuned to perform pseudoscience detection in online text. This section describes the data collection, Intelligent ReLabeling, the models used, the training of models, and the evaluation procedure.

3.1 Data Collection

The dataset was built by scraping online news sources. A total of 20 Web sites that regularly publish a wide variety of articles related to science and/or pseudoscience was scraped to build a corpus of 112,720 samples. The samples were assigned a label 0 or 1 (non-pseudoscientific and pseudoscientific, respectively) based on the credibility of the source. The credibility was determined through manual inspection of the Web site content, as well as referencing curated lists by Zimdars [23] and Aker [24]. The Web sites and their associated labels are shown in Table 2.

A separate test dataset (Manual) consisting of manually verified[2] samples to reliably evaluate the performance of the models. This dataset consists of 91 pseudoscientific samples and 91 reliable samples, adding up to a total of 182 samples was created.

3.2 Intelligent ReLabeling (IRL)[3]

The drawback of naively labeling every sample from the same source with the same label is that a significant number of samples will be mislabeled. Concretely, not *all* articles on a Web site known to publish pseudoscientific content will be pseudoscientific. Conversely, not all articles published by credible Web sites may be reliable. However, it seems unlikely that a Web site will remain credible if it consistently produces pseudoscientific content. In light of this, the samples taken from Web sites known to be reliable were considered to have been labeled correctly as

---

[2]The manual verification was performed by the authors through fact-checking against reliable sources as appropriate.

[3]We will make the code for IRL publicly available, along with experimentally demonstrating its effectiveness using the IMDB reviews dataset [25]. We artificially add noise to the labels of the IMDB reviews dataset, substantially degrading the performance of models trained on the noisy data. Then, we show that applying IRL improves the performance of the models close to or to the same levels before adding noise.

Table 2 Table of Web sites and their assigned labels

Web site	Label
www.collective-evolution.com	Pseudoscientific
www.danachildintuitive.com	Pseudoscientific
www.davidwolfe.com	Pseudoscientific
www.factcheck.org	Non-pseudoscientific
www.foodbabe.com	Pseudoscientific
www.gizmodo.com	Non-pseudoscientific
www.goop.com	Pseudoscientific
www.greenmedinfo.com	Pseudoscientific
www.greggbraden.com	Pseudoscientific
www.infowars.com	Pseudoscientific
www.mercola.com	Pseudoscientific
www.naturalnews.com	Pseudoscientific
www.pnas.org	Non-pseudoscientific
www.popsci.com	Non-pseudoscientific
www.reuters.com	Non-pseudoscientific
www.sciencebasedmedicine.com	Non-pseudoscientific
www.sciencedaily.com	Non-pseudoscientific
www.scimthsonianmag.com	Non-pseudoscientific
www.snopes.com	Non-pseudoscientific
www.thinkingmomsrevolution.com	Pseudoscientific

non-pseudoscientific. The samples from unreliable Web sites were initially labeled as pseudoscientific but were modified according to the process detailed below.

Starting from the set of all samples, S, the datasets derived from reliable sources are labeled 0 (for non-pseudoscientific) and the datasets derived from unreliable sources are labeled 1 (for pseudoscientific). Formulas (1) and (2) give the datasets of all pseudoscientific samples A and all non-pseudoscientific samples B, respectively.

$$A = \{x \in S | x \text{ belongs to class pseudoscientific}\} \tag{2}$$

$$B = \{x \text{ in } S | x \text{ belongs to class non} - \text{pseudoscientific}\} \tag{3}$$

The data can also be separated into its own set according to its source. That is, 20 datasets corresponding to the 20 Web sites from which the data was obtained.

$$S_1, \ldots, S_{20} \subseteq S \tag{4}$$

Also, by labeling according to the source,

$$S_1, \ldots, S_{11} \subseteq A \tag{5}$$

$$S_{12}, \ldots, S_{20} \subseteq B \tag{6}$$

Next, 11 new datasets are built, one for each unreliable dataset, such that they contain all pseudoscientific samples except those from one pseudoscientific Web site. In addition, all non-pseudoscientific samples are included in each of the 11 datasets (buckets).

$$D_i = \{ x \in S, i \in \{1, \ldots, 11\} | x \notin S_i \} \tag{7}$$

A RoBERTa$_{\text{BASE}}$[4] model,[5] M_i, was trained for each dataset D_i, where $i \in \{1, \ldots, 11\}$. Each model, M_i, was used to predict the labels for each dataset S_i, where $i \in \{1, \ldots, 11\}$. All samples initially assigned the label 1 (pseudoscientific), but predicted to have the label 0 (non-pseudoscientific) by the model M_i were relabeled to 0. This dataset built using IRL is referred to as S_{IRL} (Fig. 1), and a dataset retaining the original labels is referred to as S_{original}.

3.3 Models Used

All models used in this work are RoBERTa$_{\text{BASE}}$ models. RoBERTa is built on BERT but uses a better optimized training method [20]. The transformers[67]library is obtained for pre-trained models and also for fine-tuning the models.

3.4 Model Training

The datasets, S_{IRL} and S_{original}, were separated into train sets (IRL$_{\text{train}}$ and original$_{\text{train}}$) and test sets (IRL$_{\text{test}}$ and original$_{\text{test}}$) with 80% and 20% of the samples randomly assigned to each set, respectively. Note that the splitting is equivalent across both datasets, i.e., the same samples exist in the train and test splits for both S_{IRL} and S_{original}, albeit with possibly different labels.

During training, for text longer than the maximum sequence length, the "sliding window" technique is used to ensure that the model "sees" the entire text. Any text longer than the maximum sequence length of 509 WordPiece tokens[6] was separated into text of maximum length below the sequence length limit of RoBERTa models.

---

[4]See Sect. 3.3.

[5]12-layer, 768-hidden, 12-heads, 125 M parameters.

[6]https://github.com/huggingface/transformers.

[7]3 tokens reserved for special tokens.

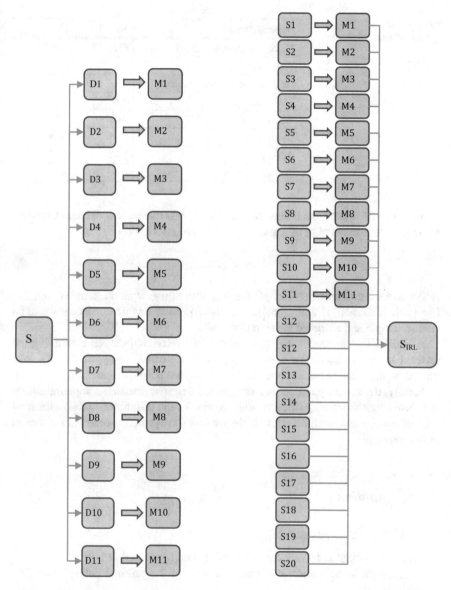

Fig. 1 Set of all samples, S, was used to derive the 11 datasets, D_1, ..., D_{11}, according to Formula (7). Eleven models, M_1, ..., M_{11}, were trained on D_1, ..., D_{11} (left). S_{IRL} (right) was built by combining the predictions (only positive to negative conversions are considered) and the unaltered data from the non-pseudoscientific sources

Table 3 Table of hyperparameter values

Hyperparameter	M_i	F
Maximum sequence length	512	512
Train batch size	4	4
Gradient accumulation steps	8	8
Weight decay	0.1	0.1
Warm-up ratio	0.06	0.06
Learning rate	5e-5	5e-5
Adam epsilon	1e-6	1e-6
Training epochs	1	5

A *stride* of 407 tokens and a maximum length of 509 tokens was defined. Consider a text sample E consisting of n tokens given in Formula (8).

$$E = \{t_1, \ldots, t_n\} \tag{8}$$

For stride s, and maximum length l, E was split into e_i separate samples such that $i = \{1, \ldots, \text{round}(\frac{n}{s})\}$ and $e_i = \{t_i, \ldots, t_{\{i+s\}}\}$. Each sample e_i was considered a unique sample and assigned the same label as E.

A RoBERTa$_{BASE}$ model, F_1, was trained on the training dataset containing the updated labels, IRL$_{train}$. For comparison, another RoBERTa$_{BASE}$ model, F_2, was trained on the training dataset containing the original labels, original$_{train}$.

Mainly, the same hyperparameters from the original RoBERTa implementation were used for the training of all models except for the difference in the number of training epochs and learning rates. Table 3 shows the hyperparameter values chosen in this paper.

3.5 Evaluation

The models were tested against three test datasets.

1. Manual—Separate test set consisting of manually verified samples
2. IRL$_{test}$—Holdout test set consisting of 20% of the original dataset with IRL applied
3. original$_{test}$—Holdout test set consisting of 20% of the original dataset (IRL not applied).

During the evaluation, samples longer than the maximum length (509 WordPiece tokens) are separated according to the "sliding window" procedure described in Sect. 3.4. The model then makes predictions on each subsample. The final prediction for the sample is taken to be the average of all predictions. In the case of a tie, the "better safe than sorry" principle was followed and the sample was predicted to be pseudoscientific.

The F_1 score and the accuracy of two models against each of the test sets given above was reported. The first model (F_1) has been trained on data with the IRL technique applied (IRL$_{train}$), while the second model (F_2) is trained directly on the original dataset with the naively assigned labels (original$_{train}$). This enables us to observe the effects of applying IRL.

4 Results

First, both models (F_1 and F_2) were observed to show satisfactory performance on all test datasets. This is in keeping with the performance of pre-trained models on a variety of NLP tasks. The most important benchmark is the performance of the models on the Manual dataset, as all samples are manually verified and known to be labeled correctly. Model F_1 outperforms Model F_2 on the Manual dataset with nearly a 3-point gap in both F1 score and inaccuracy. This reaffirms the contribution of IRL on improving the quality of datasets, leading to performance increases in models trained on the ensuing relabeled data. Better hyperparameter tuning of the models used in IRL is left to future work, where it is expected to see further improvements in models trained on the relabeled data (Table 4).

Comparing the performance of the two models on the two test datasets, original$_{test}$ and IRL$_{test}$, Model F_1 outperforms Model F_2 on the IRL$_{test}$ dataset whereas Model F_2 outperforms Model F_1 on the original$_{test}$ dataset. This is expected since the training data used for Model F_1 is more similar to the IRL$_{test}$ dataset and the training data used for Model F_2 is more similar to the original$_{test}$ dataset.

The biggest gap in performance between the two models was observed when testing on the original$_{test}$ dataset. This is likely due to the Model F_2 learning to accurately detect the general writing style of the group of sources classified as pseudoscientific rather than learning the linguistic patterns or knowledge associated with pseudoscience. There may also be various markers present in the text that offer clues to the source of the text, such as author names or locations, making it relatively easier to determine the probable source of the text.

A comparison of the performance of the two models on the IRL$_{test}$ dataset shows a similar gap in performance to the gap seen in the Manual dataset. This is particularly interesting since it may indicate that IRL has effectively changed the original$_{test}$ dataset labels to better reflect the Manual dataset which consists entirely of verified samples. This could explain why the performance gap between Model F_1 and Model

Table 4 Results of F_1 and F_2 on Manual, IRL$_{test}$, and original$_{test}$ datasets

Model		Manual		IRL$_{test}$		original$_{test}$	
Name	IRL applied	F_1 score	Accuracy	F_1 score	Accuracy	F_1 score	Accuracy
F_1	Yes	0.929	0.929	0.956	0.970	0.926	0.946
F_2	No	0.901	0.907	0.919	0.942	0.990	0.992

F_2 are similar across the Manual and IRL_{test} datasets, since the primary difference between the two models is the use of IRL on training data. However, further work is necessary to properly validate this hypothesis.

All training was performed on a single RTX 2080 GPU using mixed precision training with the Apex library.[8]

5   Future Work

Although semi-supervised learning methods exist that are capable of generating large labeled datasets from small seed datasets, their performance is greatly dependent on the quality of the seed data. IRL does not suffer from this constraint. Further analysis on the performance of IRL on datasets where one class cannot be trusted over the other is planned as future work. Particularly, more work is required to tune the hyperparameters of the models used in IRL to improve its overall effectiveness. We expect task-specific hyperparameter tuning will improve the performance of IRL. The number of buckets (the number of datasets the original dataset is split into during IRL) can also be used as a tunable hyperparameter. A comparison of splitting the dataset by source vs splitting randomly into buckets may also yield interesting insights. Due to the time and computational resources necessary for these tasks, these ideas are left for future work.

Future work also includes augmenting the dataset of pseudoscience related texts, as well as increasing the size of the manually verified dataset (Manual).

6   Conclusion

Recent developments in transfer learning with language models have opened various avenues for the application of natural language processing to real-world tasks. Our work contributes a fine-tuned RoBERTa model that can accurately detect pseudo-science in text. Its performance on a holdout test dataset, and a manually verified test dataset was demonstrated, showing the applicability of pre-trained language models in real-world tasks. Further, a lack of labeled data that can be used to train pre-trained models was observed. In response, a dataset is provided with 112,270 texts related to scientific and pseudoscientific news. Additionally, a novel technique, Intelligent ReLabeling (IRL), was proposed which can be applied to various domains to build textual datasets efficiently while minimizing mislabeled samples. The effectiveness of Intelligent ReLabeling was shown by comparing the performance of two models used to detect pseudoscience, one trained with IRL applied and the other without.

---

[8]https://github.com/NVIDIA/apex.

References

1. Boudry, M., Blancke, S., Pigliucci, M.: What makes weird beliefs thrive? the Epidemiology of Pseudoscience. Philos. Psychol. **28**(8), 1177–1198 (2015)
2. Hansson, S.O.: Science and pseudo-science (2017)
3. Hansson, S.O.: Science denial as a form of pseudoscience. Stud. Hist. Philos. Sci. Part a **63**, 05 (2017)
4. Pigliucci, M., Boudry, M.: The dangers of pseudoscience. New York Times **21**, 1735–1741 (2013)
5. Douglas, K.M., Sutton, R.M.: Climate change: why the conspiracy theories are dangerous. Bull. Atom. Sci. **71**(2), 98–106 (2015)
6. Torabi Asr, F., Taboada, M.: Big data and quality data for fake news and misinformation detection. Big Data Soc. **6**(1), 2053951719843310 (2019)
7. Lilienfeld, S.O., Ammirati, R., David, M.: Distinguishing science from pseudoscience in school psychology: science and scientific thinking as safeguards against human error. J. Sch. Psychol. **50** (2012)
8. Bond, C.F., Jr., DePaulo, B.M.: Accuracy of deception judgments. Pers. Soc. Psychol. Rev. **10**(3), 214–234 (2006)
9. Shu, K., Sliva, A., Wang, S., Tang, J., Liu, H.: Fake news detection on social media: a data mining perspective. ACM SIGKDD Explor. Newslett. **19**(1), 22–36 (2017)
10. Niall J Conroy, Victoria L Rubin, and Yimin Chen. Automatic deception detection: Methods for finding fake news. Proc. Assoc. Inf. Sci. Technol. **52**(1), 1–4, (2015)
11. Shvets, A.: A method of automatic detection of pseudoscientific publications. In Intelligent Systems' 2014, pp. 533–539. Springer (2015)
12. Figueira, Á., Oliveira, L.: The current state of fake news: challenges´ and opportunities. Procedia Comput. Sci. **121**, 817–825 (2017)
13. Hochreiter, S., Schmidhuber, J.: Long short-term memory. Neural Comput. **9**(8), 1735–1780 (1997)
14. Chung, J., Gulcehre, C., Cho, K., Bengio, Y.: Empirical evaluation of gated recurrent neural networks on sequence modeling. arXiv preprint arXiv:1412.3555 (2014)
15. Vaswani, A., Shazeer, N., Parmar, N., Uszkoreit, J., Jones, L., Gomez, A.N., Kaiser, Ł., Polosukhin, I.: Attention is all you need. Adv. Neural Inf. Process. Syst. 5998–6008 (2017)
16. Devlin, J., Chang, M.W., Lee, K., Toutanova, K.: Bert: pretraining of deep bidirectional transformers for language understanding. arXiv preprint arXiv:1810.04805 (2018)
17. Wang, A., Singh, A., Michael, J., Hill, F., Levy, O., Bowman, S.R.: Glue: a multi-task benchmark and analysis platform for natural language understanding. arXiv preprint arXiv:1804.07461 (2018)
18. Lai, G., Xie, Q., Liu, H., Yang, Y., Hovy, E.: Race: large-scale reading comprehension dataset from examinations. arXiv preprint arXiv:1704.04683 (2017)
19. Rajpurkar, P., Jia, R., Liang, P.: Know what you don't know: unanswerable questions for squad. arXiv preprint arXiv:1806.03822 (2018)
20. Liu, Y., Ott, M., Goyal, N., Du, J., Joshi, M., Chen, D., Levy, O., Lewis, M., Zettlemoyer, L., Stoyanov, V.: Roberta: a robustly optimized bert pretraining approach. arXiv preprint arXiv: 1907.11692 (2019)
21. Wu, Y., Schuster, M., Chen, Z., Le, Q.V., Norouzi, M., Macherey, W., Krikun, M., Cao, Y., Gao, Q., Macherey, K., et al.: Google's neural machine translation system: Bridging the gap between human and machine translation. arXiv preprint arXiv:1609.08144 (2016)
22. Pew Research Centre.: State of the news media 2016 (2004)
23. Zimdars, M.: False, misleading, clickbait-y, and satirical "news" sources. https://docs.google.com/document/d/10eA5-mCZLSS4MQY5QGb5ewC3VAL6pLkT53V_81ZyitM/preview, 2016
24. Aker, A., Vincentius, K., Bontcheva, K.: Credibility and transparency of news sources: Data collection and feature analysis (2019)

25. Maas, A., Daly, R.E., Pham, P.T., Huang, D., Ng, A.Y., Potts, C.: Learning word vectors for sentiment analysis. In: Proceedings of the 49th Annual Meeting of the Association for Computational Linguistics: Human Language Technologies, pages 142–150, Portland, Oregon, USA, June 2011. Association for Computational Linguistics

Identification of Music Instruments from a Music Audio File

W. U. D. Rodrigo, H. U. W. Ratnayake, and I. A. Premaratne

Abstract The goal of this research is to design an intelligent system to recognize the specific musical instrument from an audio file and construct the notation text file correspondingly for the Music Instrument Digital Interface file. The scope was limited to the identification of two different instruments of the same pitch at a time and identification is done based on the analysis of the waveform pattern of instrument sound. Ten different musical instruments from a variety of musical instrument categories were selected to observe their waveform patterns. A sampling of different tones of sound patterns of the musical instruments was carried out. For this research, 12 tons of three different instruments in three different instrument categories were selected. From these three instruments, the same pitch sound combination of two instruments at a time was used to take the samples. There are two parts in the system; first is feature extraction of the sampled waveform and the second is feature classification. In the first part, feature extraction is done with mel frequency cepstral coefficient algorithm. Thirteen mel frequency cepstral coefficient features of the sampled waveform have been extracted and then the neural network has been trained using those coefficients. In the second part, classification is done with an artificial neural network. Its functionality was evaluated using confusion matrices. Based on the performance analysis of the result, the project was successful in detecting two different instruments in two different instrument categories played simultaneous time.

Keywords Neural network · MFCC · Signal processing · Sound wave pattern · MIDI

W. U. D. Rodrigo (✉) · H. U. W. Ratnayake · I. A. Premaratne
Department of Electrical and Computer Engineering, The Open University of Sri Lanka, Nawala, Nugegoda, Sri Lanka
e-mail: udarirodrigo@gmail.com

H. U. W. Ratnayake
e-mail: udithaw@ou.ac.lk

I. A. Premaratne
e-mail: iapre@ou.ac.lk

© The Editor(s) (if applicable) and The Author(s), under exclusive license to Springer Nature Singapore Pte Ltd. 2021
S. Shakya et al. (eds.), *Proceedings of International Conference on Sustainable Expert Systems*, Lecture Notes in Networks and Systems 176,
https://doi.org/10.1007/978-981-33-4355-9_26

335

1 Introduction

Most sound processing systems available in the music industry such as digital music workstations and synthesizers are designed to use Musical Instrument Digital Interface (MIDI) format. Even though the productions are made through such workstations, the final recording is made to audio format. Furthermore, live recordings and other performances are also recorded in audio format. For some advanced music analyzers such as intelligent systems for music processing and automatic chord, progression systems need to use MIDI format to store the music data. In 2016/2017, a research project carried out at The Open University of Sri Lanka called "Intelligent time of Use Deciding system for a melody to provide a better listing experience" [15] also required the identification of the music instrument from the audio file, which was in MIDI format. In such cases, music recorded in audio format has to be converted to MIDI format. There are many types of MIDI composers available in the industry. After creating MIDI files, one can edit the file as the user wishes. The user has to select a musical instrument manually because nobody can identify the music instrument category and music instrument from the audio file. Even in the same type of real musical instruments, the waveform patterns are not very similar. Therefore, it was necessary to identify it using an intelligent system. If audio files contain more than one musical instrument, there is no basic method to recognize the instrument and convert it to MIDI format correspondingly. Therefore, this research is focused on designing an intelligent system to recognize the musical instrument and construct the notation text file correspondingly for the MIDI file.

This paper is organized as follows: Sect. 2 elaborates the literature survey, Sect. 3 is focused on methodology, Sect. 4 discusses the feature extraction and classification, Sect. 5 is about results and discussions, and Sect. 6 covers the conclusion and future work.

2 Literature Survey

There are few systems for musical instrument recognition and conversion to music instrument signal to MIDI format, recognizing different sounds and distinguishing which instrument is being played with the use of artificial neural networks [1]. In this research, recording sound clips of different notes or chords played on the guitar, piano, and mandolin, a neural network will be parameterized and have its topology constructed so that it can accurately identify the instruments based on their audio clips.

In [2], an artificial neural network has been trained to categorize musical instruments according to sound. They have transformed the audio samples into the frequency domain and then analyzed different features of the sound in both time and frequency domains in order to compare them to see how much information can be extracted. Recognition of musical instruments from isolated notes is discussed in

[3]. They have introduced a Time Encoded Signal Processing method to create simple matrices from complex sound waveforms for encoding and recognizing instrument notes. These matrices are given as input to a fast-artificial neural network to identify the instruments. Authors claim successful results in the classification of using organ sound with computational cost reduction measures.

In these research projects, the detection of instrument sound is done through the audio file, but it is not possible to detect the sound of multiple instruments at the same time. When using the MIDI file, there will be a need to identify the various sounds and their notes in the converted audio file. Therefore, as an improvement, our research identifies musical instruments and tones from an audio file that contains music of instruments played together.

2.1 Theoretical Background

Under the theoretical background, three topics are discussed. They are why different instruments have different sounds, feature extraction algorithms, and feature matching (classification).

2.1.1 Why Different Instruments Have Different Sounds

Most of the time, musical instruments vibrate strings or beams of the air usually hundreds or thousands of times per second. Instruments like piano, guitar, and violin vibrate strings, whereas flute, clarinet, and organ vibrate beams of air. When a music instrument vibrates, it alternately compresses and expands air creating sound waves. The human brain recognizes different features of the created sounds such as loudness and pitch. Even if the same note is played in two instruments, the sounds will be different as different musical instruments vibrate at different frequencies. Different frequencies of note are called harmonic, participles, or overtones. The relative pitch and loudness of these overtones give a characteristic sound to the note, which is called the timbre of the instrument [4]. The pitch, intensity, loudness, duration, and timbre characterize a tone in a musical instrument [7, 9].

2.1.2 Feature Extraction Algorithm

For any sound recognition system, components of the audio signal which are called features need to be extracted. From different sound feature extraction algorithms available, mel frequency cepstral coefficient (MFCC) feature extraction algorithm is selected for this research because the tone can only be identified by using features.

To extract the features using the MFCC feature extraction algorithm, the following steps are needed to be followed initially to divide the audio signal into short frames, calculate the spectral density estimation of the power spectrum for each short frame, apply the mel filter bank to the power spectra, sum the energy in each filter, calculate the log of all filter bank energy, calculate the discrete cosine transform of the log filter bank energy, and keep 13 discrete cosine transform coefficients and discard the rest.

These steps should be applied to each frame and one set of 13 MFCC coefficients is extracted for each frame [5, 6, 8, 10, 11]. To carry out the task of feature extraction, following methods are also important to be carried out.

Pre-emphasis
Pre-emphasis boosts the high-frequency part of a signal before transmission or recording in a storage medium. It also amplifies the importance of the high-frequency part of a signal.

Frame Blocking
This procedure divides the input audio signal into short frames of 20 ~ 40 ms. The resulting overlap will be about 50% of the frame size. In general, the frame size is made equal to the power of two so that fast Fourier transform can be performed. Framing is very important to ensure good results, especially where the variation of amplitude is great, such as in a large signal compared to a small signal.

Hamming Window
The Hamming window is used to integrate all the close frequency lines. To keep the continuity of the first and last points of the frame, all frames will be multiplied with the Hamming window denoted by $w(n)$.

Consider that the signal in a frame is represented as $x(n)$, $n = 0 \ldots N - 1$.

When using the Hamming window to the particular signal it becomes, $x(n) * w(n)$.

Fast Fourier Transform
The fast Fourier transform (FFT) converts each frame from the time domain into the frequency domain. By applying FFT for each frame, the signal frequency components in the time domain can be derived. Spectral analysis shows that different characteristics in audio signals correspond to different energy distribution over frequencies. By performing FFT, obtain the magnitude frequency response of each frame.

Triangular Band-Pass Filter (Mel Scale Filter Bank)
The magnitude frequency response that is obtained is multiplied by triangular band-pass filters to get the log energy of each triangular band-pass filter. The positions of these filters are equally spaced along with the mel frequency. The mel scale relates the perceived frequency, or pitch, of a pure tone to its actual measured frequency. In [5], it is stated that human beings can recognize small changes in pitch at low frequencies than at high frequencies.

The formula to convert the frequency to the mel scale:

$$M(f) = 1125 \ln(1 + f/700) \tag{1}$$

To convert mel scale back to frequency:

$$M^{-1}(m) = 700(\exp(m/1125 - 1)) \tag{2}$$

Logarithm Discrete Fourier Transforms

The log mel spectrum can be converted to the time domain using a discrete cosine transform (DCT). The result is called MFCC and the set of coefficients are vectors. In this step, DCT is applied to the output of the N triangular band-pass filters to obtain L mel scale cepstral coefficients.

$$C(n) = \sum Ek^* \cos(n^*(k - 0.5)^* \pi/40)$$

where $n = 0, 1, \ldots$ to N.

According to [5], here $N =$ the number of triangular band-pass filters, $L =$ number of mel scale cepstral coefficients.

In [5], $N = 40$ and $L = 13$.

2.1.3 Feature Matching (Classification)

Feature machine is an aspect of pattern reorganization. A pattern of a vector is extracted from audio input using the MFCC feature extraction algorithm. These patterns are used to classify algorithms. There are many methods to recognize instrument tones and an artificial neural network (ANN) was used here. ANN is a computer system with a number of simple, interconnected configuration components that respond to external input. In this research, the MFCC vector (13 * no of frames) is generated for each tone of the music tone matrix. Therefore, neural network method is used for classification [12–14].

First, a training dataset was created to decide the most suitable training method or the algorithm to train the ANN. Then, a graphical user interface was designed to extract music instrument sound and to identify the music instrument categories and covert to MIDI format using a trained neural network.

3 Methodology

The methodology adopted for this research is described under two phases. In the first phase, features were extracted from different music instruments (real sound) which is depicted in Fig. 1. During the second phase, an artificial neural network has been trained for the experiment is defined in Fig. 2.

Phase 01

Ten different musical instruments namely: flute, clarinet (woodwind), violin, viola (string), grand piano, piano (keyboard), tuba, brass (brass), bass drum, and sidestick (percussion) were selected, and then using digital oscilloscope, different kinds of waveform patterns were observed. For this research, 12 tones were played by three different instruments (flute, violin, and grand piano) in three different instrument categories (woodwind, string, keyboard). All sounds were generated using a standard music keyboard and sounds have been captured using a dynamic microphone. From these three instruments, the same pitch sound combination of two instruments at a time was used to take the samples. For this research, data extraction time duration was limited to 800 ms. For the purpose of observing the waveform pattern, digital storage oscilloscope (DSO) has been used. Using the DSO, recorded data has been extracted to a comma-separated value (CSV) file.

Phase 02

Then each waveform is framed into 77 number of short frames of 200 ms with a 25% overlap. Next, thirteen (13) MFCC features of the sampled waveform have been extracted using mel frequency cepstral coefficient (MFCC) algorithm and then the neural network has been trained using those coefficients. The classification was

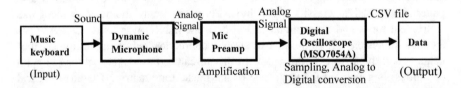

Fig. 1 Process of accrue, real data using music keyboard

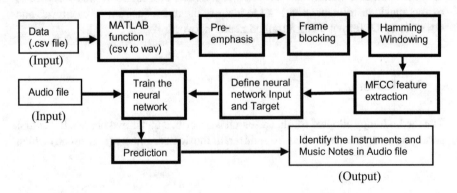

Fig. 2 Methodology of the research

done with a neural network using the MATLAB toolbox. Neural network tool in the MATLAB was used to identify the two musical instruments and notes.

4 Feature Extraction and Classification

Considering the assumption that all notes are played with the same "key press," only one note was being played during the feature extraction period and the environment of the trained dataset was not changed when recording the prediction dataset. The tone size of the trained dataset is equal to the tone size of the prediction data.

In the implementation process, the following stages are identified as initially identifying variations in different music equipment and different music categories, accrue of real data (using standard music keyboard), extracting the features, training the neural network, and identifying the Instrument and Notes.

4.1 Identifying Differences in Different Music Equipment and Different Music Categories

Woodwind, string, keyboard, brass, and percussion instrument categories were selected, and using standard music keyboard and digital oscilloscope observed different kinds of waveform patterns of twelve (12) different tones of ten (10) music instruments. The.csv files (120 number of.csv files) were generated using digital oscilloscope. CSV files were input to MATLAB function (Fourier analysis) to convert the time domain to frequency domain. Accordingly, 12 tons and 3 different instruments of three different instrument categories were selected for this research project such as woodwind, string, keyboard categories and flute, violin, grand piano (Figs. 3 and 4).

Terminology in this table.

I1—Instrument 1.

T1—tone 1.

I1T1—Instrument 1, tone 1.

I2T5, I1T5—combined the instrument 2, tone 5 and the instrument 1, tone 5 (Fig. 5).

The generated tones of Matrix A and Matrix D are the same instrument tone combination. Matrixes B and C are different instrument tone combinations. The

C5	C#5	D5	D#5	E5	F5	F#5	G5	G#5	A5	A#5	B5
T1	T2	T3	T4	T5	T6	T7	T8	T9	T10	T11	T12

Fig. 3 Selected 12 tones

Fig. 4 Instrument combination, blue and pink cells are same pitch different instrument

	Flute	**Violin**	**Grand Piano**
Flute		√	√
Violin	√		√
Grand Piano	√	√	

Fig. 5 Tone matrix

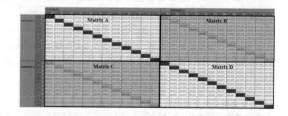

generated tones of Matrix B and the generated tones of Matrix C are equal. So, in this project, further experiments only consider Matrix C for any two-instrument combination. According to the defined scope of this research, different instruments with the same pitch are considered, which means selecting the diagonal of this matrix. Following green color 12 tones are considered, after combining any two instruments (Fig. 6).

	I1T1	I1T2	I1T3	I1T4	I1T5	I1T6	I1T7	I1T8	I1T9	I1T10	I1T11	I1T12
I2T1	I2T1,I1T1	I2T1,I1T2	I2T1,I1T3	I2T1,I1T4	I2T1,I1T5	I2T1,I1T6	I2T1,I1T7	I2T1,I1T8	I2T1,I1T9	I2T1,I1T10	I2T1,I1T11	I2T1,I1T12
I2T2	I2T2,I1T1	I2T2,I1T2	I2T2,I1T3	I2T2,I1T4	I2T2,I1T5	I2T2,I1T6	I2T2,I1T7	I2T2,I1T8	I2T2,I1T9	I2T2,I1T10	I2T2,I1T11	I2T2,I1T12
I2T3	I2T3,I1T1	I2T3,I1T2	I2T3,I1T3	I2T3,I1T4	I2T3,I1T5	I2T3,I1T6	I2T3,I1T7	I2T3,I1T8	I2T3,I1T9	I2T3,I1T10	I2T3,I1T11	I2T3,I1T12
I2T4	I2T4,I1T1	I2T4,I1T2	I2T4,I1T3	I2T4,I1T4	I2T4,I1T5	I2T4,I1T6	I2T4,I1T7	I2T4,I1T8	I2T4,I1T9	I2T4,I1T10	I2T4,I1T11	I2T4,I1T12
I2T5	I2T5,I1T1	I2T5,I1T2	I2T5,I1T3	I2T5,I1T4	I2T5,I1T5	I2T5,I1T6	I2T5,I1T7	I2T5,I1T8	I2T5,I1T9	I2T5,I1T10	I2T5,I1T11	I2T5,I1T12
I2T6	I2T6,I1T1	I2T6,I1T2	I2T6,I1T3	I2T6,I1T4	I2T6,I1T5	I2T6,I1T6	I2T6,I1T7	I2T6,I1T8	I2T6,I1T9	I2T6,I1T10	I2T6,I1T11	I2T6,I1T12
I2T7	I2T7,I1T1	I2T7,I1T2	I2T7,I1T3	I2T7,I1T4	I2T7,I1T5	I2T7,I1T6	I2T7,I1T7	I2T7,I1T8	I2T7,I1T9	I2T7,I1T10	I2T7,I1T11	I2T7,I1T12
I2T8	I2T8,I1T1	I2T8,I1T2	I2T8,I1T3	I2T8,I1T4	I2T8,I1T5	I2T8,I1T6	I2T8,I1T7	I2T8,I1T8	I2T8,I1T9	I2T8,I1T10	I2T8,I1T11	I2T8,I1T12
I2T9	I2T9,I1T1	I2T9,I1T2	I2T9,I1T3	I2T9,I1T4	I2T9,I1T5	I2T9,I1T6	I2T9,I1T7	I2T9,I1T8	I2T9,I1T9	I2T9,I1T10	I2T9,I1T11	I2T9,I1T12
I2T10	I2T10,I1T1	I2T10,I1T2	I2T10,I1T3	I2T10,I1T4	I2T10,I1T5	I2T10,I1T6	I2T10,I1T7	I2T10,I1T8	I2T10,I1T9	I2T10,I1T10	I2T10,I1T11	I2T10,I1T12
I2T11	I2T11,I1T1	I2T11,I1T2	I2T11,I1T3	I2T11,I1T4	I2T11,I1T5	I2T11,I1T6	I2T11,I1T7	I2T11,I1T8	I2T11,I1T9	I2T11,I1T10	I2T11,I1T11	I2T11,I1T12
I2T12	I2T12,I1T1	I2T12,I1T2	I2T12,I1T3	I2T12,I1T4	I2T12,I1T5	I2T12,I1T6	I2T12,I1T7	I2T12,I1T8	I2T12,I1T9	I2T12,I1T10	I2T12,I1T11	I2T12,I1T12

Fig. 6 Diagonal of the tone matrix

4.2 Accrue of Real Data (Using Standard Music Keyboard)

Using the layer function of standard music keyboard is shown in Fig. 7 selected two instruments at a time and played. The sound combination of two instruments at a time was used to take the samples. The time duration of each sample is 800 ms.

Digital oscilloscope is used for sampling. The captured sample rate is 5000 Sa/s. The sampled dataset is sent to the CSV file to use in further experiments (Fig. 8).

Figure 9 shows how to vary the spectrum of the same tone with two different instruments and a combination of two instruments. According to it, extracting the feature of the spectrum can recognize different timber of different instruments and instrument combinations.

Fig. 7 Real data acquisition setup

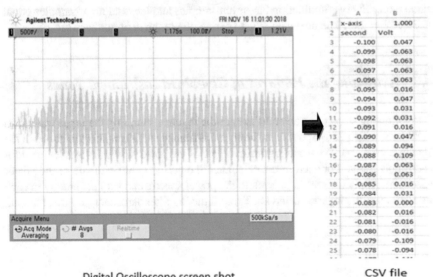

Fig. 8 Piano 5C time domain to.csv file

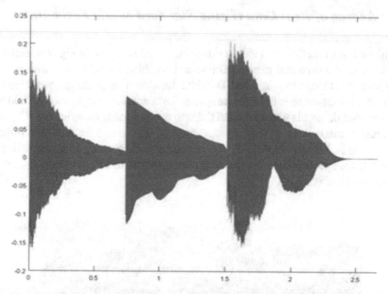

Fig. 9 Piano 5C, guitar 5C, and combination of piano and guitar

The sampled.csv files of instrument 1 and instrument 2 waveforms were generated as the above method. Using MATLAB function converted the.csv files to.wav file. One combination considered only the same pitch (same tone). So, one combination has 12 tons of the same pitch. Three combinations have $= 3 \times 12 = 36$ tones of same pitch. Every combination has seven samples and the total no of samples equal to $36 \times 7 = 252$. The next step was to extract the features of all these samples.

4.3 Extracting the Features of Combined Instrument Tones

The following steps were taken to extract the features sampling the input waveform at 800 ms, with a 25% of overlap; the signal is framed to 200 ms short frames (As the standard framing). The periodogram estimate of the power spectrum is calculated for each short frame. For each frame calculated the periodogram estimate of the power spectrum. Apply the mel filter bank to the power spectra and sum the energy in each filter. Finally, the logarithm and DCT for all the log filter bank energies are obtained.

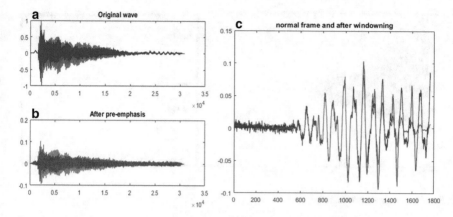

Fig. 10 **a** Pre-emphasis signal. **b** Comparison of framing and after windowing

These steps were applied to every single frame, and one set of 13 MFCC coefficients was extracted for each frame. The MFCC and preprocessing steps were done by using the melcepst () function of the voicebox tool.

Pre-emphasized signal
Since the audio signal is always changing, it is assumed that the audio signal does not change much on short-term scales and therefore the signal is framed at 50-200 ms frames.

Sampling and windowing
To avoid reducing the ripples and get a more accurate idea of the frequency spectrum of the original audio signal, the Hamming window method was used.
Fourier transform of the time domain signal

Using the MATLAB function converted the audio waveform time domain to the frequency domain and the different frequency levels shown in Fig. 10.

After the combination of one musical instrument, one music tone has one MFCC feature vector. Every sample includes 13×77 data: 13 coefficients and 77 frames (Figs. 11 and 12).

4.4 Training the ANN (Classification)

Considering the MFCC feature, a training dataset was developed to train an artificial neural network that can classify music and tone using integrated instrument tones. For pattern recognizing and classification, MATLAB neural network tool is used ("Neural Pattern Recognition app select data, create and train a network") and its functionality using confusion matrices is evaluated.

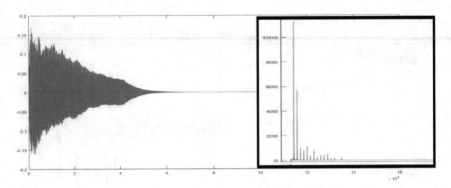

Fig. 11 Frequency domain

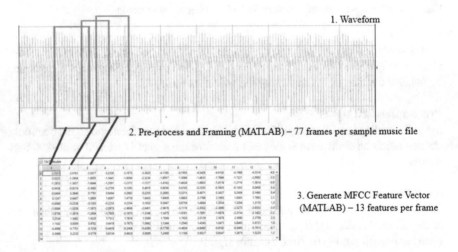

Fig. 12 MFCC feature vector

The designed ANN can classify the instrument and tone from a file of combined instrument tones. One instrument combination has 12 tones. It generates 12 tones of MFCC feature vectors. One tone combination has seven samples. After defining the input vector and the target vector, the neural network was trained to insert a different number of hidden layers.

MATLAB neural network pattern recognition app was used to train the neural network. ANN has two layers with the activation function being the sigmoid in the hidden layer and the softmax in the output layer. The training was done using the backpropagation algorithm (Figs. 13 and 14).

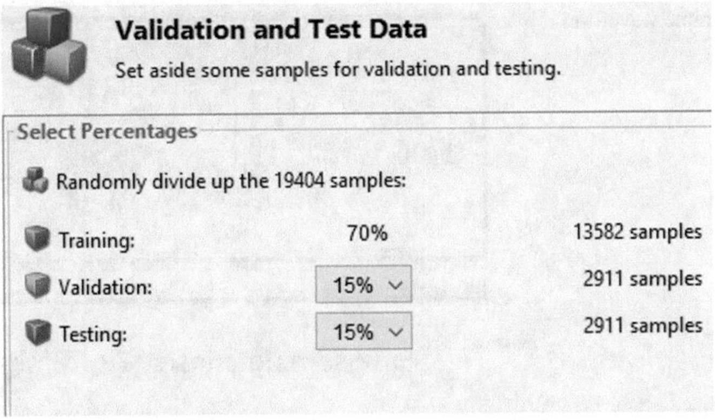

Fig. 13 Validation and test data

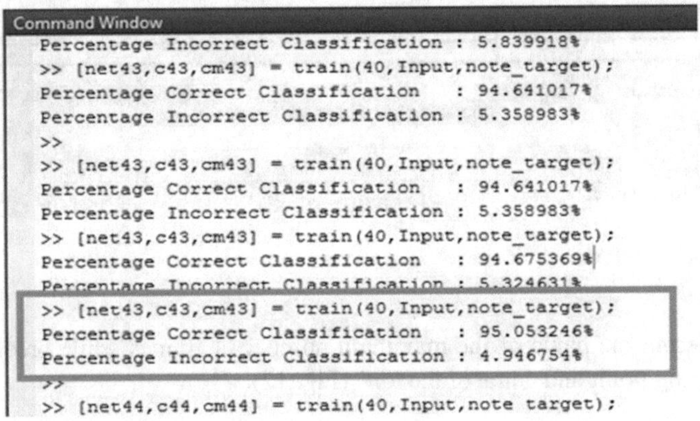

Fig. 14 Correct classification percentage

4.5 Identifying the Instrument and Notes

4.5.1 The Waveform Uses to Train Neural Network

The waveform is framed into short frames of 200 ms with a 25% overlap. Each wave was divided into 77 frames, and each frame contains 13 MFCC coefficients (Fig. 15).

4.5.2 The Waveform Uses to Input the Neural Network

The input audio waveform is framed into short frames of 200 ms with a 25% overlap. Subsequently, the instrument and instrument tone classification was carried out, considering each of the 77 frames (Fig. 16).

Fig. 15 Sample waveform

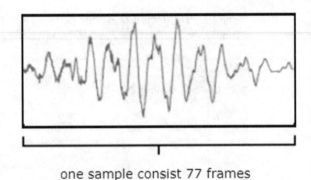

one sample consist 77 frames

one sample consist 77 frames

Tone 1
Instrument 1 and
Instrument 2

one sample consist 77 frames

Tone 2
Instrument 1 and
Instrument 2

one sample consist 77 frames

Tone 1
Instrument 1 and
Instrument 2

Fig. 16 Input waveform

Considering the name of the maximum number of frames while predicting the near changing point and name of the tone (Fig. 17).

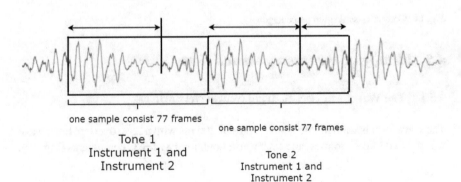

one sample consist 77 frames

Tone 1
Instrument 1 and
Instrument 2

one sample consist 77 frames

Tone 2
Instrument 1 and
Instrument 2

Fig. 17 Waveform changing point

5 Result and Discussion

The best performance of a trained neural network was obtained using the neural network training tool when 40 hidden neurons were used.

[net43,c43,cm43] = train(40,Input,note_target);

Percentage CorrectClassification:**95.053246%**

Percentage IncorrectClassification:**4.946754%**

Figure 18 shows the performance of the ANN with the training and test datasets.

According to the diagonal cells in Fig. 19, it can be seen that 95.5% of the predictions are correct and 4.5% are incorrect (Fig. 20).

When two (2) instruments are played simultaneously, it may not be possible to start playing at the same time. So, the music keyboard was used to minimize this error. Playing only one attribute when extracting the features, the environment status of the prediction dataset was made not to differ from that of the training dataset, and the duration of the tone in the training dataset was equal to the duration of the tone in the predictive data, and the wave files generated by the MATLAB function produce monosound but the predicting wave file gives the stereo sound. Therefore, prediction stereo sound needs to be converted to a monosound. Each sample signal is framed into 77 frames, and thirteen (13) MFCC features were extracted from each frame. These sets of features were used as the neural network input. Ten (10) hidden neurons were used to train the neural network first. In the MFCC calculation, 77 frames were obtained from a sample wave of 800 ms and 13 MFCC coefficients per frame. One sample wave has a sample MFCC coefficient of $77 \times 13 = 1001$ and a single tone combination has 7 sample waves. Three instruments consist of three combinations and one instrument combination consists of 12 tones. The input layer consists of 77 $\times 13 \times 7 \times 12 \times 3 = 252{,}252$ number of data. The hidden layer was increased to 40 neurons while the backpropagation algorithm with a feed-forward neural network

Fig. 18 Performance plot with 40 hidden neurons

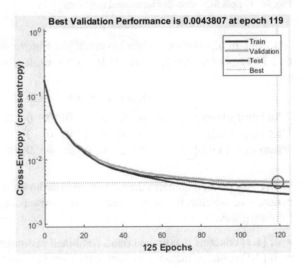

Confusion Matrix

Output Class	30	31	32	33	34	35	36	
30	0 0.0%	525 2.7%	0 0.0%	0 0.0%	0 0.0%	0 0.0%	0 0.0%	98.3% 1.7%
31	0 0.0%	0 0.0%	534 2.8%	0 0.0%	0 0.0%	0 0.0%	0 0.0%	99.6% 0.4%
32	0 0.0%	0 0.0%	0 0.0%	530 2.7%	0 0.0%	0 0.0%	0 0.0%	97.4% 2.6%
33	0 0.0%	0 0.0%	0 0.0%	0 0.0%	527 2.7%	1 0.0%	0 0.0%	96.7% 3.3%
34	0 0.0%	0 0.0%	0 0.0%	0 0.0%	0 0.0%	526 2.7%	0 0.0%	98.7% 1.3%
35	0 0.0%	0 0.0%	0 0.0%	0 0.0%	0 0.0%	0 0.0%	498 2.6%	92.2% 7.8%
36	0 0.0%	0 0.0%	0 0.0%	0 0.0%	0 0.0%	0 0.0%	0 0.0%	529 2.7%	98.7% 1.3%
	99.6% 0.4%	97.4% 2.6%	99.1% 0.9%	98.3% 1.7%	97.8% 2.2%	97.6% 2.4%	92.4% 7.6%	98.1% 1.9%	95.9% 4.1%

Fig. 19 Confusion matrix

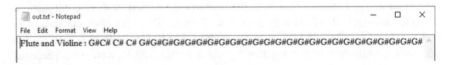

Fig. 20 Output file—contains instrument and notes

was used. In this research project to identify the instrument note, at least there must be a 77 number of frames, in one note. At the result of ANN, it generates a matrix.

$$
\begin{array}{c}
\\
\textbf{Flute and Piano} \\
\textbf{Flute and Violin} \\
\textbf{Piano and Violin}
\end{array}
\begin{array}{cccccccccccccc}
\text{C} & \text{C\#} & \text{D} & \text{D\#} & \text{E} & \text{F} & \text{F\#} & \text{G} & \text{G\#} & \text{A} & \text{A\#} & \text{B} & \text{B\#} \\
\left[\begin{array}{ccccccccccccc}
0 & 0 & 0 & 0 & 0 & 0 & 0 & 0 & 0 & 0 & 0 & 0 & 0 \\
0 & 0 & 0 & 0 & 0 & 0 & 0 & 0 & 0 & 0 & 0 & 0 & 0 \\
0 & 0 & 0 & 0 & 0 & 0 & 0 & 0 & 0 & 0 & 0 & 0 & 0
\end{array}\right]
\end{array}
\Longrightarrow
\begin{array}{c}
\textbf{Total} \\
\left[\begin{array}{c} 0 \\ 0 \\ 0 \end{array}\right]
\end{array}
$$

In every point, it contains the note count in audio file under the instrument combination. Considering the total count of every row audio file categorize under the following way,

- If [1,1] contains maximum count, combined instruments are flute and piano
- If [2,1] contains maximum count, combined instruments are flute and violin

Table 1 Real audio file generated instruments versus ANN identified instruments	Real audio file generated instruments	ANN identified instruments
	Flute and piano	Flute and piano
	Flute and violin	Flute and violin
	Piano and violin	Piano and violin

- If [3,1] contains maximum count, combined instruments are piano and violin.

  Considering the test audio file following conclusion was reached (Table 1).

6 Conclusion and Future Work

Three instruments from three instrument categories were selected and 252,252 number of data items were extracted to train the artificial neural network. Changing the number of hidden layers increased the correctness of a trained artificial neural network. The greatest accuracy obtained was 95.9%. The trained neural network facilitated music instrument identification and note identification. Instrument note identification was done and according to the time frame of this project if noise filtering technique used with proper sound recording this result would have been more accurate. Based on these results, it can be concluded that an ANN can be used to accurately detect different instrument sounds in different instrument categories.

Acknowledgments As future work, this research can be developed to identify more than two music instruments from an audio file using ANN. For greater accuracy, it is recommended to record with proper sound recording techniques which are available at recording stations.

References

1. Dr. Scott Gordon, Musical instrument recognition with neural networks (2013)
2. Toghiani-Rizi, B., Windmark, M.: Musical instrument recognition using their distinctive characteristics in artificial neural networks
3. Mazarakis, G., Tzevelekos, P., Kouroupetroglou, G.: Musical instrument recognition and classification using time encoded signal processing and fast artificial neural networks
4. Keen, P., Why do different musical instruments have different sounds? Available at https://meandering-throughmathematics.blogspot.com/2013/09/why-do-different-musical-instruments.html?m= mathematics.blogspot.com/2013/09/why-do-different-musical-instruments.html?m=, 2 Sep 2013
5. Practical Cryptography, Mel Frequency Cepstral Coefficient (MFCC) tutorial , Available at: https://practicalcryptography.com/miscellaneous/machine-learning/guide-mel-frequency-cepstral-coefficients-mfccs/ (2013)
6. Parwinder P.S., Rani, P.: An approach to extract feature using MFCC, Volume 04 available at: https://iosrjen.org/Papers/vol4_issue8%20(part-1)/D04812125.pdf (2014)

7. Seppänen, J.: Audio signal processing basics at https://www.cs.tut.fi/sgn/arg/intro/basics.html (1999)
8. Sharmila V.N., Mangal J.: Extraction of feature vectors for analysis on musical instruments (2014)
9. Fan, L., Zhang, J., Pal, N.: Automatic sign language detection at https://meandering-through-mathematics.blogspot.com/2013/09/why-do-different-musical-instruments.html
10. Jang, R.: Audio signal processing and recognization-12–2MFCC, available at https://mirlab.org/jang/books/audiosignalprocessing/speechFeatureMfcc.asp?title=12-2%20MFCC
11. Mazarakis, G., Tzevelekos, P., Kouroupetroglou, G.: Speaker identification using pitch and MFCC available at https://www.mathworks.com/help/audio/examples/speaker-identification-using-pitch-and-mfcc.html
12. Menia A.-K.: Voice-signature-based speaker recognition (2017)
13. Barua, P.: Neural network based recognition of speech using MFCC features (2014)
14. Vorgelegt, V.: Neural network base feature extraction for speech and images recognition (2014)
15. Janaka, M.S., Ratnayake, H.U.W., Premaratne, I.A.: Intelligent time of Use Deciding system for a melody to provide a better listing experience at: https://link.springer.com/chapter/https://doi.org/10.1007/978-981-13-9129-3_9. (2019)

Security in Software Applications by Using Data Science Approaches

Akkem Yaganteeswarudu, Aruna Varanasi, and Sangeet Mohanty

Abstract The clients are facing the issues once the application is deployed in a production environment because while development, developers by mistake or wrongly coded without encrypting the sensitive information or even the testing team tested the functionality of the application but not looked into the database for what is stored in place of sensitive information. The sensitive information may be hacked by hackers or due to storing sensitive information in the database the customer's important information is exposed to the outside world. To provide security in software applications various data science approaches are used like machine learning and natural language processing (NLP). In the current statement, proposing a technique uses machine learning and deep learning techniques which will scan the entire application code and databases used. The model is trained with machine learning algorithms. To identify sensitive information in all programming files in application, natural language processing term frequency–inverse document frequency (tf–idf) technique is used. By using tf–Idf, it will detect whether sensitive information is present in each document or total documents. The model will recognize the user's sensitive information like password, mobile number, account number, card verification value (CVV), and all the sensitive information which client or customer feels it is sensitive information. As per the standards of security algorithms, all the sensitive information of customers is verified. When the sensitive data is not in encrypted format, model which is trained will automatically encrypt the data with encryption algorithms. By default, data encryption standard (DES) algorithm will be used for encryption. But this tool is designed in a more convenient way so that the client will choose different algorithms for encryption or key size which used in the algorithm.

A. Yaganteeswarudu (✉) · A. Varanasi · S. Mohanty
SureIT Solutions Inc, Hyderabad, India
e-mail: yaganteeswaritexpert@gmail.com

Sreenidhi Institute of Science and Technology, Hyderabad, India

A. Varanasi
e-mail: arunavaranasi@sreenidhi.edu.in

S. Mohanty
e-mail: sangeet@nextbos.com

S. Shakya et al. (eds.), *Proceedings of International Conference on Sustainable Expert Systems*, Lecture Notes in Networks and Systems 176,
https://doi.org/10.1007/978-981-33-4355-9_27

This tool can be used as a security measure for any software application developed before deploying in a production environment.

Keywords Production environment · Sensitive information · Testing · DES · Key size · Natural language processing · Tf–idf · CVV · Data science

1 Introduction

Mainly application code is reviewed for various sensitive information regarding password, mobile number, CVV, and account number. Some applications have mandatory login which includes username and password.

Mostly observed the subsequent points are misled by the developers and they are as follows, in developed code that some developers directly storing sensitive information in databases, the developers by mistake hardcoded sensitive information values in some places of application. And also developers while retrieving values from databases without encrypting directly showing in the front-end Web pages. Test cases are written for testing also observed but the test cases focused on functionality but not testing what is storing in databases. Some developers store sensitive information in caches but not cleared the caches.

When ever any patient details stored in Database to be kept in more secured way because for example if any details of patient will be displayed out means that will create an issue for patient so we need to maintain patients details in a more secured way. Personal information of the patients could not be shown to the outsiders due to security issues or to prevent an attack from the outsiders. Like that based on the domain or client requirement, sensitive information will change.

The main idea of this paper is by using machine learning and natural language processing; the model is trained with sensitive information keywords. For training, the model used natural language processing to divide the code developed into tokens. Python NLTK [1, 2] (natural language tool kit) package is used to split the data into tokens like NLTK. The word_tokenize method is used to divide the tokens.

Machine learning model is designed to check whether the application is safe or not by using clustering algorithms. Once if anything about if the application is not secured user will be given the choice to select a security algorithm to encrypt [3, 4] the sensitive details. If the user will not specify a security algorithm by default, data encryption standard algorithm is used to encrypt the sensitive information.

2 Natural Language Processing tf–idf Score Calculation for Sensitive Details

2.1 Word Tokenization to Identify Sensitive Information

By default, natural language processing by using natural language tool kit (NLTK) [5] will provide the word_tokenize method. By using the word_tokenize method, divide the sentences in all the files into words. For example, patientdetails.js file contains patient-related information.

Procedure:

- First needs to open the file by using the open method of Python and store the sentences in one variable, let say "text" is a variable to store the text of the file.
- To identify token first, it is needed to clean the text like first to convert it into lower case and then remove the punctuation marks and remove all stop words. NLTK [6] will provide stop words so remove all stop words.
- Lower = text. Lower ().
- tokenize = words tokenize (Lower, "English").
- Once the sentences are tokenized next step is to find sensitive words that are there in the file or not.

Declare sensitive words based on user requirements. For example, patient details. Patient sensitive information = [patient name, patient id, patient address]. Account sensitive information = [account no, balance details]. Loginsensitiveinfo = [password, secret question].

The above sensitive information details vary from application to application or based on user requirement.

Once sensitive details formed, find whether the file contains sensitive information or not as below.

Sample code:

```
Final sensitive words= []
for word in Words after tokenize:
    # For patient sensitive information
    if word in Patient sensitive information:
        Finalsensitivewords.append(word)
    # For Account details sensitive information
    if word in Account Sensitive Information:
        Finalsensitivewords.append(word)
    # For login sensitive information
    if word in Loginsensitiveinfo:
        Finalsensitivewords.append(word)
```

2.2 tf–idf Score Calculation for Sensitive Words

Once word tokenization finished and sensitive words collected from a different file, the next step is to find tf–idf [6, 7] (term frequency–inverse document frequency)— To calculate tf–idf score all the JavaScript files, HTML files, core logic-related files, and database connection files are considered.

Let term frequency is "*a*".

Let inverse document frequency is "*b*".

- Final sensitive words contain all sensitive detail fields of the user.
- As per natural language processing, term frequency is defined a number of times the term present in the document. Here, the terms are the sensitive information found in all files
- Inverse document frequency (IDF) = (total number of documents)/(number of documents containing word)
- Inverse document frequency of sensitive words is calculated as how many documents containing sensitive information.
- By calculating tf–idf [8] score, it is possible to find what are the documents where sensitive details are used.

3 Machine Learning Model to Find Whether Application is Secured or not

3.1 Sensitive Keywords and Their Value Encrypted Format Testing

During the development of any software application, based on the client requirement, the programmer has to develop different Web pages and also each page will interact with different tables in the database. Many of the application's minimum requirements are a login page where the username and password are required. Some banking applications will store user account numbers, user transaction details, and so on. Some educational institutions store information about their student's names, marks, and many personal details.

Once the tf–idf score is calculated for all sensitive words, it is cleared that in which documents that are in which files sensitive information contains. Once documents identified next, the step is to find the value of the sensitive keyword is encrypted [9, 10] or not.

Procedure:

Once tf–idf is calculated for the sensitive information, the next step is to find whether values encrypted or not.

For example, the password is a key and to check whether the value of the password is present in any file.

First needs to collect all password rules used in application like.

- Password must be 8 characters
- At least 1 upper case character
- At least 1 lower case character
- At least 1 special character.

By using those rules, Python regular expression is framed and the pattern present exactly after the password key is found. If a match is found, then store the final sensitive information values so that the test can be performed with the standards of the encryption algorithm.

3.2 Machine Learning Clustering Algorithm to Test Whether Application is Secured or Not

Machine learning model is trained with K-nearest neighbors (KNN) [8, 11, 12] algorithm and agglomerative clustering algorithms. Different software applications were collected to do the clustering process. Once the tf–idf score is calculated and after testing each sensitive information [13, 14] value, the probabilities are calculated based on total sensitive information terms.

Based on user requirement, number of clusters will be formed or several neighbors will be selected.

For example, if more than 60% of sensitive information not encrypted means the application security is worst.

Cluster 1: Greater than 90% of sensitive information is encrypted, the application is more secured.

Cluster 2: Greater than 80% of sensitive information is encrypted, and then the application is semi secured.

Cluster 3: Less than 60% of sensitive information is encrypted, then the application needs to be retested as per security standards.

The number of clusters can be formed as per client requirement because based on end-user requirement rules to be framed.

Once clusters are formed and if any new application came for testing, then classify whether the application is secured [15, 16] or not by calculating the Euclidian distance between all clusters and new applications. The application is classified depending on the cluster and checks whether it is secured or not.

4 Data Encryption Algorithms

After classifying the application is secured or not, the next step is to encrypt sensitive information values.

If the application is more secured then to satisfy security standards all the sensitive information must be encrypted.

To encrypt information, there are a lot of algorithms available. Data encryption standard (DES), advanced encryption standard [10, 17] (AES), blowfish [3], two fish, triple DES, and many more algorithms are available.

The choice will be given to the user to select the algorithm to encrypt the data of sensitive information.

If the user will not select any algorithm by default, data encryption standard algorithm is used to encrypt the sensitive information.

5   Data Flow in Proposed Model

Figure 1 represents the proposed model. The data flow diagram in the proposed model functions as initially the word tokenization done on all the files. Later tf–idf score was calculated for all the terms in all JavaScript files, HTML files, and database connection files. A model trained with clustering algorithms and different clusters formed will be formed based on requirements like secured applications, unsecured applications, and semi-secured applications. A new application comes for testing whether it is secured or not. All the database files, HTML files, and input statements are given as input to the model to predict whether it is secured or not. After predicting if the application is belonging to not a secured application, the user will be given a choice to select security algorithms like DES, RSA, and so on. If the application is secured, the encryption step will be skipped.

6   Results

6.1   Hardware Requirements

Processor—Intel Xeon E2630 v4—10 core processor, 2.2 GHz with Turboboost upto 3.1 GHz. 25 MB Cache.
   Motherboard—ASRock EPC612D8A.
   RAM—128 GB DDR4 2133 MHz.
   2TB Hard Disk (7200 RPM) + 512 GB SSD

6.2   Software Requirements

Python 3.x version.
   Jupyter notebook/Pycharm.

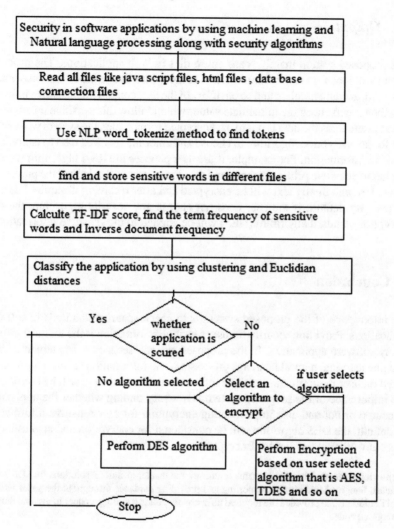

Fig. 1 Data flow

Python packages (Numpy, Pandas, Scikit-learn, NLTK).
Windows or Linux operating system.

6.3 Inputs

All the files are contained in the Web application developed. Sample files are listed below HTML files, XML files, JavaScript files, database files, Python scripts (.py files), and so on.

6.4 Output

This proposed system mainly scans entire files in Web applications. The proposed system will look for sensitive information in those files like patient name, user name, password, account number, and so on. if any of the files contain this sensitive information, then it will check the immediate values which follow this sensitive information. If the system finds the data is not in the encrypted format, then the system will ask the user for an encryption algorithm to choose and enter the required data to encrypt the sensitive information. For example, if the user chooses the RSA algorithm then the user has to enter two prime numbers, and then the system will calculate the public key, private key, and finally text will be encrypted and after receiving if required decrypt the text. By default, if the user does not choose any security algorithm, the data encryption standard algorithm (DES) will be applied to the sensitive information.

7 Conclusion

The importance of the proposed system is to check security standards in software applications. Providing security for user sensitive information is the topmost priority in any software application. In the proposed system, security is implemented with machine learning, natural language processing, and information security algorithms. Based on user requirement, sensitive information of user details will be considered. The importance of the proposed system is not only finding whether the application is secured or not and also implementing encryption for user sensitive information. By default, the DES algorithm will be considered for encryption and also users can select any other algorithm for Encryption.

Acknowledgements The author wants to convey the thanks to SureIT solutions Inc, Dr. Aruna Varanasi, Sangeet Mohanty for supporting and providing guidance throughout the paper writing. SureIT solutions Inc provided the required infrastructure during writing a paper to test and develop the required code.

References

1. Bafna, P., Pramod, D., Vaidya, A.: Document clustering: TF-IDF approach. In: International Conference on Electrical, Electronics, and Optimization Techniques (ICEEOT), Chennai, pp. 61–66 (2016)
2. Trstenjak, B., Mikac, S., Donko, D.: KNN with TF-IDF based framework for text categorization. Procedia Eng. **69**, 1356–1364 (2014)
3. Kumar, Y., Munjal, R. and Sharma, H.: Comparison of symmetric and asymmetric cryptography with existing vulnerabilities and countermeasures. (IJAFRC) **1**(6) (2014). ISSN 2348–4853

4. Mahindrakar, M.S.: Evaluation of Blowfish Algorithm based on avalanche effect by Manisha Mahindrakar gives a new performance measuring metric avalanche effect. Int. J. Innov. Eng. Technol. (IJIET) (2014)
5. Joby, P.P.: Expedient information retrieval system for web pages using the natural language modeling. J. Artif. Intell. 2(02), 100–110 (2020)
6. Nam, S., Kim, K.: Monitoring newly adopted technologies using keyword based analysis of cited patents. IEEE Access 5, 23086–23091 (2017)
7. Fan, H., Qin, Y.: Research on text classification based on improved TF-IDF algorithm. In: International Conference on Network, Communication, Computer Engineering (NCCE 2018), vol. 147 (2018)
8. Jung, Y.G., Kang, M.S., Heo, J.: Clustering performance comparison using K-means and expectation maximization algorithms. Biotechnol. Biotechnol. Equip. 28(sup1), S44–S48 (2014)
9. Mahindrakar, M.S.: Evaluation of blowfish algorithm based on avalanche effect by Manisha Mahindrakar gives a new performance measuring metric avalanche effect. Int. J. Innov. Eng. Technol.; Karthik, S., Muruganandam, A., 2014. Data Encryption and Decryption by using Triple DES and performance analysis of crypto system. (IJIET) 2(11), 2347–3878 (2014)
10. Gurusamy, V., Kannan, S.: Preprocessing Techniques for Text Mining. RTRICS, pp. 7–16 (2014)
11. Amancio, D.R., Comin, C.H., Casanova, D., Travieso, G., Bruno, O.M., Rodrigues, F.A., et al.: A systematic comparison of supervised classifiers. PloS one. 9(4), e94137 (2014). pmid:24763312
12. Manoharan, S.: Image detection, classification and recognition for leak detection in automobiles. J. Innov. Image Process. (JIIP) 1(02), 61–70
13. Kou, G., Peng, Y., Wang, G.: Evaluation of clustering algorithms for financial risk analysis using MCDM methods. Inf. Sci. 275, 1–12 (2014)
14. Raykov, Y.P., Boukouvalas, A., Baig, F., Little, M.A.: What to do when K-means clustering fails: a simple yet principled alternative algorithm. PLoS ONE 11(9), 1–28 (2016)
15. Dai, W.: Improvement and implementation of feature Weighting Algorithm TF-IDF in text classification. In: International Conference on Network, Communication, Computer Engineering (NCCE 2018), vol. 147 (2018)
16. Santhanakumar, M., Columbus, C.C.: Various improved TFIDF schemes for term weighing in text categorization: a survey. Int. J. Appl. Eng. Res. 10(14), 11905–11910 (2015)
17. Rodriguez, M.Z., Comin, C.H., Casanova, D., Bruno, O.M., Amancio, D.R., Costa, L.D.F.: Clustering algorithms: a comparative approach. PLoS ONE 14, e0210236 (2019). https://doi.org/10.1371/journal.pone.0210236

Akkem Yaganteeswarudu, Finished post-graduation (M. Tech.) in 2012. Started professional career in 2012. Worked at Honeywell, ITC infotech, Unisys and Pyramid Soft Sol. Currently working in SureIT Solution Inc (Hyderabad). Having 8+ years of experience and in Data science having 4+ years of experience. Published various IEEE papers like preventing suicides of farmers, Speaking compiler and many more. Having interest in research in health care domain. Interested to develop health care solutions by using machine learning and deep learning techniques. Having more than 10 international journals in various publications.

Aruna Varanasi, FIE, Ph.D. in C.S.E, JNTUH, M. TECH (C.S.E) Andhra University. Having total 24 years experience (6 years Govt. service and 18 years teaching). Areas of Research: Computer Networks, Cryptography, Image Cryptography, Information Security and Internet of Things. Published 1 book, 35 research papers in international journals, 5 patents filed and 2 patents got published.

Sangeet Mohanty, Technology Director and Manager at SureIT solution Hyderabad. Having huge industry experience.

Smart Agriculture Management System Using Internet of Things (IoT)

M. B. Abhishek, S. Tejashree, R. Manasa, and T. G. Vibha

Abstract Agriculture is a part of the most significant features in human civilization. In the last two decades, information and communication technology has contributed immensely to this field. Internet of Things [IoT] is an innovation in which actual material items like sensor nodes function together to render a technology-driven and information-based network that maximizes benefits with significantly reduced risks. IoT technology helps to collect online crop-monitoring information about conditions such as weather, humidity, soil, temperature, and fertility. IoT allows farmers to connect with their farms from anywhere at any time. Wireless sensor networks (WSNs) are usually used to monitor farm situation, and the process is regulated and automated with controllers. A smartphone enables farmers to stay up-to-date on the ongoing conditions of their farms from any part of the world. This paper deals with an agricultural system on the principle of Internet of Things monitoring and controlling the production process.

Keywords Smart agriculture · Monitoring and controlling · WSNs · Internet of things · Cloud computing

1 Introduction

The United Nations Food and Agriculture Organization better known as the FAO estimates that in 2025, the world population is estimated to reach eight billion and by the year 2050, the value will be 9.6 billion. There must be a 70% increase in food production globally by 2050, in order to keep pace. India is one of the largest countries in terms of agricultural production, and this has a significant impact on national food security. The farmland area per capita in the developed countries in the world is much lower than the world average level and the production value per capita and land

M. B. Abhishek (✉) · S. Tejashree · R. Manasa · T. G. Vibha
Department of Electronics and Communication,
Dayananda Sagar College of Engineering, Bengaluru, India
e-mail: abhishek.mb@gmail.com

© The Editor(s) (if applicable) and The Author(s), under exclusive license
to Springer Nature Singapore Pte Ltd. 2021
S. Shakya et al. (eds.), *Proceedings of International Conference on Sustainable Expert Systems*, Lecture Notes in Networks and Systems 176,
https://doi.org/10.1007/978-981-33-4355-9_28

yield per unit in the case of India are also much lower. Therefore, it is required to build innovative ways to achieve higher yields with limited natural resources in order to combat the challenges of food production. In most of developing countries, agriculture stands as an essential pillar in the Indian economy and accounts for 22% of gross domestic product (GDP) of the country. More than 70% of Indians live in rural areas with agriculture being the main source of livelihood of most rural residents. Agriculture offers the vast majority of the rural population not only food safety but also employment opportunities. Agriculture is, therefore, an essential part of the Indian economy, but, if asked, "What is the condition of farmers?" the response that comes to our mind first is "They lack Education, Technology, and Capital." These three things are needed to develop and live in this day and age. Indian farmers adopt conventional agricultural practices, including methods of tilling, sowing and planting that involves labor. On the other hand, modern agricultural techniques use mechanized machinery for irrigation, tilling and planting, along with hybrid seeds. Agricultural technology in India is primarily labor-intensive. It makes the work slow and imperfect to do most of the work by hand. Farmers face many challenges in agriculture without crop rotation due to climate change in the environment, inadequate rainfall for crops, and conventional methods. Agriculture is India's backbone. Agriculture is regarded as the production or cultivation of beneficial crops in the appropriate ecosystem. Farm productivity may be augmented by evaluating the variety of crops that produced the highest output under different conditions of soil, fertilization, irrigation, and climate. Agriculture is regarded as the chief source of food grains, raw materials; it is therefore termed the basis of life. It serves a significant purpose in national economic growth. It also offers people job opportunities. Agricultural growth is required to develop the country's economic condition, such as trade. Evidently, conventional agriculture techniques are still being used by many farmers, and this leads to a lower crop and fruit output. Thus, the country's trade declines; hence, the tractor was the first technology introduced into agriculture. It improved productivity and was useful to farmers as a machine. The technology advancement will help farmers boost crop yields. The emerging technologies in agriculture are the IoT, wireless sensor networks (WSNs), and Precision. Internet of things better known as IoT is a system where real-life objects are interconnected, and they appear to form many integrated networks, including fields such as electronics and sensors through which data can be reliably transmitted and received. Precision farming equipment with proper wireless connections to transmit data collected from remote satellites and ground sensors can take crop conditions into account and modify how each portion of a field is farmed. Global ICT Standardization Forum for India described the possible benefits of IoT as

- (i) Enhanced the quality, accessibility, and scalability,
- (ii) Faster and more cost-effective operation,
- (iii) Physical stream clarity and accurate status data,
- (iv) Increased capacity, reliability, flexibility, and automation.

1.1 Precision Farming

Precision farming uses a variety of technologies, e.g., sensors, GPS services, and large data transmission to maximize crop yields. ICT-based systems to facilitate decision-making, supported with real-time data, instead of replacing farmers' capabilities and notions, also give information on all features of agriculture at an unprecedented level of granularity. This makes it possible to make better decisions, leading to less waste and maximum operational efficiency. Scientific solutions, GPS innovations, technological advancements, computer-based image analysis, remote sensing, weather forecasting, ecological controls and more are the specialties and expertise now required for agriculture. Precision agriculture is often referred to as "smart farming," a general word for simpler comparison with other M2M-based applications such as smart towns, smart measurement and so on. It is based on sensor technologies that are well recognized in other sectors, e.g., environmental inspections for pollutants, eHealth monitoring of patients, building management for farm soil monitoring, etc. IT systems retrieve, compile, analyze and present the data for all M2M implementations in order to trigger a suitable response to the information obtained. An extensive range of details on soil and crop behavior, storage tanks condition, animal behavior, machinery condition is presented for action to the farmer from remote sites.

1.2 What is Smart Farming?

To various individuals, the Internet of Things (IoT) implies utilizing or incorporating modern innovations such as Google Glass, Apple watch, or a driverless car. In fact, some of the most revolutionary and effective applications occur in the industrial IoT like smart cities and smart agriculture, factories. Then, as now, IoT is transforming the agricultural sector by allowing farmers and producers to solve the overwhelming challenges they face. In a number of ways, IoT can help farmers. To allow farmers to obtain an abundance of insightful information such as the temperature of stored goods, the number of fertilizers used, the amount of water in the soil, the quantity of seed required to be planted, etc., sensors can be installed across the farm and farming machinery. When IoT is enabled, smart systems are deployed and they can track and make informed decisions on a number of environmental variables on the farm. IoT implementation in agriculture can tackle many challenges and improve the quality, yield, and cost-effectiveness of agricultural products. The Concept of Smart: Typically, a machine is considered to be smart if such a device/artifact or system does something that can only be performed by intelligent person. Any device, process, or domain that follows the six different levels of intelligence is said to be smart.

- Adapting: The term adapting refers to the changes made in order to fulfill certain requirements; in terms of smart agriculture, these changes are said to be environmental.

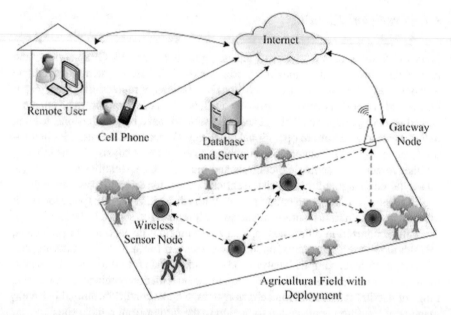

Fig. 1 Basic architecture of Internet of Things in agriculture

- Sensing: This is the ability to feel/detect and perceive any changes in the surroundings.
- Inferring: This simply refers to a conclusion based on results and findings.
- Learning: Once observations and observations have been confirmed, training can be used to consolidate previously used methodologies. It involves various types of information.
- Anticipating: This is all about thinking about something new and innovative that might happen, and it is considered to be the next phase of something.
- Self-Organizing: This level of intelligence is attributed to any device that is capable of sensing and tracking and then modifying its configurations according to the requirements (Fig. 1).

1.3 Process Flow in IoT

- Smart irrigation.
- Livestock monitoring.
- Weather monitoring and forecasting.
- Sensor-based precision farming.
- Remotely monitoring the quality of the soil.
- Smart warehousing, logistics, distribution.
- Greenhouse monitoring and automation system.

Smart farming deals with

- The environment monitoring and control subsystem includes monitoring water quality, regulation of water quality autonomously. Precise fertilization manages fertilizer. Regulate soil constituents and moisture, and also some environmental conditions like light, wind, air, etc. The agricultural asset regulation subsystem includes a smart greenhouse that will automatically adjust the temperature. A water irrigation system
- That will optimally regulate the flow of water automatically in order to conserve water. Monitoring of insects and pests.
- The production process monitoring and control subsystem includes the recognition of individual animals that ensure safe breeding. Monitoring the growth of animals and crops and product sorting ensures efficiency.
- Farm produce and food safety subsystem includes an arrangement of warehouse inventory logically. The traceability program on the farmland encourages the supply chain.
- Agricultural equipment and facility systems include remote monitoring of farm machinery, process tracking of farm machinery. Diagnosis and appropriate maintenance of farm machinery.

Since the computer age, two new concepts emerging are Internet of Things (IoT) and cloud computing. The idea of "Digital India" was put forward by India's Prime Minister Narendra Modi, which primarily stressed the growth of IoT and innovative new companies. IoT is closely linked to cloud computing as it is connected to the Internet. India is a traditional agricultural country with rice, dal, wheat, fruit, and cotton production. In India's socialist modernization, agriculture, rural areas, and farmers are of particular interest. Wikipedia's interpretation of cloud computing is this: Cloud computing is an Internet-based computing program that is used to distribute software and hardware data to computers and other devices in the network. End users don't need to have professional or even basic information about the "cloud," or directly control it. All of them need to know the type of tool they really want and how to get appropriate internet service. Cloud computing represents a modern system of implementing, utilizing, and sharing Internet-based IT services that include using the internet to deliver dynamic, expandable, and mostly virtualized resources.

2　Literature Survey

In papers [1–3], the agricultural implementation of a wireless sensor system for crop field tracking was proposed. There are two types of sensor nodes installed to these systems to determine temperature, humidity, and an image sensing node to compare and contrast information by occasionally taking images of crops. These variables perform a vital function to make good decisions about the crops' health within a period of time. Humidity, temperature, and images are these major parameters. By following these techniques, high sensor stability with low power consumption can be achieved. The agricultural field area can be monitored for a longer period of time.

Paper [4] discussed a greenhouse regulatory system on the basis of agricultural IoT with cloud computing capabilities. Inside a greenhouse, the administrator can effectively track and control various ecological parameters remotely using sensor devices such as temperature sensors, light sensors, soil moisture sensors, and relative humidity sensors. The sensors collect data on the agricultural field area at a regular interval of 30 seconds, and the data is logged and collated online with the aid of cloud computing and the Internet of Things.

Paper [5] illustrates an IoT-based irrigation automation and crop-field monitoring system. In this research, a system is designed to track crop-field using sensors and according to a server's decision based on sensed information, the system of irrigation is automatic. By using wireless communication, sensed data is transmitted to the database of the web server. When irrigation is done automatically, it ensures that the fields of humidity and temperature drop below the potential range. With the aid of a program that shows the user a web interface, the user can remotely track and manage the device.

According to Paper [6], a smart drip irrigation system is recommended. In this system, an application usually by android is used in order to reduce human intervention and for the remote tracking and managing the crop area. With the drip irrigation system, water wastage can be reduced and it works using the data obtained from water quantity sensors. To monitor environmental conditions, some additional sensors can be used.

Papers [7–9] proposed smart irrigation systems with the use of the Internet of Things. Wireless sensors are also needed to obtain the level of soil moisture and humidity. With the use of a gateway known as Generic IoT Border Router Wireless Br 1000, these detected data are transmitted through a network to a smart gateway. The data is then sent via a network to a web service from the gateway.

Paper [10] says an IoT-based smart agriculture system is built to carry out different agricultural operations such as bird and animal scaring, weeding, moisture detection, spraying, etc.

Paper [11] proposed storing the collated data, and only a database management program is required which will hold all the details about the soil. Based on the temperature sensor values, the primary focus is regulating the flow of water to the agricultural field automatically. Weather forecasting is made possible using a sensor to monitor conditions of the weather; this will be monitored via the smartphone of the farmer. In WSNs, data detected from the agriculture field is automatically analyzed with the aid of different intelligent application software, and a decision is made concerning the crops' health which is then forwarded to the farmer.

Paper [12] proposed a farming system that ensured low maintenance and high production using novel sensor technologies that are energy-efficient and eco-friendly. This paper explicitly explains automated irrigation techniques and farm monitoring which include a broad spectrum of sensors to remotely detect and regulate different conditions of the soil like temperature, fertility, and moisture and also regulate the distribution of fertilizer and water to the farmland.

Paper [13] proposed a farming system for monitoring pest insect traps by the application and distribution of image sensors. Imaging devices that use GSM and are run via a network of wireless sensors. Retrieval and relay of images are done by GSM from the trapping region and sent to the remote host station. Feedback about the accumulation of pests is relayed via call or text message to the farmer's smartphone. This technique only determines the presence of pests and doesn't proffer any way of pest control.

This paper identifies a theoretical framework and structure of a system that supports the decision for smart farming with network sensor applications to be able to achieve the desired tasks for farmers with the aid of the Internet of Things (IoT). Recommended is a Smartphone Irrigation Sensor [14]. For use on the crop field is a new invention an automated irrigation sensor and also the digital images are captured by making use of a smartphone, and with this, it can remotely monitor the crop area and also determine the level of water.

Smart Agric [15] depicts a combination of IoT and image analysis approach to evaluate the environmental or the man-made factor (pesticides/fertilizers) that directly impedes plant growth. Its decision-making system is utilized in the transmission of an evaluated framework from collated data of systems in challenging environments and the image of the leaf; it is then analyzed by software such as MATLAB with the aid of histograms for analysis.

In paper [16], the use of a monitoring system based on smart sensors for agriculture to reach decisive results is proposed. To continually monitor agricultural environments, various new technologies are being used. This serves as the gateway (FPGA) that includes various kinds of sensors, for example, soil moisture, relative humidity and temperature sensors, the field-programmable gate array, serial protocol, microcontroller, and the wireless protocol. In the dissemination of collated data kilometers away and also in an area with background noise keeping the sophistication of the network at a low level, TM radio chip is used. With the help of terrestrial microbial fuel cell, the power is supplied to the device in a zero-emission and eco-friendly manner.

Agricultural system is handled with information input, maybe humidity prediction, pH determination, and agricultural field area temperature and multi-processing can be accomplished with cloud computing, Internet of Things (IoT). Mobile computing, sensors, and big data analysis [17] are utilized. In this agricultural system, environment and soil characteristics are detected, analyzed, and occasionally sent to Agro Cloud via the Internet of Things (Beagle Black Bone). Heavy data analysis is done on Agro Cloud data for the required amount of fertilizers, total production, optimal crop cycle analysis, and current stock and market demands. The recommended system is favorable for regulating the cost of agro-products and for improvement in agricultural production with the display characteristics. The collated information is transmitted into a microcontroller in an agricultural environment with the use of the wireless Bluetooth unit.

A wireless transmitter–receiver unit pair is used to transmit and receive the data that is then fed to FPGA with the use of a serial communication protocol UART. A smart, ultra-low-power, inexpensive, and energy-neutral device with microbial fuel cells [18] to track the degree of prelatic aquifers was proposed. In [19] say that IoT technology innovation can be used to improve livestock and plan operational productivity and efficiency. The advantages of IoT and DA have been described and presented in this paper as well as open challenges. IoT is expected to offer the agricultural sector a number of benefits. Nonetheless, to make it affordable for small- and medium-sized farmers, many issues still need to be resolved. Price and security continue to be key issues. The rate of IoT adoption in agriculture is expected to increase as competition in the agricultural sector increases and favorable policies are implemented. The implementation of LPWA communication technology for agricultural purposes is one major area that is likely to attract significant research attention. Among LPWA technologies, the NB-IoT is expected to stand out. This is due to the open standard of 3GPP and telco.companies adoption.

Manishkumar Dholu et al. [20] proposed that in the agriculture domain, cloud-based IoT should be applied. Precision agriculture is essentially a concept that focuses on providing the appropriate amount of resources for the same length of time. Examples of these resources are water, light, pesticides, etc. The advantages of IoT were used in the proposed paper to incorporate precision agriculture. These agriculture parameters include light intensity, temperature, relative humidity, and soil moisture. Appropriate resolution is made on the basis of the reading received from the sensor, i.e., fogger valve (for spraying water droplet) is actuated on the basis of relative humidity (RH) readings, irrigation valve is actuated on the basis of soil moisture readings, etc. This paper suggested the integration of the sensor node to analyze all of these factors and to establish the actuation signal for all actuators. In addition, sensor nodes can also send this information to the cloud. To regulate all these parameters of farming, an android application is also designed. The proposed system is able to locally collate data and regulate the parameter, while at the same time transmit data to the cloud that can be accessed on the smartphone by the user. Such work can be done in the future by enhancing the use of mobile apps such as adding alerts when different parameters are not properly controlled. During the coding of the MCU, the set point for ambient temperature, relative humidity, soil moisture is specified in the proposed framework and to make such design more functional, the mobile app can be assigned the function of regulating the set point.

Prem Prakash Jayaraman et al. [21] proposed that optimizing productivity on the farm is important to improve activity yields profit and meet the ever-increasing food demand driven by rapid population growth worldwide. By interpreting and predicting crop production under a range of environmental conditions, farm productivity can be increased. Currently, crop recommendation is on the basis of the data compiled in field-based agricultural studies capturing plant production under a range of variables (e.g., environmental conditions and soil quality). Nevertheless, the collection of data on crop performance is actually slow, as studies on crops are mostly carried out in secluded and dispersed areas, and such data is generally acquired manually. In addition, the validity of these collated data is very low as it does not consider the earlier conditions that have not yet been encountered by the human handlers, and it is important to discard collated data that will lead to inaccurate results (e.g., solar radiation values in the midday after a short rain or overcast in the morning are inaccurate, and must not be included in the data analysis). Nascent Internet of Things (IoT) innovations like IoT devices (e.g., smartphones, cameras, wireless sensor networks, and network-connected weather stations) is useful in collating extensive amounts of information on the environmental conditions and crop performance, from temporal data retrieved from sensors, to spatial data obtained from cameras, to human observations collated and documented via smartphone applications. This data can now be analyzed to dismiss inaccurate details and evaluate personalized crop recommendations for any particular farm. This paper discusses the development of Smart FarmNet, an IoT-based system that automates the collection of data on the environment, fertilization, irrigation, and soil; automatically coordinate these data and discard inaccurate data to assess crop performance; evaluate crop forecasts and personalized crop recommendations for a specific farm. Smart FarmNet can be incorporated into almost any IoT device; this includes the ones that are available commercially like cameras, sensors, and weather stations, etc., and accumulate the collated data in the cloud for analysis of the performance of crop and also for crop recommendations.

Paper [22] says that to boost output and cost-effectiveness with the latest technologies like the Internet of Things (IoT), it is important to strengthen the efficiency of farming and agricultural operations. In general, by limiting the involvement of humans through mechanization, IoT can make processes in the farming and agricultural industries more effective. The purpose of this study is to examine newly initiated IoT applications in the farming and agricultural sectors in order to give an outline of data collected from sensors, technologies and sub-verticals like crop inspection and water conservation. It was discovered that water conservation is the most examined IoT sub-vertical under potential applications of the Internet of Things followed by crop inspection, smart farming, irrigation management, and livestock management with the same percentage. According to the analysis, the most essential measure in data collection from a sensor is relative humidity, ambient temperature, and some other sensor data are collected for IoT implementation such as soil pH (acidity/alkalinity) and moisture content of the soil. Wi-Fi has the highest demand for use in farming and agriculture, followed by mobile technology. In the farming and agricultural industries, other innovations such as RFID, ZigBee, WSN, LoRa, Raspberry pi, Bluetooth, and GPRS are of lesser interest. The agricultural sector has a higher percentage using IoT for mechanization as compared to the farming sector.

3 Analysis of IoT Hardware Requirement

3.1 Device

An IoT setup makes use of devices that perform an operation in detecting, actuating, regulating, and monitoring [23, 24]. Depending on temporal and spatial limitations (i.e., processing capabilities, memory, speeds, communication delays, and deadlines), IoT devices can share information with other linked devices and applications or collate data obtained from other devices and then send the data to the base station server and from there to the cloud server through a gateway or execute some functions locally and some other functions within the IoT structure. An IoT device can comprise multiple wired and wireless interfaces that are used to communicate with other devices [24]. These include (i) I/O interfaces for sensors, (ii) interfaces for connectivity to the internet, (iii) interfaces for memory and storage, and (iv) interfaces for audio/video.

3.2 Communication

A communication block is used to relay information across the remote servers and devices. IoT communication process typically works with the network layer, data link layer, application layer, and transport layer [24].

3.3 Services

IoT systems can be used to perform functions like device discovery, device prototyping and representation, device monitoring, data dissemination, and data statistics.

3.4 Management

Management block may perform various functions such as governing an Internet of Things system and finds the governance underneath the Internet of Things system.

3.5 Security

Security blocks can be used to provide features such as privacy, encryption, content credibility, message integrity, authorization, and data security. IoT system is also protected by the security block [21].

3.6 Application

For users, the most critical layer is most likely Application layer. This layer gives the essential modules that will manage and track the different features of the Internet of Things system. Users are allowed to view and evaluate the status of the system using applications, often predicting future prospects [22].

4 Challenges and Future Work

While the architectures described in the previous section make the IoT theory technologically conceivable, it still needs further research effort. This chapter deals with technical difficulties related to existing IoT systems. An innovational concept of IoT architecture was later developed to meet all the essential elements that are not present in the current architecture. A thorough understanding and research of industrial features and demands on conditions like privacy, security, cost, and uncertainty must be done prior to the widespread recognition and implementation of IoT in all domains. Let us discuss a few issues in this regard

- Maintaining cost—This is a far more essential variable for farmers. Researchers, therefore, concentrate on developing new IoT architecture for smart agriculture with added benefits to attain this level.
- Currently, the database management system may not be capable of handling information in real time due to the size of the data collected. It is important to idealize the appropriate solutions. Data based on IoT would be initiated at a fast rate. At the receiver end, the current RAID system is unable to handle the data collated. The data based on IoT service-centric system needs to be re-evaluated to address the issue.
- Data is an unprocessed fact that is typically not consistent with non-relevant handouts. Data plays a major role in IoT decision-making. The data value is the data pool. Data can be obtained by orienting mining, analyzing and interpreting meaningful information. A similar architectural framework can be used for mining data and analyzing them and therefore helps with decision-making operations. A large data method is integrated with data mining and analysis.
- Service-oriented architecture (SOA) for IoT is a major challenge in which service-based artifacts can face quality and cost-related problems. SOA is needed to deal with a wide range of system-connected devices with scalability issues. Challenges such as transmitting, managing, processing and storing, data become a challenge of service allocation.
- Service standard and integrity is also a major challenge. A developer needs to focus on QoS parameters to obtain an appropriate range of QoS.
- IoT envisages an unbelievably great amount of nodes. All devices and data attached to the network are recoverable. A distinctive identification is compulsory for the secure initialization of point-to-point connection. IPv4 protocol specifies a 4-byte address for each node. The supply of numbered IPv4 addresses is declining quickly, and it's approaching zero in the next few years, so the ability to identify new policy addressing IPv6 area is the field where the utmost attention is required and the adequacy of architectural skills is compulsory.

5 Conclusion

Through IoT-enabled technology, precision farming can be made more accurate and effective. In different fields of agriculture, IoT can be applied. Energy and water are one of the highly essential resources for agriculture and their prices can boost the agricultural sector or break it down. This means that water wastage has not been curbed as a result of substandard irrigation systems, inadequate field application procedures and sowing crops that require much water to grow in unsuitable farmland. Electrical energy is required for the optimum functionality of boosters, pumps, and lighting, etc. Water for agriculture can be used conservatively by controlling and adjusting the volume of water, location timing, and flow duration. Use of effective electrical energy for boosters, lighting, pumps, and other uses with the aid of IoT; the second one is crop monitoring. Applying pesticides and fertilizers based on crop and soil health and pest control is the major concern in this area. IoT can be used to make a proper decision by installing sensors and imaging systems in the crop field that is connected to the internet. IoT can be used efficiently for fertilizers and pesticides. Finally, it can be concluded that an efficient agri-IoT architecture has to be developed with low cost, the minimal power consumption of devices, optimum performance, improved decision-making action, and QoS service so that farmers can easily understand it without basic knowledge.

References

1. Liqiang, Z., Shouyi, Y., Leibo, L., Zhen, Z., Shaojun, W.: A crop monitoring system based on wireless sensor network. Proc. Environ. Sci. **11**, 558–565 (2011)
2. Zhu, Y., Song, J., Dong, F.: Applications of wireless sensor network in the agriculture environment monitoring. Proc. Eng. **16**, 608–614 (2011)
3. Jaishetty, S.A., Patil, R.: IoT sensor network based approach for agricultural field monitoring and control. IJRET: Int. J. Res. Eng. Technol. 5(06) (2016)
4. Keerthi, V., Kodandaramaiah, G.N.: Cloud IoT based greenhouse monitoring system. Int. J. Eng. Res. Appl. **5**(10), 35–41 (2015)
5. Rajalakshmi, P., Devi Mahalakshmi, S.: IoT-based crop-field monitoring and irrigation automation system. IEEEx-plore.ieee.org/iel7/7589934/7726872/07726900
6. Kaur, B., Inamdar, D., Raut, V., Patil, A., Patil, N.: A survey on smart drip irrigation system. Int. Res. J. Eng. Technol. (IRJET) **3**(02) (2016)
7. Parameswaran, G., Sivaprasath, K.: Arduino based smart drip irrigation system using Internet of Things, p. 5518. Int. J. Eng. Sci. (2016)
8. Khelifa, B., Amel, D., Amel, B., Mohamed, C., Tarek, B.: Smart irrigation using internet of things. In: 2015 Fourth International Conference on Future Generation Communication Technology (FGCT), pp. 1–6. IEEE, New York (2015)
9. Reshma, S., Babu, B.S.M.: Internet of Things (IOT) based Automatic Irrigation System using Wireless Sensor Network (WSN). Int. J. Maga. Eng. **3** (2016)
10. Gondchawar, N., Kawitkar, R.S.: IoT based smart agriculture. Int. J. Adv. Res. Comput. Commun. Eng. **5**(6), 838–842 (2016)
11. Wang, H., Liu, C., Zhang, L.: Water-saving agriculture in China: An overview (2002)

12. Srisruthi, S., Swarna, N., Ros, G.S., Elizabeth, E.: Sustainable agriculture using eco-friendly and energy efficient sensor technology. In: 2016 IEEE International Conference on Recent Trends in Electronics. Information and Communication Technology (RTEICT), pp. 1442–1446. IEEE, New York (2016)
13. Priya, C.T., Praveen, K., Srividya, A.: Monitoring of pest insect traps using image sensors and dsPIC. Int. J. Eng. Trends Tech. 4(9), 4088–4093 (2013)
14. Jagüey, J.G., Villa-Medina, J.F., López-Guzmán, A., Porta-Gándara, M.Á.: Smartphone irrigation sensor. IEEE Sens. J. **15**(9), 5122–5127 (2015)
15. Kapoor, A., Bhat, S.I., Shidnal, S., Mehra, A.: Implementation of IoT (Internet of Things) and Image processing in smart agriculture. In: 2016 International Conference on Computation System and Information Technology for Sustainable Solutions (CSITSS), pp. 21–26. IEEE, New York (2016)
16. Mathurkar, S.S., Patel, N.R., Lanjewar, R.B., Somkuwar, R.S.: Smart sensors based monitoring system for agriculture using field programmable gate array. In: 2014 International Conference on Circuits. Power and Computing Technologies [ICCPCT-2014], pp. 339–344. IEEE, New York (2014)
17. Channe, H., Kothari, S., Kadam, D.: Multidisciplinary model for smart agriculture using internet-of-things (IoT), sensors, cloud-computing, mobile-computing and big-data analysis. Int. J. Comput. Technol. Appl. **6**(3), 374–382 (2015)
18. Sartori, D., Brunelli, D.: A smart sensor for precision agriculture powered by microbial fuel cells. In: 2016 IEEE Sensors Applications Symposium (SAS), pp. 1–6. IEEE, New York (2016)
19. Elijah, O., Rahman, T.A., Orikumhi, I., Leow, C.Y., Hindia, M.N.: An overview of Internet of Things (IoT) and data analytics in agriculture: Benefits and challenges. IEEE Internet Things J. **5**(5), 3758–3773 (2018)
20. Dholu, M., Ghodinde, K.A.: Internet of things (IoT) for precision agriculture application. In: 2018 2nd International Conference on Trends in Electronics and Informatics (ICOEI), pp. 339–342. IEEE, New York (2018)
21. Jayaraman, P.P., Yavari, A., Georgakopoulos, D., Morshed, A., Zaslavsky, A.: Internet of things platform for smart farming: Experiences and lessons learnt. Sensors **16**(11), 1884 (2016)
22. Madushanki, A.R., Halgamuge, M.N., Wirasagoda, W.S., Syed, A.: Adoption of the Internet of Things (IoT) in agriculture and smart farming towards urban greening, A review (2019)
23. Abhishek, M.B., Shet, N.S.V.: Cyber physical system perspective for smart water management in a campus (2019)
24. Abhishek, M.B., Shet, N.S.V.: Data Processing and deploying missing data algorithms to handle missing data in real time data of storage tank: A cyber physical perspective. In: 2019 1st International Conference on Electrical. Control and Instrumentation Engineering (ICECIE), pp. 1–6. IEEE, New York (2019)

Offline Multilanguage Validation Analysis Using FEDSEL

T. P. Umadevi and A. Murugan

Abstract Handwriting recognition is a challenging task. In this paper, characters are first extracted from the English, Arabic, and Chinese handwriting dataset. The handwritten content recognition process comprises four stages, namely information preprocessing, segmentation, feature extraction, and dimension. The character segmentation algorithm is based on forward and backward classification. The present work implements the feature dimension sequence labeling method (FEDSEL) and semantic role labeling (SRL) for handwritten recognition. The dataset used is the HCAR database. The accuracy of the proposed method is 95.77%.

Keywords Feature dimension sequence labeling · Semantic role labeling · Recognition · Accuracy

1 Introduction

Handwritten characters are first extracted from the English, Arabic, and Chinese handwriting and then recognized. Using dynamic programming, the two strings are linked together based on a dictionary depending on multiple language analysis. Finally, some experimental results are presented to demonstrate the effectiveness and robustness of the proposed method DWT previous our research. To recognize the Chinese with English handwriting, the letters must first be extracted and then identified. In Chinese and Arabic handwriting, there is no space between two adjacent letters, which is different from English [1].

When writing the document page, loops, and arc variation, there is a gap between two adjacent words. To improve the adaptive efficiency of character feature extraction

T. P. Umadevi (✉)
Department of Computer Science, JBAS College for Women (Autonomous), Chennai, India
e-mail: Umashiva06@gmail.com

A. Murugan
Department of Computer Science, Dr. Ambedkar Government Arts College (Autonomous),
Affiliated to University of Madras, Chennai, India
e-mail: amurugan1972@gmail.com

[2], a fast method based on the stroke interval histogram is proposed to better solution FEDSEL. The handwritten stroke histogram linearization threshold is used to extract the stroke line. The finalization of the histogram of the stroke is used to extract the threshold characters [3]. The first line is the original document page. The second line is the line of extracted text, marked with a bordered rectangle. The third line is a lot of extracted characters, which is marked with a bordered rectangle, so after the English and Arabic document page is split, a character can be wrongly divided into several characters, and many characters are incorrectly grouped into one character[4], making it difficult to get the correct extraction.

Separation errors will inevitably lead to signature authentication errors. Although some characters are correctly extracted, they are misidentified [5]. Focus on the representation and analysis of real handwriting signals, not the signature set [6]. Consider the standard form of the signature path analysis of all new spelling validation performance. This is more practical than motion simulation when there is no dynamic information.

More effective identification Handwritten Arabic, English, Chinese text is challenging cursive and the uncontrolled nature of Arabic writing. Online routine signature recognition usually has to wait for the whole curve is in front of start analysis.

This delays the identification process more takes timeless character recovery very poorly. Prevent the implementation of the following advanced features input type, such as automatic words, not complete forms (Fig. 1).

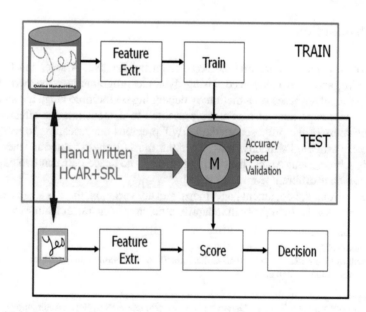

Fig. 1 Multiple language documents proposed framework SRL

2 Related Work

The paragraph level for script identification and system attributes are computed in documents. Characterization of Gabor filters [7] is mainly developed in [8], while sterile pyramid filters are used in [7]. At the word level, the properties of the system are calculated in the articles [9]. Again, Gabor filters are mainly used, while an additional feature of the unique transformations of cosines is used in [10]. Wave and radon changes are calculated in [9, 10] properties. In the article [6], various languages of Latin writing are classified using gray-level compatibility matrix, while changes based on frequency and half are used to identify eight Indian characters.

Surinta O. et al. have implemented a support vector machine classification to identify written characters using pixel intensity from input images, although variation in writing style limits writing recognition. Next, Gun Xiao Shi et al. proposed a way of recognizing handwriting based on learning checker vocabulary from non-biased places and were able to identify different characters with different fonts. However, this limits the capabilities of other languages, such as Chinese, to more than 30,000 versions. Gomti et al. suggest the extraction procedure for structural features and statistical features. For FAR and FRR, the proposed hybrid + FLMN algorithm achieved 95% higher accuracy. The most popular text locating methods for identifying text lines is discussed in, Lianwen Jin et al. Three basic experiments have been performed using authenticated text detectors. The results highlight the diversity of SCUT-HCC doctors and the challenges of understanding HCT in documentaries.

Zhaozheng Yin et al. (2018) achieve cognizance accuracies regarding one hundred percent and 93% within the half–half then the leave-one-out experiments, respectively, on the ASL benchmark dataset and achieve focus accuracies concerning a hundred percent because of both the half and the leave-one-out experiments concerning the NTU count dataset.

Chamnongthai et al. [11] computerized signal language style developed for communicating in normal-hearing people or toughness people. Due to imitation of shape variations, only some algorithm is no longer enough to generate good focus results.

Nai et al. 12 extracted applications from solely depth pix regarding randomly positioned line segments and ancient a loosely wooded area because classification, with 81.1% accuracy spoke of in their paper.

3 The Proposed Recognition System

3.1 Preprocessing

The SRL classification of male and female using loop line frequency return a set of K-spot characters datasets. Candidates male and female in the sample handwritten package provide an integrated framework for data in the following ways left and right handwritten identifiers in this processing works. Disconnected and non-uniform sample handwritten with printed or non-print handwritten samples is caused by vibration form different writing line style. This proposed method supports two choice methods one backward method using the eliminating noise other with Douglas Baker instruction. The second method is forward sharp character reorganization using the secondary section interpolation function.

3.2   Offline Segmentation

Add each offline frequency along the horizontal axis. For curved or fluctuating lines of text, the image can be divided into vertical bars. Image documents break the page into columns that determine the minimum of the histogram generated by the horizontal projection of all columns. A horizontal stroke is drawn from each minimum sequence in the column [10]. The relationship left line angle 90° to right 180° angles between these strokes allows the separation of two adjacent lines. Huff transform is used to find a straight line in an image. The best alignment of text lines corresponds to black pixels. No black pixel can pass the number of rows of pixels, and each row has a unique sequence (0, 90, 180, 360) and range pixels called accumulator polling algorithm for the converter when the line passes through a black pixel. The text line is the largest line of the aggregator [9].

Instead of splitting directly into a binary (1,0) image black with white recovery words, it is transformed into a probability reduce dimension, where each element represents the probability [11] of a line of pixel text. Probability sets are analyzed using the level set method to divide lines of text. The zero sequence of the range will automatically increase, dissolve the error code recovery loop fish line character, and stop for the final text line boundary.

Using a clustering algorithm to group multiple elements in a supervised classification method, each supervised classification method first divides the area at each point by creating a scatter similarity map and estimating the local orientation of each major component of the term. Use the width-first search algorithm to find synchronous packages on similar reduces feature maps. There is a path from every element in the collection to every other element bird's angle line character degree recovery letters that indicates a different line of text to identification HCAR new form to analyze results FEDSEL (Fig. 2).

3.3   Reduce Dimension of Feature Extraction

Text extracts the most relevant information from the image documents to help identify the text ideogram character size level of upper during the text recognition phase. This information is an attribute of each symbol that must be distinguished from other symbols. The certification phase is the final key decision stage lower ideogram

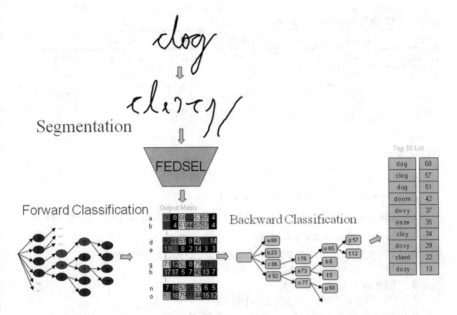

Fig. 2 Segmentation of validity of authentication

level of words recovery. This is a classification mnemonic font recognition process that identifies each unknown symbol encode key line pixel points and assigns it to a predefined class. This classification is based on the extracted attributes from the previous step. Document analysis is still an open research area due to the low quality and complexity of these documents background noise.

The segmentation is to find the identity string forward $\{I_1,..., I_n\}$, with the corresponding segments of the sequence $\{S_1,..., S_n\}$, backward $S_1 = \{z_1,..., z_{t1}\},...,$ digit matrix frameworks $S_n = \{z_{tn-1},..., z_T\}$, that best explain the sequence.

$$\{I_1^*, ..., I_n^*\} = \arg\max p(S_1, ..., S_n | I_1, ..., I_n) p(I_1, ..., I_n)$$

$$= \arg\max p(I_1) p(S_1|I_1) \prod_{i=2}^{n} p(S_i|I_i) p(I_i|I_{i-1})$$

Algorithms execution FEDSEL

Input:

Original randomly selected handwritten image document character in an existing multiple language document script of any format 5 to 8 characters or meaningful sentence actual word of the image in the dictionary (Table 1).

Table 1 FEDSEL recognition parameters

	A-Z	C-C	A-A	B-S	SB-S	SB-B
HCAR	98.3	97.3	63.6	67.3	96.8	95.6
FEDSEL	97.6	96.8	62.6	67.0	96.6	95.766
SRL	97.1	96.3	71.0	73.0	96.1	94.4
DWT	0.5%	0.5%	13.4%	9.0%	0.5%	1.3%

Algorithm

1. Begin
2. $S \rightarrow N +$ il variance $W \rightarrow E + i\,\Gamma$
3. Compute M0 is a singular point of $f(M)$ if $f'(M)$ exist at $M = M0$
4. N be a closed contour inside a simply connected region $N \subseteq M$.
5. For n = -1, each line segment integrates to $\pi/2$ integer n
6. M0 inside region bounded by M
7. End For
8. (N+2)-1 is SRL inside SRL+HCAR
9. While $f(z) < g(z)$ be LLMI in regions R < S
10. Repeat handwritten to the W - 1 HWSC and TST1 dataset / $(i - \lambda1)$ gives root $E = \lambda2$
11. Radius of convergence is frequency from M0 to nearest singularity M1
12. End while
13. $f(M)$ known on any curve segment through z0 is enough to determine $f(n)$ (M0) $\forall$ n
14. if(f(n) (M0))
15. Return (M)
16. End if
17. End

The SRL computational complexity is its $(N^3 \log N)$ matrix pixel. Embed the set Hilltop model the standardized space is easy to use Metric indexing method to solve Looking for HCAR in large collection Avoid doing more linear scanning documents all formats.

Frequency male and female handwritten and content forward and backward assessments of left and right character analysis.

The image documents are scanned for 100 users or 100 pages a situation with a lot of dataset 10 k, and the best way to select and evaluate the model is to divide the dataset into three parts: a training group, a validation group, and a test group. Training kits are used to fit the model. The test set is used to evaluate forecast errors when choosing a model. The test set is used to evaluate the prediction error of the last selected model. The Kufic dataset contains a specific character size of 70% for training and 25% for verification and testing after error 5%. The best solution is to

Table 2 Sample dataset total 10 k

# of Samples	Character position
1405	Left
1196	Right
1629	Middle
1372	Bottom

use a large custom test suite that is often unavailable. For the SRL methods presented FEDSEL here, there is not enough dataset to divide them into four parts as mentioned in Table 2.

Choose multiple characters at random. Set the change of the given number to an approximation. If multiple changes are selected, latency glyph assigns priority order to each change based on test results. Sort the selected list of changes according to their priority and use them to accumulate effects A1, A2 used as symmetry. Balance the variance of the KL deviation and the length of the code if the static length of the script code and the transformation matrix match exactly and the code length is almost the same $<d\ (T_1, T_2)$ is close to 1.

Since there are fluctuations associated with the return on both of these assets, to select α is to reduce the dimension of the overall risk and variability of our FEDSEL speculation. Hilltop method is the correct data sequence for storing defined group of points. To standardize and validate the plan of the multilanguage character database that was feature extracted from the dataset classification cursive forward classification segmentation online. The HCAR database is available for other applications HMM Conduct research in this area. The Markov chain in the transition space X is a random process (infinite array of random variables) $(x(0), x(1), x(t),...)$.

That is, the probability of a state at a particular time in history depends on sentence character top to bottom female handwritten only on time L and POS is frequency bottom to top male handwritten only.

4 Result and Discussion

The system has been trained and tested spelling and word parts were extracted HCAR database classification. This section presents the results of the multiple documents handwriting recognition system using multiple segmentation binary of extraction methods and reduces the feature map classification forward level.

The separation is supposed to be taken as input algorithm execution of validation results is based on the classification of the previous works. Divide the entire dataset using five hash strategies HMM as shown in Table 4. The development of the hash strategy illustrates how the accuracy of classifying the workbook varies according to the training set the size.

Fig. 3 Handwritten sequences

Table 3 Testing spelling validation

Configuration	Accuracy [Top 1] (%)	Validation [Top 3] (%)	Speed [ms]
High accuracy	91	96	29.9
Low latency	87	94	0.12
Fast learning	90	95.77	4.4

High accuracy: Recommended method SRL
Low delay: Avoid candidate evaluation DWT
Rapid learning: Avoid TR and indicators FEDSEL

Table 4 Comparison study for offline handwriting

Method	Results	Dataset
HMM	51% (SR)	OHASD—a self-collected dataset that includes 154 paragraphs (more than 3800 words) written by 48 writers
HMM	75%	1400 self-collected isolated character collected isolated character
Hybrid recognition system	83%	1400 self-collected isolated character
Proposed method (FEDSEL)	95.77%	2000 collection 100 pages

5 Conclusion

The classification can only produce novel forms within the variation of the training model and the performance is limited by the models used for training and testing results. Although some test results have been shown in tables and still it is unknown how to objectively evaluate synthetic scripts and compare different synthetic statistical FEDSEL models using multilanguage scanning document pixel to calculate the semantic score of a parallel text similar to the path in the reduced dimension. The dictionary used contains 2000 collection 100 pages Chinese, English, and Arabic characters and produces the results of fast learning validation 95.77%. In the future, combine high information, grammar, and semantics which produce the best result.

References

1. Simard, P.Y., Steinkraus, D., Platt, J.: Best practices for convolutional neural networks applied to visual document analysis. In: International Conference on Document Analysis and Recognition (ICDAR), IEEE Computer Society, Los Alamitos, pp. 958–962 (2003)
2. Sari, T., Sellami, M.: Cursive Arabic script segmentation and recognition system. Int. J. Comput. Appl. **27**, (2005)
3. Al-Emami, S., Usher, M.: On-line recognition of handwritten Arabic characters. IEEE Trans. (PAMI) Pattern Anal. Mach. Intell. **12**, 704–710 (1990)
4. Esposito, F., Malerba, D., Semeraro, G.: Automated acquisition of rules for document understanding. In: Proceedings of the Second International Conference on Document Analysis and Recognition. IEEE Computer Society, Los Alamitos, pp. 650–654
5. Fernandes, A.F., Oliveira, M.M.: Real-time line detection through an improved Hough transform voting scheme. Pattern Recognit. **41**(1), 299–314 (2008)
6. Doermann, D.: The indexing and retrieval of document images: A survey. Comput. Vis. Image Understand. pp. 287–298 (1998)
7. Busch, A., Boles, W.W., Sridharan, S.: Texture for script identification. IEEE Trans. Pattern Anal. Mach. Intell. **27**, 1720–1732 (2005)
8. Kundu, S., Paul, S., Bera, S.K.: Text-line extraction from handwritten document images using GAN. Expert Syst. Appl. **140** (2020)
9. Fernandes, A.F., Oliveira, M.M.: Real-time line detection through an improved Hough transform voting scheme. Pattern Recognit. **41**, 299–314 (2008)
10. Sahare, P., Chaudhari, R.E., Dhok, S.B.: Word level multi-script identification using curvelet transform in log-polar domain. IETE J. Res. (2019)
11. Pattanaworapan, K., Chamnongthai, K., Guo, J.-M.: Signer, Independence finger alphabet recognition using discrete wavelet transform and area level run lengths. J. Vis. Commun. Image Representation **38**, 658–677 (2016)
12. Nai, W., Liu, Y., Rempel, D., Wang, Y.: Fast hand posture classification using depth features extracted from random line segments. Pattern Recognit. **65**, 1–10 (2017)

Pattern Matching Compression Algorithm for DNA Sequences

K. Punitha and A. Murugan

Abstract Scientifically, the DNA contains hereditary information about the biological species. The DNA databases are considered to be the repository of a huge collection of DNA sequences. Ceaseless advancement in DNA researches has led to various key difficulties in the process of transferring, maintaining, as well as storing data. Such a huge size of DNA sequences often results in the requirement of abundant storage space which needs to be handled and minimized with the help of an effective methodology. One such technique being the data compression using which the DNA sequences can be minimized in size thus saving upon space and bandwidth requirement. There is a recommendation of Improved_Compress algorithm which aids in determining DNA compression along with biological sequence's pattern matching. The matching information can be stored in an offline dictionary. Hence, the proposed algorithm Improved_Compress leads to an average better compression ratio of 89% when compared to the existing algorithm compression ratio through the dictionary-based method and encoded ASCII value with the National Center for Biotechnology Information (NCBI) datasets.

Keywords Compression · Decompression · DNA sequences · Pattern ·
Complement · Reverse complement · ASCII

1 Introduction

There has been an apparent growth in the size of biological data which certainly requires timely analysis. The domain of molecular biology confronts such a challenge since there happens to be the regular implementation of amino acid sequences or

K. Punitha (✉)
Department of Computer Science, AgurchandManmull Jain College (Shift II), Chennai, India
e-mail: punithsathish@gmail.com

A. Murugan
PG & Research Department of Computer Science, Dr. Ambedkar Government Arts College,
Chennai, India
e-mail: amurugan1972@gmail.com

S. Shakya et al. (eds.), *Proceedings of International Conference on Sustainable Expert
Systems*, Lecture Notes in Networks and Systems 176,
https://doi.org/10.1007/978-981-33-4355-9_30

387

nucleotides for estimating biological molecules. The prime element of the biological data is the DNA sequence that comprises four main nucleotide bases which are namely Adenine, Cytosine, Guanine, and Thymine which are being represented as A, C, G, and T, respectively. Usually, the size of the DNA sequence is considerably large. For instance, the human genome comprises of nearly 3.1647 billion DNA base pairs. The initial and foremost step that is performed in a DNA based research involves pattern searching within DNA sequences databases like the alignment of DNA sequence.

There must be compression of DNA sequence for achieving reasonable storage. The lossless compression schemes comprise of three general classes namely symbol-wise substitution, dictionary-based, and context-based methods. Compression of the DNA sequence is performed via conventional compression methods like the word-based Huffman, arithmetic coding, or LZ based [1] which does not take into account the special features of biological sequences. Alternatively, method-ologies that take into account repetition structures or any sort of regularities inherent in DNA sequence performs pretty well. These include algorithms such as the BIOCOMPRESS, GENCOMPRESS, and BWT-base compress. The conventional text compression algorithms could consider only short and frequent repeats which were reflected in the DNA sequence compression too. So it is essential to determine a different set of repeats in a DNA sequence which can be useful for gaining an effective compression ratio [2]. There are many compression algorithms of substi-tution method along with the Lempel–Ziv compression approach to achieve DNA sequence compression.

The intensive research is being carried over compressed pattern matching in DNA sequences which signifies identification of overall matching patterns in the given sequence. Compressed pattern matching enables compression of DNA sequences by shrinking their size and thus reducing the time involved in searching patterns. The LZ family of compression systems is employed by the majority of the techniques, whereas the remaining ones are based upon Huffman-coded text and run-length encoding and BWT output. Chen et al. [3] have recommended d-BM (derivative Boyer–Moore) algorithm by making use of a novel compression method on the DNA sequence that yields ineffective storage space and improvised performance related to pattern searching with the compressed DNA sequence. Without making the use of a multi-core environment, the d-BM still delivers satisfactory performance in most of the scenarios. The methodology recommended in the present research deter-mines DNA compression using pattern matching and ratio for biological sequences depending upon its frequency. By making use of intrinsic DNA sequence methods, repetition structures can be carried out. The repetitive sets and mismatch informa-tion are stored in the offline dictionary. The following are the steps carried out the reading of the first four sequences from a given input sequence. Checking is done for matching the sequence with the rest other sequences as Exact, Reverse, Comple-ment, and Reverse Complement. Thereafter, the information gets stored within the dictionary table. Next reads 8, 16, 32 sequences, and so on up to the half of the length of the input sequence and computing the number of positions in the match, the frequency is determined. A match is found for the pattern or sequence concerning other sequences as Equal, Reverse, Complement, and Reverse Complement thereafter storing the frequency within the offline dictionary. The maximum frequency pattern can be stored in the dictionary table along with its position and its type of match. Using the above techniques, DNA compression and its ratio can be determined. Subsequently, any repeats present in the original sequence are being removed for producing the final parsed sequence which is then combined with the offline dictionary to produce the compressed sequence. Mismatches that result in improvised compression are taken into account and encoded into extended ASCII codes.

Following is the classification of the research: All the related work is presented in Sect. 2, the recommended system is elaborated in Sect. 3, the results generated and all other discussions are covered in Sect. 4, and lastly, Sect. 5, brings forward the conclusion and future scope of the proposed method.

2 Related Works

The huge size of DNA databases which comprises complex and logical structures often results in the requirement of abundant storage space. It is a great challenge to store and access the large size of DNA sequence [4], and these massive size DNA databases can only be handled by making use of a robust compression algorithm. General compression methods will not hold efficient in dealing with biological sequences that comprise of delicate regularities in DNA sequences. DNABIT [5] divides the sequence into blocks and then compress them based on the occurrence of the patterns.

Saada and Zhang [6] have recommended the extended ASCII encoding in the DNA compression technique. To begin with the position of the matching, Index is taken into consideration which compares the Text characters from a specified position with the pattern characters. Then, depending upon the match numbers, the frequency is calculated, and the ASCII encoding takes place.

In the compression technique, the focus lies upon shrinking the storage space through mathematical algorithms that tend to be either lossy or lossless. Since lossless compression helps in rebuilding of original sequence on decompression, it stands very significant. The technique of lossless compression is achieved by substituting the substrings of the repetition sequence from the dictionary. The repetitions are identified through the dictionary-oriented algorithms by considering the earlier sequences [7].

The concept of data compression involves data indexing too. A single genome resembles a static data, which makes pattern searching flexible due to indexing. LZ-based indexes are one such an effective indexing technique that successfully aids in eliminating redundancy [8].

The combination of reference-based and assembly-based algorithms (de novo assembly) with a statistical approach was used to achieve the DNA sequence compression. The algorithm Quip has used the probabilistic data structure for DNA compression, since amount of space consumed by memory extremely less [9].

Genome resequencing encoding (GReEn) by Pinho et al. [10] symbolizes a reference genome based tool that enables compression of genome resequencing data. As soon as the target sequence is equivalent to the reference sequence, the compression is carried out successfully by the suggested tool. The sequence size and similarity with reference sequences are the parameters for determining compression time. Resultant, anticipating encoding time for all cases is not feasible.

Emerging and expansion of biological data in the coming years remain undeniable. No doubt, there will be a generation of the massive volume of data that must move gradually for backing up the deployment of the personalized and precision medicine paradigms. Toward this, the technique of data compression will become most significant and helpful. And this has been very well elucidated by the association of ISO/IEC, Moving Picture Experts Group (MPEG), toward their attempts in genetic information processing, especially involving compression [11].

There has been an abundant volume of sequencing data generation rapidly since the launch of the high-throughput sequencing (HTS) technologies. Resultant, there is a constant revival of scientific information about genome information along with simplifying the process of diagnosis and therapy. Though the merits are noteworthy, there also lies prime issues of storage, processing, and transmission of this data. Besides, there is more expense incurred toward handling HTS data in contrast to generating sequences that are all the more aggravating [12]. These issues can be dealt with the effective technique of compression which primarily enables compressing of storage space and lessen processing expense, for instance, I/O bandwidth. It also aids in enhancing transmission speed.

The recommended CaBLASTP algorithm put forth a course database that enables the storing of distinct data fetched from the original protein sequence database. Herein, sequence segments that are being associated with prior sequences that are appended to a link index. CaBLASTP algorithm is formed by merging the two algorithms namely dictionary-based compression and sequence alignment. The dictionary-based approach relies upon ASCII replacement and arbitrary or repeated protein sequence reduction [13].

Organisms or species fundamentally rely upon protein which is an essential factor in controlling the development process. A protein sequence can be depicted using a 20-symbol alphabet of amino acids. A protein sequence tends to be complicated when compression is involved due to the presence of a large ratio of amino acids which is tangled inherently [14].

The recommended methods for exploiting the "dictionary" are ProtCompSecS formed by combining the ProtComp algorithm and a dictionary-based method that employs dictionary of protein secondary structure (DSSP) database. The proposed method encodes the protein sequences concerning their annotated secondary structure information. CaBLASTP algorithm put forth a course database that enables the storing of distinct data fetched from the original protein sequence database [15].

The heuristic method makes use of protein domain compositions. By employing evolutionary techniques of gene duplication and fusion, a hyper-graph is generated for the proteome. Further, the proteome can be compressed using a minimized cost function by utilizing a minimum spanning tree [16].

The challenges confronted can be dealt with the effective technique of compression which primarily enables compressing of storage space and less the processing expense for instance I/O bandwidth [17]. There exist two prime implementations of genomic data compression apart from minimizing storage, processing costs, and transmission speed. The performance of de novo assembly is improved in building the sections of the genome from raw sequence reads de Bruijn graphs without a reference genome by reducing the memory consumption by one order of magnitude.

DNA comprises a logical structure and maintaining a data structure for its storage, access and processing can be tedious enough. There is a ceaseless spike in DNA collection that is being fetched from the organisms [18] which arises the earnest need for storage and safe transmission. The size of every single base stands very important while storing such significant information. Yearly, the DNA order size can range from Megabyte (MB) to Terabyte (TB).

The algorithm focuses on the internal of genetic palindrome, palindrome, and reverse. It resembles a DNA sequence compression model revealing actual features of the DNA order. Results acquired from the experiment ascertain the better performance of the proposed process in contrast to the RAY compression ratio [19]. Moreover, the method aids in determining regularities in DNA sequences like crossover and mutation.

Algorithms for DNA sequence compression can either be lossy or lossless. The DNA sequence compression algorithms have modes like standard, vertical, or horizontal [20]. A standard algorithm comprises of standard text compression methods but does not consider redundancy in biological data. Gzip and bzip2 are two such algorithms which do not perform sequence compression effectively. Context tree weighting (CTW), Huffman's coding, prediction with partial match (PPM), and arithmetic coding are some of the other standard compression algorithms. Among these, Huffman's coding too stands ineffective for DNA sequence compression since there occur small variances in DNA base probabilities. However, CTW and arithmetic compression tend to be comparatively better though they are treated as slow decompression process.

To achieve DNA sequence compression, many techniques have been put forward. Compression acts as an effective remedy against the large size of sequencing data. By the means of generic techniques, compression of DNA sequences can be achieved. Compression comprises of natural depiction as a string of characters with good enough literature. Genomic data is intrinsically redundant since a significant part of the genome is being shared by similar species. Some particular biological attributes can aid in compressing DNA sequences as repeat content and association with prevailing sequences. Zhu et al. [21] presents reviews illustrating former work on DNA sequence compression and exploiting redundancy within DNA sequences, respectively.

It has been observed that just two compressors namely DELIMINATE and MFCompress stood up to the practicality threshold [22]. The compressor is stable, enables generally employed FASTA format features, and effectively handles practical processes like compressing/decompressing whole of the vertebrate genomes. Though DELIMINATE and MFCompress portray better compression ratios, they depict a slow decompression rate which restricts their capability in case of large databases.

The fast growth of the population changes the lifestyle of humans not only the periodic life but its impact in the medical field too. The various datasets are used in healthcare recommendation systems, and particularly the DNA sequence is used for the various healthcare process such as sleep pattern analysis [23], patient diet recommendation system [24].

3 Proposed System

3.1 Data Compression

The proposed method is based on both dictionary based matching method and substitution-based method. There are only four alphabet sequences (A, C, G, and T) which are used in the DNA sequence. So it is started to match with four sequences, and the substitute will be of A-00, C-01, G-10, and T-11. The prime element in biological data is the DNA sequence which can be represented as A, C, G, and T. It is assumed that the DNA sequence is arbitrary and non-compressible. Using two bits, each nucleotide base pair is being coded. There is immense significance of biological sequences since they comprise of the utmost valuable information in the creation of a life. The techniques of compression and sequence signify that there can be redundancies due to the presence of repetitions within the biological sequences, thus leading to compaction. The identification of such dependencies becomes the ground for biological sequence compression. The technique of lossless compression can be categorized into symbol-wise substitution, dictionary-based, and context-based methods. In the symbol-wise substitution, each symbol is substituted with a new code word such that symbols with frequent occurrence indicate substituted with shorter code words. This results in an overall compression. The dictionary-based methods consider the frequently occurring input sequence and based on it creates a vocabulary of symbols. It then uses a pointer for replacing the symbol's positions in the dictionary, thus achieving compression.

The dictionary can be online or offline. In an online dictionary, the text resembles a dictionary, and previously occurring symbols within the sequence are replaced with pointers with their respective prior positions. But the offline dictionary is used in the proposed system.

3.2 DNA Compression Algorithm

The following are the steps that carried out the reading of the first four sequences from a given input sequence. Checking is done for matching the sequence with the rest other sequences as Exact, Reverse, Complement, and Reverse Complement. Thereafter, the information gets stored within the dictionary table until the end of the input file, and next reads eight sequences from the input sequence. By computing the number of positions in the match, the frequency is determined. A match is found for the pattern or sequence concerning other sequences as Exact, Reverse, Complement, and Reverse Complement, and the frequency is stored in a temporary table (Temp_Dictionary1). The next eight sequences are read, and again a similar procedure is repeated. Thereafter, the matched patterns with maximum frequency move to Dictionary1 table. Dictionary1 table stores the highest frequency-matched patterns. This process is performed again for 8, 16, or 32 and up to half of the length of the input sequences to determine DNA compression through matching patterns. The above methods help in achieving DNA compression and its ratio. Mismatches result in improvised compression gain that is taken into account and recorded with the repetitive substrings using the offline dictionary.

The research work includes a text file that comprises four successive base pairs namely 'A,' 'G,' 'T,' and 'C.' The text file helps in performing the matching between the original file and the decompression file. The output file comprises ASCII value which is equivalent to the encoded binary value of the four unmatched base pairs. The architecture diagram of DNA sequence compression is illustrated in Fig. 1. The Dictionary1 (storing highest frequency pattern) is illustrated using Table 2.

Algorithm 1 defines the compression process which involves the process of calling Algorithm 2 to find the matching pattern and assigning binary codes to the unmatched nucleotides and converts those binary codes into the ASCII codes.

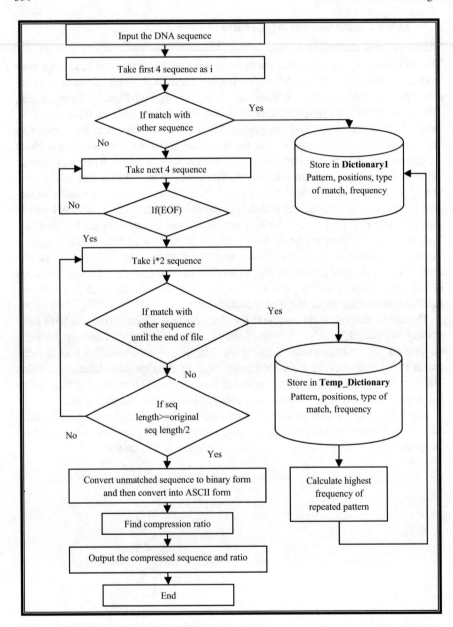

Fig. 1 Architecture diagram for DNA sequence compression

Algorithm 1: Improved_Compress(input_sequence)

Begin
{

       Read the DNA sequence from input_sequence
       Set i = 0; //assign the starting position
       Set read_seq=4;
       Do
       {

              Read 4 sequence as a pattern from i^{th} position of input_sequence
              Assign Dictionary.Name="Dictionary1"
              Out_sequence = **call** pattern_match(pattern,Dictionary)
       } **Until** i>=input_sequence.length;
       For(x=8;x<=(input_sequence.length)/2;x*2)
       {

              Do
              {
              Read x sequence as pattern from input_sequence
              Assign Dictionary.Name:="Temp_Dictionary"
              Out_sequence = **call** Pattern_Match(pattern,Dictionary,
              read_seq)
              } **Until** (i<=input_sequence.length);

       }
       Set the starting position j=0;
       Do
       {

              Read Out_sequencefromjth position

              If Out_sequence[j]='A' **then**
                  Esequence[j]=00;
              Else if Out_sequence[j]='C' **then**
                  Esequence[j]=01;
              Else if Out_sequence[j]='G' **then**
                  Esequence[j]=10;
              Else
                  Esequence=11;
              End if
       } **Until** (j<=Out_sequence.length);

  Set n=0;
  Do
       {
              Read 8 bits from Out_sequence[n] and assign it to bin_value
              Assign ASCII value equivalent to bin_value
  }**Until**(n<=Out_sequence.length);
      }

End

Algorithm 2 defines the pattern matching process that is it finds the matching starts from four sequences. If it is matched as Equal or Reverse or Complement or

Reverse Complement, then the matched sequence is directly stored in the permanent Dictionary1 table. This process continues until the end of the input file and then takes the 8 sequences,16 sequences, 32 sequences, and so on till the length of the sequence not greater than the half-length of the input file and finds the matching and stores the matched sequence in a temporary dictionary table Temp_Dictionary. The Find_Max() function finds out the maximum frequency matching sequence and store that in a permanent Dictionary1 table.

Algorithm 2: **Function Pattern _Match(pattern, Dictionary, read_seq)**

Function Pattern_Match(pattern, Dictionary, read_seq)

Begin
{
       Assign m=0; // (starting position)
       Assign frequency = 0;
       While (m<=Out_sequence.length)
       {
If pattern is match with any type (E)/(R)/(C)/
      (RC)**then**
      {
      If read_seq==4 **then**
Set Dictionary.Name="Dictionary1"
Set Dictionary1.matchpattern to pattern
      Set Dictionary1.position=m
      Set Dictionary1.type_of_match='E' / 'R' /

'C' / 'RC'
          Remove pattern from Out_sequence
      Else
      Set Dictionary.Name="Temp_Dictionary"
      Set Temp_Dictionary.matchpattern to pattern
      Set Temp_Dictionary.position=m
      Set Temp_Dictionary.type_of_match='E' /
      'R' / 'C' / 'RC'
Set Temp_Dictionary.frequency
          : =frequency+1
End if
}
         Increment m value by m value
}
      If read_seq==4 **then**
         Call Find_Max(Dictionary.frequency)
         Remove pattern from Out_sequence
      End if
      Return Out_sequence
}
End

Algorithm 3 finds the maximum frequency sequence (pattern) from the Temp_Dictionary table and return it to the Algorithm 2.

Algorithm 3: Function Find_Max(Dictionary)

Function Find_Max(Dictionary)
// Set variables max largest Record [i:j].
// Input: A subarray of array Record [1, n] with parameters I and j such that $(1 \leq I \leq j \leq n)$

Begin
{
Set i=0;
Read Record [0];   //Read the first record from the Dictionary
Do
{
If Record[i].frequency<Record [i+1].frequency **then**
        Max: =Record[i].frequency;
        j:=i;
Endif
i++;
}**Until**(i! =Dictionary.count);
Dictionary1.Pattern:=Dictionary.Record[j].pattern;
Dictionary1.Position:=Dictionary.Record[j].position;
Dictionary1.Type_of_match:=Dictionary.Record[j].type
}
End

3.3 DNA Decompression Algorithm

There is decompression of sequence which is responsible for compressing a specific value to generate the original content. The compressed file helps in creating the offline dictionary by identifying repeated patterns from the source. The dictionary size is fetched from bit blocks that depict the pattern. Then, there is fetching of all the compressed pattern blocks, recollecting the pattern type followed by patterns. Decompression algorithm which being the reverse of compression makes use of offline dictionary information for decompressing the sequence. The file matching algorithm enables determining if there is a match between the input, and output sequence thus concludes the algorithm as a lossless compression algorithm.

3.4 Compression Ratio

Compression ratio:

$$1 - \frac{(\text{Size of the Dictionary} + \text{size of the Output})}{\text{Size of the original DNA Input sequence}} \times 100 \qquad (1)$$

Table 1 Structure of the permanent Dictionary1 table

Pattern	Position	Type of Match
AACT	11	Exact
AATAACTT	3	Reverse
GAATTCCGGATCACAC	7	Exact

Table 2 Structure of the temporary table Temp_Dictionary1

Pattern	Position	Type of Match	Frequency
GAATCGTA	4,15,27,29,37,45,55,67, 17,31,60,101,250	Exact Reverse	13
AACTCCTG	11,136,373,750 187,273,441,460,543,923	Reverse Complement	10
AATAACTT	3,107,37,655,932,1108 306,2003	Reverse Reverse Complement	8

Table 1 defines the structure of the permanent Dictionary1 table, and Table 2 defines the structure of the temporary table Temp_Dictionary1.

Finding a threshold is half of the entire DNA sequence length which is the threshold value that is the length of the input sequence/2.

Encoding mismatches is a recommended method that determines overall approximate repeats in the DNA sequence and then stores them in an offline dictionary. These approximate repeats have mismatches which are then encoded into binary code. Again the binary codes are grouped into 8 bits which are encoded as ASCII code for further reducing the file size.

4 Results and Discussion

The Improved_Compression algorithm was tested and validated on a group of DNA sequences fetched from NCBI Repository. The standard sequences are Homo sapiens dystrophingene (HUMHDYSTROP), human growth hormone (HUMGHCSA), human beta globin region on chromosome 11(HUMHBB), human DNA sequence (HUMHDABCD), human hypoxanthine phosphoribosyl-transferasegene (HUMHPRTB), and vaccinia virus Copenhagen complete genome (VACCG). The testing was carried out on the above data for different sequence lengths via Improved_Compression algorithm. The research includes an implementation which is being performed on JAVA environment and the entire dataset that is retrieved from different notepad file is taken as an input and been managed using the MYSQL database for storing and retrieving the results.

Table 3 Comparison of compression ratio

Sequence	Length	Optimal seed-based algorithm (%)	Improved_Compression (%)
HUMHDYSTROP	38,770	82	90
HUMGHCSA	66,496	90	89
HUMHBB	73,308	82	89
HUMHDABCD	58,563	84	88
HUMHPRTB	56,832	84	89
VACCG	1,91,735	84	89

Table 4 Comparison of time execution [in secs]

Sequence	Length	Optimal seed-based method	Improved_Compression
HUMHDYSTROP	38,770	1.5	37
HUMGHCSA	66,495	2.5	65
HUMHBB	73,308	2.8	89
HUMHDABCD	58,864	2.2	63
HUMHPRTB	56,737	2	65
VACCG	1,91,735	4	155

The comparison of the compression ratio of the proposed algorithm Improved _compression with the other existing compression algorithms [25] is given in Table 3.

Table 4 shows the comparison of time execution (in secs) of the proposed algorithm Improved_Compression with existing algorithms [25]. The response time of the proposed algorithm is lesser than the existing algorithm.

Figure 2 describes the implementation screen of the proposed approach. This first button is used to upload the input sequence file, and the second button is used to compress the sequence which is extracted from the input sequences, and the third button is used to decompress the sequence.

Figure 3 shows details of the uploaded dataset with different types of datasets. It reads the input sequence from the selected file and displays the input sequence with its data length, i.e., size.

Figure 4 displays the compressed file along with its data length and its compression ratio.

5 Conclusion and Future Scope

The sole effort lies to put forward a technique enabling DNA sequences compression similar to that of the pattern match methods. Toward this, the recommended algorithm

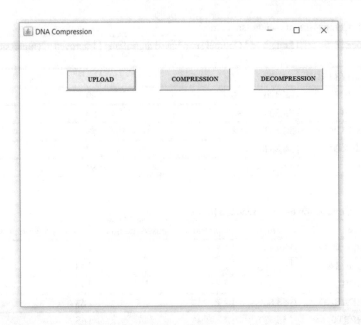

Fig. 2 Implementation screen of the proposed approach

HUMDYSTROP.txt - Notepad
File Edit Format View Help

GAATTCCGGATCACACAACAAATAACAATTAGATAACAGCTTGTTGGGAGAAAGCGAGGATCAGTGTTTGCCA
TTTGACTTTGATTTTCTCTGTTTTTATTTTTTAACTCCATTCTCAAGAAGTCTGCACATAAGAGTTTCAACATCTAC
GTGGATTTATCGTCGAGTGATTTCTTCCTGGCATTTTATCTCTGAGACCAGGATCTGGTTGCTAAGCATGTAGA
GAGCACAATGTCAATGTGATTATTTGTAATGGGGGAAAGGTTACCGGAGAATATTACACGACCATCCACATAGA
AAGTTTGTTTAGCAGCACAGAATTGAAGGAAGACAACCAGATGGTTATGAGGAAGATTCATCCAAACTATGCC
GAATCCATTTTGTGTCAGCACTTCCCAAGTTCTTGTTAATGCTACCTTAAGTTCAATTCAAACCAGGCAGCATTT
GATCTTCTAGATCAAGTTTTTATTTCAATATTTCTACCTCATTTCTGATTCTTAGGTGTTCCTTATTTCCCAATTTAT
GATTATTTATGTTATATTCTGTTTTGTGAAGTTTGACCTCTACCTAATTAAATTACATTTTCAATTGTATCTTGGATT
ATCAGACATTAGGAATTAATTATAAAAATCTCTAGATTGGTACTTGGAGCTTAAAGGAATAAGGTGGTGGAACG
CATCTTCCCACCTGCTTGATCTCTGGCATATAATTTACTATCCGTGAACCTCAGGTTTCTCTTCCGAAAAGCTGC
CAGAGCATTAATATTATTATAGACTTTTGATGATTTACACAATTTTAGCTTTTTGGCAAGACATATTTACTAGTACT
GAAAAAGAAAGTGTTTTGTTTGATAATATAGCATTATAACACTGCACAAAAAAAAAATGGTATATGCAGAGACT
TCAGACACTTCCTTTGTGTTCAATAAATTTCAATTTCCTCCTTTCCTTCAGTTCACTTCAAGAAGGACGGCAGCA
GCATTCTGTAGCAAAATGCTGTGGGTGAAAATGCCTTCCTTCTTAAGGGAATTTAGCTTCTGTAGTACCAGAAT
GAGCCCATTTTCTGGCATCTCTCTCCCTTTAATATTCCTCAAAAAGTTGGATTTTTCCTGGACTTTTCATATTACAGA
CATCCTGGCATCTTCTTTCTTCCAGACTTGTATATCCAACTGCTTCCATTCATACACTTGACCAACCTTTTAATTT
GTTCTTCTTTTGCTTCTTTGTTTCAGACAATGGCACCACCATTCTCGAGTAAGGCACGTTCATTTATCAGGTCCT
GTCATTTCTTTACCTGGGTAGCCTGCACCTTCTACCTGCATTGATTCAGCAGTCTCTTCACCACTGGCTCTCCC
TGACTTTTATTAATGCAAATATGACCTTATAACTCCCTTGCTTAAAGACCCACTCATGTTTGTCTTTGTATCCATA
AGATGAAAGAACAAGTTGTATAAATACTGAATGGTCTGATGTGCTCTTTGTTGTGTCAAGAAGGACATTTTGCA
ATTCCTCAAGAGAAGATGGATGTATTGATTCTGTATTTCAAATGACATAACTTTTGTGAAATAAGAGGCTGCCA
AAGGTAACAGGCTAAAGGGTTCAGTCTTAAACTTTCTTAAGACTGTAGTTCAGGGTTCCTATGGTGGGGCTATA
CAAGGGGTATTAGGAAAGAATCCAGGTTTGATGCAGGGAAAAATAAAAACAACTGATAATCTCTAGTGTCCCC
CTTACTTTGTCTTCTACATGTTTAAGGGAGAAAAATGAGTTAACAGAAGGGGAGGTACAGCATTTCTATTTACT
TCTCTCTCTCCCCAGCCTTCCCCCGCTTCTCTCTCTCTCTCTCTCTCTCTCTGTGTGTGTGTGTGTGTGTGTGTG
CCTTTCTTTTCTTTCAAGCATATGTTGTGGCAGAGACAAGTGTACATCAAAATTCGTGGTCCCTCTTTCATAGTA

Fig. 3 Uploaded dataset

Fig. 4 Compression result extracted from different dataset

is elaborated that aids in compressing of a DNA sequence with the average compression ratio of the results ascertain the algorithm is prompt, easy to imbibe, flexible, and caters to repetitive structure pattern matching pertaining to DNA sequences. The compression ratio for DNA sequences are clearly determined and presented. Improved_Compression method which is recommended for DNA sequence compression benefits by offering compressed file size than others. In the future, the compression ratio and the execution time can be improved by using any standard pattern matching methods.

References

1. Ziv, J., Lempel, A.: A universal algorithm for sequential data compression. IEEE Trans. Inf. Theory **23**(3), 337–343 (1977)
2. Chen, X., Kwong, S., Li, M.: A compression algorithm for DNA sequences. IEEE Eng. Med. Biol. Maga. **20**(4), 61–6 (2001)
3. Chen, L., Lu, S., Ram, J.: Compressed pattern matching in dna sequences. In: Proceedings of IEEE Computational Systems Bioinformatics Conference. IEEE, New York, pp. 62–68 (2004)
4. Zhu, Z., Zhang, Y., Ji, Z., He, S., Yang, X.: High-throughput DNA sequence data compression. Brief. Bioinform. **16**(1), 1–15 (2015)
5. Pothuraju, R., Apparao, A.: DNABIT compress—genome compression algorithm. Bioinformation **5**(8), 350–360 (2011)
6. Saada, B., Zhang, J.: DNA sequence compression technique based on nucleotides occurrence. In: International Multi-Conference of Engineers and Computer Scientists, Hong Kong, vol. 1, pp. 63–66 (2018)
7. Wandelt, S., Bux, M., Leser, U.: Trends in genome compression. Curr. Bioinform. **9**(3), 315–326 (2014)

8. Deorowicz, S., Grabowski, S.: DNA compression for sequencing data. Algorithm. Mol. Biol. **8**(1), 25 (2013)
9. Jones, D.C., Ruzzo, W.L., Peng, X., Katze, M.G.: Compression of next generation sequencing reads aided by highly efficient de novo assembly. Nucl. Acid Res. **40**(22), 171–171 (2012)
10. Pinho, A.J., Pratas, D., Garcia, S.P.: GReEn: a tool for efficient compression of genome resequencing data. Nucl. Acids Res. **40**(4), e27–e27 (2012)
11. Alberti, C., Mattavelli, M..: Investigation on genomic information compression and storage, pp. 1–28 (2015)
12. Raffaele, R.G., Rombo, S.E., FilippoUtro: Compressive biological sequence analysis and archival in the Era of high-throughput sequencing technologies. Brief. Bioinform. **15**(3), 390–406 (2014)
13. Haque, S.M.R., Mallick, T., Kabir, I.S.: A new approach of protein sequence compression using repeat reduction and ASCII replacement. IOSR J. Comput. Eng. **10**, 46–51 (2013)
14. Yu, J.-F., Cao, Z., Yang, Y., Wang, C.-L., Zhen-Dong, Su., Zhao, Y.-W., Wang, J.-H., Zhou, Y.: Natural protein sequences are more intrinsically disordered than random sequences. Cell. Mol. Life Sci. **73**(15), 2949–2957 (2016)
15. Daniels, N.M., Gallant, A., Peng, J., Cowen, L.J., Baym, M., Berger, B.: Compressive genomics for protein databases. Bioinformatics **29**(13), 283–290 (2013)
16. Hayashida, M., Ruan, P., Akutsu, T.: Proteome compression via protein domain compositions. Methods **67**(3), 380–385 (2014)
17. Bonfield, J.K., Mahoney, M.V.: Compression of FASTQ and SAM. Format Sequencing Data **8** (2013)
18. Jahaan, A., Ravi, T.N., PanneerArokiaraj, S.: Bit DNA Squeezer (BDNAS): A unique technique for DNA compression. Int. J. Sci. Res. Comput. Sci. Eng. Inf. Technol. **2**(4), 512–517 (2017)
19. Rai, D.S., Bharti, R.K., Parihar, B.: Survey of compression of DNA sequence. Int. J. Comput. Appl. **73**(6), 52–58 (2013)
20. Bakr, N.S., Sharawi, A.A.: DNA lossless compression algorithms: review. Am. J. Bioinf. Res. **3**(3), 72–81 (2013)
21. Zhu, Z., Zhang, Y., Ji, Z., He, S., Yang, X.: High-throughput DNA sequence data compression. Briefings Bioinf. **16**(1), 1–15 (2015)
22. Pinho, A.J., Pratas, D.: MFCompress: a compression tool for FASTA and multi-FASTA data. Bioinformatics **30**(1), 117–118 (2014)
23. Pandian, M.D.: Sleep pattern analysis and improvement using artificial intelligence and music therapy. J. Artif. Intell. **1**(02), 54–62 (2019)
24. Manoharan, S.: Patient diet recommendation system using K clique and deep learning classifiers. J. Artif. Intell. **2**(02), 121–130 (2020)
25. Eric, P.V., Gopalakrishnan, G., Karunakaran, M.: An optimal seed based compression algorithm for DNA sequences. Adv. Bioinform. 2016, Article ID 3528406 (2016)

Domain Specific Fine Tuning of Pre-trained Language Model in NLP

Jony, Jyoti Vashishtha, and Sunil Kumar

Abstract Text analysis is essentially the computerization process of examination of the known contents to decide the sentiments passed on it. Data mining is the automated or semi-automated process of analyzing and modeling of the large data repository to extract interesting information. This paper presents a deep hashing method-based text analytics framework that deals with a large amount of textual data. The proposed method uses a deep belief network as a pre-trained model for reducing errors in the system. Deep learning techniques with the help of the hashing method provide fast response and accurate results in terms of accuracy. The study further provides a deep comparative investigation on convolution neural methodologies used in text analysis. In the experimental part, the Twitter database is used for sentiment analysis. The characterization of sentiment data is relying upon the various degrees of investigations. A deep hashing concept is used for improvement in the accuracy of the system. The experimental results show that the proposed technique outperforms other existing techniques.

Index Terms Text analytics · Natural language processing · Deep hashing · Convolutional neural networks

1 Introduction

In the modern era, science and technology are the major components for driving a computer system. Data mining and its retrieval are increasing attention in the latest data-based technologies that are part of knowledge discovery. Data mining includes

Jony (✉) · J. Vashishtha (✉) · S. Kumar (✉)
Guru Jambheshwar University of Science and Techonology, Hisar, Haryana, India
e-mail: jonymalik.engr@gmail.com

J. Vashishtha
e-mail: jyotivst@gmail.com

S. Kumar
e-mail: skvermacse@gmail.com

© The Editor(s) (if applicable) and The Author(s), under exclusive license
to Springer Nature Singapore Pte Ltd. 2021
S. Shakya et al. (eds.), *Proceedings of International Conference on Sustainable Expert Systems*, Lecture Notes in Networks and Systems 176,
https://doi.org/10.1007/978-981-33-4355-9_31

the concept of integration of various fields like mathematics modeling, computer science algorithms, etc. It helps to provide useful data by extraction and correlation. The major methods for processing data are the retrieval of data, extraction, processing, analysis, etc.

Similarly, the search method plays an important role in statistical analysis, computer vision, and applications with the latest trends. These data are represented by high dimension vectors. This similarity is assessed by the extraction of features in text or images. Data similarity is measured by various computational functions. The similarity level is being calculated by various vectors pairs. Search is the main operation that requires extracting useful information. The main purpose of this search is to find the nearest neighbors in the given data. Many techniques have been used for this process in one form or another. The use of indexing did not work properly in large datasets. The presence of data graphs, figures, and other objects causes a challenge to search the exact form of data. After this, the semantic similarity process is applied that helped to provide the strength of semantics. This helps to find sentence similarity in a given document. This process is very useful in biomedical applications and natural language processing applications.

Text mining is the computerized or semi-robotized procedure of examining and displaying of the enormous information archive so as to remove unwanted data. It might be acted as an interdisciplinary field which includes the reconciliation of different systems. It might be characterized as the way toward finding significant data from given information by utilizing artificial intelligence.

Analytics of text is the part of the latest language processing that helps to process complete document at a time with smart technology. It uses raw documents as input and then applies to feature selection methods. The dataset is trained by a training algorithm that removes unwanted words from the text and sorted text for testing analysis in first left section as shown in Fig. 1. This data is modeled by the classifi-

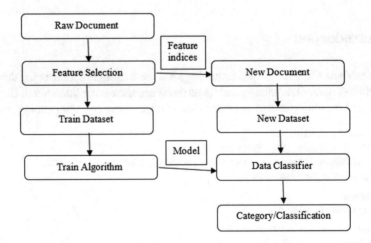

Fig. 1 Text classification

cation model and provides useful output in terms of accuracy in Sect. 2. These are the generalized model for text classification means that these steps must be followed in the process.

The remaining description of the paper is as per the following; Sect. 2 examines the surveyed work done by various authors in this field. Section 3 presents the proposed methods used during this study. The results discussion is presented in Sect. 4. The conclusions and its future scopes are described in Sect. 5.

2 Literature Survey

This section presents the recent review on various text analysis techniques. Sentiment analysis is the popular trend for social-based analytics on the Web. It covers most of the fields like consumer opinion, emotions, finance, healthcare, market strategies, and media. Sentiment analysis is based on social Web data where availability of information is in the form of imprecision, vagueness, and uncertainty. In a recent literature survey, researchers' efforts are focused on combining deep and machine learning. Table 1 presents the summary of the survey done by various authors in this field.

Atasu et al. [6] presented the matching of expressions for improving throughput in text analytics. It was done by hardware setup and produced a network with optimized output. The relation between clients and difficult modules are described as semantic connections in Bradel et al. [7] research work. It additionally summarizes the multi-modal collaboration and figured out as a perception pipeline model. Santos et al. have presented a new DCNN approach to exploit sentiment analysis by analyzing the character information on short texts. Two different domains used for this approach are: Stanford Sentiment Treebank which extracts sentences from movie reviews, Stanford Twitter Sentiment corpus that holds Twitter messages [8].

Chiranjeevi et al. [9] proposed a process of data retrieval in text documents. The retrieved data was used in the education sector with the help of data analytics. It used deep hashing with a semantic model structure for analyzing the best results.

Table 1 Overview of research analysis

References	Year	Overview
Zvarevashe et al. [1]	2018	Sentiment analysis on the hotel review data by use of Naive Bayes algorithm
Sawant et al. [2]	2018	OCR based speech conversion from printed text
Wei et al. [3]	2018	Deep learning-based text classification in legal documents
Shahareet al. [4]	2017	Sentiment analysis for the news data based on the social media
Hennig et al. [5]	2016	Social analytics of changes in consumer opinion of a television broadcaster

It presented a text engine suitable for this data retrieval by using computational techniques. The model uses the concept of representation of semantics.

Sabra et al. [10] reported analysis issues in Arabic language. Research work proposed a notation process for the vocabulary in the database and performs sentiment analysis to recognize the text slope. S. Khedkar et al. have proposed a classifier that was based on a machine learning technique. It was used to gain praise and then further analyzed the system based on performance parameters. This approach was used to understand reviews of customer that was based on their complaints [11].

Macedo Maia et al. present a financial domain prediction model based on sentiment analysis such as FinSSLx. Model uses clausal or phrasal sentence for simplification steps. Using this, a complex structure of sentences is shortened into sound independent smart and short sentences based on polarity and distant supervision of lexical acquisition [12]. Bhargava et al. discussed the possible method to extract sentimental tweets from three Indian languages. They have created thirty-nine sequential models with three different NN layers: (i) long short-term memory, (ii) recurrent neural networks, and (iii) convolutional neural network. To avoid overfitting and accumulation error, these three algorithms are used with optimal parameter settings [13]. Paolanti et al. have proposed a new approach for promote the cultural heritage sites within the territory. In this direction, an advance contribution is involved using sentiment analysis. The collection of sentiment for the monuments can be used to evaluate its positiveness to reveal its influences in public by increase its value. The sentiment analysis pictures are recognized by a special method trained using DCNN [14].

Kim et al. have shown the use of CNN for the method of sentiment classification. The experimental process using three consecutive convolutional layers is applied to three different well-known datasets, so that it works effectively with longer sentiment texts [15]. Qian Li et al. performed a comprehensive experiment that has been performed on five real-life datasets using the new approach called Bi-level based on multi-scaled masked of the CNN-RNN network. They have employed the most significant and valuable multi-grain noise tolerated patterns to find relation among word and characters in vector structure [16].

Das et al. [17] presented a parsing method and an extracted entity-based model for the company. It used the concept of text analytics for the derivation of structured data from the unstructured form of the data. It helped to integrate with a workflow system. Most companies use this technique for selection of the resume of the candidates for recruitment. Wei et al. [3] have proposed the concept of legal documentation understanding by deep learning methods. It compared the performance of deep learning results with conventional SVM method results on the different datasets. Results indicated that the performance of CNN was better as compared to SVM results.

Issues in existing research models

The main issues in existing text analytics models are the use of conventional techniques for the analysis of datasets. This cannot help to improve the accuracy of the system up to the maximum level. The investigation dependent on the sentiment concept is basically the computerization of assessment of a referred to content to choose the sentiment passed on in it. Generally, the text investigation is done at archive-based, and sentence based, in the report based, gives the positive or negative assessment in the entire record as a single element. In the sentence-based method, each sentence is to be sure, negative, or unbiased supposition in the record.

3 Proposed System

Text configuration is an information escalated process. Different subject matters and aptitudes are used in leading on each phase of the planned activities including applied structure, encapsulation plan, and gritty plan. The basic inspiration driving the examination is to make a book mining structure that can extricate mechanical knowledge from electronic content sources. This information is a prime necessity for fruitful innovation on the board. This content mining structure can assist with distinguishing innovation foundation, find covering or comparable exploration exercises, recognizing different methods for improving framework execution. The objective of the research is to obtain high script analysis accuracy in datasets. In the research of social media analysis, there are issues in mining data from such a huge dataset of Twitter. The main goal is to analyze text data.

The aim of the research is to explore the existing data mining techniques for the Twitter dataset, to design an algorithm for text analytics of the Twitter dataset, and to verify the results in terms of precision and recall.

In this work, it proposes another hashing from an alternate perspective to past ones. It saves and encodes the spatial installing of each example in the space spread over by k-bunching centroid of the preparation tests, intending to accomplish great execution with double codes and straight intricacy. In the preparation arrange, it initially packages the preparation tests into k-groups by a direct cluster-based technique.

3.1 Text Analysis Using Existing Convolutional Neural Network (CNN)

This section presents a CNN based analysis of the Twitter dataset to check the accuracy of the system. It consists of various sections like the convolutional layer, pooling section, hidden layer, and finally softmax for output.

Sentence Model: This is presented by an embedding of words in d-dimension matrix data. A comment contains n words that are concatenated with each other, provides an output matrix Y which gets after processing input with convolution matrix data.

Convolutional Layer: This layer contains m filter, levels F whose output is attached with a sliding window with length L. Then, features c are measured with X concatenation matrix that is by Eq. (1):

$$c = \sum X * F \qquad (1)$$

Max Pooling: The convolutional output is applied to reluctant activated function before pooling. This helps to generate the maximum value in a fixed set of levels.

Hidden Layer: This hidden layer mainly performed the transformation work. This provides a vector form of embedding data which is basically a hidden process.

Softmax: This is basically connected after the hidden layer. It provides the single output after receiving embedding data under probability with the largest value.

Network Parameters: After getting output, it gets trained into a neural network with X embedding words. It provides the final desired output after CNN.

3.2 Text Analysis Using Proposed Hashing Method

This work presents a concept of text analytics using a deep hashing method on the Twitter dataset. This hashing method provides a sharp response with better accuracy in the analysis system. It does not use complex neural networks like ANN and CNN because of its conventional based approaches and also high complexity. This is the major drawbacks of these systems. Due to this, deep learning-based approaches have been used nowadays to overcome these problems. Deep hashing uses a hash table values that help to preserve and encode sample data in cluster form and based on sentiment distance, and it provides output value. It uses binary coding and thresholding concepts for desired and useful output data. It performs training and testing stages. After receiving raw data, it goes for parsing and removal of unwanted words by use of a package of words concept that helps to ignore spaces and punctuation words in sentences. Then, the clustering of text data is performed that provides embedding of useful data for feature extraction. This useful data is applied to statistical and rule-based models for deep hashing results. The proposed system model for text analytics is depicted in Fig. 2.

Figure 2 depicts the system model for text analysis. The text data is processed into various processes that include the parsing of data with unwanted words removal and then processed into text filtering and clustering analysis. The feature identification is processed after clustering. This data is passed through rule-based models, and deep learning method to provide better accuracy is results. The information it utilizes for model investigation is acquired from Google, and an average of five thousand example tweets are considered for objective assortment. Before identifying the necessary information, it is essential to pre-processing to the data. Every one of the tweets is lowercased, tokenized, and labeled.

The main component is an ostensible credit comparing to the part of speech tag in its unique circumstance. This element gives morphological data of the word. There are observational proofs that abstract and target writings incorporate various

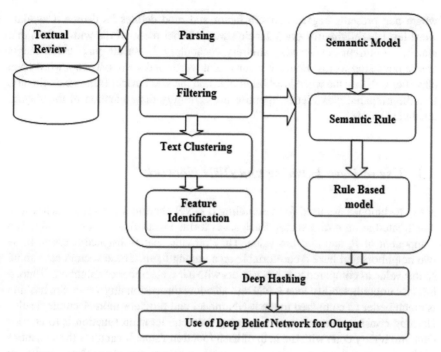

Fig. 2 Proposed model for text analysis

dissemination of labels. It pictures the extended vocabulary forces of words delegated positive and negative through word mists. This system uses the Principle of Mathematical Induction (PMI) approach for generating a relationship between words and words. The relation can be described by Eq. (2):

$$\text{PMI}(\text{word}_1, \text{word}_2) = P(\text{word}_1, \text{word}_2) \div P(\text{word}_1) * P(\text{word}_2) \quad (2)$$

The higher value of the PMI is better than the relationship between expression and their words. Here, the word is the smallest unit to present its expression. This work presents a deep hashing concept that comes closer from an alternate perspective to past ones. It saves and encodes the spatial installing of each example in the space spread over by k-bunching centroids of the preparation tests, intending to accomplish great execution with double codes and straight intricacy. In the preparation arrange, it initially packages the preparation tests into k-groups by a direct bunching technique, for example, straight ghostly grouping.

The proposed procedure can be outlined as collecting data from the objective area and also to tag each of the words from the objective assortment utilizing a grammatical feature tagger.

Calculate the influence factor for text analysis Q:

$$Q = \text{neg} \times \text{mod} \quad (3)$$

where neg presents negative adverb factor and mod defines its degree. Calculate word-level highlights for every labeled word. Label these words with a slant that matches a current hand-made extremity vocabulary. Erase all punctuation marks and remove stop words. Remove inconsistent words and spaces. Train a word-level classifier utilizing the word-level highlights and the word marks from the dictionary. Use the prepared classifier to appraise the extremity dissemination of the staying unlabeled words.

3.3 Use of Deep Hashing and DBN Network

Many techniques are used for assembling a k-NN chart to a limit that each test in X is treated as an n-dim vector. Each set is formed near to neighbor x_i; which is a component of P_i and nonzero value. They provide spatial distance between these two neighbors and have a comfortable area among them. These vectors have more sparse value as compared to k-NN vectors with a large number of neighbors. Then, it has less unpredictability than k-NN and also low dimensionality value. So, hashing is much better as compared to the k-NN model and provides more accurate results. Hashing consists of small sets of training datasets. Its main function is to convert data into binary code with the help of hash function value. It captures the semantics of data in a space which is done by a function that clustered the data together. It produces the hash functions that provide equivalent values 0 and 1 s. All bit codes are useful for the retrieval operation.

In this work, a deep belief network is also used at output to reduce the error value in the system. It uses the concept of probability in the system. The entire input is learned in this network which is not done on CNN. The symmetrical weights connect all layers. It starts from the bottom and then started moving up with tuning of weights. The use of DBN provides a network graph containing some random variables. It performs with high accuracy with minimum error. The steps of DBN are training features values that were getting from deep hashing output, and the next step is to activate the training function and perform learning features in the hidden layer. Finally, the desired output is obtained with minimum error after the output layer.

3.4 Performance Parameters of the System

The main parameters for analysis of this system are precision, recall, F-1 score, and accuracy. The accuracy is defined as the degree of closeness of standard value. It is calculated as Eq. (4):

$$\text{PMI}(\text{word}_1, \text{word}_2) = P(\text{word}_1\text{word}_2) \div P(\text{word}_1) * P(\text{word}_2) \qquad (4)$$

The precision is defined as the measure of reproducibility of measure values. It is different from accuracy in terms of reproducibility values. It can be calculated as Eq. (5):

$$\text{precision} = (\text{No. of True positive}) \div (\text{Truepositive} + \text{False Negative}) \quad (5)$$

The recall is presented as the ability of the classifier to find all positive samples. Its maximum value is 1, and the worst value is 0. It considers false negatives value also in the system. It is calculated by Eq. (6):

$$\text{recall} = (\text{No. of True Positive}) \div (\text{True Positive} + \text{False Negative}) \quad (6)$$

The F1 score is the weighted average of both precision and recall values. So, it considers all positive and negative values for scoring results. It is calculated by Eq. (7):

$$F1 \text{ score} = (2 \times \text{precision} \times \text{recall}) \div (\text{precision} + \text{recall}) \quad (7)$$

4 Results and Discussion

There is an enormous number of texts in which investigators tried to deal with many remarkable classes: fun, stun, shock, pity, and fear. Sentiments like fulfillment and issue are reasonable facets. Sentiment analysis is fundamentally stressed over distinctive positive or negative suppositions. This uses a Twitter dataset with 5000 documents and 2726 number of words of vocabulary. After cleaning of data, it just goes to 491 documents in the dataset for final processing.

4.1 Results Using Convolutional Neural Networks

This work describes the implementation of two deeper variants, namely with two and three convolutional layers. The two-layer architecture has two consecutive convolutional by adding the pooling layers followed by a hidden layer and soft-max operation. The three-layer architecture has an additional convolutional along with the pooling layer before the hidden and soft-max ones.

This work presents a novel framework for geo-spatial supposition examination of sentiment related to Twitter data objects. The data is taken from the social media Twitter site as shown in Fig. 3. After this, cleaned data is generated using sentiment analysis. The proposed system model for sentiment analysis using hashing is presented for the optimization of data. In this, the database of the Twitter site

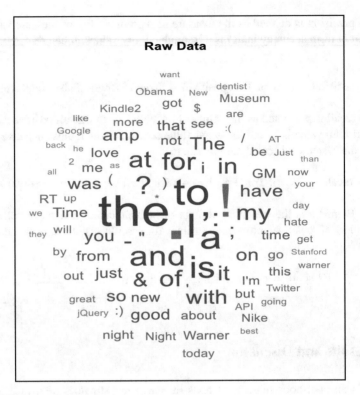

Fig. 3 Raw data used for analysis

is selected, and then, various operations are applied on it before it goes to clustering. Then, sparse embedding is applied to it for zero paddings, and removal of unwanted data from the matrix before hashing is applied for optimizing the data. The models were trained and tested on data with social media Twitter datasets. One dataset contained all sentiment data (negative, positive, and neutral), and one dataset contained all positive and negative data. Each dataset was divided into a training set and a validation set during training; the training set contained 80% of the whole dataset. Since the neural network is not handling strings, all sentences in the dataset were transformed into a list of sequences.

Each sentence had to be the same length since the neural network has an input that is fixed to a certain length. In this case, all sentences were padded to the same length as the longest one. When padding a sentence, it will be stretched out with a token that will get the value of zero in the vocabulary list. This lets the neural network know that the token does not have any impact on the literature of the sentence. As a final step in the data pre-processing, the labels of each sentence were encoded to be readable for the neural network. It covers the data obtained through the training process of the convolutional neural networks. There will be a loss graph and an accuracy graph for each type and to each graph, and there is a table of the models

Table 2 Performance parameters of system using CNN

Algorithm	Accuracy	Precision	Recall	$F1$-score
CNN	71.70%	60.29%	55.37%	57.72%

to make it easily readable. These models show that they gradually increase their accuracy percentage, meaning that they are learning from the data. Table 2 shows the performance parameters of the system using CNN. These parameters include accuracy, precision, and recall. The accuracy of system is low because of the complex structure.

4.2 Results Discussion

Matrix evaluation is very important for assessing the performance of the proposed approach. The exhibition of a recovery framework is assessed depending on a few criteria. A portion of the regularly utilized exhibition measures is normal exactness and normal review. The accuracy of the recovery is characterized as part of the recovered information that is to be sure applicable for the inquiry. The proposed results are presented by the use of hashing and deep belief networks. Hashing is a fast and accurate method for improving the accuracy of the system in analytics. The cleaned data is generated after text cleaning and removal of unwanted words from total text data as shown in Fig. 4.

The performance parameters mainly precision and recall are presented in Figs. 5 and 6, respectively. A decent recovery framework ought to have high esteems for accuracy and review. The review is the part of important information that is returned by the inquiry. Table 3 shows the performance parameters of proposed model. It is observed from the table that accuracy of the proposed model shows better improvement as compared to actual results by the use of the hashing method as given in Table 3.

Table 4 shows the performance results of the proposed system. A comparative analysis of sequential minimal optimization and Naive Bayes framework is given by Zvarevashe et al. [1], and it outperform in terms of accuracy, precision, and recall. When it comes to classifying the classes on the new data, all models are performed incredibly well by reaching around 90% accuracy for all models and datasets. The result is far better than the minimum threshold of 70%, and all models could be embed into production.

Fig. 4 Cleaned data generated after sentiment analysis

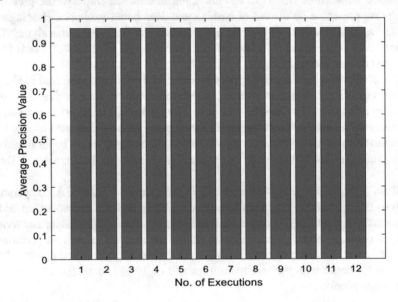

Fig. 5 Average precision values

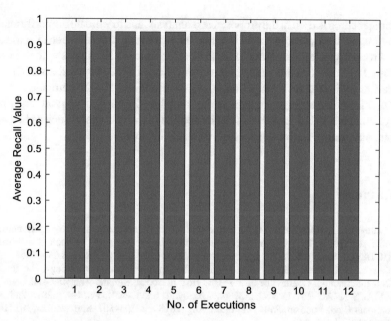

Fig. 6 Average recall values

Table 3 Results of the proposed system

Algorithm	Accuracy	Precision	Recall	F1-score	RMSE (DBN)
Hashing + DBN	89.14%	96%	95%	95.50%	0.31

Table 4 Performance comparison of different method

Parameters	SPBM (%)	CNN method (%)	Proposed method (%)
Accuracy	84	71.7	89.14
Precision	80	60.29	96
Recall	85	55.37	95
F1-score	81	57.72	95.5

5 Conclusion and Future Scope

Text analytics is a perceptive interpretation technique that deals with structured as well as unstructured data. Preparation of information and arranging the writings into bits or characters is the initial process in text analysis. Investigating or mining is used to extract the useful and essential information from large amount of data which is distributed in nature. Proposed research work provides an efficient text investigation model based on different modules and different creators. Along with mining content, digging system is evolved to obtain better knowledge from the database.

Hashing method is used in the proposed text analysis, and to validate the performance, conventional deep learning model is compared in terms of precision, recall, and accuracy. Proposed model demonstrates better performance in all aspects with minimum threshold of 70% which is much better than other models. Proposed research work limited to Twitter dataset only limitations of maximum 10,000 comments. In the future, this work can be generalized to any social media datasets without any limitation of comments. The framework can be applied to many domains such as business, design, and healthcare for the analysis of user's behavior.

References

1. Zvarevashe, K., Olugbara, O.O.: A framework for sentiment analysis with opinion mining of hotel reviews. In: 2018 Conference on Information Communications Technology and Society (ICTAS), Durban, Mar. 2018, pp. 1–4. https://doi.org/10.1109/ICTAS.2018.8368746
2. Sawant, N.K., Borkar, S.: Devanagari printed text to speech conversion using OCR. In: 2018 2nd International Conference on I-SMAC (IoT in Social, Mobile, Analytics and Cloud) (I-SMAC)I-SMAC (IoT in Social, Mobile, Analytics and Cloud) (I-SMAC), 2018 2nd International Conference on, Palladam, India, Aug. 2018, pp. 504–507. https://doi.org/10.1109/I-SMAC.2018.8653685
3. Wei, F., Qin, H., Ye, S., Zhao, H.: Empirical study of deep learning for text classification in legal document review. In: 2018 IEEE International Conference on Big Data (Big Data), Seattle, WA, USA, Dec. 2018, pp. 3317–3320. https://doi.org/10.1109/BigData.2018.8622157
4. Shahare, F.F.: Sentiment analysis for the news data based on the social media. In: 2017 International Conference on Intelligent Computing and Control Systems (ICICCS), Madurai, Jun. 2017, pp. 1365–1370. https://doi.org/10.1109/ICCONS.2017.8250692
5. Hennig, A. et al.: Big social data analytics of changes in consumer behaviour and opinion of a TV broadcaster. In: 2016 IEEE International Conference on Big Data (Big Data), Washington DC, USA, Dec. 2016, pp. 3839–3848. https://doi.org/10.1109/BigData.2016.7841057
6. Atasu, K., Polig, R., Hagleitner, C., Reiss, F.R.: Hardware-accelerated regular expression matching for high-throughput text analytics. In: 2013 23rd International Conference on Field programmable Logic and Applications, Porto, Portugal, Sep. 2013, pp. 1–7. https://doi.org/10.1109/FPL.2013.6645534
7. Bradel, L., North, C., House, L., Leman, S.: Multi-model semantic interaction for text analytics. In: 2014 IEEE Conference on Visual Analytics Science and Technology (VAST), Paris, France, Oct. 2014, pp. 163–172. https://doi.org/10.1109/VAST.2014.7042492
8. Deep Convolutional Neural Networks for Sentiment Analysis of Short Texts, p. 10
9. Chiranjeevi, H.S., Manjula Shenoy, K., Prabhu, S., Sundhar, S.: DSSM with text hashing technique for text document retrieval in next-generation search engine for big data and data analytics. In: 2016 IEEE International Conference on Engineering and Technology (ICETECH), Coimbatore, India, Mar. 2016, pp. 395–399. https://doi.org/10.1109/ICETECH.2016.7569283
10. Sabra, K.S., Zantout, R.N., Abed, M.A.E., Hamandi, L.: Sentiment analysis: Arabic sentiment lexicons. In: 2017 Sensors Networks Smart and Emerging Technologies (SENSET), Beirut, Sep. 2017, pp. 1–4. https://doi.org/10.1109/SENSET.2017.8125054
11. Khedkar, S., Shinde, S.: Deep learning and ensemble approach for praise or complaint classification. Proc. Comput. Sci. 167, 449–458 (2020). https://doi.org/10.1016/j.procs.2020.03.254
12. Maia, M., Freitas, A., Handschuh, S.: FinSSLx: A sentiment analysis model for the financial domain using text simplification. In: 2018 IEEE 12th International Conference on Semantic

Computing (ICSC), Laguna Hills, CA, Jan. 2018, pp. 318–319. https://doi.org/10.1109/ICSC.2018.00065

13. Bhargava, R., Arora, S., Sharma, Y.: Neural network-based architecture for sentiment analysis in Indian languages. J. Intell. Syst. **28**(3), 361–375 (2019). https://doi.org/10.1515/jisys-2017-0398

14. Paolanti, M. et al.: Deep convolutional neural networks for sentiment analysis ofcultural heritage. In: ISPRS—The International Archives of the Photogrammetry, Remote Sensing and Spatial Information Sciences, vol. XLII-2/W15, pp. 871–878, Aug. 2019. https://doi.org/10.5194/isprs-archives-XLII-2-W15-871-2019

15. Kim, H., Jeong, Y.-S.: Sentiment classification using convolutional neural networks. Appl. Sci. **9**(11), 2347 (2019). https://doi.org/10.3390/app9112347

16. Li, Q., Wu, Q., Zhu, C., Zhang, J.: Bi-level masked multi-scale CNN-RNN networks for short text representation. In: 2019 International Conference on Document Analysis and Recognition (ICDAR), Sydney, Australia, Sep. 2019, pp. 888–893. https://doi.org/10.1109/ICDAR.2019.00147

17. Das, P., Sahoo, B., Pandey, M.: A review on text analytics process with a CV parser model. In: 2018 3rd International Conference for Convergence in Technology (I2CT), Pune, Apr. 2018, pp. 1–7. https://doi.org/10.1109/I2CT.2018.8529772

Machine Learning Based Decision Support System for High-School Study

Pornpimol Chaiwuttisak

Abstract The objectives of this study are to investigate the correlations between personal factors, learning factors family, and economic factors affecting high-school study program selection and also to create and compare models of high-school study program selection with data mining techniques and to develop a decision support system for high-school study program selection with a data mining technique. Data were analyzed by five data mining techniques, and models of high-school study program selection were constructed. These models were then used to construct a decision support system from data mining software called RapidMiner Studio 9. The research findings were as follows personal factors, learning factors family, and economic factors affecting high-school study program selection, and from the result of high-school study program selection, the Decision Tree method, C4.5 algorithm provided the highest accuracy. Therefore, the researcher selected the forecasting model with the Decision Tree method, C4.5 algorithm together with the selection of features with the backward elimination method to create a decision support system.

Keywords Data mining · Decision support system · Model for selection of study program · Accuracy

1 Introduction

Education is a learning process to develop and build skills necessary for life and as a way to develop human resources. The development of human resources is an important activity to increase the competitiveness and survival of the organization. Furthermore, the government of the nation has the policy to promote and support education for all citizens. Especially both primary and secondary education are fundamental, as it is necessary for further education a high level and able to apply knowledge to be employed in future careers for oneself and family. Management of the education

P. Chaiwuttisak (✉)
Department of Statistics, Faculty of Science, King Mongkut's Institute of Technology
Ladkrabang, Bangkok 10520, Thailand
e-mail: pornpimol.ch@kmitl.ac.th

419

S. Shakya et al. (eds.), *Proceedings of International Conference on Sustainable Expert
Systems*, Lecture Notes in Networks and Systems 176,
https://doi.org/10.1007/978-981-33-4355-9_32

system in Thailand has affected both internal and external factors. Namely, internal factors arise from the need to develop societies to be modern and prosperous. External factors are caused by global changes in societies, economy, and politics as well as communication. As a result, the education system in Thailand must be developed in order to be the same as the standard of other developed countries. It is an important factor that helps to advance the society, economy, and politics to be stable and progress. From the information of the Secretariat of the Education Council Ministry of Education, the plan aimed to produce and develop personnel with an important goal to develop the potential of students to be as effective as the aptitude and ability of the students as much as possible.

Choosing the study program at the upper secondary level is a basic course that affects the tertiary education at the undergraduate level in various fields and also a relationship with a future career. The upper secondary education has a duration of study for a total of three years. In principle, students choose the study programs according to their aptitude and interest. Moreover, Rampudcha [1] presented that there has a strong positive relationship between personal factors and the selection of the program in higher education of Mathayomsuksa 3 Students, while there has a negative correlation between the financial factors and the educational program selection. Perrone [2] reported that the parents have a considerable influence on the educational program selection of their children. Choosing an appropriate program study for students is important. The Education Plan of the Ministry of Education, 12th edition (2017–2021) sets out educational management guidelines for learners at each level of education to develop their individual potential, to promote the quality of education of the country, to promote the production and development of manpower, including research and innovation. Machine learning and data mining techniques extract useful knowledge to support students in selecting the most suitable study program in the upper-secondary stage.

Therefore, the researcher realizes the importance and has the idea to study by using data mining techniques to classify and select the high-school education plan of students in a case study of Nawamintharachuthit School. Bangkok. The result of the research will be a guideline to help support in deciding the study program that is suitable for the aptitude of each student. The factors that will be considered: personal factors, learning factors, and family factors. Data is analyzed by using five data mining techniques, namely Decision Tree, Neural Network Method, Support Vector Machine, Naïve Bayes, and Multinomial Logistic Regression, and choose the best model to create a decision support system.

2 Data Mining Techniques

Tan, Steinbach and Kumar [3] stated that data mining is the process of dealing with large amounts of data in order to find the patterns and relationships hidden in that dataset. Currently, data mining has been applied to various applications. In the business, it helps in the decision of the executive. In science and medicine, as well as economically and socially, data mining is an evolution in data storage and interpretation. From the original with simple data storage to storage in the form of a database, it can retrieve information for use in data mining that can find the knowledge that is hidden in the data.

Data mining techniques follow standardized process procedures in the industry (Cross-Industry Standard Process for Data Mining: CRISP-DM) by Shearer [4]. According to the steps of CRISP-DM, it consists of six steps as follows: business understanding, data understanding, data preparation, modeling phase, evaluation phase, and deployment phase.

Business understanding phase is the first step in the CRISP-DM process. The goal of this step is to understand business problems or opportunities and then convert the problem to be in a suitable form for data analytics.

Data understanding phase is the process of collecting relevant information to be used in the analysis using data mining techniques when collecting data, it should be considered that it is obtained from a reliable source, the amount of data have been enough, and there are enough details and suitable information to be used in the analysis.

Data preparation phase is the longest process. Because the model obtained from data mining will give the correct results or not, it depends on the quality of data used. The data preparation process can be divided into three sub-steps:

A. Data selections executed by analyzing the data and then selecting the data that are relevant to the data analyzed earlier. Data selection is based two other process such as data cleaning and data transform. Data cleaning is a process of data preparation by separating valuable data lost and error recording due to problems during data storage

B. Data transform is the data management in the image that is suitable for use.

Modeling phase is the process of choosing the method of data analysis and choosing a suitable model.

Evaluation phase is the phase that measures the effectiveness of the results before continuing to use the results. Whether it meets the objectives set out in the first step or how much credibility which may go back to the previous step to make changes to get the desired results.

Deployment phase in the CRISP-DM work process does not stop just the results from data analysis using data mining techniques. Although the results show knowledge that is useful, it must apply the knowledge gained from these works to the real problem in accordance with Fig. 1.

2.1 Decision Tree

Decision Tree is a data mining technique based on the tree structure for data classification to support various decisions. It usually consists of rules in the form of "if <the condition> then <the result>" which is similar to the nature of the tree invert

Fig. 1 CRISP-DM
procedure (Chapman et al.
[5])

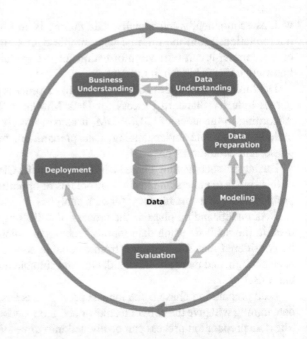

structure. The node in the first level of the tree is called "Root." Each node describes the attribute, and a branch shows the value of the attribute. The leaf node shows the class as shown in Fig. 2.

C4.5 algorithm can be applied to both continuous and discrete data. It can customize the tree for making a decision, known as pruning trees. The information gain and the entropy are calculated as Eqs. (1) and (2).

$$\text{Entropy}(S) = \sum_{i=1}^{C} -p_i \log_2 p_i \qquad (1)$$

Fig. 2 A structure of
decision tree

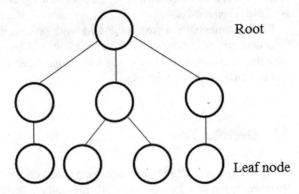

where

S is the attribute that is used to measure entropy;

P_i is the ratio of the number of members of the group i to the total number of members of the sample group.

$$\text{Gain}(S, A) = \text{Entropy}(S) - \sum_{v \in \text{Value}(A)} \frac{|S_v|}{|S|} \text{Entropy}(Sv) \tag{2}$$

where

A is an attribute;

$|S_v|$ is the number of members of attribute A with the value v;

$|S|$ is the number of members of the sample group.

2.2 Bayesian Learning

Bayesian learning applies the principle of probability by considering the probability distribution in data classification.

Simple Bayes learning is based on Bayes' rules, but it reduces complexity by adding the assumption that the properties of the data will not depend on each other. It can be said that the probability of data classified in the group C_i for data that have n attributes $(A_1, A_2, \ldots, A_n)$ and can be represented by symbols as follows: $P(C_i|A_1, A_2, \ldots, A_n)$.

From Bayes's theorem:

$$P(C_i|A_1, A_2, \ldots A_n) = \frac{P(A_1, A_2, \ldots, A_n|C_i) \times P(C_i)}{P(A_1, A_2, \ldots, A_n)}.$$

Each attribute does not depend on each other. It can be written in Eq. (3):

$$\frac{\prod_{j=1}^{n} P(A_j|C_i) \times P(C_i)}{P(A_1, A_2, \ldots, A_n)} \tag{3}$$

2.3 Logistics Regression

Logistic Regression is a machine learning classification algorithm that is used to predict the probability or the likelihood of a categorical dependent variable. In logistic regression, the dependent variable is a binary variable that contains data coded as

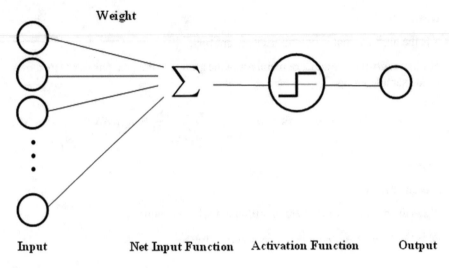

Input   Net Input Function   Activation Function   Output

Fig. 3 A structure of neural network

1 (yes, success, etc.) or 0 (no, failure, etc.). In other words, the logistic regression model predicts $P(Y = 1)$ as a function of X that are independent variables.

2.4 Neural Network

Neural Network is a simulation based on the concept of the work in human brain cells. The smallest unit of the neural network is called perceptron by accepting input and calculating these values by giving the weight of each input. The output will be calculated by errors to adjust the input weight [6]. It can be shown in Fig. 3. Firstly, each input (x_i) is multiplied by a weight (w_i). Next, all the weighted inputs are added together with a bias (b). Finally, the sum is passed through an activation function shown in Eq. (4). The activation function is the mathematical equation to turn an unbounded input into an output (y).

$$y = f(x_1 * w_1 + x_2 * w_2 + \cdots + x_n * w_n + b) \tag{4}$$

2.5 Support Vector Machine

Support Vector Machine is a supervised learning model that finds a hyperplane in an N-dimensional space to classify the data points. The dimension of the hyperplane depends on the number of features. The algorithm is processed to maximize the margin between the data points and the hyperplane. The loss function that helps maximize the margin is hinge loss of the prediction y shown in Eq. (5).

$$l(y) = \max(0, 1 - t \cdot y) \tag{5}$$

where $t = \pm 1$.

Y is the raw output of the classifier's function.

The related research started with Kanjanasamranwong et al. [7]. They studied the needs of further education in higher education of students in government schools in Songkhla province. The research studied demographic and social factors such as gender, GPA, study plans, religion, marital status, parents' occupation using the chi-square test to determine the relationships of variables. The study indicated that factors affecting the selection of further education in higher education institutions are gender, GPA, study plan, and total income, while factors affecting the selection of faculty to further study are gender, GPA, religious education plan, parents' education level, occupation of parents, and the total income of parents.

Laohawaranan, Suthiwandumoom, and Thanasophon [8] studied the classification and selection of subjects for students of the faculty of information technology. Comparative studies have been conducted using five data mining techniques, including Decision Tree, Naïve Bayesian, Neural Network, Support Vector Machine, and Logistic Regression. The data used in the research are data on gender studies and a computer aptitude test to assess competency. From the research, it is found that the most accurate technique is Naïve Bayes.

Ma et al. [9] proposed the scoring based on association (SBA) to selecting the potential students for remedial classes. It showed that it achieved much better results by using a data mining technique. Sereerat [10] studied the factors that have affected students deciding to continue their vocational education in the southern provinces of the Gulf of Thailand. The variables that are related to the decision to study further are the supporting factors/expectations of the parent, economic status of parents, quality of the school, salary or compensation, and school environment. The variable that has a large effect on the selection of further education is the supporting factors/expectations of the parents.

Sintanakul and Sanrach [11] developed the decision support system for choosing the high-school learning plan using the integration of grade 9 (O-NET) scores and multiple intelligences. Sungsri [12] improved the accuracy of the recommended courses by using the multiple classifiers, and answers from these models were calculated by majority voting. The results showed that the accuracy of the multiple classifiers is better than the accuracy of the single classifiers.

Tran and To [13] presented a model to support students' course selection using educational data mining. The proposed model experiments with educational data from the Faculty of Civil Engineering, Ho Chi Minh City University of Transport, Vietnam during the period of 2013–2016.

WadKien, Chotisukan, and Juneam [14] studied factors related to decision making in general education or vocational education of students in Mattayomsuksa 3 (or Grade 9) level at Rittiya Wannalai School. The result showed that personality factors, educational values, family support, and influence from friends are important factors for making a decision. Yassein et al. [15] identified factors affecting academic performance. Data mining was used to identify which of the known factors can give an indicator of student's performance. The result showed that the strong relationship is between practical work and success rate of courses and between student attendance and student success rate.

3 Experiment Design

The sample of 816 high-school students from Nawamintrachuthit School, Bangkok, in the academic year 2018 is given in Table 1.

From literature reviews, the variables used in the analysis are given in Table 2.

In the modeling phase, there are seven classes as you can see in Table 1. Moreover, the learning success in choosing the interested study plan is divided on a Grade Point Average (GPA) in high school into three levels as given in Table 3.

Creating predictive models using RapidMiner version 9.0 contains various processes. The screen shows the processes as shown in Figs. 4 and 5. Firstly, data collected and preprocessed in an excel file is retrieved by using the Read Excel operator. After that, select the Attributes operator that is taken to select the variables to be analyzed and manipulate data to be used to create the model by using Replace Missing Value, Set Role, Discretize, and Filter Examples operator. In the sub-step of cross-validation of both the prediction model for learning outcome and the appropriateness of selecting the study plan in which the student is interested is done. Optimize selection is the feature selection operator that selects the most relevant attributes of the given dataset. Namely, forward selection and backward elimination. Cross-validation operator in Figs. 6 and 7 is a nested operator consisting of two sub-processes: a training sub-process and a testing sub-process. Data mining algorithms are concealed in the operator. Support Vector Machine, Neural Network, and Multiple Logistic Regression cannot handle nominal attributes. Thus, nominal to numeric operator is used to convert nominal values into numeric values.

Table 1 Number of students according to the study program

No.	Study program	N	%
1	Science–Mathematic	203	24.88
2	English–Mathematic	157	19.24
3	English–Japanese	103	12.62
4	English–Chinese	142	17.40
5	English–French	37	4.53
6	English–Korea	31	3.80
7	Computer Business	143	17.52
Total		816	100.00

Table 2 Variables used in the research

No.	Variable name	Description
1	GPA_SH	Overall Grade Point Average in upper-secondary level
2	THAI	Grade Point Average of Thai language subject in the lower-secondary level
3	MATH	Grade Point Average of mathematics subject in lower-secondary level
4	SCIENCE	Grade Point Average of science subject in lower-secondary level
5	SOCIAL	Grade Point Average of social studies, religion and culture subject in lower-secondary level
6	PE	Grade Point Average of health subject in lower-secondary level
7	ART	Grade Point Average of art education subject in lower-secondary level
8	TECHNO	Grade Point Average in lower secondary level for academic, job, and career subject
9	ENG	Grade Point Average of foreign language subject in lower-secondary level
10	GPA_JH	Overall Grade Point Average in lower-secondary level
11	RELIGION_CA	Religion
12	FAT_OCC_CA	Father's occupation
13	MOM_OCC_CA	Mother's occupation
14	REVENUE_CA	Family income
15	STU_HEALTH_CA	Congenital disease
16	DISTANCE	Distance from home to school
17	PROGRAM	Study program that are interested in studying

Table 3 Code of Grade Point Average used to verify that students are suitable for their study in the selected study plan

Level of learning outcome in choosing a study plan(Grade Point Average)	Code
Grade Point Average is greater than or equal to 3.00	Good
Grade Point Average is between 2.00 and 2.99	Moderate
Grade Point Average is less than 0.2	Bad

4 Results

From Table 4, the prediction model using the Decision Tree method, C4.5 algorithm with the backward elimination feature selection, has the best efficiency of 97.81%. As same as the English–Japanese study program, the prediction model using Decision Tree method based on the C4.5 algorithm and the backward elimination feature selection provides the best accuracy of 88.11%, while the prediction model for the English–Chinese study program using Neural Network with the forward selection

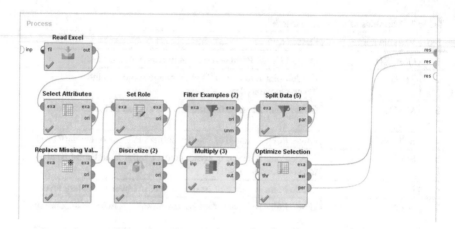

Fig. 4 Process of the model for selecting the high-school study program interested in studying using the RapidMiner software

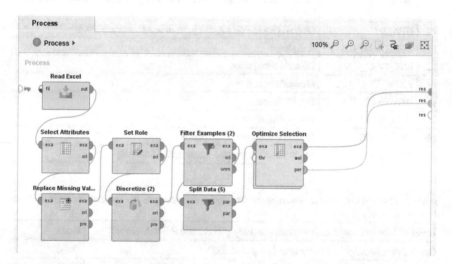

Fig. 5 Process of the model for forecasting the learning outcome in the study program interested in studying using the RapidMiner software

feature provides the best accuracy of 81.29%. The predictive model for the English–French study program using Neural Network combined with the characteristics of forward selection and backward elimination are equally accurate at 93.81%. For the English–Korean study plan, the best accuracy of the prediction model is 86.00%, and the prediction model for the business computer study program using the decision tree method and the backward elimination feature selection gives the best accuracy of 78.95%.

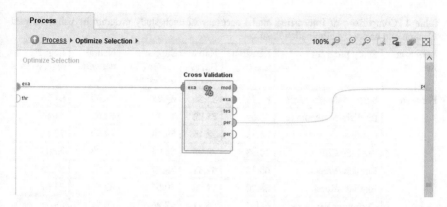

Fig. 6 Sub-steps of creating predictive model for selecting study using Decision Tree and Naive Bayes in RapidMiner software

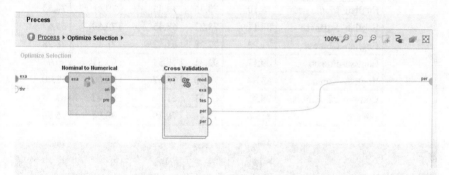

Fig. 7 Sub-steps of creating predictive model for selecting study using Support Vector Machine, Neural Network, and Multiple Logistic Regression in RapidMiner software

The C4.5 algorithm with the selection of features using the backward elimination method is employed to create a decision support system. The system for selecting high-school study programs shown in the Thai language is presented in Fig. 8.

When the student has filled out all the information and clicked the "Predict Results" button, the system will display the results page for prediction of the high-school study program that is suitable for the user and the learning outcome of the study program that the student is interested in, as shown in Fig. 9.

To evaluate the accuracy of the model by applying to the testing dataset consisting of all 82 people, the results can conclude that the efficiency in choosing the high-school study plan will be able to correctly predict 59 out of 82 students, accounting for 71.95%, while the accuracy of prediction for the suitability of the study program selected is equivalent to 73.17%, that is, 61 students can be predicted correctly from 82 students.

Table 4 Comparison of forecasting model accuracy of each study program by using forward selection and backward elimination features

Selection method	Study program	Decision tree (%)	Naïve Bayes (%)	Support vector (%) machine	Neural network (%)	Logistic regression (%)
Forward	Science–Mathematic	84.76	83.56	67.78	78.69	84.76
	English–Mathematic	76.60	75.22	73.77	78.69	74.46
	English–Japanese	74.33	72.16	59.18	79.59	72.16
	English–Chinese	70.38	78.95	53.94	81.29	79.05
	English–French	68.33	92.31	78.57	93.81	90.95
	English–Korea	68.00	72.00	70.67	82.67	74.67
	Computer Business	66.65	66.71	47.26	69.75	63.51
	Average	72.72	77.27	64.45	80.64	77.08
Backward	Science–Mathematic	97.81	82.19	73.62	85.24	84.16
	English–Mathematic	78.10	73.45	73.45	79.41	74.53
	English–Japanese	88.11	73.22	63.45	79.59	65.67
	English–Chinese	71.92	77.38	61.82	78.12	78.28
	English–French	89.17	92.31	78.57	93.81	90.95
	English–Korea	63.33	75.33	78.57	86.00	72.00
	Computer Business	78.95	68.31	61.82	64.42	63.60
	Average	81.06	77.46	70.19	80.94	75.60

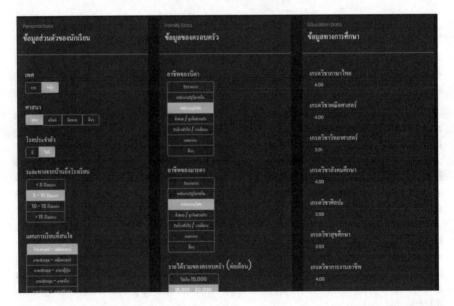

Fig. 8 Components of the window to input data for supporting the decision making in choosing the high-school study program

Fig. 9 Results of the prediction of the appropriate study program and the suitability of the study program selection

5 Conclusion

In Thailand, selecting the study program in high school is important for entering the program of higher education which plays a crucial role in the transition of students to work. The purpose of this research is to study the relationship of the personal, learning process, family, and economic factors affecting the selection of study programs and to build models for the development of a decision support system for further study in the upper secondary education program. There are five data mining techniques for the classification of each study programs namely the Decision Tree method, Naive Bayes, Support Vector Machines, Neural Network, and Multiple Logistic Regression. The selection features are considered in the study: the forward selection and backward elimination methods. The models are built for the selection of the appropriate study program and the learning outcome of the study program that is interested in studying.

The results show that the C4.5 algorithm give the best accuracy on 11 variables: Grade Point Average at the lower secondary level in Thai language, Grade Point average at the lower secondary level in mathematics, Grade Point Average at the lower secondary level in science subject, Grade Point Average at lower secondary school level in social studies, religion, and culture, Grade Point Average at the lower secondary level in art education, Grade Point Average at the lower secondary level academic in job and career, Grade Point Average at the lower secondary level in foreign language subjects, Grade Point Average at the lower secondary school level, gender, occupation, and family income.

From the analysis, it can be seen that no predictive modeling techniques work best with all data groups. However, the result of the C4.5 algorithm shows good accuracy overall. The author chooses the prediction model using the C4.5 algorithm with the selection of the backward elimination method to create a decision support system in the Thai language.

References

1. Rampudcha, P.: Factors affecting the selection of higher education of Mathayomsuksa 3 students in educational service area 1. NakhonRatchasima Province, Master Thesis of Education Program, Ramkhamhaeng University (2004)
2. Perrone, P.: Values and occupational preference of junior high school girls. Personnel Guidance J. **44**, 254–257 (1965)
3. Tan, P.-N., Steinbach, M., Kumar, V.: Introduction to Data Mining. Pear Addison Wesley, Boston
4. Shearer, C.: The CRISP-DM model: The new blueprint for data mining. J. Data Warehousing **5**(4), 13–22 (2000)
5. Chapman, P., Clinton, J., Kerber, R., Khabaza, T., Reinartz, T., Shearer, C., Wirth, R.: CRISPDM 1.0 step-by-step data mining guide. Technical report, CRISP-DM (2000)
6. Craven, M.W., Shavlik, J.W.: Using neural networks for data mining. Future Gener. Comput. Syst. **13**(2–3), 211–229 (1997)
7. Kanjanasamranwong, P., Urawong, K., Kongrod, T., Salaeh, A.: The needs of attending in graduate institute of Grade 12 students In Public High Schools, Songkhla Province. J. Educ. Res. **11**(1), 75–90 (2017)
8. Laohaharanan, J., Limsutwanphum, R., Thanasophon, B.: The use of data mining techniques in classifying and selecting fields for students of the faculty of information technology. Lat Krabang Information Technol. J. **4**(2), 45–53 (2015)
9. Ma, Y., Liu, B., Wong, C., Yu, P., Lee, S.: Targeting the right students using data mining. In: The Sixth ACM SIGKDD International Conference on Knowledge Discovery and Data Mining, pp. 457–464 (2000)
10. Sereerat, P.: The Discriminating Factor for Decision Making of Students to Study in Vocational Colleges in Southern Region (Gulf of Thailand) Provincial Cluster. Hat Yai University, Educational Administration Program (2017)
11. Sintanakul, T., Sanrach, C.: A model of decision support system for choosing high school learning plan using students' O-NET score and multiple intelligence. Int. J. Inf. Educ. Technol. **6**(7), 555–559 (2016)
12. Sungsri, T.: The use of multiple-classifier data mining technique for study guideline. Int. Journal of Inf. Educ. Technol. **8**(1), 22–25 (2018)
13. Tran, T.Y., To, B.L.: Educational data mining for supporting students' courses selection. Int. J. Comput. Sci. Network Security **19**(7), 106–110 (2019)
14. WadKien, P., Chotisukan, S., Juneam, P.: Factor that related to the decision making of Ritthiyawannalai. J. Graduate Stud. Valaya Alongkron Rajabhat **7**(1), 89–98 (2013)
15. Yassein, N., Helali, R., Mohomad, S.: Predicting student academic performance in KSA using data mining techniques. J. Inf. Technol. Software Eng. **7**(5), 1–5 (2017)

A Guided Visual Feature Generation and Classification Model for Fabric Artwork Defect Detection

W. M. M. Botejue and L. Ranathunga

Abstract Defect detection of fabric artworks on garments is a manual, error-prone, subjective, and a time-consuming process. A novel guided visual feature generation and classification-based approach is proposed in this study which can automatically detect the defects in the printed artworks embossed on garments. The system consists of four major processes, namely background removal, artwork segmentation, quality analysis, and defect report generation. The proposed is to analyze the process which focuses on detecting the shape, size, rotation, placement, and color defects in a printed artwork. This system was introduced to increase the productivity of the existing manual defect detection process in the apparel industry. The defect recognition of the quality analysis module proved 90.95% accuracy along with 0.7917 of recall and 0.9282 of precision.

Keywords Defect detection · Hu moments · Image moments · Curvature · Color descriptors

1 Introduction

In the apparel industry, the quality of a garment is analyzed at various levels. Prior to the production, the raw fabric undergoes a quality checking process. Subsequently, each part of the garment is checked for damages before the final stitching process. Furthermore, printed artwork or embroideries on the garments are also checked to capture the existing defects. Finally, the sewed apparels undergo a separate quality checking process. However, all the above-mentioned quality analysis processes are conducted manually even in well-established garment factories that work on mass

W. M. M. Botejue (✉) · L. Ranathunga
Faculty of Information Technology, University of Moratuwa, Moratuwa, Sri Lanka
e-mail: mbotejue60@gmail.com

L. Ranathunga
e-mail: lochandaka@uom.lk

© The Editor(s) (if applicable) and The Author(s), under exclusive license
to Springer Nature Singapore Pte Ltd. 2021
S. Shakya et al. (eds.), *Proceedings of International Conference on Sustainable Expert Systems*, Lecture Notes in Networks and Systems 176,
https://doi.org/10.1007/978-981-33-4355-9_33

production. Therefore, an extensive amount of labor is required for a single production process. Moreover, due to different vision conditions and eye exhaustion, the quality decisions of humans become subjective. A defect for a person may seem as a non-defect to another person.

Different artworks such as cartoon characters, abstract arts, logos, and mottos can be printed on fabrics. Preceding the sewing process the colors of the artwork, the artwork formation on the garment is finalized and informed to the production team. Tolerance measures of the print deviations are also discussed in this stage as it is not possible to have exactly similar printed artworks for each and every garment. In the quality checking (defect identification) process, each printed artwork is analyzed against the accepted artwork to detect the deviations. Maintaining the correct shape, size, and colors of a print is very crucial when printing characters and logos of the famous companies with a standardized brand. Moreover, it is important to detect any print defects before the final stitching process in order to minimize the production wastage. Through this study, automated printed artworks defect detection system is introduced to overcome the mentioned drawbacks while reducing the required manpower for the manual process. The system is designed by analyzing the flow of the manual print defect detection routine. Since the different production orders have different artworks, it is not suitable to use machine learning as it will consume large learning samples and time whenever a new artwork is introduced.

This paper comprehensively discusses the quality analysis process. In the quality analysis module, artwork features such as size, shape, rotation, placement, color, and boundary pattern are extracted from the images that captured the artworks. Approaches of extracting image features were gathered from researches conducted on content-based image retrieval (CBIR). CBIR is a solution that uses image features such as colors, shapes, and textures in order to support fast and accurate retrieval of content from digital repositories [1]. Identified approaches were examined and modified to create suitable descriptors for the features examined by the quality analysis module. The experiments conducted validate the quality analysis module and its individual feature matching stages to be accurate.

Figure 1 depicts a correct and defected cat print. In (b), one facial hair of the cat is damaged. This can be categorized as size, shape, or a color defect. The proposed system will identify such defects and graphically present it to the quality checking personnel. Thereafter, the decision is made whether to accept or reject the defected print.

The rest of the paper is structured as follows; Sect. 2 briefly explains the motivation for the proposed system. Section 3 explains several feature extraction methods in CBIR solutions. Section 4 describes the methodology used in the quality analysis module. The performance of the module is referred to in Sect. 5. Sections 6 and 7 express the discussion and conclusion of the paper, respectively.

(a) (b)

Fig. 1 **a** Correct print and **b** defected print

2 Motivation

The defect detection process of prints in printed garments is a manual process at the industrial level. The print analysis is based on human vision and thus leads to subjective decisions. On average, printed garment orders placed on a considerably large factory exceed 15,000 pieces. When the quantity analyzed increases, human quality controllers' eyes get familiar with the print. Hence, the accuracy of detecting errors decreases with the increase in a number of pieces. Additionally, due to various vision conditions of each quality inspector, the color variations may not be detected accurately, which may cause erroneous decisions. Moreover, due to various negligence mistakes, the color combination needed for the print may not acquire by the manual paint mixing process. These factors were proven during the field visits to the apparel factories and printing firms.

Defects in the prints of branded garment orders may cause severe legal issues if the customers complain about the defects. It is important for high-end brands to maintain consistency and standard in every print of their garments. Additionally, when printing trademarked characters and logos, clients will look thoroughly for the correct color usage and character representation. Improper representations of the trademarks can even lead to legal measures.

With the use of image processing, an image can be further processed to incur valuable insights. Prints can be also captured as images under a good and stable capturing environment to examine and gain insights. In this manner, unbiased, less erroneous, and accurate decisions about consistency and standards of the prints can be made by the quality controllers.

3 Review of Literature

Images can be retrieved by searching for the shape features of their content. Fourier descriptor is a common shape descriptor used for shape-based content retrieval. It is well known for its robustness, ability to capture perceptual features [2]. Nevertheless, it only considers the magnitude of the shape. Hence, it can provide the same descriptor value for the contours that are different in shape. A shape descriptor that is invariant to scaling, occlusion, translation, and rotation is proposed in [3]. This descriptor outsmarts the others since it answers more challenges such as intra-class variation and nonlinear deformation. However, the computational effort needed for the method proposed in [3] is high in the context of this research as the quality analysis module consists of many feature extraction processes. Hu moments are geometric moments that are invariant to scale, rotation, and transformation [4]. Therefore, it is extensively used as descriptors in pattern recognition. Moreover, Hu moments require relatively low computational effort compared to other shape descriptors. Hu in [5] has introduced seven Hu moments. The latter moments are more sensitive to the shape changes as they use higher degrees in their equations.

Boundary extrema-based descriptors are useful in describing features for object detection. Curvature scale space is such a descriptor that iteratively detects local extrema at different scales of shape [6]. Gaussian smoothing is used to obtain different scales of the boundary. At each scale, new extrema points are discovered. These steps are continued until there are no zero-crossing points. However, sometimes the curvature scale space algorithm detects false local maxima due to the jagged nature of the boundary lines. An enhanced solution to overcome this issue is suggested in [7]. It suggests checking the curvature of a point against a threshold. If the curvature is less than the threshold, then that particular point is not chosen as a maximum. The solution suggests a special threshold named adaptive local threshold. This threshold is defined based on the neighboring points' curvature values.

Color can be successfully used in statistic generation without depending on the geographical differences of an image [8]. Humans initiate object identification by grouping similar colors together. Following this notion, Soo-Chang Pei and Ching-Min Cheng proposed a color feature extraction method based on the DSQ approach [9]. In the DSQ approach, a color palette is generated for an image by quantizing the colors of the true image to a limited color range. The approach consists of two major components, namely the recursive binary MP thresholding and the bit allocation. A study in [10] proposes an equation to allocate bits to each color component. To obtain the required thresholding level, recursive binary MP thresholding is used instead of the multilevel thresholding stated in [11]. A method for image retrieval based on color histograms and color moments is proposed in [12]. Histogram-based content retrieval approaches contain several disadvantages as the histograms are sensitive to noise and their inability to express the spatial information of the image components in the image space. The solution proposed through [12] mitigates such drawbacks. In this method, region-wise calculations for average, skewness, and variance are done for the components hue, saturation, and value separately. A location-based color vector is proposed in [13] which use a combination of color correlogram (CC) and dominant color to construct the final descriptor. The method shows a high accuracy than the histogram intersection descriptor, color correlogram descriptor, and dominant color descriptor. A study proposed in [14] uses color features of the salient regions of an image and their spatial relationship to facilitate content-based image retrieval. By using the global color contrasting method [15], the object areas, i.e., the salient regions are identified. Then, the original image is cropped using the grab cut method [16], in order to isolate the identified salient regions. The rough object shapes can also be analyzed using these maps. The use of this color descriptor is proven to be better than the classical histogram-based method.

4 Methodology

The flow of the system is performed through four major stages, namely background removal, artwork segmentation, quality analysis, and defect report generation.

There are two types of raw image inputs to the system such as reference image and test images. The reference image garment is with the correct print artwork. The test images of the garments are with the printed artworks under defect detection.

Background removal module isolates the printed artwork from the fabric background and feeds it to the segmentation module. The segmentation module uses a color-based segmenting method to segment the artwork. The quality analysis module extracts the segment-wise features from the reference image and generates a reference model. And then it extracts features from the segments of the test images and compares them with the corresponding features in the reference model in order to capture the defects. Finally, the identified defects are graphically presented to the quality supervisors through a defect report. Segment-wise defect detection of the mentioned features enables the system to achieve more detailed detection results.

The software in the proposed system is designed using Open CV libraries and C++ programming language. The user interface is created using QT, an open-source toolkit, to design interfaces. The system can process both uniform and textured fabric types. However, in the current implementation, a sample image of the fabric should be given to the system if the fabric is textured. Samples of textured and uniform fabric types are given in Fig. 2.

4.1 Quality Analysis Module

Reference model generation and inspection are the two main stages of the quality analysis module. Figures 3, 4, and 5 depict the flow of this module. This module is capable of detecting the shape, size, rotation, placement, boundary extrema, and color defects in extracted segments of the test images. Therefore, this module performs a

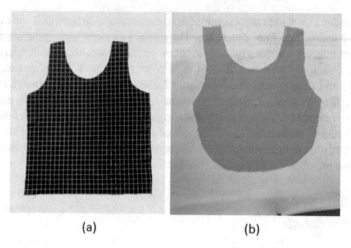

(a)   (b)

Fig. 2 **a** Textured fabric and **b** uniform fabric

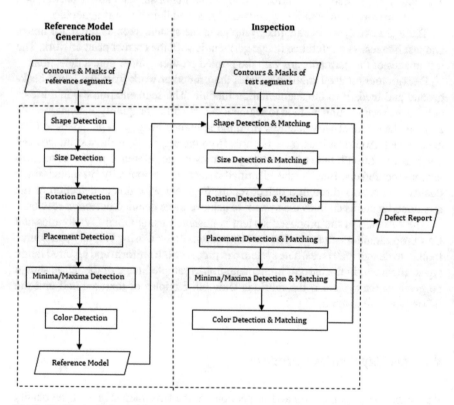

Fig. 3 Quality analysis module

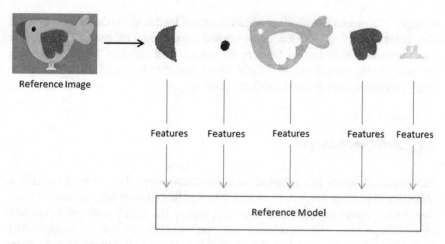

Fig. 4 Reference model generation

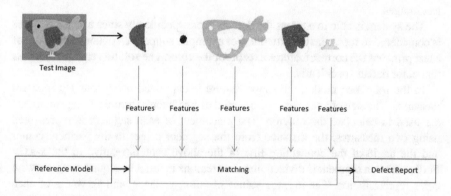

Fig. 5 Inspection stage

feature extraction to obtain the above six features from each segment of the given print.

Figure 4 represents the reference model generation process of the quality analysis module. The bird print is segmented by the artwork segmentation module. Segment-wise features are then extracted in order to generate the reference model.

The high-level depiction of the inspection stage is represented in Fig. 5. In this stage, the features of the print segments in the test image are extracted and compared with the reference model to detect the defects. The defect in the wing segment of the bird will be detected in this stage.

"Features" term in Figs. 4 and 5 refers to the six main feature set explained earlier. The size of the reference model depends on the number of segments that are extracted from the segmentation module.

To reduce the computational effort, the flow of the inspection stage is optimized. A conditional matching mechanism is used for the feature matching approach. For example, if a certain test segment includes a considerable shape defect, it affects the size, color, placement, rotation, and edge and color feature measures. Therefore, in such scenarios, the rest of the features are not matched and only the shape defect is presented to the user. A single notable defect in a print is a valid reason to fail the quality analysis in the manual checking process.

4.2 Reference Model

The reference model is a template that characterizes the defect-free print. It is a collection of per segment feature vectors extracted from the correctly printed artwork. One feature vector includes measures representing six main features of a segment, namely shape, size, rotation, orientation, color, boundary minima, and maxima. The feature vector measures are recorded as double-type numerical values in order to be used in comparisons performed against the corresponding segment features of the test images.

The system is able to express the defects more specifically since a set of features is considered to represent the artwork. For example, suppose an obtained segment of a test print has the correct features except for the color. The solution can then address that color defect specifically.

In the reference model, the shape feature is expressed using four Hu moment measures. The size of a segment is expressed using Green's formula. Image moments are used to calculate the rotation. The placement of each segment is represented using two measures, the distance from the segment center to the garment center and the angle of the connecting line of the above centers relative to the x-axis. Furthermore, a segment is divided into four regions in order to acquire a region-wise color representation. The average values of color, variance, and skewness of each region are calculated for hue, saturation, and value channels separately. Therefore, the color of a segment is represented by 36 feature values as a region requires 9 values to represent the color. The above-mentioned features acquire 43 values in a segment feature vector. However, the total number of values for a feature vector varies with the number of zero-crossing points of the shape boundary. The system is able to express the defects more specifically since a set of features is considered to represent the artwork. For example, suppose an obtained test segment has the correct features except for the color. The solution can then address that color defect specifically.

4.3 Shape Defect Detection

The shape of an object is independent of its scale, location, or rotation. Therefore, a scale, translation, and rotation invariant shape descriptor is required to represent the shape measure. In literature, Hu moment shape descriptors are described to be simple and invariant to scale, translation, and rotation [4]. However, in some cases, a moment that is scale-invariant indirectly implies transform invariance. Selecting the necessary Hu moments which are suitable for the study minimizes the computational complexity as well as the redundancy. Therefore, to define a shape descriptor for the segments, a compressed version of Hu moments is used. Out of the seven moments, only the first three are used to obtain the shape descriptor. The latter Hu moment calculations use large orders in the equations than in the initial ones. Therefore, even a slight change in the shape will generate a considerable variation in the values compared to the first three moments [5]. However, in the context of this research, a minor shape deviation might be acceptable. The seventh Hu moment checks the mirroring of a shape. If a test print segment is a mirror reflection of its reference segment, it is considered to be a defect. Hence, the seventh moment is also measured and recorded as the module can directly reject a test segment if it is flipped.

Equation (1) is used to transform the moments into a format suitable for numerical comparisons. H_i is the Hu moment value, and $S(H_i)$ is the sign of the obtained value.

$$H_i = -1.S(H_i).\log_{10}|H_i| \tag{1}$$

In the inspection phase, the seventh Hu moment is first calculated for a selected test segment and compared with the corresponding value in the reference model. If the test segment is not flipped, the other three Hu moments are calculated. The deviation of a particular test segment is determined by (2). If a calculated deviation is larger than Ts, the module decides that the shape of the test segment is highly damaged and can be detected effortlessly by the naked human eye. In this study, the experimentally decided threshold (Ts) for shape matching is 0.15. In (2), H_n = Hu moments of the reference segment, H'_n = Hu moments of the test segments, and d = deviation.

$$d = \sum_{n=1}^{3}(H_n - H'_n)^2 / \sum_{n=1}^{3} H_n \tag{2}$$

4.4 Size Defect Detection

If a particular test segment has a substantial shape variation, it also implies that it has a size defect also. Therefore, if a shape defect is captured in the shape matching stage that segment also has a size defect. Even though two segments match in shape they might have different size measures. In such occurrences, the Hu measures return the same or approximately similar values for shapes. Therefore, the objective of size defect detection is to detect the considerable size variations in the test segments which have the desired correct shape.

For each segment, the number of nonzero pixels is calculated using the corresponding binary mask in order to obtain a size measure. The test segment size measure is compared with the corresponding measure in the reference model using (3). If the deviation is less than or equal to the specified threshold (Ta), the size of the test segment is decided to be correct or almost correct. The experimentally proved threshold (Ta) which is used in this research is 0.15. In (3), S = reference segment size, S' = test segment size, and d = size deviation.

$$d = |S - S'|/S \tag{3}$$

4.5 Rotation Defect Detection

It is possible for the printed artwork segments to be rotated from a certain angle, deviating from its expected appearance while having the correct shape and size. Therefore, the segments that have passed as defect less through the size and shape matching phases are inspected in order to detect the rotation defects. Following image, moment calculations are used upon the segment contours to obtain a rotation measure.

$$m\_11 = 2.m[i].m_{11} - m[i].m_{00}.(\bar{x}^2 + \bar{y}^2) \tag{4}$$

$$m\_02 = m[i].m_{02} - m[i].m_{00}.\bar{x}.\bar{y} \tag{5}$$

$$m\_20 = m[i].m_{20} - m[i].m_{00}.\bar{x}.\bar{y} \tag{6}$$

m_11, m_02, and m_20 of Eq. (7) are obtained using (4), (5), and (6) equations, respectively. Let $(\bar{x}, \bar{y})$ be the center point of a chosen segment. $m[i]$ is the moment calculation for the segment i and is calculated using (9).

$$= \tan^{-1}(m\_11/m\_20 - m\_02)/2 \tag{7}$$

The rotation measure is obtained by using (7). Equation (8) is used to obtain a test segment rotation deviation. That deviation is compared against a specific threshold (Tr). The experimentally proven threshold for this phase is 0.05.

$$d = |r - r'|/1 + r + r' \tag{8}$$

r = rotation measure of the reference segment and r' = rotation measure of the corresponding test segment. If the rotation deviation of a test segment is less than or equal to the 0.05 threshold, the module decides the test segment preserves the correct rotation.

4.6 Placement Defect Detection

An artwork segment may have the expected shape, size, and rotation features but it might be printed in a different location. Placement defect detection is executed in such segments. The quality analysis module defines the placement measure of a segment using a combination of two values: the center of the garment and the contour enter of the segment.

The center of the garment block is calculated by using the garment's bounding box. This idea is depicted in Fig. 6. This center acts as a fixed position relative to which the distance to each segment contour center is measured. The width of the bounding box is obtained from the minimum and maximum x pixel coordinates difference of the garment area. Similarly, the height is obtained from the maximum and minimum y pixel coordinates difference.

Since each garment block is checked and verified before printing the artwork, the size and shape variances of the reference garment block and test garment blocks are negligible. The garment blocks are fixed to frames by the quality supervisors in order to maintain the correct formation. Moreover, images are captured from the same distance and with a fixed camera focus to obtain the same scale. Therefore, bounding boxes drawn for each garment boundary has the same scale.

However, the bounding box method is not used to calculate the segment centers. A test segment with a shape defect can give the same bounding box as its matching reference segment if the defect does not change the maximum and minimum x, y coordinates of the segment boundary. Figure 7 displays such instance.

Therefore, to avoid such scenarios, the segment center is obtained using (10) and (11). In (9), (i, j) is a contour point and m_{ij} is the moment of that point. $I(x, y)$ represents the pixel intensities of a grayscale image. $(\overline{X}, \overline{Y})$ is the center point of a

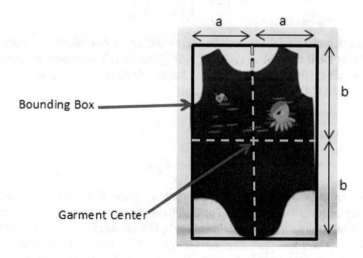

Fig. 6 Determining the garment center using the bounding box

Fig. 7 An instance where
the actual segment and the
defected segment results in
the same bounding box

segment.

$$m_{ij} = \sum_x \sum_y x^i y^j I(x, y) \tag{9}$$

$$\overline{X} = m_{10}/m_{00} \tag{10}$$

$$\overline{Y} = m_{01}/m_{00} \tag{11}$$

The distance between the two center points is then calculated using the Pythagoras theorem. However, based on a point, the same distance can be marked in many directions. Therefore, in addition to the distance, the angle between the *x-axis* and the line connecting the two centers is also calculated to measure the placement. In (12), $[X_c, Y_c]$ is the contour center and $[X_g, Y_g]$ is the garment center.

$$\theta = \tan^{-1}\left(\frac{|Y_c - Y_g|}{|X_c - X_g|}\right) \tag{12}$$

In the reference model, the positive angle measure is recorded. θ is modified to get a positive angle if it is not an acute angle in quadrant 1. To compare the placement measures, the angle deviation (da) and distance deviation (d_d) are first calculated using (13) and (14).

$$d_a = (p - p')^2/p \tag{13}$$

$$d_d = (a - a')^2/a \tag{14}$$

Then, (15) is used to generate the placement deviation. If this deviation is less than or equal to the specified threshold (Tp), the placement is correct or closely acceptable. The experimentally accepted threshold (Tp) in this module is 0.5. In (13), (14), (15), p = reference segment placement measure, p' = test segment placement measure, a = reference segment angle measure, a' = test segment angle measure, and d = placement deviation.

$$d = \sqrt{d_a - d_d} \tag{15}$$

4.7 Boundary Defect Detection

A particular test segment with a minor shape deviation includes defected boundary points. Therefore, to elaborate more on identified minor shape defects, the quality analysis module identifies the defected boundary points as well. This is achieved using the maxima and minima points of the segment boundary line.

When damage occurs in a particular region of a segment boundary, that damage either increases or decreases the segment area. In either way, the damage leads to new maxima or minima creation on the boundary.

The detection of the extrema points is a challenge due to the jagged nature of the boundaries. A method was required to avoid detecting the small spikes or valleys as maxima or minima. Therefore, before detecting these special points, the contour boundary is approximated to the nearest polygon. However, the approximation should not alter the original shape strongly as it reduces the number of extrema points.

In literature, zero-crossing points are numerically represented by their curvature measure. As depicted in (16), curvature K of the boundary point (x, y) is obtained from the $\dot{x}(t)$, $\dot{y}(t)$ first derivatives and $\ddot{x}(t)$, $\ddot{y}(t)$ second derivative values of a point t.

$$|K| = \frac{[\dot{x}(t)\ddot{y}(t)\dot{y}(t)\ddot{x}(t)]}{[\ddot{x}^2(t) + \ddot{y}^2(t)]^{3/2}} \tag{16}$$

However, if the neighboring points have defected it can affect the curvature value. Therefore, a modified version of the curvature calculation stated in (17) is used in this solution to minimize the impact of the defected zero-crossing points. The zero-crossing points of each test segment are identified and the curvature values of those points are determined. These curvature values are compared with the corresponding reference segment curvature measures to detect any deviations. At this step, the shape matching phases have already identified a segment to have the desired shape. Therefore, this phase supports detecting the subtle anomalies in the shape boundary which are hard to notice at the first glance.

$$|K| = \frac{\dot{x}(t) - \dot{y}(t)}{[\dot{x}(t) + \dot{y}(t)]^{\frac{3}{2}}} \tag{17}$$

The binary masks of the test and the reference segments are overlapped and subtracted to identify deviations at the test segment boundary area. The matched zero-crossing points identified in the inspection stage are used to overlap the masks with each other. The center point of the test segment does not necessarily match the

Defect Type	Original Image	Test Image	Output
Edge			

Fig. 8 Presenting the edge defect in the defect report

reference segment center if their shapes do not match exactly. Therefore, the segment center is not suitable to overlap the two segments.

The location of the garment panel in the captured image can differ from image to image. Therefore, a certain extrema point of a test image to match with a reference segment point but not to have the same location coordinates. Hence, the location offset of the above two matching points is measured using (18) and (19) before overlapping the masks.

$$\text{offset}X = x' - x \tag{18}$$

$$\text{offset}Y = y' - y \tag{19}$$

Here, (x, y) is a reference segment zero-crossing point and (x', y') is the matching zero-crossing point in the equivalent test contour. To present the edge defect in a meaningful manner, the deviated pixels are marked in the test image to present the output as illustrated in Fig. 8.

4.8 Color Defect Detection

Each segment is divided into four region of interests (ROIs) before gathering the color information. Thereafter, region-wise color details of those segments are measured and recorded in their corresponding feature vectors. This method allows the module to identify the area of wise color defects.

The binary masks along with its respected segment are passed to the quality analysis module as inputs. These masks are used to identify the ROIs mentioned earlier. As depicted in Fig. 9, the bounding box of a segment is detected using its mask. Since the mask is a binary image of the reference image, the identified pixel coordinates map to the same coordinates in the reference image. The color information is then extracted from those coordinates in the reference image. The reference image is converted from RGB to HSV color channel before obtaining the measures as the HSV channel is used to achieve a perception similar to the human vision.

Mask   Bounding box area   ROIs

Fig. 9 ROI determination

The color distribution information is calculated using three color moments: average, variance, and skewness. These three moments are calculated for each channel. Therefore, for each region, there exist nine values to represent the color. Since there are four regions per segment, a segment feature vector will represent color details using 36 values.

Figure 10 depicts an instance where the defect report presents a color defect in the wing segment of the test image. The identified color defect exists in the upper left region of the wing segment. Due to the region-wise color detection method, the system can highlight the defected area rather than the entire segment.

The color deviation of each region calculated using (20). In here, n = color channel, H = hue measure, A = average measure, S = skewness measure, and d = color deviation. d is compared to a specified threshold (Tc). If the deviation is less than or equal to Tc, the color deviation is negligible. The experimentally specified threshold for color matching in this study is 0.015. The following formula calculates the deviation of the color measures of each channel.

$$d = \sum_{n=0}^{2} \left[\left(H_n - H_n'\right)^2 + \left(A_n - A_n'\right)^2 + \left(S_n - S_n'\right)^2 \right] \tag{20}$$

Defect Type	Original Image	Test Image	Output
Color			

Fig. 10 Presenting the color defect in the defect report

5 Evaluation

Performance of the quality analysis module was conducted using a dataset of printed garment block images that contained 1200 defected segments. 200 defected segments per each defect type mentioned in Table 1 were included in the chosen dataset. As explained in previous sections, the quality analysis module detects shape, size, rotation, placement, and color defects. Hence, individual performance evaluation for each inspection phase was carried out. Moreover, the performance of the quality analysis module as a whole was evaluated as well. Accuracy, precision, and recall metrics are used for the performance evaluation. 0.15, 0.15, 0.05, and 0.015 threshold values are used in shape, size, rotation, placement, and color matching stages, respectively. The evaluation results are presented in Table 1.

The evaluation results suggest that the quality analysis module has performed well as a whole. However, the shape module underperformed when recognizing shape defects in complex segments that have a considerable number of peaks and valleys in the boundary. An example of such shapes is the complex tentacle design in an octopus print. Moreover, the color defect detection falsely identifies defects for segments with glitter prints. It is noticed that the quality analysis module's performance highly depends on the output of the segmentation module. If the reference image segments and the test image segments do not match properly, the false positive rate of the identified defects will increase.

The capturing environment contained controlled illumination conditions. The images were captured using a 12 MP camera that has a 1.8 aperture. The distance from the camera to the image kept unchanged during the image capturing process.

Defect report is the final output of the defect identification. Figure 11 depicts the method that the defects in a given defected print "Test Image" is visualized in the defect report. Under each category of a defect, the defected areas are highlighted on a copy of a test image. The birdwing segment has been detected with a shape defect. Therefore, for the wing segment, other defect types are not analyzed. However, due to the wing shape defect, a color defect is detected in the body segment. No defects were found in the beak, eye, and leg segments.

Table 1 Evaluation results of the quality analysis module

	Precision	Recall	Accuracy (%)
Quality analysis module	0.9282	0.7917	90.95
Shape	0.944	0.5967	73.52
Size	0.9193	0.8507	96.04
Placement	0.9426	0.8862	90.275
Rotation	0.9516	0.7866	82.35
Color	0.84	0.7078	90.37

Reference Image

Test Image

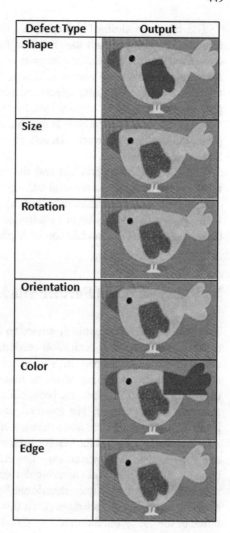

Defect Type	Output
Shape	
Size	
Rotation	
Orientation	
Color	
Edge	

Fig. 11 Defect report representation of a given test image

6 Discussion

The printed artwork defect detection system for garments is a novel approach that automates the manual process of artwork defect detection. The solution uses an image processing-based approach since a machine learning-based solution requires the system to be trained each time a new artwork is introduced to the production line. This approach consists of three main modules, and the quality analysis module is the final module that detects existing shape, size, rotation, placement, color, and edge defects in the print.

The current industrial procedure of detecting defects in printed artworks is a tedious manual process that requires extensive labor power. Moreover, due to different vision capabilities and the rates of eye exhaustion, defect detection becomes subjective to the quality inspector. Hence, automating this process reduces the erroneous decisions on detecting defects and minimizes the effort needed for the detection process. Usually, in the apparel industry, a printed garment is rejected even if one defect is detected. The module is capable of detecting six different types of defects presented in printed artworks. Hence, there is less chance for a defected print to be labeled as a correct one.

The illumination conditions and the capturing device features may affect the results of the background removal and segmentation modules. In such situations, the quality analysis module may output unexpected results. Therefore, the images used in the system were captured in a controlled environment. Improvements for the flow of the quality analysis module can be applied in the future.

7   Conclusion and Further Work

The quality analysis module discussed in the paper has yielded good performance metrics as stated in the evaluation section. The module as a whole and individual defect detection stages perform well under the chosen thresholds. This module will assist the quality checking teams to make an unbiased and accurate decision on the print quality. Therefore, the probability of passing defected prints to the sewing process can be reduced. The lowered rate of completing garments with defected prints will save more time and monetary costs of the factories.

It was noticeable that the system performance is sensitive to the capturing device and the illumination environment. Therefore, the system can be improved further by using a more powerful capturing device and a well-controlled environment. The results of the quality analysis module can be further improved by experimenting with different thresholds in each defect detection step. Additionally, the current implementation of the background removal module is unable to isolate the print for a textured fabric without using a fabric sample image. An automated background removal process for textured fabrics will increase the value of the proposed system. Moreover, the entire system can be improved to analyze prints with colorful sequences or glitter. In such prints, the color similarity in all the prints is not considered. The proposed automated system is an excellent initiative to influence the process optimizations in the apparel industry. Such optimizations will assist in reducing the costs of production while increasing the quality of the products.

References

1. Kotoulas, L., Andreadis, I.: Colour histogram content-based image retrieval and hardware implementation. IEE Proc. Circuits, Devices Syst. **150**(5), 387 (2003)
2. Zhang, D., Lu, G.: Content-based shape retrieval using different shape descriptors: A comparative study. Gippsla
3. Xu, H., Yang, J., Yuan, J.: Invariant multi-scale shape descriptor for object matching and recognition. In: 2016 IEEE International Conference on Image Processing (ICIP), pp. 644–648 (2016)
4. Huang, Z., Leng, J.: Analysis of Hu's moment invariants on image scaling and rotation, pp. V7-476–V7-480 (2010)
5. Ming-Kuei, H.: Visual pattern recognition by momentinvariants. Inform. Theory, IRE Trans. **8**, 179–187 (1962)
6. Pillai, C.S.: A survey of shape descriptors for digital image processing. IRACST – Int. J. Comput. Sci. Inf. Technol. Security (IJCSITS), **3**(1) (2013) ISSN: 2249-9555
7. He, X.C., Yung, N.H.C.: Curvature scale space corner detector with adaptive threshold and dynamic region of support. In: Proceedings of the 17th International Conference on Pattern Recognition. ICPR 2004, vol. 2, pp. 791–794 (2004)
8. Pei, S.-C., Cheng, C.-M.: Extracting color features and dynamic matching for image database retrieval. IEEE Trans. Circuits Syst. Video Technol. **9**(3), 501–512 (1999)
9. Pei, S.C., Cheng, C.M.: Dependent scalar quantization of color images. IEEE Trans. Circuits Syst. Video Technol. **5**, 124–139 (1995)
10. Duda, R.O., Hart, P.E., Stork, D.G.: Pattern Classification, 2nd ed. Wiley (2001)
11. Trans Foster, I., Kesselman, C.: The Grid: Blueprint for a New Computing Infrastructure. Morgan Kaufmann, San Francisco (1999)
12. MangijaoSingh, S., Hemachandran, K.: Image retrieval based on the combination of color histogram and color moment. Int. J. Comput. Appl. **58**(3), 27–34 (2012)
13. Fierro-Radilla, A.N., Nakano-Miyatake, M., Perez-Meana, H., Cedillo-Hernandez, M., Garcia-Ugalde, F.: An efficient color descriptor based on global and local color features for image retrieval, pp. 233–238 (2013)
14. An, J., Lee, S.H., Cho, N.I.: Content-based image retrieval using color features of salient regions. In: 2014 IEEE International Conference on Image Processing (ICIP), Paris, France, pp. 3042–3046 (2014)
15. Amanatiadi, V.G., Kaburlasos, A., Gasteratos, S.E., Papadakis: Evaluation of shape descriptors for shapebased image retrieval. IET Image Process **5**(5), 493–499 (2011)
16. Rother, C., Kolmogorov, V., Blake, A.: Grabcut: Interactive foreground extraction using iterated graphcuts. ACM Trans. Graphics **23**, 309–314 (2004)

A GPS-Based Methodology for Analyzing Bicyclist's Behavior in Different Road Environment

Lemonakis Panagiotis, Botzoris George, Galanis Athanasios, and Eliou Nikolaos

Abstract The bicyclist's behavior under different road and build environment in the road network of Volos, a mid-sized Greek city, is recorded and analyzed. The developed methodology was based on the use of GPS technology embedded in instrumented vehicles for the acquisition of behavioral and performance data parameters such as speed, position, and lateral/longitudinal accelerations. These parameters are crucial for two-wheeler road users compared to the four-wheeler. This study concludes that it is feasible to record bicyclists' speed and acceleration profiles with accuracy and speed, and moreover supports that there is a strong indication that a bicycle is a rather controllable and predictable transport mode. However, the generalization of the conclusions drawn in a wider proportion of road users demands the conduction of more experiments including a greater number of participants and road sections.

Keywords Intelligent embedded system · Instrumented bicycle · Bicyclist behavior · Road safety

L. Panagiotis (✉) · E. Nikolaos
Department of Civil Engineering, University of Thessaly, Pedion Areos, 38334 Volos, Greece
e-mail: plemonak@uth.gr

E. Nikolaos
e-mail: neliou@uth.gr

B. George
School of Civil Engineering, Democritus University of Thrace, Kimmeria Campus, 67100 Xanthi, Greece
e-mail: gbotzori@civil.duth.gr

G. Athanasios
Department of Civil Engineering, International Hellenic University, Terma Magnesias, 62124 Serres, Greece
e-mail: atgalanis@ihu.gr

© The Editor(s) (if applicable) and The Author(s), under exclusive license to Springer Nature Singapore Pte Ltd. 2021
S. Shakya et al. (eds.), *Proceedings of International Conference on Sustainable Expert Systems*, Lecture Notes in Networks and Systems 176,
https://doi.org/10.1007/978-981-33-4355-9_34

1 Introduction

Bicycle is a common transport mode for numerous commuters worldwide who travel daily in urban and suburban areas. Bicyclists are vulnerable road users resulted from the available data of registered accidents and the higher risk of their road and personal safety [1]. Nowadays, more citizens are willing to bike daily for utilitarian or non-utilitarian trips in short- and medium-distance trips in urban areas [2]. The achievement of a safer, more convenient, attractive, and accessible built environment for bicyclists will encourage commuters to ride their bikes instead of using their vehicles [3]. Bicyclists' built environment differs among cities and/or districts of an urban area. Built environment and bicycling psychological factors affect the acceptable bicycling distance of rural residents [4]. Various factors can influence bicycle usage [5]. Trip purpose, trip distance, and cycling infrastructure can influence cycle use, as well as specific participant characteristics and attitudes [6]. Furthermore, the sustainable city transport challenge can be addressed in the context of selecting electric bicycles [7].

Bicyclists examine several factors when they plan their routes in the urban road network. Bicyclists usually travel either across an existing bikeway or in the roadway in mixed traffic conditions. Commuters are willing to ride their bikes with safety and convenience to reach their destinations in the entire urban area. Municipality officials should improve the bike-ability level of their city if they want to promote bicycle use and sustainable transportation with relative benefits for their city [8, 9]. The presence of a bikeway network in the city has to be well designed and maintained to be functional; otherwise, bicyclists may not select to use it as planned [10].

Various manuals and guidelines worldwide classify and describe the features of bicycle facilities to be functional and attractive to bicyclists [11, 12]. Furthermore, naturalistic driving studies have examined the driving behavior of motorists [13].

The present study examines the bicyclists' riding behavior under normal traffic conditions in the road network in the city of Volos, a mid-sized port city, situated midway on the Greek mainland in the region of Thessaly. For the successful implementation of the study, a new methodology was developed based on the use of GPS technology embedded in instrumented vehicles for the acquisition of behavioral and performance data parameters such as speed, position, and lateral/longitudinal accelerations. All these parameters are more crucial for 2-wheeler road users compared to the 4-wheeler ones. This study proposes a methodology that is suitable to evaluate the bicyclists' behavior under various traffic conditions. A first attempt to develop this methodology was made with an earlier study in the same area with a limited number of recorded data [14].

2 Methodology and Data Collection

The appropriate research method sought among the most popular ones that are being used by the researchers to investigate behavioral aspects of road users. These methods are studies based on accident records, field operational tests carried out on simulators or test tracks, naturalistic studies with the use of instrumented vehicles, and questionnaire studies [15].

The third method met the requirements of the present study, and consequently a naturalistic riding study was planned and implemented in different road environments using an instrumented bicycle based on GPS technology this is displayed in Fig. 1.

2.1 Participants

The bicyclist who participated in the experiment had to be an experienced one, with a low center of mass and enhanced sports adult profile. That was due to the additional load that should stand (the equipment weights approximately 7 kg). The candidates were tested and their performances were compared to other typical riders in order to choose the one whose behavior was more similar to the typical rider as possible. For the same reason, the gender of the participant was chosen to be male while in order to mitigate the accident risk during the implementation of the experiments,

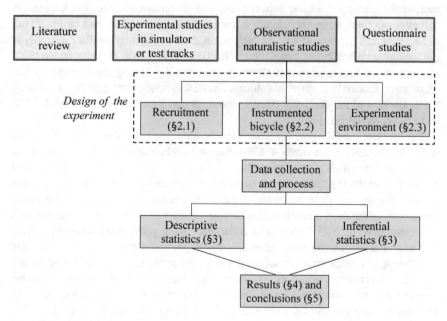

Fig. 1 Scheme of the proposed methodology

the candidate had to be no older than 50 years old, and in good physical condition, although it was unclear the correlation between age and accident risks.

Moreover, the candidates should have a blank accident record and they should consent to be equipped with the appropriate safety gear such as a reflective vest and helmet. Another important issue was the successful operation of the data logging equipment and hence, the candidates should be familiar with its handling. That was achieved with a training process using the manuals of the equipment. Taking into account the above-mentioned, a male bicycle rider was chosen without impaired riding ability and place of residence close to the experiment routes.

2.2   Bicycle and Equipment

The common practice of obtaining driving performance data is the use of a radar gun as well as speed detectors. However, many authors expressed the opinion that these recording equipment introduce either measuring errors or can affect road users' behavior who might perceive that equipment and consequently, adjust their speed and lateral position to the legal ones. In order to overcome these considerations, there is a new approach to record riding parameters, based on GPS technology.

Data loggers based on GPS technology are able to capture simultaneously video and vehicle data. Proper modifications to the instrumented 2-wheeler vehicle make their use ideal for motorsport experiments due to their small weight, size, and increased accuracy. Thus, it is more feasible to measure speed, track position, distance, lap time, lateral acceleration, longitudinal acceleration, and height. It has to be noted that the exploitation of such parameters can be very useful especially in circumstances where riding behavior is not approached exclusively in speed terms but also in terms of acceleration to the three dimensions. Apparently, such an approach is particularly crucial in 2-wheeled vehicles. In order to conduct the experiment utilized the equipment of the company Race Technology and specifically the Video 4, DL1, and Speedbox devices [16].

For this study, a bike type is selected that is appropriate to travel in the urban road network. The selection process was based on the fundamental idea of the naturalistic cycling studies which is to simulate the conditions of the experiment to be as close as possible to the typical transport standards for utilitarian trips. The first step was to select the type between a mountain or city bike, based on the most popular trends nowadays. Ultimately, a combination between those two options was achieved with a bike equipped with 28-inch wheels and a suspension, providing a smoother riding. The use of a suspension was necessary, due to the maintenance level of the road pavements. The price range of the bike was based on the average amount of money that a Greek commuter could afford. There is no official data for this index, therefore, to make a hypothesis based on the average bike types presented in the city. The aluminum frame constitutes a popular option for city bicycles, because of its low weight, which is crucial as the steel frame and the measuring equipment significantly increase the total weight. This could lead to an inaccurate cycling profile because of distorted measurements. The hydraulic brakes used in this specific bicycle provide sharp and accurate breaking, a common feature of mountain bikes.

The next step was the installation of the equipment on the bicycle. Specifically, incorporate a large rack into the original design of the bicycle and added metal bars with the welding technique. In this way, a permanent stable is created as base for the waterproof case that included the equipment and the necessary wiring. Outside of this case, the antenna section was placed at a considerable distance from the bicycle rider in order to receive the GPS data unaffected. For equilibrium reasons (there was a significant increase in weight at the rear with the probability of concentrating the center of mass closer to the drive wheel, thus indirectly affecting the driving profile), a smaller rack was placed on the front of the bicycle with the optional choice for positioning the battery of the equipment. Finally, continue to the final step of the equipment installation process that was the operational testing and calibration of the equipment on the field.

2.3 Study Area, Road Type, Time and Weather Conditions

The next step regarding the experimental design was the selection of the study area and specifically the study routes. They were chosen according to four criteria: location, road type, weather conditions, and the date and time of data collection. The study routes should have ensured the continued operation of the recording equipment. Additionally, for practical purposes, the study routes should have been close to the participant's residence. Taking into account the above-mentioned, the routes that could meet the standards of this research and the requirements of the experiment in the city of Volos are presented in Fig. 2.

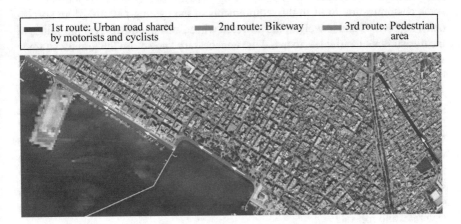

| ▬ | 1st route: Urban road shared by motorists and cyclists | ▬ 2nd route: Bikeway | ▬ 3rd route: Pedestrian area |

Fig. 2 Study area

In a more in-depth investigation, three routes were determined as most appropriate to conduct the measurements. The first one was a typical urban road simultaneously used by motorists and cyclists (first route, from 39° 21′ 22.2″ N 22° 57′ 59.4″ E to 39° 21′ 37.0″ N 22° 57′ 46.4″ E), the second one was a road segment with a designated bikeway separated from the road network and the pedestrian area (second route, from 39° 21′ 25.6″ N 22° 57′ 03.1″ E to 39° 21′ 39.8″ N 22° 56′ 31.9″ E), and finally, a pedestrian area where bicyclists are allowed to use (third route, from 39° 21′ 19.4″ N 22° 57′ 21.3″ E to 39° 21′ 25.4″ N 22° 57′ 03.2″ E). All coordinates are derived from Google Earth.

For safety reasons, the recording was stopped when the pavement was even slightly wet, while all measurements were conducted at the same time each day aiming to ensure similar traffic conditions. The level of natural light had to be invariant and hence, during cloudy or foggy days the measurements were ceased as well. Generally, the instructions that were initially set forbidden the conduction of the experiment in case any environmental or traffic conditions might divert the regular riding behavior. The bicycle rider conducted 36 measurements (12 repetitions for each route) for the needs of this study.

3   Data Process

The data process was based on two software; the Analysis V8.5.369 of Race Technology which accompanies the recording and data collection devices and Microsoft Excel. This selection was based on the rejection of the common way of analysis, which consists of the data conversion into comma-delimited files (extension *.csv). The original data files were recorded at the SD card (extension *.run) and opened using Analysis V8.5.369. The program analysis platform (layout) was modified after some customization to fit in the requirements of the cycling pattern. More specifically, adjust the vehicle weight, the powered wheel, and smoothed over the values of the measurement variables, to avoid the unnecessary "noise" on the graphs. In some specific cases where the variable smoothing was not enough, the sample rate was reduced from 100 to 10 Hz to achieve the desired effect.

In the presentation process, the tools are used to demonstrate the vehicle mobility patterns are common x, y graphs, single or multiple with the combination of different laps in the same area. Especially, the presentation of a specific variable was based on color changing bar imprinted on the cycling route of the experiment. The map where the routes were located was instantly downloaded from Google Earth into the Analysis program layout. Finally, the Analysis program calculates the summary statistics of the specific variables that are used to approach the urban cycling pattern, such as maximum/minimum and average values of speed, longitudinal acceleration, yaw rate, and roll rate.

At the end of each measurement, rows of data were generated in 100 Hz frequency, including the following physical quantities such as time (sec), speed (km/h), distance traveled (m), longitudinal acceleration (g), yaw rate (deg/sec), and roll rate (deg/sec).

The recording of these quantities initiated as soon as the rider inserted the SD card into the relevant slot of the data logger and ended after he pulled it out. However, between these two actions, the rider did not necessarily ride the bicycle. For instance, he might try to fit the body on the saddle right before he started to pedal or he might try to bring the vehicle to a halt and balance himself and the bike on a proper position in order to be able to remove the SD card. Therefore, many recordings at the start and end of each measurement had to be withdrawn since they do not represent net riding time. The criterion to do so was the speed of 2 km/h. All rows of data in which speed was less than this number at the start and end of the measurements were removed from the data processor.

4 Results

The descriptive analysis of the raw data of all measurements concluded to the aggregated values of Table 1. For each route individually, the table contains the maximum speed (U_{max}), the mean speed (U_{mean}), the minimum longitudinal acceleration, and the range of the recorded yaw and roll values. The longitudinal acceleration is a reliable quantity regarding the rapid braking of the rider in case of an incident or near accident while the range of the yaw and roll rate values indicates how convenient the riding was. Taking into account that all three routes consist of straight lines and smooth bends (Fig. 2), the more the range of the yaw and roll rate is, the more the riding is tensed and inconvenient.

It has to be mentioned that according to the statistics, the average speed of vehicles on local A' roads in England across 24 h in the year ending March 2017 is estimated to be 25.2 miles/h (or 40.6 km/h) [17], while the average pedestrians' crossing speed at signalized intersections in Izmir (Turkey) is found to be 1.31 m/s [18]. Also, the bicyclists' speed in roads with mixed traffic is approximately 17.0 km/h in Bologna, Italy [19]. Hence, by comparing the referred values with the corresponding ones in Table 1, it is inferred that the use of bicycles instead of a car in urban roads double

Table 1 Aggregated results of the three routes

Route	U_{max} (km/h)	U_{mean} (km/h)	Min longitudinal acceleration (g)	Yaw rate range (deg/sec)	Roll rate range (deg/sec)
Urban street	24.2	19.4	−0.19	38.75 to − 40.32	40.72 to − 44.11
Bikeway	25.5	16.4	−0.25	67.36 to − 103.26	84.27 to − 70.03
Pedestrian area	26.9	22.6	−0.32	88.22 to − 49.89	49.80 to − 67.81

the travel time. Nevertheless, this travel time compared to the pedestrians' one is approximately four times smaller.

On the other hand, the disturbance due to the presence of other users on the cyclist's way impels him to travel slower than in the pedestrian area, although as mentioned in research that dealt with the bicycle travel speed and disturbances on off-street and on-street facilities, the presence of pedestrians adjacent to the rider's trajectories is the primary factor that determines the average speed [19]. However, in the present research, the pedestrian area provides the opportunity to the rider to speed more than in any other route, probably because that the density of the pedestrian volumes was not very high. Although the rider managed to maintain the highest mean speed at the pedestrian area, he had to break harsh occasionally and that is why, according to Table 1, the maximum absolute value of the longitudinal acceleration was recorded at the pedestrian area.

Another element of the recordings is that the highest range of yaw and roll rates observed primarily on the bikeway and secondarily on the pedestrian area. Therefore, the movement of the rider along these traffic environments is rather tensed and inconvenient. On the contrary, the small range of yaw and roll rate values recorded in the urban street indicate the cautious and stable riding behavior of the participant. The sense of fear for an eventual incident makes him ride very carefully choosing to decrease the speed, steer smoothly, and avoid intense maneuvering and thus to roll the bike less intensively compared to the normal riding behavior which would result in rapid change of the center of mass. Moreover, the motorists forced him to adopt a stressed riding behavior. He is constantly checking backward before any maneuver and slow down very often due to the presence of vehicles ahead. Moreover, it is also noticeable the fact that the intense variation in the speed profile implies an abnormal riding attitude. Figures 3, 4, and 5 illustrate the speed profile of one indicative measurement for each route.

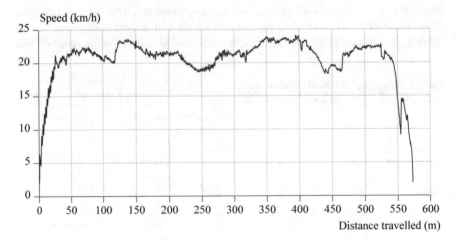

Fig. 3 Speed profile of an indicative measurement at the urban street (first route)

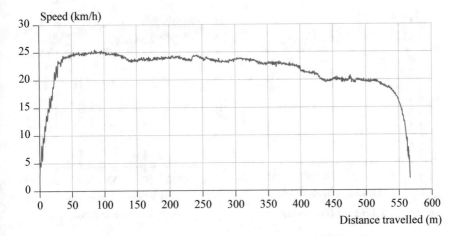

Fig. 4 Speed profile of an indicative measurement at the bikeway (second route)

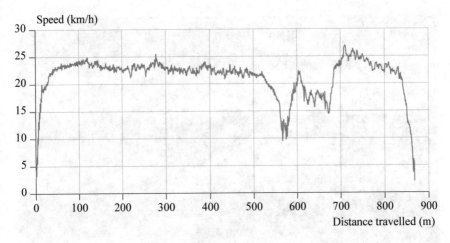

Fig. 5 Speed profile of an indicative measurement at the pedestrian area (third route)

Further, in-depth analysis of the measurements emerged that two incidents occurred while the rider was riding along the pedestrian area and the bikeway. These are depicted with rapid drops in speed profile as shown in Fig. 6. The fact that no incidents recorded on the urban street together with the fact that the speed profile was the smoothest one compared to the other two routes suggest that the rider did not have to cope with hazardous situations.

Figure 7 illustrates the trajectories and the average speed variation for the three measurements of each route with color gradation. The intense variation in the speed profile implies an abnormal riding attitude.

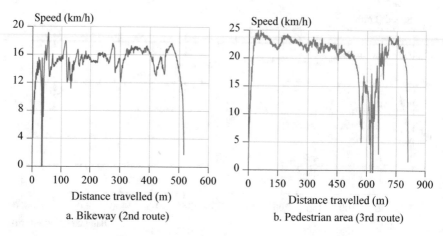

a. Bikeway (2nd route)   b. Pedestrian area (3rd route)

Fig. 6 Speed profiles of the first measurement at the bikeway (**a**) and the third measurement at the pedestrian area (**b**)

Fig. 7 Average speed profiles for the three routes (complete study area) [16]

5 Conclusions

This study aims to provide a methodology to investigate cyclist's behavior in urban road environments and to trigger the execution of additional experiments including a greater sample of riders, road sections, weather conditions, equipment, and vehicles to generalize the conclusions drawn. The core element of the proposed methodology is the exploitation of GPS technology because it ensures accurate recordings and unaffected riding behavior. Indeed, the use of pneumatic road tubes and radar guns to investigate road users' behavior poses a great risk of validity to the recorded data because drivers tend to change their attitude since they perceive the said equipment as means of traffic violations monitoring.

The conclusions that derived from the process of the behavioral cycling data of the participant who carried out the measurements are the average speed at the pedestrians/cyclists area was the highest one (22.6 km/h) compared to the other two routes, followed by the urban street (19.4 km/h) while the lowest speed recorded on the bikeway (16.4 km/h). All three routes present similar acceleration profiles. At the beginning of the route, the bicycle rider strives to reach a certain speed (increased acceleration) and then he merely tries to maintain this speed (decreased acceleration). At the pedestrians/cyclists area (third route), the bicycle rider performs consecutive maneuvers and hence the raw/roll recordings are the highest of all the three routes. Furthermore, these values are reduced at the cyclists/motorists area (first route). That is probably because the bicycle rider is overwhelmed of fear and insecure and hence he is very cautious in performing maneuvers.

Bicycle is a common transport mode used for daily trips in the urban road network. Bicyclists use the existing road infrastructure traveling across designated bikeways and/or mixed traffic conditions with motorists, resulting to uncertainties for their road safety and travel convenience. However, analyzing the behavioral parameters of a rider, bicyclist is rather predictable and consistent in terms of driving behavior. Indeed, the speed, acceleration, and yaw/roll rate data are particularly homogenous and hence bicycle could be classified as a transport mode with controllable and predictive mobility pattern. More research in this area is needed to record and analyze bicyclists' behavior under various traffic conditions and urban/suburban routes.

References

1. European Road Safety Observatory, Traffic Safety Basic Facts 2018—Cyclists. https://ec.eur opa.eu/transport/road_safety/sites/roadsafety/files/pdf/statistics/dacota/bfs20xx_cyclists.pdf
2. Galanis, A., Botzoris, G., Profillidis, V., Lantitsou, K., Eliou, N.: The Greek ECONOMIC CRISIS AS AN OPPORTUNITY FOR THE PROMOTION AND THE ENHANCING OF SUSTAINABLE URBAN TRANSPORT. In: 8th International Congress on Transportation Research ICTR 2017, Thessaloniki, Greece (2017)
3. Eliou, N., Galanis, A., Proios, A.: Evaluation of the Bikeability of a Greek city: Case study 'City of Volos.' WSEAS Trans. Environ. Dev. 5, 545–555 (2009)
4. Wang, Y., Ao, Y., Zhang, Y., Liu, Y., Zhao, L., Chen, Y.: Impact of the built environment and bicycling psychological factors on the acceptable bicycling distance of rural residents. Sustainability 11(16), 4404 (2019)
5. Lourontzi, E., Petachti, S.: Cycling: Utilitarian and symbolic dimensions. Transp. Res. Proc. 24, 137–145 (2017)
6. Konstantinidou, M., Spyropoulou, I.: Factors affecting the propensity to cycle—The case of Thessaloniki. Trans. Res. Proc. 24, 123–130 (2017)
7. Sałabun, W., Palczewski, K., Wątróbski, J.: Multicriteria approach to sustainable transport evaluation under incomplete knowledge: Electric bikes case study. Sustainability 11(12), 3314 (2019)
8. Profillidis, V., Botzoris, G., Galanis, A.: Traffic noise reduction and sustainable transportation: A case survey in the cities of Athens and Thessaloniki, Greece. In: Nathanail, E., Karakikes, I.

(eds.) Data Analytics: Paving the Way to Sustainable Urban Mobility. AISC, vol. 879, pp. 402–409. Springer, Cham (2019)

9. Bakogiannis, E., Siti, M., Christodoulopoulou, G., Karolemeas, C., Kyriakidis, C.: Cycling as a key component of the Athenian sustainable urban mobility plan. In: Nathanail, E., Karakikes, I. (eds.) Data Analytics: Paving the Way to Sustainable Urban Mobility. AISC, vol. 879, pp. 330–337. Springer, Cham (2019)

10. Tilahun, N., Levinson, D.M., Krizek, J.K.: Trails, lanes or traffic: The value of different bicycle facilities using an adaptive stated preference survey. Transp. Res. Part a: Policy Pract. **41**, 287–301 (2007)

11. AASHTO Guide for the Development of Bicycle Facilities. American Association of State Highway and Transportation Officials, Washington DC (2012)

12. NACTO Urban Bikeway Design Guide, National Association of City Transportation Officials, Washington, DC (2011)

13. Laporte, S.: Literature Review of Naturalistic Driving Studies, 2-BE-SAFE, Deliverable 4; Work package 2; Activity 2.1 (2010)

14. Zavitsanos, K., Galanis, A., Lemonakis, P., Botzoris, G., Eliou, N.: Investigation of bicyclists' riding behaviour under normal traffic conditions in the road network of a mid-sized Greek City. In: 5th Conference Economics of Natural Resources & the Environment, pp. 211–219, Volos, Greece (2018)

15. Eliou, N., Misokefalou, E.: Driver's distraction and inattention profile in typical urban high speed arterials. In: Stanton, N.A. (ed.) Advances in Human Aspects of Road and Rail Transportation, pp. 252–260. CRC Press, London (2012)

16. Race Technology, Hardware. https://www.race-technology.com/wiki/index.php/Hardware/Index

17. Department for Transport, Travel time measures for local 'A' roads, England: April 2016 to March 2017. https://www.gov.uk/government/statistics/travel-time-measures-for-local-a-roads-england-april-2016-to-march-2017

18. Onelcin, P., Alver, Y.: The crossing speed and safety margin of pedestrians at signalized intersections. Transp. Res. Proc. **22**, 3–12 (2017)

19. Bernardi, S., Rupi, F.: An analysis of bicycle travel speed and disturbances on off-street and on-street facilities. Transp. Res. Proc. **5**, 82–94 (2015)

Hybrid International Data Encryption Algorithm for Digital Image Encryption

Anupkumar M. Bongale, Kishore Bhamidipati, Arunkumar M. Bongale, and Satish Kumar

Abstract Image security is one of the prominent research areas in the field of information security. In this paper, a new image encryption method called hybrid international data encryption algorithm (HIDEA) is proposed. In the proposed methodology, image encryption takes place in different phases. In the first phase, the input image is transformed into a scrambled image by using Arnold transformation. Secondly, the scrambled image is encrypted using 128-bit key based on international data encryption algorithm (IDEA) encryption algorithm. Same process is applied in the reverse order to obtain the original image from the cipher image. The encryption algorithm has been thoroughly validated based on key sensitivity analysis and additive noise attack analysis and found that the encryption methodology is stable and robust.

Keywords Image encryption · Image security · Histogram · Arnold transformation

1 Introduction

Cryptography is an art of securing the information by converting the original form into unreadable or unrecognizable form and vice versa. The original information is called plain text. The unreadable message is called ciphertext. The process of cryptography involves key generation, encryption, and decryption. Cryptography is broadly classified into two categories, namely symmetric and asymmetric cryptog-

A. M. Bongale (✉) · A. M. Bongale · S. Kumar
Symbiosis Institute of Technology, Symbiosis International (Deemed University),
Lavale, Pune, Maharashtra, India
e-mail: anupkumar.bongale@sitpune.edu.in

A. M. Bongale
e-mail: ambongale@gmail.com

K. Bhamidipati
School of Electrical Engineering and Computer Science, Oregon State University, Corvallis, USA
e-mail: bhamidik@oregonstate.edu

© The Editor(s) (if applicable) and The Author(s), under exclusive license   465
to Springer Nature Singapore Pte Ltd. 2021
S. Shakya et al. (eds.), *Proceedings of International Conference on Sustainable Expert
Systems*, Lecture Notes in Networks and Systems 176,
https://doi.org/10.1007/978-981-33-4355-9_35

raphy. In asymmetric cryptographic systems, two keys will be used. Out of the two keys, one key is accessed by all parties and is called as public key. Other key is called as private key and is shared with authentic users only. Usually, secret key (or private key) is used for decrypting the ciphertext. On the other hand, symmetric cryptosystems use only a single key for encryption and decryption [3, 4, 9, 16, 18].

Multimedia security is the prime concern in information security paradigm. Over the communication network, the multimedia information is transmitted and shared among the connected users. Even though the communication channel is secured, enforcing a secure transmitted data within itself is still remaining as an important aspect. Image security is a broad research area, and image encryption/decryption is considered as one of the subfields of image security. Several image encryption methods are available, but most of them are based on single encryption technique. Review article [11] describes such methods of image encryption, and there are other existing secure cryptic algorithms available to enhance the image, video, and audio security [13, 17]. This paper is targeted toward the provision of security for digital images through a novel hybrid cryptographic algorithm.

The organization of this article is as follows. Brief literature review of the related articles on image encryption and decryption is described in Sect. 2. The proposed image encryption and decryption method, namely HIDEA, is presented in Sect. 3. Section 4 discusses the detailed result analysis with justification of the proposed encryption and decryption technique. Finally, the article is concluded in Sect. 5.

2 Literature Review

Over the last two decades, many encryption algorithms have been developed for image security. Researchers have proposed secure symmetric and asymmetric image encryption techniques. In this section, a brief overview of recently presented approaches is specified.

Yong and Xinghuo in [6] have proposed invertible two-dimensional map, called line map, for image encryption and decryption. The main idea of line map is "to map an array of pixels and again mapping it to same-sized image. The algorithm uses two phases called as left line map and right line map. The advantage lies in performing permutation and substitution simultaneously. The authors have identified other advantages such as lossless image encryption and decryption, and key can be determined arbitrarily without any limitations."

In [14], a new cryptic algorithm is taken up by mixing the Duffing chaotic algorithm with the Lorenz chaotic algorithm to generate new six-dimensional chaotic cryptographic algorithm. The Duffing and Lorenz chaotic algorithms have simple structure and are easy to decipher. In the technical paper [19], a new image encryption algorithm based on the permutation–diffusion architecture is presented using "chaotic cat map and logistic map." The method works in two steps. First one is about permutation of chaotic pixel position, and second step is diffusion of chaotic pixel values. Here, the pixels are exchanged based on the number generated by logistic map. Later phase is to process the image by cat map. Kamlesh and Sanjay [5] have proposed "elliptic curve cryptography (ECC) with knapsack for image encryption." The benefits are low encryption time, less power consuming, and more reliable. Jain and Khunteta have proposed encryption and decryption method for color images in [7] and evaluated the proposed method using mean square error. Another color image encryption method is proposed by Jithin et al. in [8] that shows the application of combined DNA operations and chaotic image encryption. Image security is not only limited to encryption and decryption, but image steganography is also an important field of study. Vinothkanna in [20] has proposed secure stenography method for different file formats. Security can also be imposed to at different stages of video encoding as well [10].

In this paper, a new image encryption algorithm is proposed for digital images. The algorithm works in two phases. First is input image is transformed using Arnold transformation. Later transformed image is encrypted using IDEA encryption algorithm. The methodology is explained in the next section.

3 Methodology

3.1 Arnold Transforms

Arnold transform for cat image was proposed by Vladimir Arnold by using Chaos theory [2]. The basic Arnold transformation is defined by using 2 matrix for the 2D images of size $N \times N$ pixels as per the equation (1):

$$\begin{bmatrix} X_1 \\ Y_2 \end{bmatrix} = \begin{bmatrix} 1 & 1 \\ 1 & 2 \end{bmatrix} \times \begin{bmatrix} X_n \\ Y_n \end{bmatrix} \quad (\text{mod } N) \tag{1}$$

where (X_n, Y_n) are the pixel arrangements before the transformation and (X_1, Y_1) are the pixel organizations after the Arnold transformation. The transformation process continues several rounds iteratively for all (X_n, Y_n) pixels.

3.2 International Data Encryption Algorithm (IDEA)

IDEA is a private key encryption algorithm designed by Lai and Massey [12]. IDEA uses modified Feistel (MF) structure with eight rounds. The IDEA is a block cipher and symmetric key algorithm. The key size used in IDEA is 128-bit long. The basic operation of IDEA uses three operators:

- Bit-by-bit XOR operation
- Addition modulo of 216
- Multiplication *modulo* of $(2n + 1)$

IDEA uses S-box of structure called multiplication–addition (MA) structure. This takes two 16-bit inputs and produces two 16-bit output strings by using two keys in each round.

Key generation: The IDEA algorithm uses 52 keys to perform encryption of plain text in eight rounds. In IDEA, primary key is divided into eight 16-bit words like $K = (K1, K2, K3, K4, K5, K6, K7, K8)$. Then, this primary key K is routed 25 bits left and again divided into eight 16-bit words. This process of generating keys will continue until getting 52 keys for eight rounds of encryption operation in IDEA. Same keys are used in reverse order to perform the decryption operation. Figure 1 shows general structure of IDEA algorithm with 8 rounds.

Fig. 1 General structure of IDEA algorithm with 8 rounds

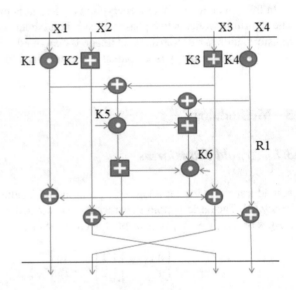
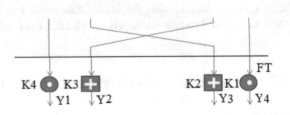

3.3 Anti-Arnold Transforms

The basic anti-Arnold transformation is the inverse of Arnold transformation, which is defined by using 2 matrices for the 2D images of size $N \times N$ pixels as in Eq. (2).

$$\begin{bmatrix} X_1 \\ Y_2 \end{bmatrix} = \begin{bmatrix} 1 & 1 \\ 1 & 2 \end{bmatrix} \times \begin{bmatrix} X_n \\ Y_n \end{bmatrix} \quad (\text{mod } N) \times \frac{2}{N} \tag{2}$$

3.4 Proposed Algorithm

The proposed image processing steps are mentioned in Fig. 2. Each step and its operations are described below. Each step shows the intermediate image processing results. The novelty of the proposed method lies in the combination of Arnold transformation and application of IDEA algorithm on the transformed image.

- *Step 01:* Original 2D image is taken as input for Arnold transform to get the output in the form of scrambled image as shown in Fig. 2a.
- *Step 02:* The scrambled image is used for encryption with the help of IDEA encryption algorithm of using 128-bit key and eight rounds as shown in Fig. 2b.
- *Step 03:* The cipher image is decrypted in receiver end using the reverse process of IDEA algorithm as shown in Fig. 2c.
- *Step 04:* The output of IDEA decryption algorithm is a scrambled image. This is given as input for the anti-Arnold transform process. The output of this process is the plain 2D image as shown in Fig. 2d.

4 Result Analysis

To implement the proposed cryptosystem, MATLAB software is used. The computer system with 64-bit Windows 10 operating system and 8GB RAM configuration is used. The strength of the proposed algorithm is explored and validated through series of experiments based on working of encryption/decryption, histogram analysis, key sensitivity analysis, and additive noise attack. For validating the proposed work, standard dataset maintained in [1] is used. The general and most of popular images of Lena, mandril, cameraman, lake, etc., are utilized from the standard database. The database has huge set of images, and all images are of dimension 512×512 size. But the dimensions of these images are reduced to 256×256 resolution grayscale mode for our experiment purpose. Detailed result analysis of the presented cryptosystem is described in next subsections.

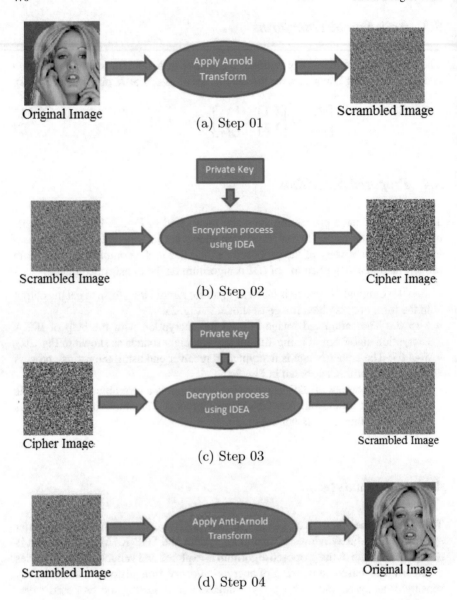

Fig. 2 Proposed method

4.1 Encryption/Decryption Validation

The first experiment is to check the correctness of the encryption quality. Figures 3, 4, and 5 show result of application of hybrid IDEA algorithm on standard image dataset to obtain encrypted and decrypted images. Here, along with the encrypted image, intermediately image obtained by Arnold transformation is shown in Figs. 3b, 4b, and 5b. Visually, encrypted image appears to hide all the information of the original image. Decryption leads to successful retrieval of the original image.

(a) Original Image   (b) arnold transformation   (c) Cipher Image   (d) Deciphered Image

Fig. 3 Different stages of hybrid IDEA algorithm use Mandril.jpg

(a) Original Image   (b) arnold transformation   (c) Cipher Image   (d) Deciphered Image

Fig. 4 Different stages of hybrid IDEA algorithm use WomanBlonde.jpg

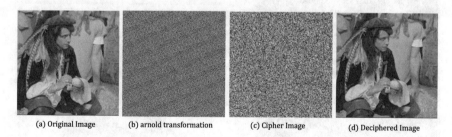

(a) Original Image   (b) arnold transformation   (c) Cipher Image   (d) Deciphered Image

Fig. 5 Different stages of hybrid IDEA algorithm use Pirate.jpg

4.2 Histogram Analysis

As per Sect. 4.1, the image encryption and decryption are successful. But in many cases, the encrypted image can reveal information of original image if the encryption strategy used is not very strong. The reason for this is that pixels of an image are statistically related to its neighboring pixels. The adversary may attack the encrypted image to get information of original image, provided encryption algorithm is weakly designed. So, the encrypted image should be having low correlation and low statistical similarity among the pixels. One way to understand the quality of the encryption is to analyze the cipher image based on histogram plots.

Figures 6, 7, and 8 show the histogram plot of "mandril" image. The histograms are shown for plain image, transformed image, and encrypted image. The histogram

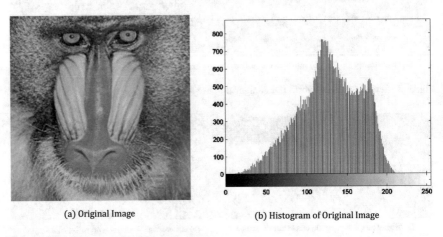

(a) Original Image   (b) Histogram of Original Image

Fig. 6 Mandril image and its histogram

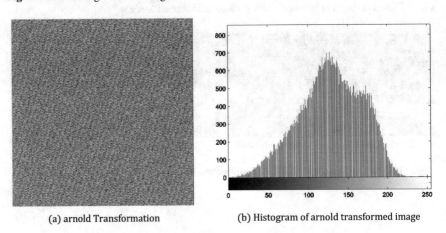

(a) arnold Transformation   (b) Histogram of arnold transformed image

Fig. 7 Arnold transformed mandril image and its histogram

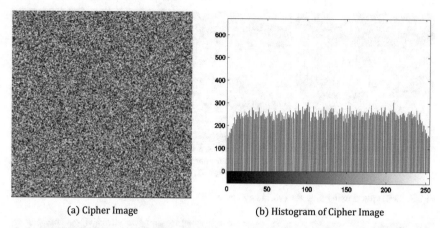

(a) Cipher Image   (b) Histogram of Cipher Image

Fig. 8 Mandril cipher image and its histogram

plot of the original plain image shown in Fig. 6b and its original image is shown in Fig. 6a. It can be observed from the histogram that statistical relation among the pixels in original image has resulted in a pattern which indicates that the pixels are statistically related to each other.

Figure 7b represents the histogram plot of Arnold transformed image shown in Fig. 7a. Even though the transformed image is visually non-readable, its histogram plot is almost similar to histogram plot of original image shown in Fig. 6b. This shows that intermediate transformed image alone is not enough to provide security.

Finally, Fig. 8b signifies the histogram plot of the encrypted image shown in Fig. 8a. Here, encryption is performed on Arnold transformed image. From the histogram, it is clear that there is no presence of visually revealing features that can be exploited by an attacker. Thus, our proposed algorithm ensures that final encrypted image will not have any correlation among the neighboring pixels, which is obviously a necessary quality of any image encryption strategy.

4.3 Key Sensitivity Analysis

Key sensitivity analysis is about checking the encryption quality of the image with respect to slight modification in the key. Adversary may try to tamper around with the encrypted image by brute force approach by trying to decrypt the image with random keys.

It can happen that randomly generated keys may be very close to actual used key. If the key is slightly different from the original key, encryption algorithm should not get compromised. In the presented encryption technique, 128-bit key is used. To test the strength of the proposed encryption method, a valid key $K1 = 5f2e6656772b366e4f795c7c74$ and invalid key $K2 = 5f2e6656772b366e4f795$

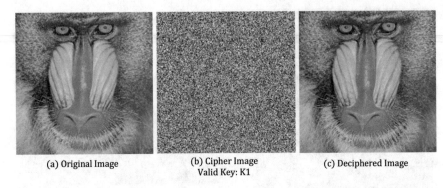

(a) Original Image   (b) Cipher Image   (c) Deciphered Image
                              Valid Key: K1

Fig. 9 Decryption using Key K1 (valid key case)

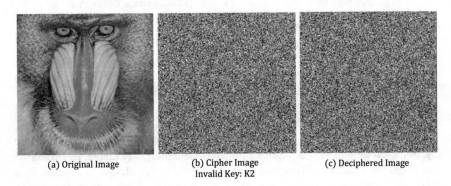

(a) Original Image   (b) Cipher Image   (c) Deciphered Image
                              Invalid Key: K2

Fig. 10 Decryption using Key K2 (invalid key case)

$c7cAA$ are considered. $K1$ and $K2$ are slightly different only by last two digits. As shown in Fig. 9, the decryption is successful for $K1$. Figure 10 shows decryption using key that for $K2$ leads to an unreadable image. This shows that only with the valid key decryption is successful.

4.4 Additive Noise Attack Analysis

Cryptanalyst can tamper the encrypted images by inducing extraneous noise. In [15], noise attacks have been discussed. To check the robustness of the proposed work, three popular noise attacks like induced salt and pepper noise, speckle noise, and Gaussian noise are considered. The density of salt and pepper noise is set to 0.1, and variance is set 0.05 for Gaussian noise attacks while evaluating the strength of the proposed approach. The decryption result as shown in Figs. 11, 12, and 13 shows that the algorithm can handle external noise attack.

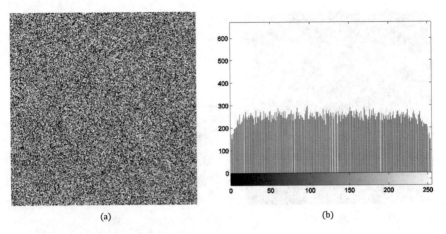

(a)   (b)

Fig. 11 Decryption under salt and pepper noise attack

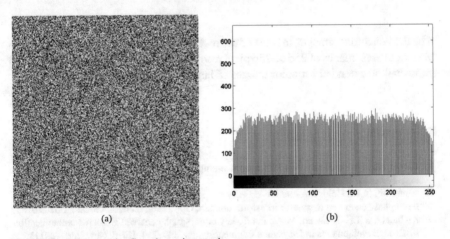

(a)   (b)

Fig. 12 Decryption under Gaussian noise attack

5 Conclusion

In this research article, a new image cryptographic algorithm is proposed. In the proposed approach, the encryption and decryption of images include mainly four steps. For encrypting the image, hybrid IDEA algorithm is proposed and also an intermediate stage called Arnold transform is introduced. The benefit of the presented approach is that original image is Arnold transformed and then encrypted using 128-bit key. Due to this process, the encrypted image is capable of counteracting the well-know adversary attacks such as histogram analysis, key sensitivity analysis, and additive noise attack analysis. The encryption algorithm has performed well for the considered sample test images of Lena, mandrill, pirate, etc., which are considered

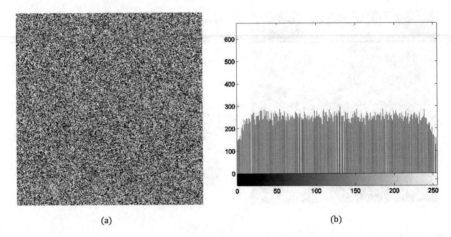

(a) (b)

Fig. 13 Decryption under speckle noise attack

to be the benchmark images in most of the research work. Current work presented in this article uses images of 256 × 256 pixel resolution size. In the future, the presented work shall be extended for color images of larger dimensions.

References

1. Standard test images. http://www.imageprocessingplace.com/root_files_V3/image_databases. htm. Accessed 19 Aug 2020
2. Abuturab, M.R.: Color information security system using Arnold transform and double structured phase encoding in gyrator transform domain. Opt. Laser Technol. **45**, 525–532 (2013)
3. Al-Janabi, S.T.F., Rasheed, M.A.: Public-key cryptography enabled Kerberos authentication. In: 2011 Developments in E-systems Engineering, pp. 209–214. IEEE, New York (2011)
4. Biswas, G.: Diffie-Hellman technique: Extended to multiple two-party keys and one multi-party key. IET Inf. Secur. **2**(1), 12–18 (2008)
5. Gupta, K., Silakari, S.: Performance analysis for image encryption using ECC. In: 2010 International Conference on Computational Intelligence and Communication Networks, pp. 79–82. IEEE, New York (2010)
6. Huang, F., Lei, F.: A novel symmetric image encryption approach based on a new invertible two-dimensional map. In: 2008 International Conference on Intelligent Information Hiding and Multimedia Signal Processing, pp. 1340–1343. IEEE, New York (2008)
7. Jain, S., Khunteta, A.: Color image encryption by component based partial random phase encoding. In: 2018 International Conference on Inventive Research in Computing Applications (ICIRCA), pp. 144–148. IEEE, New York (2018)
8. Jithin, K., Sankar, S.: Colour image encryption algorithm combining Arnold map, DNA sequence operation, and a Mandelbrot set. J. Inf. Secur. Appl. **50**, 102428 (2020)
9. Kim, Y., Sim, M., Moon, I., Javidi, B.: Secure random phase key exchange schemes for image cryptography. IEEE Internet Things J. **6**(6), 10855–10861 (2019)
10. Kumar, T.S.: A novel method for HDR video encoding, compression and quality evaluation. J. Innov. Image Process. (JIIP) **1**(02), 71–80 (2019)

11. Kumari, M., Gupta, S., Sardana, P.: A survey of image encryption algorithms. 3D Res. **8**(4), 37 (2017)
12. Lai, X., Massey, J.L.: A proposal for a new block encryption standard. In: Workshop on the Theory and Application of of Cryptographic Techniques, pp. 389–404. Springer, Berlin (1990)
13. Li, H., Wang, Y., Zuo, Z.: Chaos-based image encryption algorithm with orbit perturbation and dynamic state variable selection mechanisms. Opt. Lasers Eng. **115**, 197–207 (2019)
14. Liu, J., Lv, H.: A new Duffing-Lorenz chaotic algorithm and its application in image encryption. In: 2012 International Conference on Control Engineering and Communication Technology, pp. 1022–1025. IEEE, New York (2012)
15. Loukhaoukha, K., Chouinard, J.Y., Berdai, A.: A secure image encryption algorithm based on Rubik's cube principle. J. Electr. Comput. Eng. **2012** (2012)
16. Luciano, D., Prichett, G.: Cryptology: From Caesar ciphers to public-key cryptosystems. College Math. J. **18**(1), 2–17 (1987)
17. Rashaideh, H., Shaheen, A., Al-Najdawi, N.: Real-time image encryption and decryption methods based on the Karhunen-Loeve transform. Int. J. Intell. Eng. Inf. **7**(5), 399–421 (2019)
18. Sharma, R., et al.: A novel approach to combine public-key encryption with symmetric-key encryption. Int. J. Comput. Sci. Appl. **1**(4) (2012)
19. Song, T.: A novel digital image cryptosystem with chaotic permutation and perturbation mechanism. In: 2012 Fifth International Workshop on Chaos-fractals Theories and Applications, pp. 202–206. IEEE, New York (2012)
20. Vinothkanna, M.R.: A secure steganography creation algorithm for multiple file formats. J. Innov. Image Process. (JIIP) **1**(01), 20–30 (2019)

Pre-trained Deep Networks for Faster Region-Based CNN Model for Pituitary Tumor Detection

Shrwan Ram and Anil Gupta

Abstract Due to drastic changes in the field of technology and computing power for the last decade, it has become very easy to implement the convolutional neural networks for the classification and detection of objects from the large volume of images. Nowadays, the various deep networks with hundreds of layers are developed and implemented by the researchers for the classification of images and object detection inside the images. The Faster region-based convolutional neural network (R-CNN) is a widely used state-of-the-art approach that belongs to R-CNN techniques that were first time developed and used in 2015. Different R-CNN object detection approaches are developed and implemented by the researchers. Three approaches are developed and implemented on different platforms, and these approaches are R-CNN, fast R-CNN, and faster R-CNN. The efficiency and accuracy of the approaches are tested for various object detections inside the different images. Algorithms based on region proposals are used in R-CNN approaches to generate the bounding boxes or the actual location of the objects inside the images. The ground labels are generated through image labeling approaches. These ground truth labels are stored in a file. The features are extracted by pre-trained deep networks or the convolutional neural networks using the ground truth labeled images. The classification layer of the convolutional neural networks predicts the class of the object to which it belongs. The regression layer is used to create the relevant coordinates of the bounding boxes accurately. In this research paper, the faster R-CNN approach with retrained deep networks is used for the detection of pituitary tumor. The tumor detection performance of the detectors trained with three pre-trained deep networks is compared in the proposed approach of tumor detection. Three pre-trained deep networks such as Googlenet, Resnet18, and Resnet50 are used to train the tumor detector with ground truth labeled images.

S. Ram (✉) · A. Gupta
Department of Computer Science and Engineering, MBM Engineering College, Jai Narain Vyas University, Jodhpur, Rajasthan, India
e-mail: shrawanbalach@jnvu.edu.in

A. Gupta
e-mail: anilgupta@jnvu.edu.in

479
S. Shakya et al. (eds.), *Proceedings of International Conference on Sustainable Expert Systems*, Lecture Notes in Networks and Systems 176,
https://doi.org/10.1007/978-981-33-4355-9_36

Keywords Convolutional neural networks · Classification and detection · Deep networks · Region proposals · Tumor detector · Efficiency · And accuracy

1 Introduction

Automatic brain tumor detection is still a very challenging task in the field of medical image processing. Various approaches for object detection are proposed and implemented by researchers and academicians. In the last eight years, there are drastic changes that have been accrued in the field of machine learning. The subfield of machine learning in the form of deep learning and computer vision has emerged as a new paradigm of image processing [1]. The deep convolutional neural networks (CNNs) are designed and trained with very large image datasets such as Imagenet. Deep networks such an Alexnet were designed and implemented by Alex Krizhevsky with a total of 25 layers. He published the paper with Ilya Sutskever and Geoffrey Hinton in the year 2012 [2]. Other deep networks such as Resnet18, Resnet50, Googlenet, and Resnet101 are designed and implemented by the researchers for image classification and object detection. Academicians are continuously working in the field of medical image processing, particularly in the area of disease diagnosis from MRI images.

Nowadays complex healthcare systems are developed using the deep learning approaches for real-time patient recovery monitoring. The advancement in the field of artificial intelligence and machine learning has opened the door for the development of medical expert systems [3]. Magnetic resonance images (MRIs) have become the most important source of disease diagnosis [4]. The brain tumors are detected by the neurosurgeons and neurophysicians manually with the help of magnetic resonance images (MRIs). The manual process of tumor detection is not an efficient and accurate practice as found in many cases. The disease diagnosis systems developed with deep learning [1] and approaches are playing a vital role in the area of the healthcare system.

Convolutional neural network is one type of artificial neural network that is playing the dominant role in the field of computer vision [2]. This computer vision approach is creating enthusiasm among the researchers and academicians in the domain of image classification object localization and object detection. The working of convolutional neural networks is inspired by biological processes, based on the research done by Hubel and Wiesel, related to the vision in cats and monkeys [5]. The idea of the connection between layers in CNN came from the research based on the organization of the visual cortex system. In the vision system, separate neurons are responsible for visual stimuli focusing on the receptive field of selective region. Other regions are partially overlapped in such a way so that the whole area of vision is covered. The working principle of CNN is based on the research of Hubel and Wiesel [6].

The convolutional neural networks (CNNs) architectures are capable to extract the interesting features automatically from the large volume of image datasets. The networks adaptively learn features with the learning approach such as backpropagation by using the layered structure of CNN [7]. The extracted features are not trainable; they are learned during the training time of the network on a batch of images. This fundamental approach makes the convolutional neural network models more efficient and highly accurate for computer vision applications [7]. The main layers are the convolution layer, rectified unit layer (ReLu), pooling layer, softmax layer, fully connected layer, and classification/regression layer [7].

The pre-trained deep networks or convolutional neural networks (CNNs) are the backbones of faster region-based convolutional neural networks (R-CNN). The pre-trained deep networks are designed and trained with a large volume of images from the ImageNet classification task with replacing the last pooling layer by the region of interest (ROI) pooling layer [8]. The fully connected layer is converted into two branches such as a softmax layer and the regression layer for bounding boxes. The batch of images is fed into the main convolutional neural networks. The features from each image are extracted by the layered architecture, and finally, the feature map is obtained from the last layer of the convolutional neural network. The size of the feature map obtained from CNN is comparatively small as the size of the input image. The size of the features map usually depends on the size of the stride used in convolutional operation and max-pooing operations. The size of the stride is depending on the designed parameters of the deep n networks decided by the deep network designers. The faster R-CNN is working efficiently because the convolution operation is performed only once with each image and a feature map is generated from it. In the case of the R-CNN, 2000 region proposals are fed to the convolutional neural network every time. Classifying the images into different classes or categories is a process of taking an image input to the CNN layers and producing the relevant output class of the image such as an image with a tumor or image without a tumor. The final layer of CNN classifies the category/class of the image with a probability that the input image belongs to a particular class [6].

Pituitary tumors are the neurological brain disorders generally formed in the pituitary gland that affects the normal working of the human brain. It is a small gland associated with the brain and called the "Master gland" [6]. The pituitary gland is a very important organ of the human body that controls and regulates the many types of hormones for the normal functioning of the entire body [6]. A recent study is found that 10,000 pituitary tumor patients are diagnosed every year in the USA [9]. The actual number of patients suffering from the pituitary tumor may be much higher as predicted [9].

The introduction part of this paper focused on the faster R-CNN based on deep learning approaches. This part also focused on the pituitary tumor and its effect on the patients suffering from that tumor. The remaining part of the paper is divided into four sections, such as related work, research methodology, experimental setup, and results and the last section is related to the conclusion and future work. The related work sections explain the research work done in the domain of tumor classification and detection.

The full paper is organized in the following manner, and the related work in the domain of tumor classification and detection are explained in Sect. 2. The research methodology is focused on the approaches and methods used to classify and detect brain tumors which are described in Sect. 3. The experimental setup and the result analysis are explained in Sect. 4 along with the implementation details of the proposed model for pituitary tumor detection. Section 5 is related to the outcomes of the proposed model and the future research direction.

2 Related Work

Researchers and academicians are working in the field of image classification and object detection for the last decade using the advanced machine learning approach such as deep learning [1]. Nowadays, object detection has become a very interesting and challenging task for computer scientists [10]. The various object detection and classification models are proposed and implemented by the researchers using the recent deep learning approaches. Through the literature survey, it is found that so many research papers, review articles, book chapters, and textbooks are published in the last five years, infighting the field of medical image analytics. The disease diagnosis systems have become efficient and accurate due to the use of deep learning approaches. The healthcare system developed with such kind object detection and classification algorithms is proving to be more useful for finding interesting patterns and hidden valuable information related to early diagnosis of disease. This research paper is based on the detection of a pituitary tumor through deep pre-trained networks. The related research work based on tumor detection through a deep learning approach is included in the background study. It is found through the literature survey that the following research paper is more relevant to the proposed research approach for the detection of a pituitary tumor through pre-trained deep networks.

Raheleh Hashemzehi et al. proposed a model for brain tumor detection from the magnetic resonance images using a hybrid approach based on convolutional neural network and neural autoregressive distribution estimation (NADE) [11]. The model was trained and tested on contrast-enhanced T1 weighted magnetic resonance images of pituitary tumor, meningioma tumor, and glioma tumor. Images are resized from 512×512 to 64×64, and the model was trained on resized images. In the validation, process the authors used a sixfold cross-validation technique and ADAM optimizer. They achieved a 95% classification accuracy with the hybrid model. The detection accuracy is not mentioned in this research paper.

Ali ARI and Davt HANBAY proposed the deep learning-based approach for the brain tumor classification and detection, and the authors proposed a method with three stages, with image preprocessing, extreme learning machine local receptive field (ELM-RRF)-based tumor classification, and the tumor region extraction through image processing. The cranial magnetic resonance images were classified into two classes as benign or malignant. They achieved 97.18% accuracy of classification [12]. The detection accuracy is not mentioned in the results.

Rikiya Yamashita et al. published the research article based on the application of convolutional neural networks in radiology [13]. The article focused on the fundamental concepts of convolutional neural networks and the application of this approach in the area of radiology. The authors explained that data augmentation and transfer learning are a more efficient way of medical image processing the dataset with limited MRI images. The overfitting can be avoided through the use of data augmentation. Transfer learning is a more effective approach to train the pre-trained deep networks on a limited image dataset. The pre-trained deep networks are publically available online. Some examples of these networks are Alexnet [2], Resnet18 [14], Resnet50 [14], Resnet101 [14], and Googlenet.

3 Research Methodology

The selection of suitable research methodology is an important step in every search in many domains. In this research paper, the research methodology adopted for the detection of pituitary tumors from the magnetic resonance images is explained. The pituitary tumor detection process is carried out step by step, from the selection of the dataset to preprocessing the images. The quality of the MRI images should be very good because the detection accuracy most of the time depends on the image quality. Deep learning approaches performed well with a number of images in a dataset. The previous implementation of pre-trained networks for the detection of multiple sclerosis brain lesions is that, the detection accuracy is relatively higher than the previous outcomes of the research carried out by the researchers. The second part of the methodology is to select suitable pre-trained networks for the detection of brain tumors. It is also found by us that the object detection accuracy of Resnet18 and Resnet50 is very high as compared to AlexNet and Googlenet.

After selecting the pre-trained deep networks, the image labeling process is completed to get the ground truth labeled images for the training of deep networks. The image labeling step is a very important task. The images are labeled correctly, and then the correct ground truth labels will be generated. After completing the labeling task, three DAGNetwork are selected to train the pituitary tumor detector. Figure 1 depicts the proposed implementation strategy for tumor detection.

3.1 The Dataset Collection

The dataset of pituitary tumors containing 930 T1 weighted contrast-enhanced images downloaded from [15, 16]. These images are originally collected from 233 patients. It is found through preprocessing of the images that 914 images are good in quality and can be used for implementing the tumor detection model. These images are the collection of axial plane images, segital plane images, and coronal plane images.

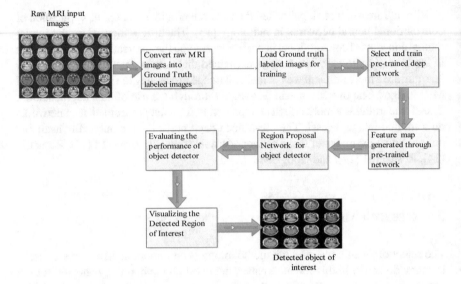

Fig. 1 Proposed steps for pituitary tumor detection

3.2 Preprocessing of MRI Images

The image datasets originally collected from the sources were of .mat file formats. All the images are converted to a PNG file format with size 512×512. The contrast and brightness of the images are further increased. After data conversion into the PNG file format and enhancing the quality of images, a data store of images is created through an image processing approach of MATLAB image processing toolbox [17]. All the MRI images are labeled using the image labeler approach of the MATLAB [17]. A ground truth data store is created with a label and anchor box as "Pituitary_Tumor". Few sample images are shown in Fig. 2.

3.3 Pituitary Tumor Detection Approaches

In this research paper, the transfer learning approach is applied to detect the pituitary tumor from the T1 weighted magnetic resonance images (MRIs). convolutional neural network approaches are playing a major role in the field of medical image analytics. To detect the infected area within the MRI image is a challenging task. The detection process requires the suitable CNN architecture for feature extraction from each image [13]. For feature extraction from image datasets, pre-trained deep networks are more suitable architectures. These networks are trained on very large image datasets such as ImageNet [2]; therefore, the deep networks are retrained based on learnable weights of pre-trained networks that were trained on very large datasets,

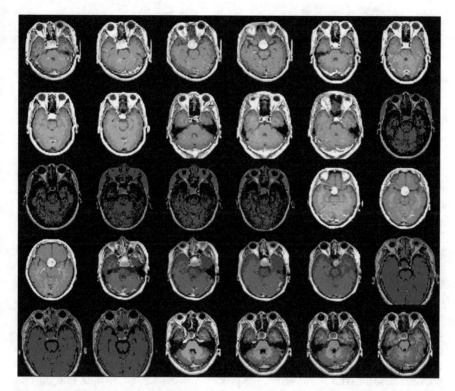

Fig. 2 Pituitary tumor sample images [15]

for small datasets. The faster R-CNN approach is implemented with the help of three pre-trained deep networks, such as Googlenet, Resnet18, and Resnet50. First of all, a batch of labeled images with label name and anchor boxes in the form of ground truth values is fed into the pre-trained network.

The training options, such as stochastic gradient descent with momentum (SGDM) as optimization function, scheduling the learning as piecewise, learning rate drop factor, the number of epochs, mini-batch size, an initial learning rate, are defined before the staring of the training process. All these parameters are known as hyperparameters. The input images are in the form of a multidimensional array passing through the pre-trained deep networks to build the convolutional feature map [18]. The feature extraction process is shown in Fig. 3. The process worked as a feature extractor for the next stage of object detection [19]. Now another important network comes in the picture, known as region proposal network used to find pre-defined regions (bounding boxes), defined at the time of the labeling process. These pre-defined regions contain the desired object[20].

Generating the list of variable length bounding boxes through the use of region proposal network (RPN) is a very typical task. The anchors are used to solve the

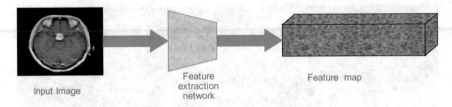

Fig. 3 Feature map generation through pre-trained deep networks [19]

variable length bounding box problem in RPN concerning reference bounding boxes drawn uniformly all through the original image data [19].

3.4 Performance Evaluation of the Proposed Approach

The performance of the proposed approach is generally evaluated based on accurate object detection, and for this purpose, use the performance evaluation approaches such as precision, recall, mini-bat accuracy, RPN mini-batch accuracy, mini-batch loss, and mini-batch root mean squared error. Formulas 1 and 2 are used to calculate precision and recall. During the training period of the pre-trained network, mini-batch accuracy is calculated for every mini-batch at the given iteration. Using stochastic gradient descent with momentum, the process divides the entire data into disjoint mini-batches [17]. The network gradient is calculated for each mini-batch to the iteration. The stochastic gradient descent momentum (SGDM) approach is used to update the weights and biases to minimize the loss function with the help of a small learning step at every epoch in the opposite direction of gradients. Equation 1 [17] is used for this purpose.

$$\theta_{\ell+1} = \theta_\ell - \alpha \nabla E(\theta_\ell) \tag{1}$$

where the ℓ is the iteration number, α is step size or learning rate which is always greater than zero, θ is the parameter vector, and ∇E is the loss function.

$$\theta_\ell + 1 = \theta_\ell - \alpha \nabla E(\theta_\ell) + \gamma (\theta_\ell - \theta_{\ell-1}) \tag{2}$$

In Eq. 2 [17], γ is the quantity obtained from the previous gradient step for the next gradient step. Other parameters as similar to Eq. 1.

The process of an epoch is relevant to move through each available mini-batch.

$$\text{Precision} = \frac{\text{True Positive}}{(\text{True Positive} + \text{False Positive})} \tag{3}$$

The precision of the model is based on the correct prediction/detection of the objects to incorrectly detected objects inside the image. It is helpful when the cost of false positive predicted results is high. If the precision of the model is low, it predicts the disease incorrectly and diagnoses the healthy person as a patient.

$$\text{Recall} = \frac{\text{True Positive}}{(\text{True Positive} + \text{False Nagetive})} \quad (4)$$

The quantity recall is helpful when the cost of a false negative is very high. A false negative detection is very harmful which helps to develop and deploy the model in the field of health care, particularly for disease diagnosis. Another one of the important performance measurement factors is the f1_score, which is used to measure the overall accuracy of the model based on precision and recall. The higher value of f1_score indicates that the value of false positive and false negative is very low. The model with f1_score one is considered as the best model, and with zero value of f1_score considered as the total failure of the model.

4 Experimental Setup and Result Analysis

The proposed tumor detection model is implemented in the MATLAB environment through the use of MATLAB version 9.8.0.1417392 (R2020a) [21]. The stochastic gradient descent momentum (SGDM) optimizer is used to calculate the gradients, the learning rate is scheduled as piecewise, the learning rate drop factor is set to 0.2, and it indicates the dropping of learning rate for epochs. 80 epochs are set for training; mini-batch size is set 1. The model is trained on HP Z6 Workstation having Windows 10 pro Workstation operating system, Intel Xeon Silver 4110 two CPUs of 2.1 GHz with 32 GB RAM. The Z6 workstation is equipped with Quadro P5000 GPU having 16 GB GDDR5X GPU Memory, 2560 NVIDIA CUDA Cores, and 8.9 Teraflops Computing Power. The MRI image dataset of a pituitary tumor contains a total of 914 images. The MRI images are labeled using MATLAB's image processing toolbox. The image labeler app of the MATLAB is very useful for generating the ground truth labels, in the form of rectangular regions of interest. After labeling the pituitary tumor images, ground truth labels are saved in the .mat file format. The labeled ground truth image data is loaded in the MATLAB programming environment. The ground truth labeled images are loaded with the three data values that are generated by evaluating of .mat file. These data values are data source, label definition, and ground truth label data stored in the form of a table. The dataset with ground truth labels is randomly split into training and test set. The training dataset consists of 548 ground truth labeled image and the test dataset consists of 366 ground truth image dataset. The training dataset is used for feature extraction through the pre-trained deep networks. The feature maps obtained through the training process of pre-trained deep networks are used as the input for region proposal network (RPN) [18]. The region proposal is generated by the RPN. These region proposals are either an object or background.

The trained object detector is used to detect the object of interest within the unseen MRI images [18].

4.1 Results Were Generated Through the Pre-trained Deep Network

All the steps as described in Sect. 4 are repeated to train the Googlenet, Resenet18, and Resnet50 pre-trained deep network.

4.1.1 Results Generating Through Googlenet Pre-trained Network

Using ground truth labeled data, the Googlenet pre-trained network is trained. Through the training, process features are extracted as shown in the proposed steps in Fig. 1. The pre-trained deep network is trained on an HP Z6 workstation, equipped with graphics processing unit (GPU) with configuration as described above. Values for negative overlap range and positive overlap range are defined in the training process. The detector is obtained after the long training process, using the faster R-CNN object detection function [20]. The pre-trained deep networks were trained on a very big dataset of RGB images with the help of high-end computing power of graphics processing units. In the proposed research, the pre-trained deep networks are retrained on a small dataset. The training process is completed in 38 h and 10 min and a table is obtained with the main fields as epochs, iteration, time elapsed, mini-batch loss, mini-batch accuracy, mini-batch root mean squared error, region proposal network mini-batch accuracy, region proposal mini-batch root mean squared error and base learning rate. Figure 4 shown below displays a graph plotted between precision and recall. The average precision graph between precision and recall depicts how a detector's performance is varying with levels of recall.

The graph between mini-batch accuracy and mini-batch loss is shown in Fig. 5, and it is clear from the plotted graph that mini-batch accuracy is increasing concerning to decay in the loss.

Figure 6 shows that the mini-batch root mean squared error and region proposal squared errors both are decreasing concerning to each other.

Figure 7 describes the region proposal network mini-batch accuracy and mini-batch accuracy both are almost following to each other with a very small variation. Both of the plots are overlapping with each other.

50 images are used for testing the detection accuracy of the object detector, and Fig. 8 shows the part of 50 images with a bounding box and detection probability printed inside each image through the testing of the object detector network trained with Googlenet pre-trained network. The detector precisely detected the infected area with high accuracy as provided in Fig. 8.

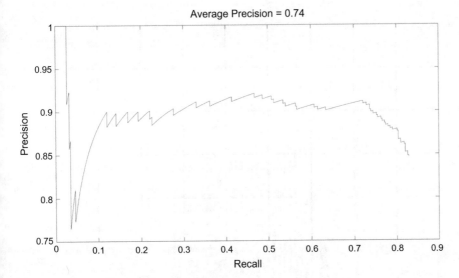

Fig. 4 Graph between precision versus recall

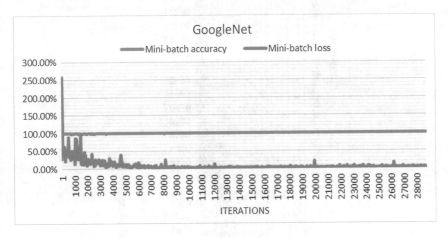

Fig. 5 Mini-batch accuracy versus mini-batch loss

4.1.2 Results are Generated Through the Resnet18 Pre-trained Network

The pre-trained deep network, Resnet18, is retrained on grayscale MIR images of pituitary tumor. The ground truth labeled image dataset is used to train the pre-trained deep network with the same method as used in the training process of Googlenet. The training process is completed in 17 h and 5 min. An execution table is generated after the successful training of the detector. The table attributes are mentioned above in the results section of Googlenet. The graph plotted between precision and recall

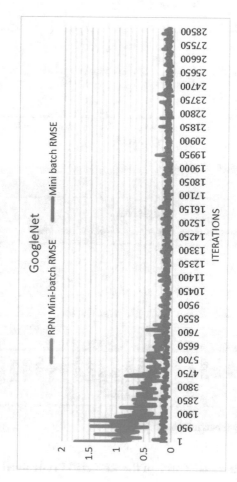

Fig. 6 RPN mini-batch RMSE versus mini-batch RMSE

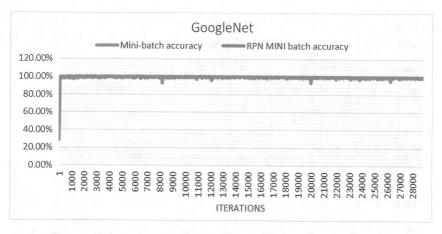

Fig. 7 Mini-batch accuracy versus RPN mini-batch accuracy

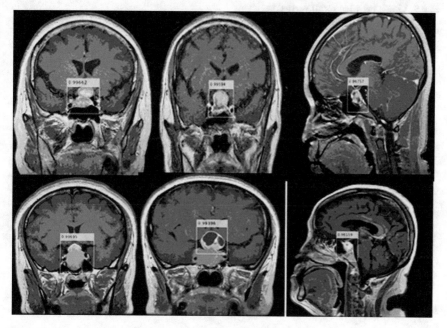

Fig. 8 Tumor area detected by the detector (Googlenet) [15]

is shown below in Fig. 9. The average precision of the Resnet18 deep network is obtained as 0.85.

The average precision of the Resnet18 pre-trained network is higher than the Googlenet deep network. The mini-batch accuracy is increasing to the mini-batch loss decreasing. Figure 10 as shown below is a graph plotted between mini-batch accuracy and mini-batch loss.

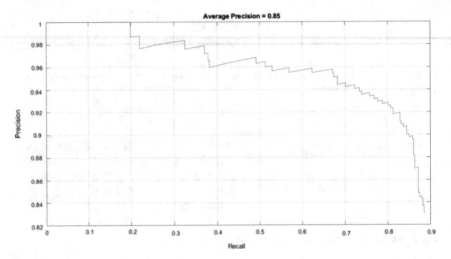

Fig. 9 Graph between recall versus precision

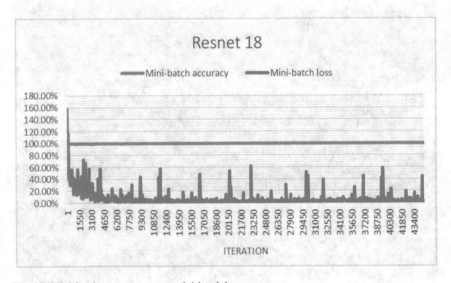

Fig. 10 Mini-batch accuracy versus mini-batch loss

Figure 11 is plotted between mini-batch root mean squared error and region proposal network mini-batch root mean squared error.

Figure 12 is plotted between mini-batch accuracy and region proposal mini-batch accuracy. Both the accuracy cures are almost similar because there is a very difference between the two quantities, that is why the curves seem to be to overlap with each other (Figs. 13 and 14).

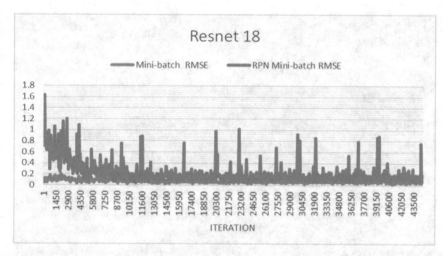

Fig. 11 Mini-batch RMSE versus RPN mini-batch RMSE

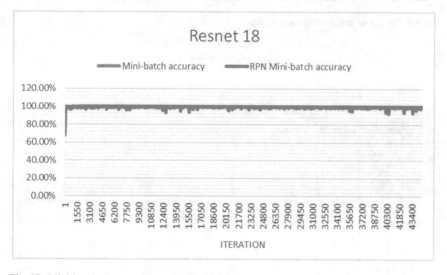

Fig. 12 Mini-batch accuracy versus RPN mini-batch accuracy

50 images are selected for finally testing the accuracy of the tumor detector model, and few images are shown above with the detected area. The bounding boxes with detection accuracy are printed inside each image

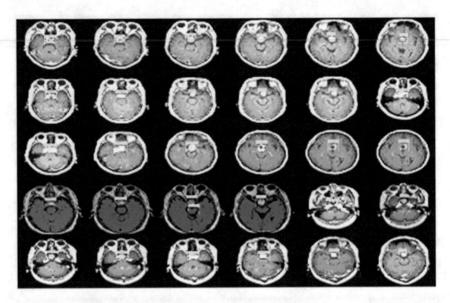

Fig. 13 Tumor detected results (Resnet18)

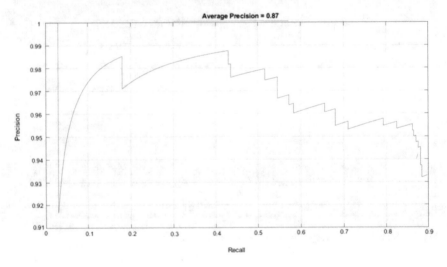

Fig. 14 Graph between precision and recall

4.1.3 Results Generated Through Resnet50 Pre-trained Network

The Resnet50 pre-trained network which consists of 177 layers is retrained on grayscale MRI images of pituitary tumors. The network is trained on the HP Z6 work-station equipped with Quadro p5000 graphics processing unit. The execution process

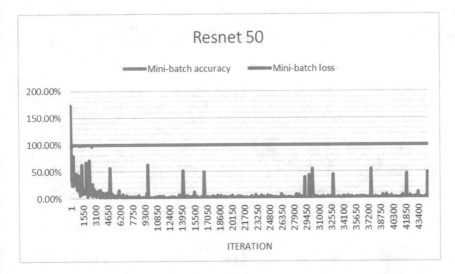

Fig. 15 Mini-batch accuracy versus mini-batch loss

is completed in 60 h and 59 min. The execution table is generated through the execution process with 9 attributes as mentioned above in the results section Googlenet. The graph shown below is plotted between precision and recall as described in the methodology section of this paper. The average precision of the pre-trained network is higher than the Googlenet and Resnet18 pre-trained networks. The average precision of the Resnet50 network is 0.87.

Figure 15 shown below is a graph plotted between mini-batch accuracy and mini-batch loss.

The detection accuracy of the model lies between 99 and 100; therefore, the curve of accuracy seems to a straight line.

Figure 16 shown below is a graph plotted between mini-batch root mean squared error (RMSE) and RPN mini-batch root mean squared error (RMSE). The region proposal mini-batch squared error (RMSE) is almost equal with very fewer variations.

Figure 17 as shown below is a graph plotted between mini-batch accuracy and RPN mini-batch accuracy (Fig. 18).

50 images are selected for testing the accuracy of the detector trained through the Resnet50 pre-trained deep network. The above images are showing the detected accuracy with the bounding box with the detected score.

The tumor detection performance of the Resnet50 pre-trained deep network is higher as compare to Googlenet and Resnet18 deep network as shown in Table 1.

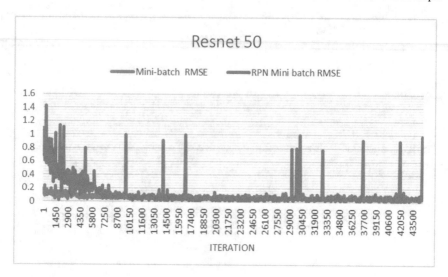

Fig. 16 Mini-batch RMSE versus RPN mini-batch RMSE

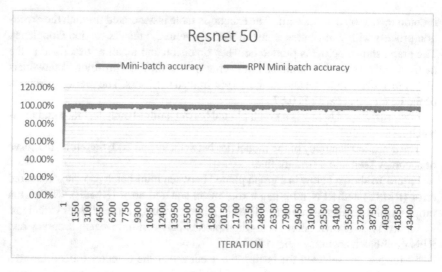

Fig. 17 Mini-batch accuracy versus RPN mini-batch accuracy

5 Conclusion and Future Scope

Three pre-trained deep networks are retrained and the pituitary tumor detection model is built through the transfer learning approaches. Three tumor detectors are built for pituitary tumor detection. The average precision of the detector trained with the help of Resnet50 is higher than the Googlenet and Resnet18. The detection accuracy of

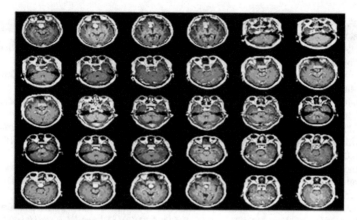

Fig. 18 Tumor detected results (Resnet50)

Table 1 Performance comparisons

	Googlenet	Resnet18	Resnet50
Average precision	0.897114	0.95353	0.957778
Average recall	0.438754	0.497117	0.482276

the model built with Resnet50 is also higher than the Googlenet and Resnet18. İt has been seen through the training process of each pre-trained deep network that the high-speed computing power is required for the training. Graphics processing unit with higher Cuda cores and memory is an essential part of the training. The quality of the MRI images should be very good for training the pre-trained deep networks. In the future, the deep networks can be used for various types of MRI images for the detection of the abnormal area inside the images. The detection accuracy of the models is higher as compared to the existing research work. Most of the research work was carried out in the field of image classification.

References

1. LeCun, Y., Bengio, Y., Hinton, G.: Deep learning. Nature **521**, 436–444 (2015)
2. Girshick, R.: Fast R-CNN. In: Proceedings of the 2015 IEEE International Conference on Computer Vision. Santiago, Chile, Dec. 2015, pp. 1440–1448
3. Vijayakumar, T.: Neural network analysis for tumor investigation and cancer prediction. J. Electron **1**(02), 89–98 (2019)
4. Faster R-CNN: Down the rabbit hole of modern object detection, available [online]. https://try olabs.com/blog/2018/01/18/faster-r-cnn-down-the-rabbit-hole-of-modern-object-detection
5. The Complete Beginner's Guide to Deep Learning: Convolutional Neural Networks and Image Classification, available [online]. https://towardsdatascience.com/wtf-is-image-classification-8e78a8235acb
6. Pituitary Tumors available [online]. https://www.mayoclinic.org/diseases-conditions/pituitary-tumors/symptoms-causes/syc-20350548

7. Ari, A., Hanbay, D.: Deep learning based brain tumor classification and detection system. Turk J Electr Eng Comp Sci **26**, 2275–2286 (2018). https://doi.org/10.3906/elk-1801-8
8. Fast R-CNN for object detection, available [online]. https://towardsdatascience.com/fast-r-cnn-for-object-detection-a-technical-summary.
9. Pituitary Tumors, available [online]. https://www.aans.org/en/Patients/Neurosurgical-Conditions-and-Treatments/Pituitary-Gland-and-Pituitary-Tumors
10. Introduction to basic object detection algorithms, available [online]. https://heartbeat.fritz.ai/introduction-to-basic-object-detection-algorithms-b77295a95a63
11. Hashemzehi R, et al.: Detection of brain tumors from MRI images base on deep learning using hybrid model CNN and NADE. Biocybern Biomed Eng (2020). https://doi.org/10.1016/j.bbe.2020.06.001
12. Krizhevsky, A., Sutskever, I., Hinton, G.: ImageNet classification with deep convolutional neural networks. In: Proceedings of the Advances in Neural Information Processing System, vol. 25, 1090–1098 (2012)
13. Yamashita, R., Nishio, M., Do, R.K.G., et al.: Convolutional neural networks: an overview and application in radiology. Insights Imaging **9**, 611–629 (2018). https://doi.org/10.1007/s13244-018-0639-9
14. He, K., Zhang, X., Ren, S., Sun, J.: Deep residual learning for image recognition. In: Proceedings of the IEEE Conference on Computer Vision and Pattern Recognition, pp. 770–778 (2016)
15. Key Statistics About Pituitary Tumors, available [online]. https://www.cancer.org/cancer/pituitary-tumors/about/key-statistics.html
16. Cheng, J.: Brain tumor dataset. Figshare. Dataset. https://doi.org/10.6084/m9.figshare.1512427.v5 (2017)
17. Precision and recall, available, [online]. https://deepai.org/machine-learning-glossary-and-terms/precision-and-recall
18. Girshick, R., Donahue, J., Darrell, T., Malik, J.: Rich feature hierarchies for accurate object detection and semantic segmentation. In: Proceedings of the 2014 IEEE Conference on Computer Vision and Pattern Recognition. Columbus, OH, June 2014, pp. 580–587
19. He, K., et al.: Spatial pyramid pooling in deep convolutional networks for visual recognition. Lecture Notes in Computer Science
20. Ren, S., He, K., Gershick, R., Sun, J.: Faster R-CNN: Towards real-time object detection with region proposal networks. IEEE Trans. Pattern Anal. Mach. Intell. **39**(6), 1137–1149 (2017)
21. MATLAB R2020a, The MathWorks, Inc., Natick, Massachusetts, United States

Wind Energy Productivity Evaluation Using Weibull Statistical Method

Omar Mahmood Jumaah, Mustafa W. Hamadalla,
Karam Hashim Mohammed, Ehssan M. A. Ibrahim, Mohammed N. Yousif,
and Sohaib N. Abdullah

Abstract Due to the increasing demand for energy and the negative effects of traditional energy systems on the environment, many researchers around the world are
trying to discover an alternative, non-consuming, harmless, inexpensive, and sustainable energy system. Wind power is the most widely used source of electricity. This
study aims to utilize wind energy by wind turbines in the Nasiriyah governorate to
obtain electricity. Different values of wind speed were taken and statistically analyzed
from the monthly data recorded for 2016. In the analysis, the average wind speed
and energy density were determined for each month. Weibull statistical distribution
method has been adopted in this study. The results obtained showed that the highest
average intensity of wind energy in 12 months was in July; in addition, the average
value of wind speed in 2016 was higher than 3 m/s.

Keywords Wind turbine · Weibull distribution · Wind velocity · Wind power

O. M. Jumaah (✉) · M. W. Hamadalla · K. H. Mohammed · E. M. A. Ibrahim · M. N. Yousif ·
S. N. Abdullah
Northern Technical University, Mosul, Iraq
e-mail: Omarmahmood803@ntu.edu.iq

M. W. Hamadalla
e-mail: mustafawadd_82@ntu.edu.iq

K. H. Mohammed
e-mail: karam.hashim@ntu.edu.iq

E. M. A. Ibrahim
e-mail: Ehssan1980@gmail.com

M. N. Yousif
e-mail: airengineer1983@gmail.com

S. N. Abdullah
e-mail: Souheebred@yahoo.com

S. Shakya et al. (eds.), *Proceedings of International Conference on Sustainable Expert
Systems*, Lecture Notes in Networks and Systems 176,
https://doi.org/10.1007/978-981-33-4355-9_37

1 Introduction

Energy is one of the most important inputs that humanity uses and is indispensable at every stage of the production of goods and services. In this context, the need for electric power is increasing due to the development of societies. The development of industry, population growth and the diversity of machinery and equipment introduced by modern technologies increase the need for electric power. This need is met through hydroelectric and thermal power plants, natural gas plants, and nuclear power plants. Due to the environmental damage caused by these plants and the positive characteristics of renewable energy sources, the use of wind energy in recent years has spread widely and reached significant levels. On the other hand, fossil fuels that meet a large part of the world's energy needs to have a certain age. Besides, when the population increases and the area of use expands, periods of depletion may be shortened [1].

The energy obtained from limited fuels directly interacts with international political developments, making it even more important to generate electricity from the country's territory in terms of strategic and national security. Because fossil fuel reserves such as oil and coal are very limited, each country is turning to renewable energy resources to protect them for longer periods. Although the share of renewable energy in the total energy supply is small today, every "kilowatt hour" obtained from this source increases the period of depletion of other resources in the world. Moreover, environmental damage from traditional sources and climate change will increase environmental awareness in the coming years [2].

A statistical analysis of the probability of wind energy was conducted on the campus: (September 12) at the University of Anadol in July, August, September, and October of 2005, based on wind speed data measured at distances of 15 h [2]. The Weibull distribution has been used in potential wind energy research in the region. The Weibull distribution coefficients used to estimate the average velocity, standard deviation, energy, and energy density were determined by the maximum probability method. As a result of this preliminary research, it was found that the standard deviation is between 0 and 3 m/h, wind speed and wind power are, respectively, greater than 3 m/h and greater than 50 watts/m2 and therefore it was determined that the probability of wind energy on campus is statistically suitable for electric power production. Weibull, which is popular in wind speed modeling and lognormal distribution functions not previously tested in the region, has been used to investigate potential wind energy in the region.

Danook et al. [3] have studied the wind energy system utilization as an alternative to traditional energy sources. It focused on the effect of humidity on the air density and the wind kinetic energy converting to electrical energy. The wind turbine frequency against the speed of wind with Kirkuk climatic conditions has been observed. Results indicated that the humid air density was lower than that of dry air. The assessment of data analysis of wind is the key requirement to achieve wind energy in any region. Statistical modeling for wind speed statistical features studying in Algeria was demonstrated by Ettoumi et al. [4]. Results showed that the data of wind for three hourly are well presented by Markov chains means. Furthermore, the fit measurements speed of wind monthly frequency distribution and the Weibull probability distribution function have been indicated.

Krewitt and Nitsch [5] developed the nature conservation criteria analysis affected the potential of wind energy in quantitative terms. The wind energy cost per kWh assessment for electricity power production utilizes three types of wind electrical conversion systems for many regions performed by Rehman et al. [6]. It was developed to estimate the cost of power generated per kWh from three wind machines selected.

In this study, the wind speed data are measured daily by the management of the weather station in Nasiriyah city between 2002 and 2009. The probability of wind speed was statistically analyzed. The maximum probability evaluation and the least squares method (LSM) were used as a technique for determining the Weibull and lognormal propagation coefficients to estimate the average velocity and energy. The analysis and evaluation of the determination factor (R^2) and the square root of the total error rate that is the root mean square error (RMSE) were performed. The probability of wind energy in Nasiriyah city is statistically encouraging in terms of electricity production.

2 Methodology

2.1 The Energy Production in the Turbine and Wind Speed

For utilizing wind energy, the probability of wind energy must first be determined. Given that the wind column is a machine that rotates by changing the torque of the wind-block particles that run in the sweep of the wing, the wind energy in the swept area is proportional to the size of the area and the density of the air and its velocity square. Accordingly, the energy obtained from the wind can be represented as follows [3]:

$$P_{wind} = \frac{1}{2}\rho.A.v^3 \tag{1}$$

where.

P_{wind}: Mechanical energy for wind flow (air) [Watts].

ρ Air density: Air density [kg/m^3].

Note: At 1 atmospheric pressure (sea level) and temperature 15 °C, $\rho_0 = 1.225$ kg/m^3.

A: the area sweeping by the wings of the rotor (the area of the passage of the wind) [m^2]; Here, consider that at 15.6 °C and a pressure of 1 atmosphere and at sea level the air density is $\rho_0 = 1.225$ g/m^3, then the corrected air density at elevation (H_m) above sea level according to other site information is as follows:

$$\rho = \rho_0 - 1194.10^{-4}.H_m \tag{2}$$

The most important input to determine the energy potential is wind speed. Although wind direction is not effective for potential, it plays an important role in the development of wind energy conversion systems. As will be seen from the equal energy, the change in the expression of energy depending on the wind speed is proportional to the speed cube.

The initial parameters and assumptions of this estimation method are the particle objective function that is calculated and evaluated, and the best particle local position is evaluated best within all of the particles; according to the initial step, the particles are updating and also if the global best is the output termination criterion is stop.

2.2 *Mechanical Energy Obtained from Wind*

The wind that cuts the turbine blades does not turn into mechanical force in the rotor. The rotor efficiency should be calculated for the expression of mechanical energy obtained from the kinetic energy of the wind. The actual amount of energy obtained by rotor blades is the difference between the kinetic energy between the wind channel inlet and the airflow at the wind channel outlet [2].

$$P_k = \frac{1}{2}\dot{m}\left(v_i^2 - v_o^2\right) \tag{3}$$

P_k Mechanical energy obtained by rotor blades (turbine energy).

V_i Wind speed at the inlet of the rotor blades.

v_c Wind speed at the outlet of the rotor blades.

v_k. Wind speed at rotor blade level (wing v)

$$v_k = \frac{v_i + v_o}{2} \tag{4}$$

The movement along the wind channel is not continually constant (wind speed from v_i' to v_o'). Thus, the mass flow rate (the amount of mass flowing per unit time) can be obtained from the air moving through the rotor blades by multiplying the rate of velocity by air density (or taking into account the speed in the rotor sweep field at the rotor level).

$$P_k = \rho . A \frac{v_i + v_o}{2}. \tag{5}$$

Instead of the energy equation P_k it would be

Fig. 1 Changing turbine productivity [5]

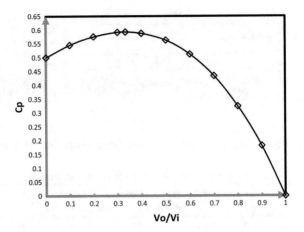

$$P_k = \frac{1}{2} \cdot \left(\rho.A \frac{v_i - v_o}{2} \right) \cdot \left(v_i^2 - v_o^2 \right). \tag{6}$$

In the expression is simplified, the following is obtained:

$$P_k = \frac{1}{2} \cdot \rho \cdot A v_i^3 \frac{\left(1 + \frac{v_o}{v_i} \right)\left[1 - \left(\frac{v_o}{v_i} \right)^2 \right]}{2} = \frac{1}{2} \cdot \rho.A.v_i^3.C_p \tag{7}$$

C_p is known as rotor productivity. Similarly, when defined as $v_i = v$ and $\lambda = v_o/v_i$, the maximum value of the rotor yield can be obtained when $\lambda = 0.33$ and its value is $C_{p\,max} = 0.5926$ or 59.26 percentage. This productivity is called the "Betz" or Betz law. In practical applications, however, this value remains below 0.5. The turbine productivity [4] is changed by λ which is depicted in Fig. 1.

2.3 Statistical Analysis of Wind Speed

A monthly average is needed when analyzing the wind characteristics of the area and thus the standard deviation, turbulence, and cumulative distribution of speeds can be seen. For 2016 data, the monthly wind speed data in Nasiriyah were represented by Fig. 2.

Wind speed distributions have been formed. To make statistical analyzes easier, they are organized by frequency distribution as shown in Table 1. Besides, Table 2 describes the wind speed was first divided into seasons and the frequency of wind was determined in each range of the wind class.

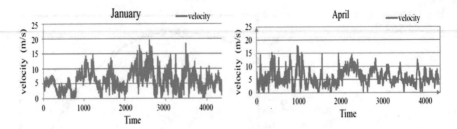

Fig. 2 Sample of month wind speed distributions to Nasiriyah data in 2016

Table 1 Frequency distributions of wind speed per month

J	V_j (m/s)	January (iteration)	February (iteration)	March (iteration)	April (iteration)	May (iteration)	June (iteration)
1	0–1	142	105	256	95	41	26
2	1–2	318	260	302	171	209	136
3	2–3	292	308	253	251	250	185
4	3–4	560	506	582	688	679	637
5	4–5	592	612	627	796	890	768
6	5–6	415	382	380	424	496	439
7	6–7	631	556	497	518	501	632
8	7–8	345	325	398	392	412	440
9	8–9	351	330	227	271	295	260
10	9–8	289	279	326	313	307	274
11	10–11	159	119	259	184	176	196
12	11–12	183	150	131	81	53	96
13	12–13	91	70	105	72	46	91
14	13–14	26	22	35	21	28	33
15	14–15	43	6	36	26	21	40
16	15–16	12	0	18	9	10	17
17	16–17	10	0	10	5	10	9
18	17–18	4	0	16	3	12	8
19	18–19	0	0	4	0	9	3
20	19–20	1	0	0	0	9	2
21	20–21	0	0	0	0	8	5
22	21–22	0	0	0	0	2	7
23	22–23	0	0	0	0	0	8
24	23–24	0	0	0	0	0	8
SUM		4464	4030	4464	4320	4464	4320

Table 2 Frequency distributions of wind speed per month

J	V_j (m/s)	July (iteration)	August (iteration)	September (iteration)	October (iteration)	November (iteration)	December (iteration)
1	0–1	10	38	59	138	139	78
2	1–2	46	66	178	471	501	480
3	2–3	107	137	251	476	521	436
4	3–4	361	448	723	858	874	789
5	4–5	638	707	795	683	700	739
6	5–6	522	572	421	331	385	419
7	6–7	625	759	581	413	501	609
8	7–8	632	630	501	356	358	428
9	8–9	367	341	249	191	150	176
10	9–8	483	356	300	207	119	183
11	10–11	302	249	159	128	40	66
12	11–12	164	85	47	42	13	26
13	12–13	141	48	38	58	9	23
14	13–14	36	11	12	28	6	7
15	14–15	24	9	5	24	4	5
16	15–16	6	6	0	13	0	0
17	16–17	0	1	0	5	0	0
18	17–18	0	1	1	7	0	0
19	18–19	0	0	0	10	0	0
20	19–20	0	0	0	5	0	0
21	20–21	0	0	0	7	0	0
22	21–22	0	0	0	8	0	0
23	22–23	0	0	0	2	0	0
24	23–24	0	0	0	3	0	0
SUM		4464	4464	4320	4464	4320	4464

2.4 Weibull Distribution

The Weibull distribution has been used to calculate wind energy potential in many studies [11–15]. It is generally known that wind data correspond to this distribution and in some areas; it is not compatible with the Weibull binary coefficient distribution of wind data [6]. Various methods have been developed to calculate the shape and coefficients of the Weibull distribution scale [7]. And one of the ways is the chart method.

The probability density function for the Weibull distribution with coefficients $f_w(v)$ is as follows:

Fig. 3 Weibull distributed
by different k values

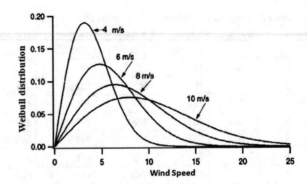

$$f_w(v) = \left(\frac{k}{c}\right)\left(\frac{v}{c}\right)^{k-1}\exp\left(-\left(\frac{v}{c}\right)^k\right) \tag{8}$$

Here v is the wind speed (m/s), k and c are the coefficients of the shape without dimensions and scale. The Weibull distribution of the cumulative probability dsity function is:

$$F_w(v) = 1 - \exp\left[-\left(\frac{v}{c}\right)^k\right] \tag{9}$$

Note the above Weibull distribution function. To clarify the transactions in more detail: k (coefficient of shape): A coefficient showing the frequency of wind. If the wind speed in a given land does not vary significantly, i.e., the wind speed blows at an almost constant speed (can be low or fast) then the coefficient k is large. The Weibull distribution [8] is observed with an average speed of 8 m/s but the values of k are different which is depicted in Fig. 3.

C (scale factor): shows the frequency of the cumulative wind speed of the relevant. In simpler terms, the coefficient C varies according to the rate of velocity. If the speed rate is high, the coefficient C is also high.

To find the Weibull coefficients (k and c), ground wind data are needed. To find these coefficients, analytical and empirical equations are used. These equations are shown below [9]

$$k = \left(\frac{\sigma}{\bar{v}}\right)^{-1.086} \quad 1 \le k \le 10 \tag{10}$$

$$c = \frac{\bar{v}}{\left(1 + \frac{1}{k}\right)} \tag{11}$$

The average wind speed and standard deviation of wind speed are calculated, respectively, from Eqs. (12) and (13).

$$\bar{v} = \frac{1}{n}\left(\sum_{i=1}^{n} v_i\right) \tag{12}$$

$$\sigma = \left[\frac{1}{n-1}\sum_{i=1}^{n}(v_i - \bar{v})^2\right]^{0.5} \tag{13}$$

$$E = \int_{0}^{\infty} P_k * f_w(v) \tag{14}$$

As a result, after a detailed review of the wind data and completion of the calculations described above, Weibull coefficients are obtained [4].

The Rayleigh density function can be expressed as follows:

$$f_R(v) = \left(\frac{\pi v}{2\bar{v}^2}\right)\exp\left(-\left(\frac{v}{\bar{v}}\right)^k\right) \tag{15}$$

Another distribution function used to determine the wind speed potential is the Rayleigh distribution. This distribution is a special case of the Weibull distribution, and Rayleigh distributions are given when the case is corrected by assuming that the Weibull coefficient without dimensions is equal to the distribution of the second probability density and its cumulative function [10].

$$F_R(v) = 1 - \exp\left[-\left(\frac{\pi}{4}\right)\left(\frac{v}{c}\right)^2\right] \tag{16}$$

As shown in Eq. (16), the only unknown coefficient in the Rayleigh distribution is the velocity coefficient. As a result of all these evaluations, there are two alternatives regarding the formation of the wind speed probability function. Henceforth, in Eq. (8), these values can be established and approximate annual energy production values can be found. Furthermore, to obtain more accurate results, the equation can be evaluated in the manner of calculating the integration in Eq. (8) and written in different codes by computer (Visual Basic, Excel, and Pascal).

The code is written in Table 3. In Excel, observe the annual energy production values calculated using Weibull distributions. The Earth's wind speed data used for these calculations are:

2.5 Graphical Method

The graphical method requires data of the wind speed to be in the cumulative format of a frequency distribution. Firstly İt should be sorted the data of time series into bins. In this study, the data of the wind speed have been interpolated by a horizontal line,

Table 3 Weibull coefficient estimates and adjusting speed

Distribution	k	C	$\bar{v}$	Σ
January	2.287176	7.374166	6.532482	3.04951999
February	2.473891	7.159603	6.350868	2.75806384
March	2.10437	7.326333	6.488799	3.27062079
April	2.542981	7.222542	6.411111	2.71449584
May	2.440378	7.291786	6.466174	2.84362821
June	2.437485	7.755616	6.877315	3.02774142
July	3.340456	8.444763	7.579749	2.4964854
August	3.340449	7.763425	6.972446	2.2714337
September	2.865737	6.998271	6.237269	2.36573729
October	1.935631	6.413192	5.687724	3.09620196
November	2.528582	6.680105	5.041204	2.24566023
December	2.586478	6.135159	5.448477	2.27116438

utilizing the rule regression of least squares [11–13]. The foundation of transformation logarithmic is the method of Weibull distribution. Calms observations have been omitted from the data to generate the line of best fit. The data analysis method has been shown in Fig. 4.

Fig. 4 Data analysis method

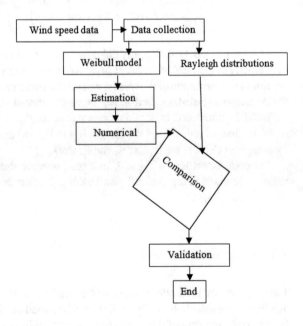

3 Result and Discussion

In this study, the equations were used to find the Weibull distribution coefficients, and therefore the coefficients, velocity, and estimated forces for the 12th month of 2016 are shown in Table 3. Figure 5 shows the probability density functions related to their wind speeds for 12 months according to Nasiriyah data for 2016. It can be seen the maximum wind speed probability density is in the range of velocity between 5 and 10 m/s, while, zero above 15 m/s. In Fig. 5, there is a similarity between the probability density functions related to wind speed for November and December and the probability density functions related to wind speed for June and July. This situation is noted in Table 4, which shows the similarity in the energy estimate. Changes in average wind speed and wind strength are shown in Table 4 and Fig. 5. Therefore, it is possible to say that there is a possibility of rising energy in July compared to the rest of the months in terms of wind speed and strength.

In Table 3, it is noted that the Weibull distribution form factor was between 1.935631 and 3.340456 and the scale factor was between 5.041204 and 7.579749. The standard deviation of wind speed data should be between 0 and 3 m/s. If the standard deviation is in any field, the wind system in that field is maximized [11]. The standard deviation measured in this region is between the values shown. Although the measurements in the region are 30 m, the average wind speed per month is greater than three. This situation is promising regarding energy production even if it is short-term research.

In this study, the potential of wind energy, which is a renewable energy source, was analyzed in the Nasiriyah region and the wind characteristics in the area were studied by statistical methods using the data of the average monthly and hourly wind

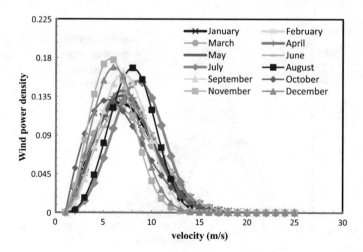

Fig. 5 Functions of wind speed probability density over 12 months according to Nasiriyah data for 2016

Table 4 Energy estimates for months (by Weibull and Rayleigh distributions)

Distribution	Weibull W/m^2	Rayleigh W/m^2
January	289	274.08
February	249	239.04
March	304	257.03
April	251	246.58
May	266	253.95
June	320	311.07
July	355	425.99
August	274	323.97
September	214	226.46
October	223	226.46
November	123	112.12
December	153	144.44

intensity in the governorate of Nasiriyah during 2016 and the average wind speed and energy density were calculated as shown in Fig. 6.

To find the Weibull distribution coefficients, the Weibull and Rayleigh distributions method has been adopted. It has been evaluated in the standard deviation of wind speed data being among the expected values. During the twelve months of study, the highest average wind speed was measured in July, so the largest wind power was achieved during the same month. The average wind speed for all months was above 3 m/s. This case shows that the use of wind energy to produce electricity is initially appropriate.

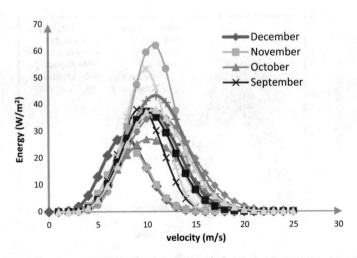

Fig. 6 Average energy speed during 12 months under to Nasiriyah data for 2016

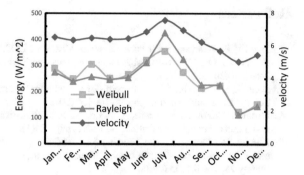

Fig. 7 Changes in the rate of wind speed and strength during 12 months according to the data of Nasiriyah for 2016 (By Weibull and Rayleigh distributions)

Such studies, conducted to obtain approximate information and undertake the necessary expansion of wind power potential in the Nasiriyah area, will be expanded and continued in detail with permanently taken measurements. The related studies accelerate the trend toward such energy sources in Iraq for great importance to contribute and expand production. Figure 7 shows a comparison among three methods to evaluate the wind energy.

It can be noted that the Weibull distribution is between the Rayleigh distribution and velocity methods which may be adopted for energy evaluation during 12 months and the Iraqi region. It can be seen from Table 4 the deviation between these two methods is approximately 6%.

It was observed that the data of the Weibull method results of this study are agreed to the data of [16] with a deviation of not more than 4% due to the environment conditions of the region.

4 Conclusion

The use of a wind energy system to traditional sources of energy as an alternative has been investigated in this study. The results of this study are concluded that the maximum value of rotor productivity is 59% at $\lambda = 0.33$. The Weibull distribution with an average speed of 8 m/s, and however, the values of k are different. It is between the Rayleigh distribution and velocity methods which may be adopted for energy evaluation for 12 months and the Iraqi region. The Weibull distribution is required for the computation of the wind available energy, and it is representing a good approximation of the actual probability frequency of wind speed according to the high coefficient of correlation values and used for the wind turbine characteristics determination. The largest wind power was achieved during the same month. The use of wind energy to produce electricity is initially appropriate.

References

1. Patel, M.P.: Wind and Solar Power Systems. CRC Pres LLC, New York (1999)
2. Vardar, A., Onur, T.: Renewable energy sources and Turkey. Int. J. Energy Power Eng. **3**, 245–249 (2014)
3. Danook, S.H., Jassim, Kh.J., Hussein, A.M.: The impact of humidity on performance of wind turbine. Case Stud. Therm. Eng. **14**, 100456 (2019)
4. Ettoumi, F.Y., Mefti, A., Adane, A., Bouroubi, M.Y.: Statistical analysis of solar measurements in Algeria using beta distributions. Renew. Energy **26**(1), 47–67 (2002)
5. Krewitt, W., Nitsch, J.: The potential for electricity generation from on-shore wind energy under the constraints of nature conservation: a case study for two regions in Germany. Renew. Energy **28**, 1645–1655 (2003)
6. Rehman, H.S., Halawani, T.O., Mohandes, M.: Wind power cost assessment at twenty locations in the Kingdom of Saudi Arabia. Renew. Energy. **28**, 573–583 (2003)
7. Tony, B., David, S., Nick, J., Ervin, B.: Wind Energy Handbook (2001)
8. Akpınar, E.K., Akpınar, S.: Determination of the wind energy potential for Maden-Elazig. Turkey. Eng. Convers. Manag. **45**, 2901–2914 (2004)
9. Kurban, M., Kanta, M.K., Fatih, O.H.: Weibull Dagılımı Kullanılarak Rüzgar Hız ve Güç Yogunluklarının İstatistiksel Analizi. Afyon Kocatepe Üniversitesi Fen ve Mühendislik Bilimleri Dergisi. **7**(2), 205–218 (2007)
10. Ahrens, C.D.: Meteorology Today: An Introduction to Weather, Climate, and the Environment (9th Edition). Brooks/Cole Cengage learning, USA (2009)
11. Brown, D., Yong, S.: A distributed density-grid clustering algorithm for multi-dimensional data. In: IEEE CCWC 2020: 10th Annual Computing and Communication Workshop and Conference. Jan (2020)
12. Baraneetharan, E., Selvakumar, G.: MPPT based ZVS resonant converter for grid connected system with SOS optimization algorithm. J. Adv. Res. Dyn. Cont. Sys. **1**, 79–87 (2017)
13. Ufuk, E., İbrahim, U.: Rüzgâr Türbinleri. Çeşitleri Ve Rüzgâr Enerjisi Depolama Yöntemleri. Süleyman Demirel Üniversitesi, Isparta, TÜRKİYE (2014)
14. Jaramillo, O.A., Borja, M.A.: Windspeedanalysis in La Ventosa, Mexico: a bimodal probability distribution case. Renew. Energy **29**, 1613–1630 (2004)
15. Shata, A.A., Hanitsch, R.: Evaluation of wind energy potential and electricity generation on the coast of Mediterranean Sea in Egypt. Renew. Energy **31**, 1183–1202 (2006)
16. Jiang, Y., Yuan, X., Cheng, X., Peng, X.: Wind potential assessment using the Weibull model at the Inner Mongolia of China. Energy Expl. Explo. **24**(3), 211–221 (2006)

COVID-19 Social Distancing with Speech-Enabled Online Market Place for Farmers

D. Mohan, K. Anitha Sheela, and P. Sudhakar

Abstract Corona virus (COVID-19) has been rampant in most countries across the world, leading to a global crisis, and it has been so dangerous because of its ability to spread through any medium especially when there is physical contact of any object. So, this puts many people including the farmers and the customers at a bigger risk when they sell or buy products physically in a market place. This is a major challenge to an Indian farmer today to sell the product at the right price at the right time. Moreover, social distancing is the need of the hour, and hence, it is an even bigger challenge to monitor the market place to avoid physical contact. Also, many initiatives by governments, local bodies, and cooperative societies to eliminate the intermediaries are not turning effective due to the non-awareness, non-accessibility, ignorance, and ease of usage difficulties. Keeping the current pandemic situation in mind and social distancing being the need of the hour, this paper proposes a speech-enabled interactive voice response (SEIVR) wherein the farmer and the customer can sell and buy products online, i.e., without any physical contact. In the current technology era, online market platforms like Amazon, Flipkart, and Olx are effectively cutting short these supply-chain overheads by establishing a direct connection between buyer and seller. Hence, the proposed implementation of a speech-enabled interactive voice response (SEIVR)-based online market place for farmers which even ensure social distancing will help to reduce the spread of the virus. The objective of this work is to build an agriculture-based mobile application with speech-enabled interactive voice response (SEIVR) based on speech recognition application with the Telugu language as a case study.

Keywords COVID-19 · Corona virus · Agricultural information system · Voice-enabled mobile application · Speech recognition · Speech-enabled IVR · Sphinx · Acoustic model · Language model · Photic dictionary

D. Mohan (✉) · P. Sudhakar
ECM Department, Sreenidhi Institute of Science and Technology, Hyderabad, India
e-mail: mohan.aryan19@sreenidhi.edu.in

K. A. Sheela · P. Sudhakar
ECE Department, JNTUH College of Engineering, Kukatpally, Hyderabad, India
e-mail: kanithasheela@jntuh.ac.in

© The Editor(s) (if applicable) and The Author(s), under exclusive license
to Springer Nature Singapore Pte Ltd. 2021
S. Shakya et al. (eds.), *Proceedings of International Conference on Sustainable Expert Systems*, Lecture Notes in Networks and Systems 176,
https://doi.org/10.1007/978-981-33-4355-9_38

1   Introduction

COVID-19, a deadly virus, has created a pandemic situation with its widespread outburst and put the entire world in a crisis. It has been so deadly and severe because of its ability to spread through any medium especially when there is physical contact involved. This has pushed us into a situation where social distancing has becoming a mandatory norm. With social distancing being the need of the hour, it has changed the way businesses are running especially in the field of agriculture where farmers sell products physically in a crowded market place. To avoid physical contact, it is imperative to adopt digital technology into the field of agriculture to avoid the spread of the deadly virus. Keeping the current pandemic situation in mind and social distancing being the need of the hour, this paper proposes a speech-enabled interactive voice response (SEIVR) wherein the farmer and the customer can sell and buy products online, i.e., without any physical contact. In this context, the adoption of information communication technology-enabled information support systems for supportable development in the agricultural expansion is very important for developing countries such as India, where agriculture is still the Indian economy's most significant field. Given the country-wide expansion of telephone and mobile networks, with easy access to smart mobile devices and access mobile data services, initiatives are now being introduced by both the government and the private sector [1] to develop and fund smart agriculture-related solutions.

With a similar context, the work is presented on developing a speech-based mobile application for "*Speech-Enabled IVR-Based Online Market Place for Farmers*" in the Telugu language. A robust automatic speech recognition (ASR) engine automatically recognizes voice queries in the form of spoken commodity names in the Telugu language at runtime, to generate text outputs for commodity variety collection, product searching, etc.

The application developed serves end-users both in rural and urban areas, especially those who are not computer literate or find it hard to access the same data during busy working hours. In reality, the rural end-users of this application include mostly semi-literate, non-tech savvy farmers, and agro-producing sellers who mainly earn their livelihood by selling farm-grown products on local markets. In comparison, price knowledge for customers helps smart customers to have equal offers when purchasing agricultural commodities. This speech-based mobile application is therefore unique in its way to encourage both farmers, and consumers through one single application.

This paper is organized as follows, and Sect. 2 explains the related work of the existing system; Sect. 3 discusses proposed work and key issues on application design, and it describes the application overview and architecture. Section 4 deals with the framework for designing a speech recognition system. Section 5 illustrates a detailed result analysis and assessment study. Finally, conclusions were made from the obtained results which are depicted in Sect. 6, and future directions for the expansion of this work presented in Sect. 7.

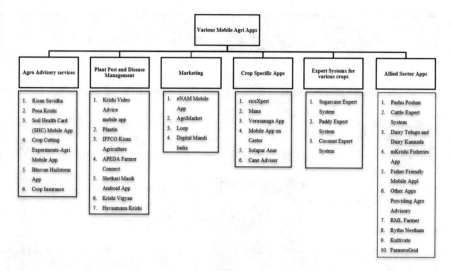

Fig. 1 Available mobile applications in indian agriculture and allied sectors [2]

2 Related Work

Currently available mobile applications in the Indian agriculture sector, the government of India, for the benefit of farmers and other stakeholders, has launched a range of online and mobile-based applications to disseminate information on agricultural-related activities, free of charge. Theses in the "mkisan.gov.in" Web site. There are also applications established by agricultural institutions, private sector, non-governmental organizations.

All these existing applications are menu-driven/touch-tone keypad-based systems, which is not easily handled by low literacy farmers (Fig. 1).

3 Proposed Work

The proposed work is primarily focused on building the user interface for agricultural-related technologies to support agriculture activities and conversational systems for low-literate users. More broadly it implements the speech recognition system, and its output is evaluated using created speech database by using real-time Google speech recognition engine as well as on CMU's sphinx speech recognition engine (Fig. 2).

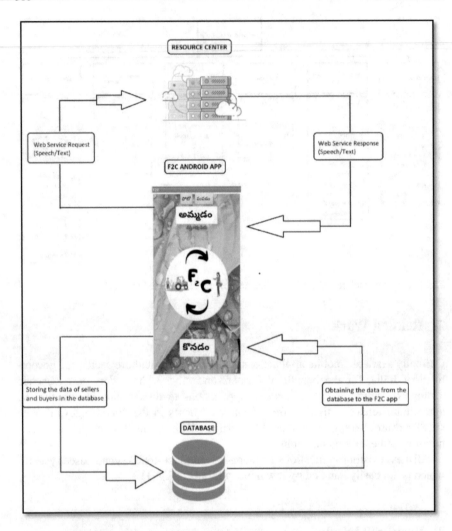

Fig. 2 General block diagram of the application

3.1 Key Issues on Application Design

Major issues being discussed and decided during the design of the developed application are as follows, and connecting stakeholders are a simple, responsive interface with an efficiently configured core framework allows for easy use by all stakeholders (government department, farmer, consumer) and therefore communicating with each of them equally. Access modality in various access modalities like touch-type-listen, speak-see, and speak-listen are included in the design to attract literate as well as semi-literate end-users. User preferences for language, input–output modality, service

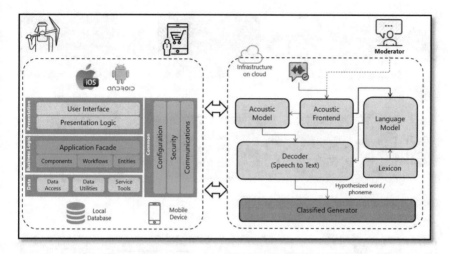

Fig. 3 Technical architecture

type, etc., are recorded while user registration and maintained until the next change. The content quality design should ensure completeness, reliability, and timeliness (important for the agriculture domain) of the information to be provided. The core technology such as application-based colorful user interface is underlying client–server architecture, and the latest ASR technology in the application core to add flexibility and ease of access.

3.2 APPLICATION Overview and Architecture with Detailed Process Flow

3.2.1 Technical Architecture

The overall design of the mobile application depends on three basic components, namely Android-based user interface software, Web service to which the application communicates, and An authentic online information source. Figure 3 demonstrates the entire framework architecture with data flow within the core modules.

3.2.2 Application Flow

When the farmer opens the Application –home screen with a mike button (ready to supply message), then by one-click—application collects details (product name, quantity, and price). Next in interactive mode farmer responds to simple questions in native language (one-word answers). The application processes speech inputs

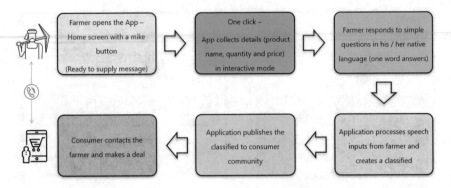

Fig. 4 Application flow

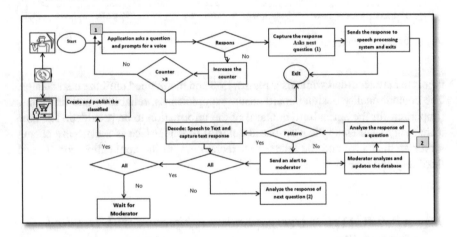

Fig. 5 Detailed process flow

from the farmer and creates a classified then publish the classified to the consumer community. The consumer contacts the farmer and makes a deal (Figs. 4 and 5).

4   Frame Work for Designing Speech Recognition System

CMU's Sphinx has been primarily used as a proof of concept [3] for our proposed designed mobile Application, i.e., "Speech-Enabled IVR-Based Online Market Place for Farmers mobile Application" (Fig. 6).

Stages involved in the design of speech recognition system [4] are:

Stage 1: Preparation of Speech database

Stage 2: Design of the Vocabulary

Stage 3: Design of the Grammar Rule

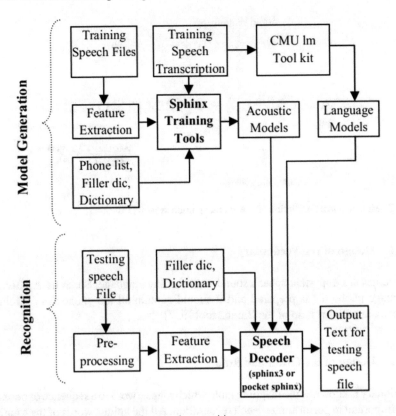

Fig. 6 Model generation diagram for speech recognition system

Stage 4: Generation of Acoustic Model

Stage 5: Generation of Language Model.

The acoustic model (AM) and language model (LM) are generated with the help of a training database using sphinx training tools. The sphinx decoder (Sphinx-3) performs speech recognition.

4.1 Database Preparation

The agricultural data collection process is for data collection primarily aims to collect isolated words such as rice, wheat, vegetable names, quantity, and price (Fig. 7).

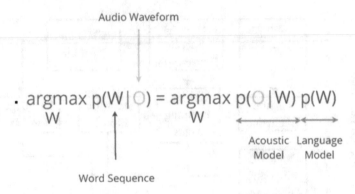

Fig. 7 Mathematical description of speech recognition system [15]

4.1.1 Design of the Vocabulary

Phonemes are important speech sound units for any language. So, to have a Telugu language phone list is prepared and it should contain all the phonemes of Telugu varnamala with the help of the Baraha tool [6, 7].

4.1.2 Design of the Grammar Rule

Dictionary is known as the mapping table which maps a word to a sequence of phones. It is important in performing speech recognition. All the unique words of the training and testing speech file transcriptions are included in the phonetic dictionary [8, 9].

4.2 Feature Extraction

The process of extracting vocal tract parameters form the speech signal is known as feature extraction. The main purpose of extraction of the feature is to generate the sequence of vectors describing the speech signal's spectral and temporal nature.

Generating AM and LM model during the training, mel frequency cepstral coefficients (MFCC) are used [10–12].

4.3 Acoustic Model (AM)

The acoustic model can be known as the phoneme model because the output of the acoustic model is "phone sequence." The acoustic model gives the likelihood of various phonemes at different time instant [4, 13, 14].

4.4 Language Model (LM)

The language model gives the probability of appearing one phone after another phone [4]. This work language model is generated using the CMU SLM tool kit [13–15].

5 Results

This work has been carried with real-time Google voice search as well as with CMU's Sphinx speech recognition system. The testing was carried out on 20 farmers of which 10 were male farmers and 10 were considered female farmers. The speech recognition accuracy of the system is measured with the following performance metrics such as accuracy of recognized words and the rejection rate. The male and two female speaker's speech samples were used in the deployed system's testing. The way to carry out the speech recognition test is tabulated with few test samples (from the available speech database consider only a few speech samples for demonstration purpose).

(a) **Accuracy of recognized words is defined as follows**:

$$\text{Accuaracy} = \frac{\text{No. of correctly recognized speech speech samples}}{\text{Total no. of test samples}} \times 100$$

The experimental results of only a few samples of speech are given below. They correctly recognized samples are denoted with T (True), and those speech samples that are not recognized are labeled as F (False). The experimental results of only a few samples of speech are given below. They correctly recognized samples are denoted with T (True), and those speech samples that are not recognized are labeled as F (False) (Tables 1 and 2).

(b) **Rejection Rate**: The error rate of total test samples that are being falsely recognized is calculated as follows: (Table 3)

$$\text{Accuaracy} = \frac{\text{No. of correctly recognized speech speech samples}}{\text{Total no. of test samples}} \times 100$$

The graph of correct acceptance rate (CAR) versus correct rejection rate(CRR) shows that the word recognition rate is greater than the word rejection rate. The graph of the CAR CRR shows that the word recognition rate is greater than the word rejection rate (Fig. 8).

Table 1 Speech recognition test results

Words	No. of speech samples	1 st Speaker	2nd Speaker	3rd Speaker	4th Speaker	5th Speaker
గోధుమలు Word1	1	T	T	T	T	T
	2	T	T	T	T	F
	3	T	T	F	T	T
	4	T	T	T	T	T
	5	T	T	T	T	T
ఐదు కిలోలు Word2	1	T	T	T	T	T
	2	T	T	T	T	T
	3	T	T	T	T	T
	4	T	T	T	F	T
	5	T	T	T	F	T
వంద రూపాయలు Word3	1	T	T	T	T	T
	2	F	T	T	T	F
	3	T	F	F	T	T
	4	T	T	T	T	T
	5	F	T	T	T	T

Table 2 Correct recognition accuracy

Words	1 st Speaker (%)	2nd Speaker	3rd Speaker (%)	4th Speaker (%)	5th Speaker (%)	Average accuracy (%)
గోధుమలు Word1	100	100	100	80	80	92
ఐదు కిలోలు Word2	100	100	100%	60	100	92
వంద రూపాయలు Word3	60	80	100	100	100	88

5.1 Results Analysis Using CMU's Sphinx-3

The performance is evaluated with CMU's Sphinx-3. Speech recognition is carried out using a speech database of 20 farmers (Table 4).

M = Male; F = Female

Table 3 Rejection rate

Words	1 st Speaker (%)	2nd Speaker (%)	3rd Speaker (%)	4th Speaker (%)	5th Speaker (%)	Average accuracy (%)
గోధుమలు Word1	0	0	0	20	20	8
ఐదు కిలోలు Word2	0	0	0	40	0	8
వంద రూపాయలు Word3	40	20	0	0	0	12

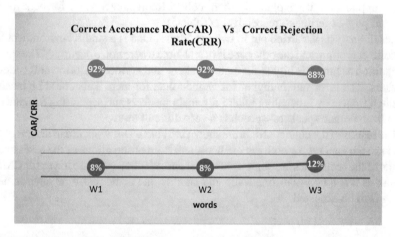

Fig. 8 Graph -CAR/CRR versus Words

Table 4 Speech recognition accuracy and rejection rates using CMU's Sphinx-3 on male and female speech samples

Training	Testing	Accuracy of recognized words (%)	Rejection rate (%)
10 (M + F)	12 (M + F)	95	5
10 (M + F)	7F	90	10
10 (M + F)	8 M	97	3

Table 5 Speech recognition accuracy and rejection rate on male and female speech data

Training	Testing	Accuracy of recognized words (%)	Rejection rate (%)
10 (M + F)	12 (M + F)	92	8
10 (M + F)	7F	88	12
10 (M + F)	8F	95	5

5.2 Results from the Field Study with Real-Time Google Voice Search

Initially, the farmers are trained on the application for a few minutes, where the application's target and the task are demonstrated with the help of a real-time voice search application are explained. Then, collect the feedback from formers for the pre- and post-experiment, respectively (Table 5).

Result analysis is carried out in two conditions. One is a studio environment like no-noise condition, and the other one is in real-time where noise occurred. The speech recognition accuracy is higher for male speakers, and normally, it is lower for female speakers. The error rate is higher for women than for men. This could be because the pitch frequency is usually higher for female speakers and separating formant and pitch from female speakers' speech is a very difficult task.

The results of the collected data are shown in Tables. This work is carried out on the same speech data set for both the methods, and from observations of the assessment, results of the accuracies are somewhat higher and rejection rates are lower for CMU's Sphinx as it is a non-real-time setup and another one is the real-time environment, which is to be expected.

6 Conclusion

In this research work, an application is developed with the Telugu speech recognition module for the online market place for farmers. The application allows farmers to sell their products with a voice search system which even ensures social distancing will help to reduce the spread of the virus. Test results show more than 90 percent accuracies for both real-time and non-real-time scenario. Investigation results display that this application performs well for isolated word queries even in noisy field conditions.

7　Future Scope

This research can be expanded to process continuous word voice queries so that applications of live voice chat related to agriculture such as soil preparation and crop advisory can be built in a similar line as well.

Acknowledgements　This is the collaborative research work of JNTUH and SNIST and funded by JNTUH under the scheme of "Collaborative research project under TEQIP-III."

References

1. Saravanan, R., Bhattacharjee, S.: Mobile phone applications for agricultural extension in India. In: Worldwide mAgri Innovations and Promise For Future. New India Publishing Agency, pp. 1–75. (2013)
2. Mobile App Empowering Farmer-Published by the Director General on behalf of the National Institute of Agricultural Extension Management (MANAGE), Hyderabad, India. Vol.1 No.2 (2017)
3. CMU Sphinx project by Carnegie Mellon University, Pittsburgh, USA (2018). https://cmusph inx.sourceforge.net/
4. Picone, J.W.: Signal modeling techniques in speech recognition. Proc. IEEE **81**, 1215–1247 (1993)
5. Website https://www.ffmpeg.org/
6. https://www.speech.kth.se/wavesurfer/
7. Website https://www.baraha.com/
8. Ramana, A.V., Laxminarayana, P., Mythilisharan, P.: Real time ASR with HMM word models for Telugu. In: Proceedings of the International Conference on Recent Advances in Communication Engineering, Osmania University, December 20–23 (2008)
9. Deller, J.R., Hansen, J.H., Proakis, J.G.: Discrete Time Processing of Speech Signals. Prentice Hall, NJ (1993)
10. Benesty, J., Sondhi, M.M., Huang, Y.A.: Handbook of Speech Processing. Springer, New York (2008)
11. Volkmann, J., Stevens, S., Newman, E.: A scale for the measurement of the psychological magnitude pitch. J. Acoust. Soc. Am. **8**, 185–190 (1937)
12. The Carnegie Mellon University Statistical Language Modeling (SLM) Toolkit. https://www.speech.cs.cmu.edu/SLM_info.html
13. Aker, J.: Dial "A" for agriculture: using information and communication technologies for agricultural extension in developing countries. In: Confernce on Agriculture for Development Revisited, University of California at Berkeley (2010)
14. Ram Reddy, M., Laxminarayan, P., Ramana, A.V.: Transcription of Telugu TV news using ASR. Published In: 2015 International Conference on Advances in Computing, Communications and Informatics (ICACCI) (2015)
15. https://medium.com/@Alibaba_Cloud/interspeech-2017-series-acoustic-model-for-speech-recognition-technology-c7cea164d654

Machine Learning Concept-Based Prognostic Approach for Customer Churn Rate in Telecom Sector

Deepika Ghanta, Guru Sree Ram Tholeti, Sunkara Venkata Krishna, and Shahana Bano

Abstract One of the leading telecommunication service providers in India has made a substantial sum of a couple of billion dollars in three months, whereas other competing service providers are finding it hard to improve their revenues. Here comes an important factor called retaining the high potential customers and knowing the risks of them by switching to a new service provider. Building a standard customer churn prediction model rather than just advertising about their service can make a huge difference in profits obtained by any telecom service companies. Because as it is said retaining the high potential, old customers are much easier than getting completely new customers. The fact that can find the customers at risk of leaving the service which provides an offer in favour to retain them. So, the models like logistic regression, decision trees, neural networks and naive Bayes are used to review the accuracies and compare each model to build a standard prediction system. An accuracy of 73.54% for decision trees, 76.33% for naive Bayes, 75.01% accuracy for neural networks and a whooping 80.88% for logistic regression.

Keywords Naive bayes · Logistic regression · Decision tree · ANN · Churn · Telecom

D. Ghanta (✉) · G. S. R. Tholeti · S. Bano
Department of CSE, Koneru Lakshmaiah Education Foundation, Vaddeswaram, India
e-mail: deepikaghanta15@gmail.com

G. S. R. Tholeti
e-mail: usertgsr@gmail.com

S. Bano
e-mail: shahanabano@icloud.com

S. V. Krishna
Department of CSE, SRM University, Amaravati, AP, India
e-mail: Venkatsunkara45@gmail.com

S. Shakya et al. (eds.), *Proceedings of International Conference on Sustainable Expert Systems*, Lecture Notes in Networks and Systems 176,
https://doi.org/10.1007/978-981-33-4355-9_39

527

1 Introduction

Generally, serving providers refers the loss of subscribers when users switch from one service to other for better rates and benefits which reduce the share values and organization profit [1]. This term is referred as churn, and it is an important factor for customer management [2]. The duration acquired by the user from initial time to end by an organization is defined as attrition rate, whereas churn defines the customer preference for an organization and its not only limited for communication industries [3]. Since churn is the key factor towards profit of an organization, telecommunication industries pay more attention towards churn to maintain the customer base [4]. This could be possible with the help of data mining technologies which effectively predict the customer churn. Algorithms such as decision trees, regression trees, neural network and other models help to predict the customer churn [5]. Based on the collected information, these prediction processes are performed but generally raw information has diverse features and no-churners [6]. Quality of service is an essential factor in telecommunication to provide better voice and data service with affordable cost will maintain the customers for an organization in long term [7]. If these services are managed efficiently, then customers will remain in the same service for long duration. In order to provide such service, providers use different carrier-based plans. Simultaneously, the utilization of mobile phones increases rapidly and churn consequences also increases in short duration [3]. So it is essential to develop a standard model to obtain customer churn and increase the organization profit without losses.

2 Related Works

Various churn prediction models are evolved in the recent times. Decision tree and logistic regression-based churn prediction are reported in Preeti et al. [8] research model. Proposed churn prediction model uses to test the dataset in order to obtain the accuracy of collected information through decision tree and regression process. The high probability customer list is generated from the collected information which is predicted in the research model effectively.

Multiple classifiers and regression models are reported in Kriti Mishra and Rinkle Rani [4] research work. Proposed model compared the synthetic minority oversampling technique, bagged classification and regression trees, conventional classification and regression trees, and partial decision tree classifiers to validate the performance of featured dataset. Among all the classifiers, classification and partial decision tree classifier yield maximum accuracy of 77% which is much better than other models.

Comparative analysis of churn dataset is performed in Vijaya and Sivasankar [6] research work using SVM, K-NN, SVM, Naïve Bayes and LDA models in which the SVM stands high accuracy with 91.39% due to its hyperparameters.

In the proposed work, the churn data which is extracted from Kaggle and used ANN, naive Bayes, decision tree model and logistic regression model for classification and achieved 80.88% accuracy for logistic regression and achieved 72.79% ROC score and 80.2% precision.

3 Methodology

3.1 Logistic Regression

Initially, the churn data is imported into our environment. The data is split into train and test set with a split ratio. Then the hypothesis equation; i.e. $\theta^t x$ is passed through a sigmoid function. The sigmoid function generates a probability, and the data instance is classified into churn or no-churn based on this value. The best parameters for theta are calculated by using a gradient descent optimizer, and this process is iterated until find the optimized values. Then the classified results are displayed (Fig. 1).

3.2 Decision Tree

The churn data is imported into our environment. The data is split into train and test set with a split ratio. Initially, the entropy for the whole dataset is calculated. The input attributes are taken, and information gain is calculated for each of them like gender, and senior citizen. The attribute with maximum information gain will be selected and made as a node in the decision tree. The branches become the outcomes of the decision. The expansion of the tree is continued until there are no more attributes left (Fig. 2).

3.3 Neural Network

The churn data is taken into our environment. The churn data is split into training and testing sets. An ANN model is built from sequential by taking three layers. The weights and biases of the network are initialized with random values. First, the pattern obtained by initial values is presented and output values are calculated. This process is continued until the maximum epochs are mentioned by updating the weights and biases of the network. Once the above process is completed, results will be displayed and the process is stopped (Fig. 3).

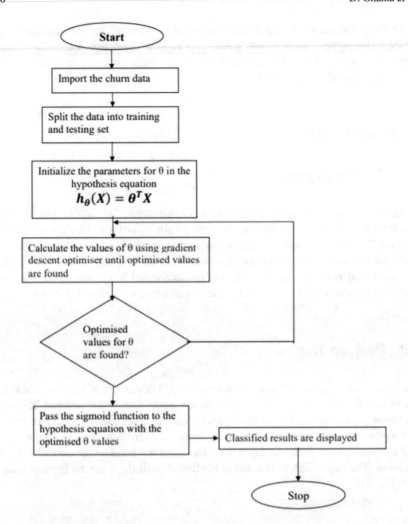

Fig. 1 Logistic regression flowchart

3.4 Naïve Bayes

The churn data is imported into our working environment and split into train and test set. Naive Bayes classifier uses Bayes theorem to compute the conditional probability of an event. Then Gaussian probability density function is used to estimate the probability of the attribute by using its mean and standard deviation estimated from training data. The class with a high probability for an attribute is selected and classified into it (Fig. 4).

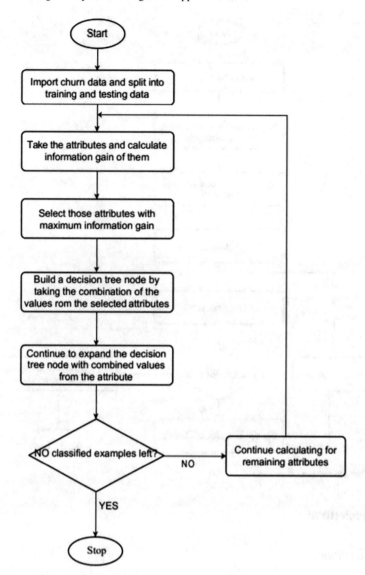

Fig. 2 Decision tree flowchart

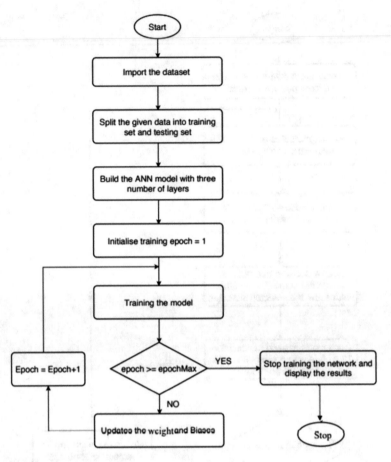

Fig. 3 Neural network flowchart

4 Procedure

4.1 Dataset

The telecommunication churn data is taken from Kaggle. It contains a total of 21 features with 20 input features some of them are customer ID, gender, senior citizen, partner, dependents monthly charges, total charges and churn.

There are a total of 7043 records with no missing values. The feature customer ID is removed as it will not have any impact on deciding whether a customer will churn or not. So, the total number of input features becomes 19. The test dataset contains 2113 records, and the train dataset contains 4930 records. There are few categorical columns which are converted into integers. Having 'tenure' which is highly indirectly proportional and which is directly 'monthly charges' proportional

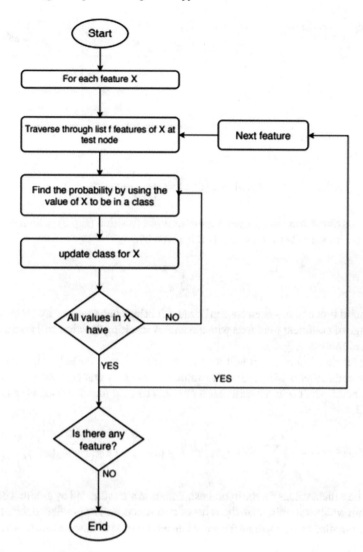

Fig. 4 Naive bayes flowchart

4.2 Logistic Regression

Logistic regression is a classification algorithm used to classify discrete data. The churn data contains a total of 20 features with 19 input features and one output feature. The logistic regression uses logs of odds of the dependent variable. The hypothesis function (Eq. 1) of the logistic regression becomes as follows,

Fig. 5 Sigmoid curve

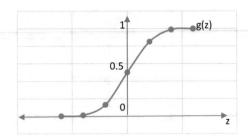

$$h_\theta(x) = g(\theta^T x) = \frac{1}{1 + e^{-\theta^T x}} \tag{1}$$

The activation function g used here is sigmoid function (Eq. 2), which will map the given real value between 0 and 1. It is given by,

$$g(z) = \frac{1}{1 + e^{-z}} \tag{2}$$

Sigmoid function has been chosen because it brings out nonlinearity to the model. The usage of nonlinear functions with a complex resulting function will give us more cultivated models (Fig. 5).

As z reaches infinity, $g(z)$ will tend to the value 1 and similarly when z reaches negative infinity $g(z)$ will tend to the value 0. Based on this $g(z)$ value decision is made whether the customer will churn or not. The cost function used here is given by Eq. 3,

$$J(\theta) = -\frac{1}{m} \left(\sum_{i=1}^{m} y^{(i)} \log h_\theta(x^{(i)}) + (1 - y^{(i)}) \log(1 - h_\theta(x^{(i)})) \right) \tag{3}$$

The best theta values for the hypothesis equation are estimated by gradient descent algorithm which will minimize the value of cost function. The gradient descent value can be computed by performing the partial differentiation of the cost function as Eq. 4,

$$\frac{\partial J(\theta)}{\partial \theta_j} = \frac{1}{m} \sum_{i=1}^{m} (h_\theta(x^{(i)}) - y^{(i)}) x_j^\wedge(i) \tag{4}$$

The cost of the proposed model is 0.437, and the accuracy for our model is 80.88%.

4.3 Naïve Bayes Classifier

Naive Bayes is a classification algorithm that is based on Bayes theorem with an assumption that there exists no dependency between predictor variables. Bayes theorem gives the conditional probability of an event happening given a prior knowledge or evidence related to the event, and it is given by Eq. 5,

$$p(H/E) = \frac{p(E/H)p(H)}{p(E)} \tag{5}$$

where $P(H|E)$ is a posterior probability, $P(H)$ is prior probability, $P(E|H)$ is the likelihood of event H occurring given E is true and $P(E)$ is marginal probability. Summary including mean and standard deviation is calculated for each attribute by class value. The Gaussian probability density function is used to estimate the probability of the attribute given its mean and standard deviation estimated from training data. The function is as Eq. 6,

$$p(y) = \frac{1}{\sigma\sqrt{2\pi}} e^{\frac{(y-\mu)^2}{2\sigma}} \tag{6}$$

where μ is the mean, σ is the variance and y is a continuous variable. With the probability of each class, i.e. churn or no-churn, the probability of entire data instance by combining the probabilities of all attributes of that data instance. It is calculated for each class and the class having a high probability value is selected, and the data instance is classified into it. The accuracy measured by our model is 76.33% with a recall score of 76.39.

4.4 Decision Tree Classifier

Decision tree portrays the outcome of a decision in each branch in tree-like representation with leaf nodes representing the class label. The method used in our model is information gain, which measures the decrease in the entropy after split the data in accordance with each attribute. Information gain is given by Eq. 7,

$$\text{Gain} = \text{Entropy (System)} - \text{Entropy (Attribute)} \tag{7}$$

Initially, the system entropy is calculated wherein entropy is used to measure the level of impurity in the data. Entropy is given by Eq. 8,

$$\text{Entropy} = \sum_{i=1}^{n} -p_i \log_2 p_i \tag{8}$$

where p_i is the probability of the class is 'i'. The entropy for each attribute is calculated. The difference between system entropy and for each attribute the entropy is calculated. The attribute which is having the highest value of gain is selected as the root node, and the data is split into two according to the root node. On each level, the nodes with maximum gain value, and their values become the leaf nodes. A tree-like structure is formed with the decision on each level. The accuracy that our model has given is 73.54% with a roc score of 67.08.

4.5 Neural Network

An artificial neural network is used to classify the data into a specific class; that is, whether the customer will churn or not. Totally 19 independent features are considered for the dataset as the input for the network. These input features are passed through two hidden layers with 150 neurons in each layer. The inputs are multiplied by appropriate weights and along with biases are pre-processed by passing through an activation function. The activation function used in our model is a sigmoid function, and it is given by Eq. 9,

$$F(s) = \frac{1}{1+e^{-s}} \tag{9}$$

This function ranges between 0 and 1 and is used here as a need to predict churn probability and classify them as an output. The function used to minimize the error is given by Eq. 10,

$$\text{Loss}(y, \hat{y}) = \sum_{i=1}^{n}(y - \hat{y})^2 \tag{10}$$

This loss function is optimized in our dataset by using a standard gradient descent function which is given by Eq. 11,

$$W(\text{new}) = W(\text{old}) - \eta * \frac{\partial \text{loss}}{\partial \text{weight}} \tag{11}$$

The initial weights of the network are changed by the slop value at that particular point. The loss value obtained after optimizing is 0.6793. The accuracy of the proposed training model is 75.01%.

5 Results

Training and testing sets with a split ratio of 70:30, i.e. 70% of the data is taken as a training set and 30% as a testing set is shown in Figs. 6 and 7.

customerID		gender	SeniorCitizen	Partner	Dependents	tenure	PhoneService	MultipleLines	InternetService	OnlineSecurity	OnlineBackup
5925	1840-BIUOG	Male	0	No	No	20	Yes	Yes	DSL	No	Yes
4395	5502-RLUYV	Female	0	Yes	Yes	69	Yes	Yes	Fiber optic	No	Yes
1579	9391-EOYLI	Male	1	Yes	No	12	Yes	No	Fiber optic	No	No
1040	1236-WFCDV	Male	1	No	No	14	Yes	No	Fiber optic	No	No
1074	2111-DWYHN	Male	0	No	No	1	Yes	No	No	No internet service	No internet service
...	...	...	...	...	...	...	...	...	...	...	...
905	0781-LKXBR	Male	1	No	No	9	Yes	Yes	Fiber optic	No	No
5192	3507-GASNP	Male	0	No	Yes	60	Yes	No	No	No internet service	No internet service
3980	8868-WOZGU	Male	0	No	No	28	Yes	Yes	Fiber optic	No	Yes
235	1251-KRREG	Male	0	No	No	2	Yes	Yes	DSL	No	Yes
5157	5840-NVDCG	Female	0	Yes	Yes	16	Yes	No	DSL	Yes	Yes

4930 rows × 20 columns

Fig. 6 Training set

```
In [13]:  x_test
Out[13]:
```

	gender	SeniorCitizen	Partner	Dependents	tenure	PhoneService	MultipleLines	InternetService	OnlineSecurity	OnlineBackup	DeviceProtection
3381	0	0	0	0	41	1	0	0	2	0	2
6180	0	1	0	0	66	1	2	1	2	0	0
4829	0	0	0	0	12	1	0	0	0	0	0
3737	0	0	0	0	5	1	2	0	0	0	0
4249	0	0	1	1	10	1	0	0	0	2	2
...	...	...	...	...	...	...	...	...	...	...	...
3934	0	0	0	1	10	0	1	0	2	2	0
1351	1	0	1	1	11	1	0	2	1	1	1
2048	1	1	0	0	21	1	0	1	0	0	2
6218	0	0	1	1	70	0	1	0	0	2	2
4297	0	0	0	0	45	1	2	0	2	0	0

2113 rows × 19 columns

Fig. 7 Testing set

5.1 Decision Tree

The accuracy obtained for the decision tree model as shown in Fig. 8 depicts that 73.54% of the test data instances are classified correctly into their corresponding classes, i.e. churn or no-churn. ROC score is 67.08% and precision 75.1%, which is greater than 50% shows us that can trust the model. In the confusion matrix, consider 1268 customers who belong to the 'no-churn' category which is correctly classified by the model. 317 customers belong to the 'no-churn' category but the model classified them into the 'churn' category. Whereas 242 customers belong to 'churn' but the model classified them into the 'no-churn' category and similarly 286 customers belong to 'churn' and classified correctly into that particular category.

```
print("\nACCURACY - %f" %(accuracy_score(y_test,dt_ypre) * 100))
print("RECALL        - : %f" %(recall_score(y_test, dt_ypre) * 100))
print("ROC SCORE. - : %f\n" %(roc_auc_score(y_test, dt_ypre) * 100))
print("CONFUSION MATRIX -\n")
print(confusion_matrix(y_test, dt_ypre))
```

```
ACCURACY - 73.544723
RECALL       - : 54.166667
ROC SCORE. - : 67.083333

CONFUSION MATRIX -

[[1268  317]
 [ 242  286]]
```

Fig. 8 Result decision tree

5.2 Logistic Regression

The accuracy obtained for the decision tree model as shown in Fig. 9 depicts that 80.88% of the test data instances are classified correctly into their corresponding classes, i.e. churn or no-churn. ROC score is 72.79% and a precision 80.2%, which is greater than 50% shows us that can trust the model. In the confusion matrix, consider 1410 customers belong to the 'no-churn' category which is correctly classified by the model. 175 customers belong to the 'no-churn' category but the model classified them into the 'churn' category. Whereas 229 customers belong to 'churn' but the model classified them into the 'no-churn' category and similarly 299 customers belong to 'churn' and classified correctly into that particular category. The minimized cost value after getting the optimized parameters of the theta is 0.437261.

```
print("\nACCURACY - %f" %(accuracy_score(y_test,lr_ypre) * 100))
print("RECALL        - %f" %(recall_score(y_test, lr_ypre) * 100))
print("ROC SCORE  - : %f\n" %(roc_auc_score(y_test, lr_ypre) * 100))
print("CONFUSION MATRIX -\n")

print(confusion_matrix(y_test, dt_ypre))
print("\nCOST          - %f"%np.sqrt(((lr_ypre - y_test) ** 2).mean()))
```

```
/Users/deepikaghanta/anaconda3/lib/python3.7/site-packages/sklearn/linear_model/logistic.py:433: FutureWarning: Defau
lt solver will be changed to 'lbfgs' in 0.22. Specify a solver to silence this warning.
  FutureWarning)
```

```
  ACCURACY - 80.880265
  RECALL       - 56.628788
  ROC SCORE  - : 72.793889

  CONFUSION MATRIX -

  [[1410  175]
   [ 229  299]]

  COST         - 0.437261
```

Fig. 9 Result logistic regression

```
print("\nACCURACY -: %f" %(accuracy_score(y_test,nb) * 100))
print("RECALL    - : %f" %(recall_score(y_test, nb) * 100))
print("ROC SCORE. - : %f\n" %(roc_auc_score(y_test, nb) * 100))
print("CONFUSION MATRIX -\n")

print(confusion_matrix(y_test, nb))
```

```
ACCURACY -: 76.336962
RECALL    - : 76.515152
ROC SCORE. - : 76.396377

CONFUSION MATRIX -

[[1209  376]
 [ 124  404]]
```

Fig. 10 Result naïve bayes

5.3 Naïve Bayes

The accuracy obtained for the decision tree model as shown in Fig. 10 depicts that 76.33% of the test data instances are classified correctly into their corresponding classes, i.e. churn or no-churn. ROC score is 76.39% and a precision 80.9%, which is greater than 50% shows us that can trust the model. In the confusion matrix, consider 1209 customers belong to the 'no-churn' category which is correctly classified by the model. 376 customers actually belong to the 'no-churn' category but the model classified them into the 'churn' category. Whereas 124 customers belong to 'churn' but the model classified them into the 'no-churn' category and similarly 404 customers belong to 'churn' and classified correctly into that particular category.

5.4 Artificial Neural Network

The accuracy obtained for the decision tree model as shown in Fig. 11 depicts that 75.01% of the test data instances are classified correctly into their corresponding classes, i.e. churn or no-churn; ROC score is 74.34% and precision 80.1%. The loss obtained is 0.6793.

The results of four algorithms used are shown in Table 1.

```
1  score = model.evaluate(x_test, y_test,verbose=1)
2  print(score)
```

```
2113/2113 [==============================] - 1s 540us/sample - loss: 0.6793 - acc: 0.7501
[0.6792810350247486, 0.7501183]
```

Fig. 11 Result ANN

Table 1 Results

S. No.	Algorithm	Accuracy	Roc score	Precision
1	Logistic regression	80.88	72.79	80.2
2	Decision tree	73.54	67.083	75.1
3	Naïve bayes	76.33	78.39	80.9
4	Neural network	75.01	67.93	80.1

6 Conclusion

The proposed work built a standard churn model using algorithms like logistic regression, neural networks, naive Bayes classifier, and decision trees. Out of all the models, logistic regression provided us the highest accuracy. A standard churn prediction model will help many telecommunication companies to increase their sales and revenue. The customers can be better understood with the help of this model to know the possibility of churn of each customer. Once the customers with high risk are found, possible measures like discounts, offers and advertisements regarding such type of customers will be brought into action to retain them. By doing this almost, all the telecommunication companies can be benefited.

References

1. Anurag B., Sumit S.:A robust model for churn prediction using supervised machine learning. In: 2019 IEEE 9th International Conference on Advanced Computing
2. Ma, H., Qin, M.: Research method of customer churn crisis based on decision tree. In: 2009 International Conference on Management and Service Science, Wuhan, pp. 1–4. (2009). https://doi.org/10.1109/icmss.2009.5305403
3. Roshini, T., Sireesha, P.V., Parasa, D., Shahana, B.: Social media survey using decision tree and naive bayes classification. In: 2019 2nd International Conference on Intelligent Communication and Computational Techniques (ICCT) (2019)
4. Kriti, M.*, Rinkle R.:Churn prediction in telecommunication using machine learning. In: International Conference on Energy, Communication, Data Analytics and Soft Computing (ICOEI 2017)
5. Vijayakumar, T.: Comparative study of capsule neural network in various applications. J. Artif. Intell. 1(01), 19–27 (2019)
6. Vijaya, J., Sivasankar, E.:Improved churn prediction based on supervised and unsupervised hybrid data mining system. In: Lecture Notes in Networks and Systems (2018)
7. Hong-Yu, H.: Research on customer churn prediction using logistic regression model. In: Advances in Intelligent Systems and Computing (2019)
8. Preeti, K.D., Siddhi, K.K., Ashish, D., Aditya B., Kanade, V.A.: Analysis of customer churn prediction in telecom industry using decision trees and logistic regression. In: 2016 Symposium on Colossal Data Analysis and Networking

Simulation of CubeSat Detumbling Using B-Dot Controller

Rishav Mani Sharma, Rohit Kawari, Sagar Bhandari, Shishir Panta, Rakesh Chandra Prajapati, and Nanda Bikram Adhikari

Abstract CubeSats tumble with high angular velocity after it is deployed into orbit due to the asymmetric force of deployment. This initial spin is not suitable for precise attitude control. Therefore, it is necessary to reduce the angular velocity of CubeSat to some acceptable value before switching to attitude control strategy. This process of damping spin, known as detumbling is often performed by utilizing the Earth's magnetic field. In this paper, numerical simulation of CubeSat detumbling by using a variant of B-Dot controller with angular velocity feedback is presented. The rotational dynamics of CubeSat is simulated using the RK4 integration of equations of rigid body kinematics and Euler's second law of motion. CubeSat is propagated in low Earth orbit (LEO) using Kepler's equation of orbital motion. The international geomagnetic reference field (IGRF) model is used to find the magnetic field strength in the body frame as the satellite propagates in orbit, spinning on its way. Cube-Sat is equipped with three magnetorquers, three-axis magnetometer, and three-axis gyroscope.

Keywords CubeSat · Detumbling · B-Dot · Magnetorquer · IGRF model · Simulation

This research is supported by University Grants Commission (UGC), Nepal and Higher Education Reform Project (HERP), TU in a part. The authors deeply appreciate the technical support from ORION Space, Nepal.

R. M. Sharma · R. Kawari · S. Bhandari · S. Panta · N. B. Adhikari (✉)
Department of Electronics and Computer Engineering, Pulchowk Campus, Institute of
Engineering, Tribhuvan University, Lalitpur, Nepal
e-mail: adhikari@ioe.edu.np

R. C. Prajapati
ORION Space, Madhyapur Thimi, Nepal

541

S. Shakya et al. (eds.), *Proceedings of International Conference on Sustainable Expert
Systems*, Lecture Notes in Networks and Systems 176,
https://doi.org/10.1007/978-981-33-4355-9_40

1 Introduction

The idea of utilizing the Earth's magnetic field to control the attitude of satellites equipped with current-carrying coils was proposed in 1961 [1]. Thirty-eight years later professors Jordi Puig-Suari and Bob Twiggs presented their CubeSat standard and four years after that CubeSats were launched for the first time in 2003 [2, 3]. QUAKESAT 1 (3U) was deployed with passive magnetic control in 2003 and magnetorquers were used for the first time onboard HITSAT (1U) which was launched in 2006 [3]. Most of the CubeSats with mission requirements of active precise attitude control use magnetorquers for detumbling. The main aim of this paper is to present an architecture for simulation of CubeSat detumbling by considering the design parameters of CubeSat as well as magnetorquers.

A magnetorquer is an electromagnet, with or without a ferromagnetic core, that is used as an actuator to detumble the spacecraft using Earth's magnetic field. The magnetic moment generated by magnetorquer interacts with the magnetic field of Earth to produce control torque to the satellite that is used to detumble or control the attitude of CubeSat. This torque relies on the strength of the dipole moment generated by magnetorquer, intensity of Earth's magnetic field at that location in orbit, and relative orientation of these two fields. The generated magnetic torque is maximum if the moment generated by magnetorquer and the field of Earth is perpendicular.

Magnetorquers are suitable actuators due to their simpler construction, higher reliability, lesser weight, and lower cost compared to other class of actuators [4, 5]. Moreover, they do not have moving mechanical parts, do not need propellant, consumes less power and can be operated as long as the solar panel or battery can supply power [5, 6].

This technology has several limitations. Magnetorquers can only generate torque on the plane perpendicular to the magnetic field of the Earth. So they can produce only two control torque components [4]. This shows CubeSat equipped with only magnetorquers is underactuated (components of control torque than the degrees of freedom of satellite) [7]. So, it must be used with other actuators for full three-axis attitude control. Its performance is dependent on the strength of Earth's magnetic field at the orbit, so it is more effective for LEO compared to higher orbits [6]. It has a slow response time compared to actuators such as reaction wheels or propulsion. Hence, it is not a good choice for missions with fast and accurate pointing applications [6].

The actuation of magnetorquer for detumbling requires some control law which must ensure the damping of angular velocity. B-Dot controller is a simple and widely used control algorithm to detumble satellites using magnetorquers. This controller computes the magnitude and direction of current to be passed through magnetorquers in order to detumble the CubeSat. The B-Dot algorithm used in this simulation was proposed by [5] and this particular algorithm was selected due to its proof of global convergence of angular velocities to zero for underactuated spacecrafts. This implies that the use of this control law detumbles the angular velocity of satellite to zero for any initial values of angular velocity and orientation.

2 Simulation Architecture

The numerical simulation presented in this paper follows the architecture as shown in Fig. 1. The flow of the simulation starts with the generation of the spherical coordinates of CubeSat in ECI frame using the Kepler's equation of orbital dynamics. This coordinate is used by IGRF model to compute the magnetic field of Earth in ECI frame which is then transformed to body frame. Thus, generated magnetic field strength, with the angular velocity obtained using equations of satellite dynamics, is used to implement detumbling control algorithm.

2.1 Mathematical Model of Magnetorquer

The magnetorquer generates a magnetic dipole whose magnitude is proportional to the current passed through it. The magnitude also depends on the number of turns of the coil, its geometry, and the nature of the core used in the coil. The direction of the dipole moment depends on the direction of the current. The mathematical model of both air core and ferromagnetic core magnetorquer is presented.

The mathematical model assumes three identical magnetorquers aligned such that the magnetic moment produced by them align with the principal axes. The magnetic moment, considering air core magnetorquer, is given by the equation below [6],

$$\boldsymbol{\mu}_M = n A \mathbf{i}_M \tag{1}$$

where $\boldsymbol{\mu}_M \in \mathbb{R}^3$ consists magnetic moment generated by each magnetorquer, n is the number of turns of the coil of each magnetorquer, A is the area of the magnetorquer, and $\mathbf{i}_M \in \mathbb{R}^3$ is a vector consisting the magnitude of current through each magnetorquer.

In the case of magnetorquer with ferromagnetic core, Eq. (1) now has an extra term ξ which accounts for the magnetic permeability and the geometry of the core.

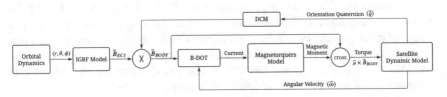

Fig. 1 Architecture of simulation of CubeSat detumbling

$$\boldsymbol{\mu}_M = \xi n A \mathbf{i}_M \tag{2}$$

where ξ is defined as below [8],

$$\xi = \left(1 + \frac{\mu_r - 1}{1 + (\mu_r - 1)N_d}\right). \tag{3}$$

Here, μ_r is the relative permeability of the ferromagnetic core and N_d is demagnetizing factor which is written as [8],

$$N_d = \frac{\left[4\ln\left(\frac{l_c}{r_c} - 1\right)\right]}{\left(\frac{l_c}{r_c}\right)^2 - 4\ln\left(\frac{l_c}{r_c}\right)} \tag{4}$$

where l_c is the length and r_c is the radius of core. The torque produced by the magnetorquers due to interaction with Earth's field is given by [9],

$$\boldsymbol{\tau}_M = \boldsymbol{\mu}_M \times \mathbf{B} \tag{5}$$

where $\boldsymbol{\tau}_M \in \mathbb{R}^3$ is the torque due to magnetorquers and $\mathbf{B} \in \mathbb{R}^3$ is magnetic field vector in body frame. Equation (5) models magnetorquers by computing the magnetic torque experienced by the CubeSat using the dipole moment generated by magnetorquers and the local magnetic field of the Earth as input.

2.2 B-Dot Controller

B-Dot controller is a popular control law used to detumble CubeSats. It is written as [9],

$$\boldsymbol{\mu}_M = -k\dot{\mathbf{B}} \tag{6}$$

where $\boldsymbol{\mu}_M$ is a vector with its components corresponding to the magnetic moment command to magnetorquers aligned on each of the principal axes, k is control gain, and $\dot{\mathbf{B}}$ is the derivative of magnetic field vector in body frame. Input current to the magnetorquer is computed using Eq. (1) for air core and (2) for ferromagnetic core magnetorquers, respectively.

This control law offers two major advantages [10]: (1) $\dot{\mathbf{B}}$ is perpendicular to $\mathbf{B}$ for high sensor sampling frequency compared to angular velocity of CubeSat. So the $\boldsymbol{\mu}_M$ produced by magnetorquer will be perpendicular to $\mathbf{B}$, hence, producing maximum control torque. (2) $\dot{\mathbf{B}}$ is proportional to minus the angular velocity vector of satellite $-\boldsymbol{\omega}$. This shows that the control law acts in a way to decrease the angular velocity of the satellite which is in fact the desired effect for detumbling.

The control law used in the simulation is the modified form of B-Dot controller which requires the angular velocity of CubeSat as feedback. Detumbling is straight-forward in a fully actuated system as they can perform three-axis control. In such case, the zero angular velocity is asymptotically achieved if the control torque is in the form $\tau_c = -k\omega$ [5]. However, CubeSat with only magnetorquers is underactu-ated so the same case does not apply to it. It is proved analytically that control law presented in this paper strictly decreases the kinetic energy of CubeSat, which means that it will converge to zero monotonically. This control law is written as,

$$\mu_M = -\frac{k}{||\mathbf{B}||^2}(\mathbf{B} \times \omega) \tag{7}$$

where ω is the angular velocity vector of the satellite. It can be seen that this equation does not look like the 6 for this control law to be called B-Dot controller in strict sense. However, the authors have also shown this controller being similar to the B-Dot controller where magnetorquer dipole moment command is defined as $\mu_M \propto (\omega \times \mathbf{B}) \approx -\dot{\mathbf{B}}$. Avanzini and Giulietti [5] also presents an equation to find the reasonable value of gain.

$$k = 2n(1 + sin\zeta)I_{min} \tag{8}$$

where n is mean motion of satellite, ζ is the inclination of the orbit with respect to the geomagnetic equator, and I_{min} is the value of minimum moment of inertia of the satellite.

2.3 IGRF Model

The IGRF model is used to determine the magnetic field at the coordinate of CubeSat in its orbit. The IGRF is updated every five years and the one used in this simulation is 13th generation (2020–2025). This model computes the strength of magnetic field for given polar coordinates (r, θ, ϕ), where r is the geocentric distance to the satel-lite, θ is the co-latitude (90-latitude), and ϕ is the east longitude from Greenwich. In simulation, this coordinate is obtained from the Kepler's equation of orbital dynam-ics. In case of the actual CubeSat on orbit, this information can be obtained from a GPS sensor. The model is implemented based on the detailed algorithm presented in [11] and [12].

The magnetic field strength generated by the IGRF model is in cartesian ECI frame, but the one needed for B-Dot is in body frame. So, coordinate transformation is performed for the magnetic field vector in ECI ($\mathbf{B}_{ECI}$) to body frame ($\mathbf{B}$) using direction cosine matrix constructed using the orientation of satellite in quaternion (q_0, q_1, q_2, q_3)[13].

$$\mathbf{B} = \begin{pmatrix} q_0^2 + q_1^2 - q_2^2 - q_3^2 & 2(q_1q_2 + q_0q_3) & 2(q_1q_3 - q_0q_2) \\ 2(q_1q_2 - q_0q_3) & q_0^2 - q_1^2 + q_2^2 - q_3^2 & 2(q_2q_3 - q_0q_1) \\ 2(q_1q_3 + q_0q_2) & 2(q_2q_3 - q_0q_1) & q_0^2 - q_1^2 - q_2^2 + q_3^2 \end{pmatrix} \mathbf{B}_{ECI} \quad (9)$$

2.4 Orbit Generation: Keplerian Dynamics

The Kepler's equation is solved to generate the coordinates (r, θ, ϕ) required by IGRF model. The orbit and the position of satellite in the orbit are parameterized using six Kepler elements. They are (1) specific angular momentum (h), (2) inclination (i), (3) right ascension of ascending node (Ω), (4) eccentricity e, and (5) argument of perigee (ω) and true anomaly (θ). It should be noted that the specific angular momentum and true anomaly are often replaced with semimajor axis (a) and mean anomaly (M), respectively [11].

Kepler's equation is used to update the true anomaly which represents the position of satellite as it propagates in its orbit. The equation is,

$$M = E - e \sin E \quad (10)$$

where M is mean anomaly and E is eccentric anomaly. This transcendental equation is solved for E using Newton's iterative method. Using this value of E, true anomaly (θ) is computed using the relation,

$$\tan^2 \frac{E}{2} = \sqrt{\frac{1-e}{1+e}} \tan \frac{\theta}{2} \quad (11)$$

It is important to note that the initial value of E is not determined using 10 but 11.

Equations 10 and 11 are run in each time step of simulation, and the evolving values of θ represent the position of CubeSat in orbit. However, the Earth is not perfectly spherical in shape and is bulged in the equator. Hence, the gravity of Earth is not perfectly directed to its center. So, for LEO satellites, the gravitational field is not only dependent on its distance from center of the Earth. The second zonal harmonic is the parameter that quantifies this effect and has dimensionless value of $J_2 = 1.08263 \times 10^{-3}$. This effect causes the right ascension Ω and the argument of periapsis ω to change with time which must be considered for LEO satellites. This change is expressed in equations below,

$$\dot{\Omega} = -\left[\frac{3\sqrt{\mu} J_2 R^2}{2(1-e^2)^2 a^{\frac{7}{2}}} \right] \quad (12)$$

$$\dot{\omega} = -\left[\frac{3\sqrt{\mu} J_2 R^2}{2(1-e^2)^2 a^{\frac{7}{2}}} \right] \left(\frac{5}{2} \sin^2 i - 2 \right) \quad (13)$$

where R and μ are the radius and gravitational parameter of the Earth, respectively. a and e are the semimajor axis and eccentricity of the orbit, respectively, and i is the orbit's inclination. The orbit is generated by propagating equations (10), (12), and (13) in each time step of the simulation.

Equations (10), (12), and (13) are used to update the Kepler elements as the satellite moves in orbit. The coordinates that are required is not in terms of Kepler elements but in spherical coordinate in ECI. So, in each time step, the updated Kepler elements are converted to the cartesian position vector of CubeSat in ECI frame. Then, this position vector is transformed into spherical coordinates (latitude, longitude, and height) which is used by the IGRF model as input. Curtis [11] presents a detailed algorithm on these transformations.

Transformation of Kepler elements to the position vector in ECI frame. First of all, the Kepler elements are converted into the position vector in perifocal frame [11].

$$\mathbf{r}_{PQW} = \frac{h^2}{\mu} \left(\frac{1}{1 + e\cos\theta} \right) \begin{pmatrix} \cos\theta \\ \sin\theta \\ 0 \end{pmatrix} \tag{14}$$

Then, the coordinate transformation of position vector of satellite in perifocal frame $\mathbf{r}_{PQW}$ to ECI frame $\mathbf{r}_{ECI}$ is performed using direction cosine matrix Q.

$$\mathbf{r}_{ECI} = Q^T \mathbf{r}_{PQW} \tag{15}$$

The 3-1-3 rotation sequences from ECI to perifocus frame are: right ascension of ascending node (Ω), inclination (i), and argument of perigee (ω). The direction cosine matrix (Q), which maps vectors in ECI into vectors in perifocal frame, is defined as [13] [11],

$$Q = \begin{pmatrix} c\theta_3 c\theta_1 - s\theta_3 c\theta_2 s\theta_1 & c\theta_3 s\theta_1 + s\theta_3 c\theta_2 c\theta_1 & s\theta_3 s\theta_2 \\ -s\theta_3 c\theta_1 - c\theta_3 c\theta_2 s\theta_1 & -s\theta_3 s\theta_1 + c\theta_3 c\theta_2 c\theta_1 & c\theta_3 s\theta_2 \\ s\theta_2 s\theta_1 & -s\theta_2 c\theta_1 & c\theta_2 \end{pmatrix} \tag{16}$$

where $c\theta_i = \cos\theta_i$ and $s\theta_i = \sin\theta_i$ are the shorthand notations, and $\theta_1 = \Omega$, $\theta_2 = i$, and $\theta_3 = \omega$ are the rotation sequences.

Conversion of position vector in cartesian to spherical coordinates. The position vector in ECI in cartesian coordinates is converted to spherical coordinates (r, θ, ϕ) using the following relations [11].

$$r = |\mathbf{r}_{ECI}| \tag{17}$$

$$\theta = \sin^{-1} n \tag{18}$$

$$\phi = \begin{cases} cos^{-1}(l/cos\theta), & \text{if } m > 0 \\ 2\pi - cos^{-1}(l/cos\theta), & \text{if } m \le 0 \end{cases} \tag{19}$$

Here, l, m, and n are the direction cosines of the position vector in ECI frame. r, θ, and ϕ are the geocentric height, latitude, and longitude of the satellite which are input to the IGRF model.

2.5 Satellite Rotational Mechanics

The rotational dynamics of CubeSat is modeled using the rigid body kinematics and the Euler's equation of motion. Both of these are coupled differential equations and are integrated numerically using the RK4 method to find the orientation and the angular velocity of the satellite in the next time step.

Kinematics. The rotation of the CubeSat is numerically simulated by integrating the kinematic differential equation for the quaternions as presented in [13].

$$\begin{pmatrix} \dot{q}_0 \\ \dot{q}_1 \\ \dot{q}_2 \\ \dot{q}_3 \end{pmatrix} = \frac{1}{2} \begin{pmatrix} 0 & -\omega_1 & -\omega_2 & -\omega_3 \\ \omega_1 & 0 & \omega_3 & -\omega_2 \\ \omega_2 & -\omega_3 & 0 & \omega_1 \\ \omega_3 & \omega_2 & -\omega_1 & 0 \end{pmatrix} \begin{pmatrix} q_0 \\ q_1 \\ q_2 \\ q_3 \end{pmatrix} \tag{20}$$

Here, (q_0, q_1, q_2, q_3) is the orientation of the satellite in quaternion, and w_1, w_2, and w_3 are the angular velocities of the CubeSat about each of body axes. This coupled differential equation describes the evolution of the orientation of CubeSat.

Kinetics. The angular velocity of the satellite is propagated using the differential equation obtained from the Euler's second law of motion on the rigid body [13, 14],

$$\dot{\mathbf{H}} = \tau_{ext} \tag{21}$$

Here, $\mathbf{H}$ is the angular momentum in inertial frame and τ_{ext} is the external torque acting on the satellite. In this case, the source of torque is magnetorquer, so $\tau_{ext} = \tau_{mt}$, where τ_{mt} is torque generated using magnetorquers. By applying the called transport theorem [13] and simplification, it is obtained as,

$$\dot{\omega} = I^{-1}(-\omega \times I\omega + \tau_{mt}) \tag{22}$$

where I is the inertia matrix of the CubeSat and it is assumed to be diagonal, ω is the angular velocity about the body axis of the satellite.

Table 1 Orbital parameters, simulation parameters, satellite parameters, and the initial conditions for simulation

Orbital parameters, CubeSat configs and initial conditions	Values	Units
Principal moment of inertia	(0.33, 0.37, 0.35)	kg m^2
Max magnetic moment	2.0	A m^2
No. of turns of coil	2000	Turns
Cross section of coil	0.02	m^2
Relative magnetic permeability of core	5500	–
Length of core	0.0800	m
Radius of core	0.0050	m
Simulation time	0.2684	days
Step time	1	secs
Semimajor axis	6978	km
Eccentricity	0	–
Inclination	56	degrees
Argument of periapsis	0	degrees
Longitude of the ascending node	0	degrees
True anomaly	0	degrees
Initial roll, pitch, and yaw	(0.0, 0.0, 0.0)	rad
Initial angular velocities	(0.6, −0.5, −0.4)	rad/sec

3 Simulation and Results

The simulation is performed with the architecture presented in Fig. 1. The CubeSat is assumed to be at the circular orbit of 56 degrees inclination at the height of 6978 km. The control law (7) is used and the gain is estimated using (8). All of the initial conditions and the parameters are shown in Table 1.

Figure 2 shows the nadir projection of CubeSat on the Earth as it revolves around the Earth.

Figure 3 is the magnetic field as observed from the ECI frame. However, the magnetic field in the body frame as seen in Fig. 4 is more chaotic than the field in ECI. This is because the observer (magnetometer fixed to the satellite body) itself is rotating.

It is seen that the field in the body frame gets less random and periodic as time passes. This is because the angular velocity of the satellite is slowly decreasing with time as shown in Fig. 5.

Figure 5 illustrates the detumbling of CubeSat. The magnitude of the angular velocities in each axis is converging to zero with time. Moreover, it is also seen that the angular velocities are converging to zero as claimed by [5]. This shows that the CubeSat is no longer spinning and is ready for fine pointing control strategy. It is observed that the magnetic field in body frame after the CubeSat detumbles is smooth

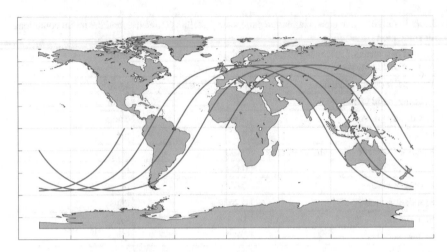

Fig. 2 Groundtrack of the CubeSat

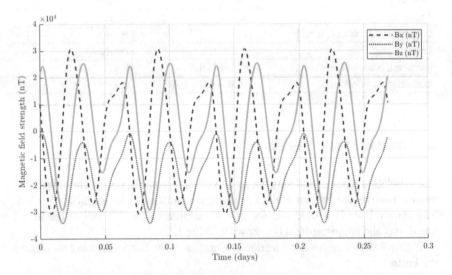

Fig. 3 Magnetic field strength in ECI frame

and periodic. This is because the sensors will measure similar value as the CubeSat travels around the orbit and comes back to same spot without spinning.

Similarly, Fig. 6 shows that the orientation of the CubeSat is gradually being settled to some value as the satellite detumbles. It must be noted that this orientation to is not necessarily the desired orientation. In satellite missions, after successful detumbling, the fine attitude control system is activated to point satellite to the desired orientation.

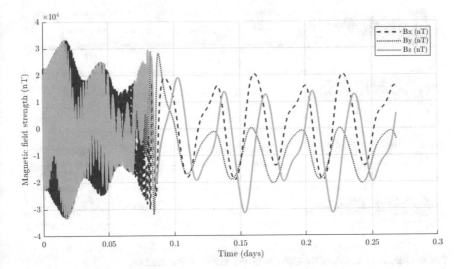

Fig. 4 Magnetic field strength in body frame

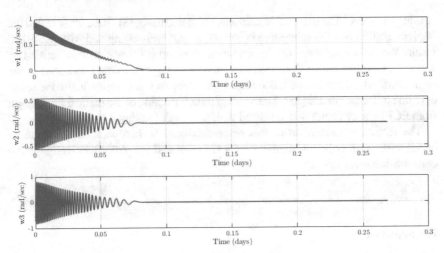

Fig. 5 Angular velocity profile of CubeSat as it detumbles

4 Conclusions

The numerical simulation of magnetic detumbling is performed in a CubeSat spinning on its way around the Earth in circular orbit using a form of B-Dot controller. The CubeSat is fitted with three magnetorquers with ferromagnetic core and three-axis sensors to measure magnetic field and angular velocity. The reason behind selecting this variant of control law is its analytical proof of the global convergence of angular velocity for underactuated satellite. Figure 5 shows the angular velocity of satellite

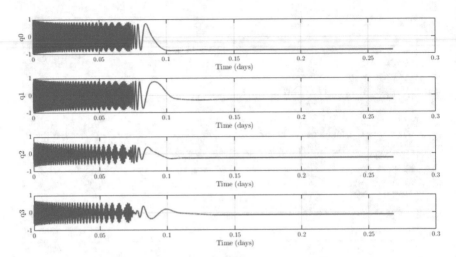

Fig. 6 Orientation profile of the CubeSat in quaternions as it detumbles

converged to zero. The reasonable gain was estimated using (8). To make sure that the implementation of each subsystem is correct, results were compared with published results and online sources. The implementation of orbit generation was verified by generating the ground track of Fig. 4.23 of [11] with given initial conditions in the book. Similarly, the result of IGRF model was compared and verified with the result of online IGRF calculator [15]. The result of transformation of magnetic field strength from ECI to body frame was validated with the result presented in [9].

The further advancement of this research would be hardware implementation of the system on onboard computer of a CubeSat and magnetorquer design using optimization techniques.

References

1. White, J.S., Shigemento, F.H., Bourquin, K.: Satellite Attitude Control Utilizing the Earth's Megnetic Field. Ames Reasearch Center, NASA, Washington (1961)
2. Swartwout, M.: The first one hundred Cubesats, a statistical look. J. Small Satellites **2**(2), 213–233 (2013)
3. Polat, H.C., Virgili-Llop, J., Romano, M.: Towards the thousandth Cubesat, a statistical overview. J. Small Satellites **5**(3), 513–530 (2019)
4. Celani, F.: Robust three-axis attitude stabilization for inertial pointing spacecraft using magnetorquers. Acta Astronaut. **107**(3), 87–96 (2015)
5. Avanzini, G., Giulietti, F.: Magnetic detumbling of a rigid spacecraft. J. Guid. Control Dyn. **35**(4), 1326–1334 (2012). https://doi.org/10.2514/1.53074
6. Bellini, N.: Magnetic actuators for nanosatellite attitude control. Master's thesis (2014)

7. Coverstone-Carroll, V.: Detumbling and reorienting underactuated rigid spacecraft. J. Guid. Control Dyn. **19**(3), 708–710 (1996). https://doi.org/10.2514/3.21680
8. Mehrjardi, M.F., Mirshams, M.: Design and manufacturing of a research magnetic torquer rod. Contemp. Eng. Sci. **3**, 227–236 (2013)
9. Monkell, M., Montalvo, C., Spencer, E.: Using only two magnetorquers to de-tumble a 2u Cubesat **62**, 3086–3094 (2020). https://doi.org/10.1016/j.asr.2018.08.041
10. Leomanni, M.: Comparison of control laws for magnetic detumbling. Research Gate (2012)
11. Curtis, H.D.: Orbital Mechanics for Engineering Students. Butterworth-Heinemann, Oxford (2020)
12. Davis, J.: Mathematical modeling of earth's magnetic field (2010). https://hanspeterschaub.info/Papers/UnderGradStudents/MagneticField.pdf
13. Schaub, H., Junkins, J.L.: Analytical Mechanics of Aerospace Systems. American Institute of Aeronautics and Astronautics Inc, Virginia, USA (2009)
14. Wise, E.D.: Design, analysis, and testing of a precision guidance, navigation, and control system for a dual-spinning Cubesat. Master's thesis (2013)
15. International geomagnetic reference field (IGRF), 13th generation calculator. http://www.geomag.bgs.ac.uk/data_service/models_compass/igrf_calc.html

A System of Vehicular Motion Sensing and Data Acquisition over Thapathali–Kupondole Bridge and Impact Prediction and Analysis Using Machine Learning

Amit Paudyal, Nirdesh Bhattarai, Shiva Bhandari, Nabin Rai, Ram Prasad Rimal, and Nanda Bikram Adhikari

Abstract The analysis of traffic load and its impact on road bridges has become an important method to ensure the longevity of the bridges. Heavy and rapid moving vehicles cause huge strain and deflection of bridges. The study of the dynamic response of bridges and their trend over a period allows the continuous assessment of their deteriorating conditions. Visual inspections and surveys only give an outlook on the condition of the exterior of the bridge. A network of wireless sensors measuring dynamic characteristics such as strain, deflection, and vibration can assist in understanding the overall bridge structure conditions. In our study, an array of high-precision inertial sensors has been deployed on both the old and the new sections of the Bagmati bridge. Huge amounts of data are extracted from each sensor and then preprocessed to remove corrupt data, outliers, and obtain clean datasets. This process is followed by visualization of the preprocessed data to derive initial observations. Statistical analysis is carried out to study the nature of obtained data, and machine learning algorithms have been applied to predict the future trends.

Keywords Structural health monitoring · Bridge dynamic response · LSTM · Inertial sensors · Visual analysis · Time domain forecasting · Statistical analysis · RNN · ARIMA

The authors deeply acknowledge the Higher Education Reform Project (HERP), TU, Kirtipur for providing us with the research grants without which this research would not have been possible.

A. Paudyal · N. Bhattarai · S. Bhandari · N. Rai · N. B. Adhikari (✉)
Department of Electronics and Computer Engineering, Pulchowk Campus, Institute of Engineering, Tribhuvan University, Lalitpur, Nepal
e-mail: adhikari@ioe.edu.np

R. P. Rimal
Ramlaxman Innovations, Kathmandu, Nepal

S. Shakya et al. (eds.), *Proceedings of International Conference on Sustainable Expert Systems*, Lecture Notes in Networks and Systems 176,
https://doi.org/10.1007/978-981-33-4355-9_41

1   Introduction

According to Nepal Bridge Standards-2067, "all the permanent bridges shall be designed for a design life of minimum 50 years". But there are bridges across the country that are in use even after their life expectancy has ended. Reasons such as significant environmental damage, varied load, and traffic conditions, corrosion, cracking and other structural damages cause the design life not to match the overall practical life expectancy. This also has a huge impact on the bridge's load-carrying capacity that further diminishes the actual lifetime of the bridge. Moreover, in the context of Nepal, these bridges are used past their life expectancy without proper and periodic monitoring and examining parameters that continuously make bridges structurally deficient. Specifically, the core problem resides within the Kathmandu valley. Mostly, the crowded and highly demanding populations dwell in this constricted region of the country. The more the population increases, the more robust transportation infrastructure is needed. One of the heavily jammed places is the Kupondole–Thapathali bridge. To add to the condition, there are dynamic loads, that means the bridges suffer from varying loads of vehicles falling under different categories according to their weight. This all relates to the fact that there should be periodic monitoring or inspection. Thus, the project features one of the highly busy and heavily loaded bridges of Kathmandu valley, Thapathali bridge, which links two districts Kathmandu and Lalitpur. There are two bridges at Thapathali, an old one and a new one. This project focuses on real-time sensory data collected from these bridges, their storage, and big data-based sensory data analysis and visualization to classify anomaly parameters.

The nationwide bridge survey uses deep learning techniques [1], and a model-free damage detection approach uses machine learning techniques [2] to evaluate the bridge serviceability in accordance to real-time sensory data or archived bridge-related data such as traffic status, weather conditions, and bridge structural configuration. Applying deep learning [3] and machine learning perspective to SHM consists of training algorithms so that it is able to classify the structure state of health-based only on data about a given damage-sensitive feature [4], and a wireless sensor network (WSN) for SHM is designed [5]. LSTM NN can achieve the best prediction performance in terms of both accuracy and stability [6]. The algorithms for vehicle detection and axle estimation based on the findings are derived, and their performance is evaluated and discussed[7]. Deep learning and machine learning tools have been used for big data time series forecasting [8–13]. Moreover, hybrid models of ARIMA and ANNs are often compared with mixed conclusions in terms of the superiority in forecasting performance [14, 15].

2   Methodology

Implementation of inertial sensors in structures like bridges is necessary to observe the inner conditions which cannot be seen by visual inspection. Visualization of traffic trends can help develop necessary regulations to restrict heavy loads and ensure a long life of the bridge. Prediction of future trends can help decide optimal time for repair and improvement as well as limit catastrophic failures. The purpose of this project is to develop an efficient data acquisition and processing system that allows inspectors and researchers to monitor the status of the bridge in near real-time. This data can be used to develop appropriate regulations on the vehicle weight limit, traffic density, as well as on repair and upgrade.

So far, structural health monitoring in Nepal has not been implemented on real road bridges. The conditions of road bridges have been monitored through visual inspection which is time-consuming and causes traffic movement restrictions and disturbances. The data obtained from such methods may be misleading or incomplete and may not indicate the true status of the bridge.

2.1 System Block Diagram

The major components of the project model are depicted in a block diagram as shown in Fig. 1, and the processes involved are described briefly in subsections below.

Data Extraction The recorded data is extracted from the SD card on the sensor module.

Fig. 1 System block diagram

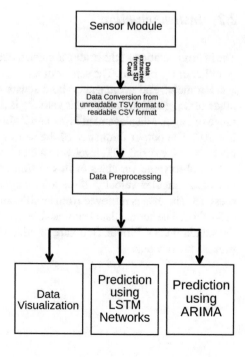

Data Conversion The data is originally stored in a compressed unreadable format as a .tsv (tab-separated values) file. This is converted into a readable .csv (comma separated values) file using recompilation software provided by the sensor manufacturer.

Data Preprocessing The data retrieved from the sensor is first converted into a readable format using the recompilation software provided by the sensor manufacturer. All the corrupted values like missing values and Not a Number (NaN) are removed. The retrieved data is originally retrieved at a sample period of 200 milliseconds. The data is resampled to a rate of 5 s for clear visualization and reducing the training time. Only the required columns are extracted, and a time series data set is formed by using the sensor's recording time as a time index.

Data Visualization The preprocessed data is visualized using a plotting library in Python. Visual inspection shows the periodic trend of vibrations increasing during the day and decreasing during the nights. The overall amplitude is lower during the holidays.

Prediction Using LSTM Networks The preprocessed data is split into training and test set and used for prediction using LSTM networks.

Prediction Using ARIMA The time-series data set is also split into training and test set and used for prediction using the ARIMA model.

2.2 Instrumentation

The instrumentation of the bridge is carried out using an array of three sensors placed on either of the bridges. The sensor module shown in Fig. 2 is a high-precision gyro, accelerometer, and geomagnetic field sensor with Bluetooth 4.0 transmission. The range of acceleration and angular velocity is ± 16 g and $\pm 2000°$/s, respectively. The range of angular displacement in X- and Z-direction is $\pm 180°$ and in the Y-direction is $\pm 90°$. The output frequency of the sensor is in the range 0.1–100 Hz. The data interface is the serial TTL level with a baud rate of 115,200.

The observable variables of this customized sensor are three-axis acceleration, three-axis angular velocity, three-axis magnetic field angle, temperature, and air pressure. The data is retrieved from an SD card in the form of a tab-separated values (TSV) file. The source data is not usable and thus requires processing to obtain it in the standard CSV format. Software provided by the sensor manufacturer is used to process the raw data.

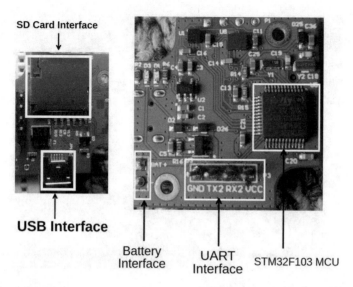

SD Card Interface

USB Interface

Battery
Interface

UART
Interface

STM32F103 MCU

Fig. 2 Sensor module

2.3 Observations

There are two beam-type bridges, each one having a length 184 m and a width 6 m where the construction date is 1967 AD and 1995 AD for the old one and the new one, respectively. Before the installation of the array of sensors, it was essential to identify and locate the sections for optimum vibration recording and collection. So testing of possible sites for sensors was done at five different locations on the bridge road surface. Figure 3 depicts the test points for observation before selecting the sensor deployment location.

With structural experts' advice, six sections along with the sensor's orientation arrangement were selected for sensor deployment on the Thapathali–Kupondole bridge as depicted in Fig. 4. The sensor nodes are fixed on the structure under the bridge. The main sensor is enclosed inside a waterproof box which is bolted on the steel frame under the road. The data is extracted manually from the SD card on the sensor. The sensor box consists of a button to turn on the sensor and LED indicators to show if the sensor is turned on or off. The sensor nodes have high flexibility, i.e. they can easily be installed and uninstalled in any position. The sensor installation line was selected under the bridge as shown in Fig. 5.

Fig. 3 Test points before sensor installation

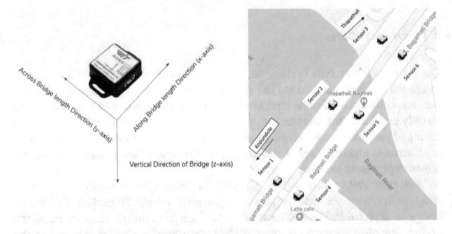

Fig. 4 Illustration of Sensors installation in the Thapathali–Kupondole bridge

Fig. 5 Sensor installation site

2.4 Time Series Modelling Using LSTM

One of the approaches in time series forecasting is the use of recurrent neural networks (RNNs). However, long short-term memory (LSTM) has proven to be better than RNN as it solves long-term dependencies and vanishing and exploding gradient problems. The basic unit of LSTM is its cell or cell state which provides memory to LSTM so it can remember the sequence. It consists of three gates: input gate, forget gate, and output gate. The following are the relevant equations for these gates:

$$i_t = \sigma(w_i[h_{t-1}, x_t] + b_i) \tag{1}$$

$$f_t = \sigma(w_f[h_{t-1}, x_t] + b_f) \tag{2}$$

$$o_t = \sigma(w_o[h_{t-1}, x_t] + b_o) \tag{3}$$

where i_t, f_t and o_t represent input gate, forget gate and output gate; σ represents sigmoid activation function; w_x represents the weight for respective gate neurons; h_{t-1} represents output of the previous LSTM block; x_t represents input of current time stamp and b_x represents biases for respective gates.

The main components of LSTM are cell states and gates. The cell state carries relevant information throughout the long sequence. The task of adding or removing the information to the cell state in the chain of sequences is of the gates. Gates are actually responsible for regulating information flow in an LSTM cell. Basically, an LSTM cell consists of three varieties of gates. The forget gate determines what information is to be thrown away or kept from the previous steps. The sigmoid function takes information from the previous hidden state as well as information from the current input state and then outputs values between 0 and 1. The value close to zero means to forget, and the value close to one means to keep. Then, there is an input gate which functions to update the cell state. It consists of sigmoid activation where the previous hidden state and current input are passed and the input gate determines what values to be updated as values between 0 and 1 means not important and 1 means important. In short, it determines what information is relevant to be added from the current step. At last, there is an output gate that determines what the next hidden should be. Actually, the hidden state carries information on the previous inputs. It can be said that the gating mechanism enables LSTM to model the long-term dependency so well.

Basically, the sigmoid function is used for gates so that the gate is able to give positive values and thus distinct answers whether to keep or discard a particular feature. Similarly, there is tanh activation function that limits the value between -1 and 1. The equations used for the cell state and the final output are as follows:

$$\hat{c}_t = \tanh(w_c[h_{t-1}, x_t] + b_c) \tag{4}$$

$$c_t = f_t * c_{t-1} + i_t * \hat{c}_t \tag{5}$$

$$h_t = o_t * \tanh(c_t) \tag{6}$$

where $\hat{c}_t$ represents candidate for cell state or memory at time stamp t; c_t represents cell state or memory at time stamp t and h_t represents output of the current LSTM block; w_c represents the weight for cell state or memory and bc represents bias for cell state.

2.5 Time Series Modelling Using ARIMA

Auto-regressive integrated moving average (ARIMA) model has been well fitted for time series forecasting problems based on its own lags and lagged errors or past values to predict future values. Generally, ARIMA models are defined by three parameters, namely p, d, and q which are all non-negative integers depicting the model as ARIMA(p, d, q). The further description of these parameters is as follows:

p: refers to the number of lags of time observations to be used as predictors also called lag order and is also known as the order of "auto-regressive" term.

d: refers to the number of times the raw observations are differenced to make time-series stationary and is known as the degree of differencing.

q: refers to the number of lagged forecast errors that should go in the ARIMA model and is also known as the order of the "moving average" term.

If there are seasonal patterns in a time series, then seasonal terms are to be added, and thus, it becomes seasonal ARIMA (SARIMA). The general equation of an ARIMA model becomes:

$$Y_t = \alpha + \beta_1 Y_{t-1} + \beta_2 Y_{t-2} + \ldots + \beta_p Y_{t-p} + \epsilon_t + \phi_1 \epsilon_{t-1} + \phi_2 \epsilon_{t-2} + \ldots + \phi_q \epsilon_{t-q} \tag{7}$$

where Y_t represents the predicted value of y; p is the order of AR term and $\beta_1, \beta_2, \ldots, \beta_p$ represent the auto-regressive parameters; q is the order of MA term and $\phi_1, \phi_2, \ldots, \phi_q$ represent the moving average parameters; ϵ_t represents the error term at timestamp t and α represents a constant.

3 Results and Discussions

3.1 Statistical Analysis

The data collected for this project is the measurement of inertial parameters and temperature collected over a period of 10 days from 2020-02-18 to 2020-02-28 with a sampling rate of 200ms. The data set consists of 13 fields covering acceleration, angular velocity, angular displacement, and magnetic field in three direc-

Table 1 Vibration data distribution for sensor 1

	Count	Mean	SD	min	25%	50%	75%	max
az(g)	4248290	1.01574	0.00863	−0.8276	1.0151	1.0156	1.0166	1.6987

Table 2 Vibration data distribution for sensor 4

	Count	Mean	SD	min	25%	50%	75%	max
az(g)	209290	1.00044	0.0579	−1.3691	1.002	1.0024	1.0029	1.1816

tions and a temperature parameter. However, it mainly discusses and analyses the vibration(acceleration) of the bridge in the vertical direction(az) in this section. Due to interruptions and failure in the sensors, a significant amount of data has been extracted only through sensors 1 and 4.

There are several statistical tests known as normality tests that can be used for inferring whether the data appears as drawn from a Gaussian distribution. For this purpose, the D'Agnostino's K^2 test was used. In this test, the p-value can be interpreted as follows:

If $p <=$ alpha; then reject the null hypothesis H0; not normally distributed
If $p >$ alpha; then accept the null hypothesis H0; normally distributed

The value of alpha was fixed to 0.05, and then, this test was applied to the data sets of sensor 1 and sensor 4. The result of the statistical test for sensor 1 data was found to be as follows: statistics = 14917213.591, $p = 0.000$.

From the results, it is concluded that the data for sensor 1 data was not normal and thus reject the null hypothesis H0. Similarly, the results of the statistics test for sensor 4 data were found to be as follows: Statistics = 559346.244, $p = 0.000$. Also, the data for sensor 4 data was not normal and thus reject the null hypothesis H0. Moreover, kurtosis and skewness were calculated. The skewness defines how much distribution is pushed left or right, and kurtosis defines how much of the distribution is in the tail. The following results were obtained for sensor 1 data: kurtosis = 6488.68643 and skewness = −68.81102, and the following results were obtained for sensor 4 data: kurtosis = 1088.71485 and skewness = −32.88305.

Since the kurtosis of our sensor 1 and sensor 4 data was greater than zero, it implies that the data is heavily tailed. Similarly, the skewness of our sensor 1 and sensor 4 data was less than zero which implies that the data is highly skewed on the left side.

The summary of data obtained from sensor 1 and sensor 4 is shown in Tables 1 and 2, respectively.

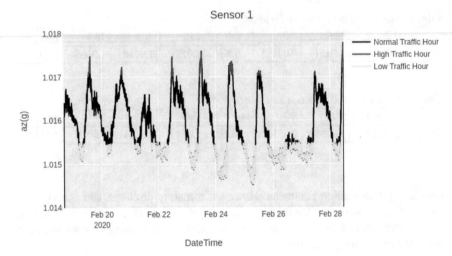

Fig. 6 Plot of az for 10 days from sensor 1

3.2 Analysis of Vehicular Motion on Bridges

Visual analysis of the vertical acceleration plot of sensor 1 as shown in Fig. 6 shows a seasonal trend occurring over a period of a day. This indicates that there is not much difference in traffic density on normal weekdays. The overall acceleration amplitude is lower on holidays due to less movement of traffic. The time series has been resampled at a rate of five minutes for better visual clarity.

The trend of vertical acceleration in sensor 5 is the same as that of sensor 1. Due to sensor failure, the data has been recorded for only three days. This indicates that the traffic density follows the same trend in both the old bridge and the new bridge.

From the vertical acceleration plot from Fig. 7, a maximum impact that occurs on the bridge is observed during midday and minimum impact occurs during the early hours of the day. This indicates that there is peak traffic density at midday and least during midnight. Similarly, the vertical vibration on the bridge at peak traffic hour and low traffic hour has been illustrated in Fig. 8.

The overall vibration is higher during normal days and low on holidays as seen in Fig. 9. From the comparative plot as shown in Fig. 10, it is observed that the z-axis acceleration provides vertical vibration of the bridge as the value varies about 1g. The y-axis acceleration is found to be the lateral vibration of the bridge as it has a significant amplitude but not maximum. The x-axis acceleration is the longitudinal vibration direction of the bridge as it has the lowest amplitude.

Since the most impact is along the z-axis, so the plots of z-axis vibration data are mainly considered.

Modeling Using LSTM Prediction of vertical vibrations for the next time periods was carried out using the long short-term memory (LSTM) sequence model. The

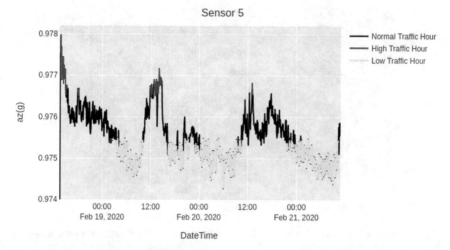

Fig. 7 Plot of az for 3 days obtained from sensor 5

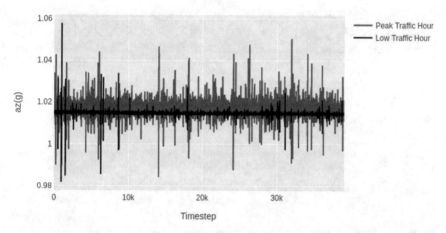

Fig. 8 Vertical vibrations in peak traffic hour vs low traffic hour as seen in the data obtained from sensor 1

data set obtained after preprocessing was first resampled into a period of 5 seconds and then split into train and test sets in ratios of 80% and 20%, respectively. An LSTM model consisting of 128 units was designed, and the model was fit for 20 epochs with a batch size of 32. The loss function used was mean square error (MSE). The optimizer used for fitting was Adam with a learning rate of 0.001. After fitting, the model was used for prediction and compared with test set values. The test set root-mean-squared error (RMSE) was found to be 0.00138 g.

The root-mean-squared error (RMSE) for both train and test data was found to be extremely low. This is because, in large structures, there is not much variation in their response for a large period of time, and the vibration data was found to be very

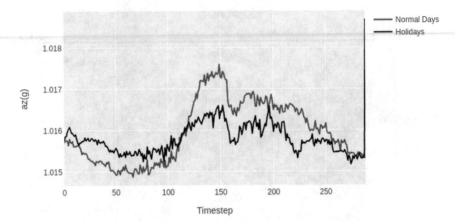

Fig. 9 Vertical vibrations during normal days vs holidays as seen in the data obtained from sensor 1

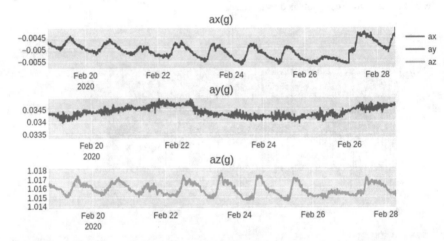

Fig. 10 Acceleration in three different axes as seen from the data collected from sensor 1

similar. The forecast by this model is depicted in Fig. 11, and train vs test error plot is shown in Fig. 12.

Modeling Using ARIMA An ARIMA(3,0,0) model was used for the modeling of time series data. The data set used was the vertical vibrations obtained in sensor 1. The data set was resampled to a period of 30 min which would result in nearly 48 points each day. In total, 471 points were obtained indicating the measurement period of nearly 10 days. The data set was split into a training set and test set in the ratio of 66% and 33%. Hence, the data obtained over a period of a week was used to forecast for the next 3 days. The forecast produced by the model is as shown in Fig. 13. It is observed that the forecast follows the actual data very closely. This is

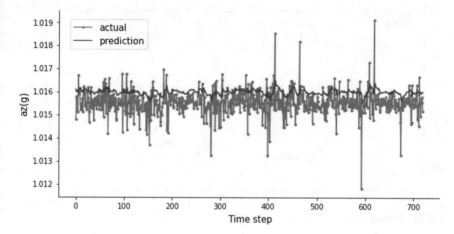

Fig. 11 LSTM prediction versus actual test set

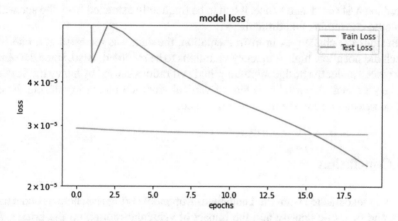

Fig. 12 Train loss and test loss

also because the sensor records vibrations in g, and due to the position of sensors, the variation is extremely little.

4 Limitations

There were a number of challenges encountered during this research. Most of the challenges occurred in the instrumentation process. During the first deployment, the sensor box was lost because of its improper placement and installation. A number of sensor modules encountered frequent power failures that led to improper measurements and the generation of corrupted data. The data recorded by the sensor was

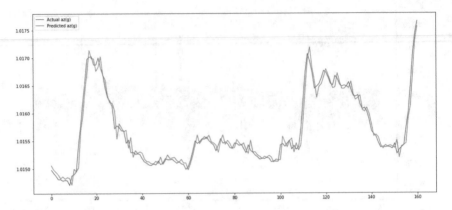

Fig. 13 Arima prediction versus actual test data

stored in an SD card, and hence, it had to be frequently extracted from the site which was time-consuming and difficult.

Besides the challenges in instrumentation, the data was recorded at a rate 5 Hz which did not allow high-frequency vibrations to be recorded. Also, since the sensor was placed under the bridge, high-amplitude vibrations caused by heavy traffic could not be recorded. A periodic pattern of vertical vibration increasing during the day and decreasing during the night was obtained.

5   Conclusion

The data set collected from the Thapathali–Kupondole bridge can help us understand the trend of traffic density and the impact of vehicular motion on the bridge. The root-mean-squared error (RMSE) of the LSTM model used for forecasting was found to be 0.00138 g. The ARIMA model also generates a very close forecast with respect to the actual data. The vibrations collected in g have a very narrow range. This is because the change in response of the structure is slow and may take years to show significant changes. The collection of data over 10 days can help us understand the trend of traffic density but may not provide significant input in predicting the future response of the bridge.

For better accuracy in vehicular motion sensing, the sensors must be placed on the surface of the road. The fusion of vibration and magnetic deflection can help detect the type and the weight of the vehicle. The period and shape of the magnetometer signal can help predict the speed and direction of the vehicle. Further, the verification of the obtained data can be done by the means of the camera. This allows us to differentiate between actual vehicular impact and sensor noise.

Since this is an ongoing project, it can be further improved by experimenting with different sensor locations. Finite element analysis can provide the expected response of the bridge and that can be compared with our observations. Theoretical analysis and data-driven analysis can be used together to build a more accurate and efficient solution.

References

1. Liang, Y., Wu, D., Liu, G., Li, Y., Gao, C., Ma, Z., Wu, W.: Big data-enabled multiscale serviceability analysis for aging bridges. Digital Commun. Networks **2** (2016). https://doi.org/10.1016/j.dcan.2016.05.002
2. Neves, A., Gonzalez, I., Leander, J., Karoumi, R.: A new approach to damage detection in bridges using machine learning, pp. 73–84 (2018). https://doi.org/10.1007/978-3-319-67443-8_5
3. Azimi, M., Pekcan, G.: Structural health monitoring using extremely-compressed data through deep learning. Comput.-Aided Civil Infrastructure Eng. (2019). https://doi.org/10.1111/mice.12517
4. Khouri Chalouhi, E., Gonzalez, I., Gentile, C., Karoumi, R.: Damage detection in railway bridges using machine learning: application to a historic structure. Proc. Eng. **199**, 1931–1936 (2017). https://doi.org/10.1016/j.proeng.2017.09.287
5. Kim, S., Pakzad, S., Culler, D., Demmel, J., Fenves, G., Glaser, S., Turon, M.: Health monitoring of civil infrastructures using wireless sensor networks, pp. 254–263 (2007). https://doi.org/10.1109/IPSN.2007.4379685
6. Ma, X., Tao, Z., Wang, Y., Yu, H., Wang, Y.: Long short-term memory neural network for traffic speed prediction using remote microwave sensor data. Transp. Res. Part C: Emerg. Technol. **54** (2015). https://doi.org/10.1016/j.trc.2015.03.014
7. Hostettler, R.: Traffic counting using measurements of road surface vibrations (2009)
8. Yuankai, W., Tan, H., Ran, B., Jiang, Z.: A hybrid deep learning based traffic flow prediction method and its understanding. Transp. Res. Part C: Emerg. Technol. **90** (2018). https://doi.org/10.1016/j.trc.2018.03.001
9. Torres, J., Galicia de Castro, A., Troncoso, A., Martínez-Álvarez, F.: A scalable approach based on deep learning for big data time series forecasting. Integrated Comput.-Aided Eng. **25**, 1–14 (2018). https://doi.org/10.3233/ICA-180580
10. Galicia de Castro, A., Talavera-Llames, R., Troncoso, A., Koprinska, I., Martínez-Álvarez, F.: Multi-step forecasting for big data time series based on ensemble learning. Knowl.-Based Syst. (2018). https://doi.org/10.1016/j.knosys.2018.10.009
11. Ni, F., Zhang, J., Noori, M.: Deep learning for data anomaly detection and data compression of a long-span suspension bridge. Comput.-Aided Civil Infrastructure Eng. (2019). https://doi.org/10.1111/mice.12528
12. Tang, Z., Chen, Z., Bao, Y., Li, H.: Convolutional neural network-based data anomaly detection method using multiple information for structural health monitoring. Struct. Control Health Monitor. (2018). https://doi.org/10.1002/stc.2296
13. K., K., Raj, J.: Big data analytics for developing secure internet of everything. J. ISMAC **01**, 49–56 (2019). https://doi.org/10.36548/jismac.2019.2.006
14. Zhang, P.: Zhang, g.p.: Time series forecasting using a hybrid Arima and neural network model. Neurocomputing **50**, 159–175 (2003). https://doi.org/10.1016/S0925-2312(01)00702-0
15. Khashei, M., Bijari, M.: A novel hybridization of artificial neural networks and Arima models for time series forecasting. Appl. Soft Comput. **11**, 2664–2675 (2011). https://doi.org/10.1016/j.asoc.2010.10.015

Semantic Interoperability for Development of Future Health Care: A Systematic Review of Different Technologies

Gajanan Shankarrao Patange, Zankhan Sonara, and Harmish Bhatt

Abstract World is getting smaller and smaller in a digital era but accurate and precise health care depends upon high-quality information. Attributes of patients, manufacturers, medicines, treatment systems, laboratory standards, hospitals, etc., are not identified in a real sense in many cases of development of healthcare data But due to unavailability of authentic information, it cannot be integrated for use of interoperability. Furthermore, an exchanged data must be understood by a user which does not exist in the present time. This paper aims to identify the role of core elements in health care. This paper provides a platform for identifying the role of each element by conducting a survey and analyzing it for semantic interoperability which is a prerequisite of Web Semantic. Survey reveals that interoperability in health care obtained only when the role of each element is identified in a holistic approach.

Keywords Semantic interoperability · Web semantic · Health care · Holistic approach · EHR

1 Background and Rationale

Electronic health records (EHRs) are nothing but real-time, patient-centered records that make actual information available readily and in a secure manner to authorized users. It contains a patient's medical history along with required data such as diagnoses, medications, and treatment plans [1].

Semantic interoperability is nothing but the ability of computer systems to transfer explicit data to information systems. Semantic interoperability is, therefore, it is concerned not just with the packaging of data, but the simultaneous transmission of the user data which is a requirement of the Semantic Web. This can be accomplished by the addition of relevant data which is requirement of linking of data in a system [2].

G. S. Patange (✉) · Z. Sonara · H. Bhatt
Department of Mechanical Engineering, Faculty of Technology and Engineering, Charotar University of Science and Technology, Changa Ta, Petlad Dist Anand, Gujarat, India
e-mail: gajananpatange.me@charusat.ac.in

© The Editor(s) (if applicable) and The Author(s), under exclusive license
to Springer Nature Singapore Pte Ltd. 2021
S. Shakya et al. (eds.), *Proceedings of International Conference on Sustainable Expert Systems*, Lecture Notes in Networks and Systems 176,
https://doi.org/10.1007/978-981-33-4355-9_42

Fig. 1 Road map of
interoperability(5)

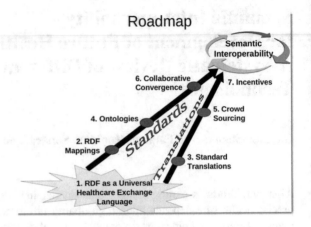

Semantic interoperability is necessarily based on ontology which is a shared specification of a conceptualization. Semantic Web is an extension of the current one, in which information is given well-defined meaning. Meaning is expressed by the resource description framework [3, 4]. Figure 1 indicates the road map of interoperability [5].

Thus, Semantic Web provides a common framework which allows data to be shared and reused in different stages during the integration of information [6, 7].

Interoperability remains a huge burden to the developers of health system because IoT devices are heterogeneous in terms of different domain communication protocols. Due to the lack of worldwide acceptable standards interoperability, tools remain limited [8].

The current limitations in healthcare data standardization are lack of patient data in layman's language and its linkages to authoritative sources for further information.

For patients to be able to enter and access medical data, they need to be able to understand the technical aspects as well as the contents of their record [9–11].

In the current situation, the challenges for interoperability include capabilities of IoT components for processing as well as interpreting exchanged data. So ontologies are essential for one to one alignment and central alignment because conceptualization and implementation may be entirely different due to human nature [12].

In every domain, standards may act as major units of reliability as they reflect the quality and significance. In the healthcare sector, it is restricted only to machine to machine. In reality, health needs some information transfers from institutions, systems, machines, and people. This is because data channelization is required to understand the various levels of interoperability and their significance to identify the limitations of the standards [13].

The World Health Organization (WHO) has given a list of twenty mandatory items which are useful for health study records. EHR information can be used if it contains accuracy, consistency, timeliness as well as trustworthiness. The biggest disadvantage of the resource description framework (RDF) is flexibility and most of the ontologies based on RDF [14].

The medical domain contains different clinical concepts which have ambiguous terms which result in different interpretations at every stage of EHR. If a case admitted in one hospital needs to be referred to any other hospital, every record related to that case has to be transferred to the other hospital. These reports may or may not be as per the semantics understood by other hospitals. Interoperability needs to be resolved such issues efficiently [15].

For healthcare system, two approaches are important one is EHR standards and other is vocabulary, however, many systems do not follow any standards which violets the object of interoperability [16, 17].

Ontology plays a central role in different domains and useful in different models through semantic interoperability for knowledge representations [18].

Half of the trials in clinical project fail due to EHR routinely collected data. Routinely collected data is known as real world data because it is a direct data taken from real cases. The gap between syntactic and semantic interoperability can be reduced by correct interpretation of each layer [19].

The stockholders of health systems include clinicians, healthcare organizations, patients, device, or manufacturers. Establishment of semantic interoperability needs the involvement of all stockholders [20, 21].

For better understanding, there is a need for integrating three interoperabilities for the holistic approach as shown in Fig. 2 [22, 23]. Semantic Web technologies may resolve many issues regarding the representation as well as extraction of information by many smart applications such as meteorology and environment observation.

Thus, the Internet of things (IoT) facing a challenge how to achieve interoperability between various healthcare platforms and it is very important how ontologies and semantic data processing system can be integrated to facilitate interoperability across the process to produce same interpretations.

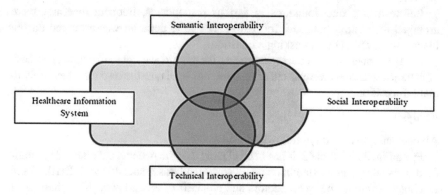

Fig. 2 Relationship among different types of interoperability [22, 23]

In this context, the most notable existing developments lagging in uncovering of information from different platforms such as doctors, patients, co-staff, manufacturer, and pharmacist.

Thus, there is a need for the practical aspects of identifying the role of each domain that gathers healthcare data from various sources in lemans language. The information can be used further by integrating with adopted standards without violating object of interoperability.

2 Methodology

Therefore, to know ground reality in healthcare process and environment, it was decided to gather such information with the help of survey which identifies role of each domain to align the objective of semantic interoperability.

The survey is carried out intentionally in developing country India which includes all the necessary elements in electronic health record (EHR) in a heterogeneous manner.

This study was designed to find heterogeneous data found among the different elements in healthcare systems and to identify the role of each one for effective implementation of semantic interoperability.

Data Collection

After permission from the authority of hospitals, manufacturer, and pharmacist, 300 participants were selected. The participants were doctors' co-staff such as technician, nurse, paramedical staff, pharmacist, and manufacturer.

Criteria for survey in the study included oral consent from the participant, able to read and understand in English at a basic level. Basic questions were asked as in Table 1.

Before asking questionnaires to various respondents, frequent meetings were arranged for getting precise information. The time duration was arranged during lunch and traveling during asking a question.

Local language is used during a process for getting accurate information; in case of patients, questions were asked after some informal questions so that they reply to real information.

Analysis

All data analyses yield the following results.

From Fig. 3 and Table 2, it has been cleared that the patients only have 21% standard vocabulary knowledge in contexts with standards associated with EHR. There is a huge difference between doctors and co-staffs of around 60%. The pharmacist and manufacturers which are important elements during the transformation of input into output in a predetermined sequence found to be imperfect. One of the critical elements in EHR is doctor and found to be lagging by 11% which may lead to serious effects in some crucial conditions.

Table 1 Survey questions (authors' contribution)

Respondent	Questions
Patients	Are you confident for writing and filling forms by yourself?
	Are you confident at saying your consultant that he will know everything in a first visit?
	Are you open at being to talk to your consultant?
	Are you confident at understanding back what your consultants tell you in his own words?
	Are you confident at understanding your medicine prescriptions?
	Are you confident at understanding your medicine side effects?
Doctors	Are you confident at writing the standard name of disease by symptoms?
	Are you confident at being to talk to your patient?
	Are you confident at understanding back what your patients tell you in their own words?
	Are you confident that your patients take medicine in the same prescriptions?
	Is there any mechanism that your patients take medicine in same as per your prescriptions?
	Are you aware of health standards update?
Co-staff	Are you confident at understanding what your consultants tell you precisely?
	Are you open at being to talk to your consultant if you don't understand what is to be done?
	Are you open at being to talk to your consultant if you don't understand how is to be done?
	Are you open at being to talk to your consultant if you don't understand when is to be done?
	Are you open at understanding precision and accuracy of your machines?
Pharmacist	Are you confident at getting the exact prescription of your consultant?
	Are you confident at being to talk to your consultant if you don't understand what is the exact spelling of prescriptions?
	Are you confident at understanding standard name and contents of medicine?
	Are you confident at understanding handling procedure of medicine?
Manufacturer	Are you able to manufacture the medicine as per standard?
	Are you confident at understanding precision and accuracy of your contents?

3 Results and Discussion

Significant differences were found between various layers of information from consultant to end user-patient during the healthcare information system which is a base of semantic interoperability.

It was found that most the doctors met the requirements of accuracy and reliability in most the cases as per medical standards. But during distributions of medicine pharmacist did not know the prescriptions, for example, a medicine which needs

Fig. 3 Meeting
requirements (survey results)

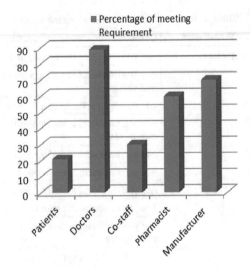

Table 2 Survey analysis
(authors' contribution)

Respondent	Percentage of meeting requirement
Patients	21
Doctors	89
Co-staff	30
Pharmacist	60
Manufacturer	70

to maintain refrigerated whole time due to lack of information. Patients did not know the doctor's prescriptions, for example, exact quantity of consumption. Co-staff are not aware of standard procedure of taking a sample, operating machine, etc. Manufacturers are not aware of quality of process including quality standards.

Collection of more and more data creates a learning health system environment which again helps to monitor and adopt dynamic interoperability standards.

Current standards sometimes do not match with semantic interoperability so they need the continuous inputs from various elements and they must play a role for innovative interoperability initiatives.

In a developing country like India, there is an adoption of own independent information system ignoring adherence to the standards which inclined toward missing interoperability.

Medical domains contain many concepts which create different interpretations. Thus, the semantic data may be good understandable through a user interface.

This paper sought to explore precise and required roles of different domains in relation to semantic interoperability effectiveness. Understanding of role can reduce the heterogeneity of system and integration in the healthcare environment to assist semantic interoperability.

The gap between syntactic and semantic interoperability can be reduced by correct interpretation at each stage from source to destination. In electronic health record, the source is varied from time to time it may be from consultant to patients or vice versa.

While finalizing conclusion the world is facing COVID-19 issue. It was observed that there is no authentic information available(from three interoperabilities semantic, technical and social) till date, and yet, no platform is developed regions among at all the stages so that accurate and precise information can flow from one source to other, and objective of semantic interoperability will be met for future prosperity.

To achieve precise and accurate semantic interoperability, adoption of a standard is required by the exchange of data among different layers which is discussed in this paper. It can be achieved with unambiguously translating the dataset in desired layman's language which varies from one stage to other. As diseases like COVID-19 growing exponentially, the evolution of knowledge is essential in the healthcare domain. Good healthcare system depends so much upon having high-quality information about a patient and interrelated system. Role of semanticinteroperability is nothing but the ability of a system to exchange data with some another system as well as to understand precise descriptions. Semantic interoperability is necessarily based on ontology which can share expected specification of a current health data for EHR which is a key requirement and will be merged as a backbone of current health conditions for reliable data. And it can be achieved in an interoperable manner for the development of standard-based.

Table 3 represents the role of each element in EHR for use of the Semantic Web for interoperability.

Study Limitations

The study was conducted in a developing country India and one region. The results may differ by some amount in other regions.

4 Conclusion

The need for entire elements communication in health care is very crucial for the end result. However, the accuracy of such a process depends on the ability of different elements to map different terms to shared semantics, or meaning for integrating a heterogeneous data in one coherent whole.

Semantic interoperability seeks more understanding rather than only transferring the data for a holistic approach.

Table 3 Role of domain (authors' contribution)

Doctor	Patient	Co-staff	Pharmacist	Manufacturer
Study the use of behavior of patients in new situations like COVID-19 by fostering the semantic interoperability	Semantic harmonization services by supporting consultant, co-staff, pharmacist, and manufacturer	Accreditation of the clinical research process	Adoption of relevant as well as usable semantic interoperability standards	Accredit manufacturing process
Presentation of patient data in layman's language	Involvement and cooperation with the rest of domain	Accreditation of clinical research machines	User-friendly EHRs	Adopt a model of value demonstrations
Adoptive listening for monitoring long-term interoperability	Interpretation of medical terminologies correctly	Facilitating required documentation at the point of care	Ontology mapping	Legal compliance for the benefit of society
Adopt medical standards to reduce medical errors	Avoid variations in treatments for a shorter time	Classification of the information in the applications for future use	Focus on achieving technical and semantic interoperability	Quality assurance building
Derive health data from outcome all other domain	Provide complete health data	Recognition at the receiver's ontology	Enrichment of the local information	Deployments of best practices
Collating evidence from health processes	Understanding of data in a specific context to avoid misinterpretations	Support doctor in interpreting data	Produce findable, accessible, interoperable, and reusable data	Fostering ecosystem standards to achieve the interoperable process
Use of data for legitimate use	Stay active with semantic connection	Maintain the security in IoT-based health devices	Analyze trustworthiness of various entities included in the EHR process	To implement the interoperable EHR process
Educate co-staff to have precise outcomes	Do not infer general facts to express the interoperability of IoT data	Inspects the communications between the Web server and smart devices	Avoiding malware practices	Maintain the specifications of services and tools for conformance testing

References

1. U.S. Department of Health and Human Services : https://www.healthit.gov/faq/what-electr onic-health-record-ehr Accessed on 5 Sept 2020
2. Martínez-Costa, C., Menárguez-Tortosa, M., Fernández-Breis, J.T.: An approach for the semantic interoperability of ISO EN 13606 and OpenEHR archetypes. J. Biomed. Inf. **43**(5), 736-746 (2010)
3. Booth, D., Chute, C.G., Glaser, H., Solbrig, H.: Toward easier RDF. In: W3C Workshop on Web Standardization for Graph Data, Berlin, Germany (2019)
4. Jaulent, M.C., Leprovost, D., Charlet, J., Choquet, R.: Semantic interoperability challenges to process large amount of data perspectives in forensic and legal medicine. J. Forensic Leg. Med. **57**, 19–23 (2018)
5. Dataversity Semantic Interoperability: The Future of HealthCare Data Retrieved from https://www.dataversity.net/semantic-interoperability-future-healthcare-data/ on 15 Jan 2020
6. Schulz S., Stegwee R., Chronaki C.: Standards in healthcare data. In: Kubben, P., Dumontier, M., Dekker, A. (eds.) Fundamentals of Clinical Data Science. Springer, Cham (2019)
7. Balas, V.E., Solanki, V.K., Kumar, R., Khari, M. (Eds.): In: Handbook of Data Science Approaches for Biomedical Engineering, Academic Press (2019)
8. Ullah, F., Habib, M.A., Farhan, M., Khalid, S., Durrani, M.Y., Jabbar, S.: Semantic interoperability for big-data in heterogeneous IoT infrastructure for healthcare. Sustain. Cities Soc. **34**, 90–96 (2017)
9. Jaulent, M.-C., Leprovost, D., Charlet, J., Choquet, R.: Semantic interoperability challenges to process large amount of data perspectives in forensic and legal medicine. J. Forensic Legal Med. Elsevier (2016)
10. Lau, F.: Infrastructure and capacity building for semantic interoperability in healthcare in the Netherlands. Build. Capacity Health Inf. Future **234**, 70 (2017)
11. Ganzha, M., Paprzycki, M., Pawłowski, W., Szmeja, P., Wasielewska, K.: Semantic interoperability in the internet of things: an overview from the INTER-IoT perspective. J. Netw. Comput. Appl. **81**, 111–124 (2017)
12. Villanueva-Miranda, I., Nazeran, H., Martinek, R.: A semantic interoperability approach to heterogeneous internet of medical things (IoMT) platforms. In: IEEE 20th International Conference on e-Health Networking, Applications and Services (Healthcom), pp. 1–5. IEEE (2018)
13. Janaswamy, S., Kent, R.D.: Semantic interoperability and data mapping in EHR systems. In: IEEE 6th International Conference on Advanced Computing (IACC), pp. 117–122. IEEE (2016).
14. Leroux, H., Metke-Jimenez, A., Lawley, M.J.: Towards achieving semantic interoperability of clinical study data with FHIR. J. Bio. Semant. **8**(1), 41 (2017)
15. Sachdeva, S., Batra, S., Bhalla, S.: Evolving large scale healthcare applications using open standards. Health Policy Technol. **6**(4), 410–425 (2017)
16. do Espírito Santo, J.M., Medeiros, C.B.: Semantic interoperability of clinical data. In: International Conference on Data Integration in the Life Sciences, pp. 29-37. Springer, Cham, (2017)
17. Roehrs, A., da Costa, C.A., da Rosa Righi, R., Rigo, S.J., Wichman, M.H.: Toward a model for personal health record interoperability. IEEE J. Biomed. Health Info. **23**(2), 867-873 (2018)
18. Soualmia, L.F., Charlet, J.: Efficient results in semantic interoperability for health care. Yearbook Med. Info. **25**(01), 184–187 (2016)
19. Kalra, D., Stroetmann, V., Sundgren, M., Dupont, D., Schlünder, I., Thienpont, G., Coorevits, P., De Moor, G.: The European institute for innovation through health data. Learning Health Syst. (2016).https://doi.org/10.1002/lrh2.10008
20. Bhartiya, S., Mehrotra, D., Girdhar, A.: Issues in achieving complete interoperability while sharing electronic health records. Procedia Comput. Sci. **78**(C), 192–198 (2016)

21. Min, L., Tian, Q., Lu, X., An, J., Duan, H.: An open EHR based approach to improve the semantic interoperability of clinical data registry. BMC Med. Inform. Decis. Mak. **18**(1), 15 (2018)
22. Garlapati, R., Biswas, R.: Interoperability in healthcare: a focus on the social interoperability (2012)
23. Patel, A., Jain, S.: Present and future of semantic web technologies: a research statement. Int. J. Comput. Appl. (2019).https://doi.org/10.1080/1206212X.2019.1570666

A Study on Multimodal Approach of Face and Iris Modalities in a Biometric System

P. Sai Shreyashi, Rohan Kalra, M. Gayathri, and C. Malathy

Abstract Biometrics is an ever-emerging field which is now being utilized in many applications and systems. Face biometrics are quite prevalent and have already been deployed in many devices and applications. Iris biometrics though already studied and deployed in certain areas are considered very effective due to their lesser chances of changing with time and uniqueness. Face detection and iris detection for verification have been employed in various biometric applications already. Unimodal systems of face and iris are secure, but for enhanced levels of security, multimodal biometric systems are being created. Multimodal biometric systems are those systems which use the various biometric traits simultaneously, in order to authenticate a person's identity. There is a direct enhancement in the verification accuracy due to such systems. The fusion of these two biometrics in consideration to make the process of unauthenticated or unauthorized intervention complex because it will be difficult and inefficient to get multiple biometric traits simultaneously in real time. In order to draw a contrast between a unimodal approach and multimodal approach for the biometrics, this paper presents a survey on using individual face classifiers or individual iris classifiers and compares it with the combined classifier results. After analysing the various techniques, it is concluded with the directions towards the future scope.

P. Sai Shreyashi (✉) · R. Kalra · M. Gayathri · C. Malathy
Department of Computer Science and Engineering, SRM Institute of Science
and Technology, Chennai, India
e-mail: sai.shreyashi@gmail.com

R. Kalra
e-mail: rohankalra97@gmail.com

M. Gayathri
e-mail: gayathrm2@srmist.edu.in

C. Malathy
e-mail: malathyc@srmist.edu.in

© The Editor(s) (if applicable) and The Author(s), under exclusive license 581
to Springer Nature Singapore Pte Ltd. 2021
S. Shakya et al. (eds.), *Proceedings of International Conference on Sustainable Expert
Systems*, Lecture Notes in Networks and Systems 176,
https://doi.org/10.1007/978-981-33-4355-9_43

1 Introduction

There are various kinds of biometrics which are available and are being researched extensively to develop better and almost foolproof methodologies. Face and iris multimodal biometric systems, although not foolproof, can conceptually provide effective levels of performance. The proposed research work will mostly be looking at various methods and techniques used for verification purposes (one to one matching). This usually results in accept or reject results.

1.1 Biometric Verification

Biometrics is the field which deals with analysing characteristics of humans and the systems which use the biometric data have the capability of identifying a human uniquely. This data is recognizable and verifiable through appropriate techniques and systems being developed. The methodologies and usage of biometrics go back to the 1860s and hence it is not a new concept. Biometrics can be categorized as physiological traits and behavioural traits. Physiological traits include fingerprints, hand shape, vein pattern, iris, retina, and face shape. Biological traits within this can be, DNA, blood, saliva, or urine. Behavioural traits include signature dynamics (speed of pen movement, accelerations, pressure exertion, inclination), keystrokes, gait, and gestures. Biometric verification is the process of verifying a person by their biometric characteristic, it can be an image of their face, an image of their fingerprint, and even a recording of their voice. This data of the person is then compared to data already collected and stored in a database for real-time verification systems.

1.2 Face Detection

The face is one of the primarily used and implemented biometrics in recognition systems. Facial verification is a biometric technology which helps in uniquely identifying a person by detection, comparison, and analysis of patterns based on the various features of the face-like contours. These applications are generally used in security systems and for related purposes. This technology has a high rate of usage in various areas related to law enforcement and other enterprises due to its various advantages. The main advantage is the ease of usage by the users and that it is non-contact in nature. Images of a person's face can be taken from a distance and hence analysis can be carried out without much physical interaction. It is a highly suitable technique for security, tracking, and attendance applications. There are certain drawbacks associated with this technique too. Some of these are lighting issues (illumination) and facial expressions.

1.3 Iris Verification

Iris is an effective physiological biometric trait which is used for recognition and authentication purposes. Iris verification is a biometric technology which uses high-quality images of the irides of the eye which uniquely identify people. Iris is an internal organ and remains protected from everyday damage and is uniform under most conditions. This is implemented by applying appropriate algorithms and matching techniques and this was developed by John G. Daugman, Ph.D., OBE. Iris verification produces exceptional results with a high percentage of accuracy as compared to many other biometric traits and related technologies. This again has the advantage of being a non-contact form of the verification system and ease of usage for the users. These systems have been inculcated in many systems for security purposes and in enterprises.

1.4 Multimodal Biometric Verification: Face and Iris

The face verification system is a unimodal system as is the iris verification system. When there is a single trait for information in the verification system, it is a unimodal system. A multimodal system is one in which, there is more than one trait for information (as inputs) in the verification system. The multimodal systems increase the scope and level of accuracy, especially in terms of security. The major reasons for usage and implementation of multimodal systems are reliability due to multiple traits, a higher level of accuracy, unavailability of one trait can be fulfilled by the presence of other traits, and better handling of spoofing attacks. Face and iris biometrics real-time data (live samples) are integrated or fused preferably at an early stage for better accuracy. There are various techniques of fusion of multimodal data, like feature-level and rank-level fusion. The templates formed after fusion after necessary transformation are then compared accordingly and a final decision is reached.

The major advantages received out of such a multimodal system is a reduction in error rates and prevention against security attacks like spoofing. Some disadvantageous facts being higher computational and storage needs as well as processing periods.

2 Modules in the Basic Methodology

2.1 Face Detection (Followed by Iris Detection) and Cropping Block

Face detection block is one of the main phases of the verification system. This block is responsible for detecting the face. The basic architecture is designed in a way as to locate the face and additionally discover the iris in it (Fig. 1).

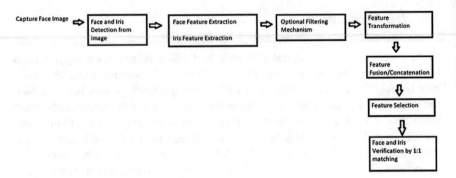

Fig. 1 System architecture design: block diagram

2.2 Pre-processing Block (within Face and Iris Detection Module)

Face image goes through a series of pre-processing steps to reduce the impact of factors that can affect face and iris verification algorithm detrimentally. Facial pose and illumination of the face are the most critical of the factors. These issues have been stated and noted beforehand. Pre-processing is a crucial and understated step for the further processing, extraction, and verification steps. In the case of faulty or low-quality pre-processing data, the results obtained will result in a low level of performance.

2.3 Feature Extraction Block (for Face and Iris), Filtering, and Feature Transformation

Feature extraction from the different modalities is carried out to find robust local and global features from them to make the verification process more accurate and also includes an essential step of dimensionality reduction. This is usually done on either of the following levels or a combination from these levels which are: Pixel-level features, local features, and global features. Feature extraction from the face for verification process also is followed by a filtering step to convolve the image using filter(s). These filters can either be based on a predetermined pattern or a pattern can be learnt from the training data for the filtering process. Filtering can be done using image patches and even in a multi-level structure. Encoding is another important part of feature extraction which usually generates a feature vector or a histogram. Spatial pooling block of feature extraction is used for further compression of features by two main methods: average pooling and max pooling. Average pooling where the average value is utilized and max pooling where the maximum value is utilized. Furthermore, there are methods like template matching using pixel values [1], eigenface method

[2], and fisherface method [3] which help in the representation of the face. Iris feature extraction is usually based on the colour, texture, and shape. For iris feature extraction, it can be done in mainly four different methods [4] namely: texture-based, phase-based, zero crossing-based, intensity variation-based. Features are then transformed using the original features by linear or nonlinear combinations. For iris verification, the segmentation of iris from the image is also an important step. This is the process of locating pupil and iris from the raw data. This displays challenging due to noises in the images introduced due to various factors. For this, one approach seen in some works use Viterbi's algorithm and full width at half maximum (FWHM). FWHM is applied after contour estimation to remove occlusion noise. Next module for iris recognition is normalization which is the transformation of iris area into scale and pupil-dilation invariant band. One method which was used [5] was the rubber sheet model by Daugman [6].

2.4 Feature Fusion/Concatenation

This module involves the fusion of information of multiple modalities for the verification purpose and its overall better performance. Fusion can take place at various levels namely: signal-level, feature-level, score-level, rank-level, decision-level. The basic descriptions are given later for each of these fusion methods.

2.5 Feature Selection

This module has the function of selecting features, those which were extracted in the previous steps to maximize the performance of the biometric system. This step of feature selection helps in taking into account those features which will create the most impact in terms of verification accuracy. Some of these techniques are correlation-based feature selection (CFS) and mutual information (MI) [7].

2.6 Face and Iris Verification Block

In this block, the comparison or the calculation of similarity between two subjects(can be the feature encodings) takes place. Among the most commonly used recognition techniques for this block is eigen identification technique. The standard verification methodology uses the Euclidian separation distance to coordinate features. Example of the simplest features used in face verification is geometrical relations and distance between vital focuses on the face. An evaluation technique is taken against each and every current show in the database and the most solidly relating model is settled. On the off chance that these are satisfactorily close, an affirmation case is enacted.

3 Survey of Techniques and Methods

3.1 Face and Iris Detection Techniques

There are various face detection techniques which are as follows.

3.1.1 Viola–Jones Face Detection Algorithm

It is a real-time object detection framework. It results in effective rates of detection. The basic steps involved in clear detection of the face in an image are:— Feature computation from an image illustration. —Classifier using AdaBoost algorithm for few important feature selection from a larger set of features. –Classifier combination for better processing of the image for clearer detection.

3.1.2 Local Binary Patterns (LBP)

It mainly utilizes image texture features. In this technique, every pixel is assigned a value which can then be combined with a target. LBP patterns help in realizing key points in the target region. Then, there is the formation of a mask for colour and texture-based feature selection.

3.1.3 AdaBoost Algorithm

This algorithm results in the creation of a strong learner by utilizing weak learners. Strong learners are those which are strongly correlated to the true classifier while weak learners are minimally correlated to the true classifier. In the various training iterations, weak learners are added methodically and adjustment of the weight vector is performed taking into consideration all the examples which were wrongly classified in past iterations. This results in a classifier which has better accuracy than all the classifiers based on the weak learners.

3.1.4 Neural Network-Based Method

In this method, there are mainly two stages. In the first, a filter is applied to all regions of the image. An input of a fixed size pixel region is received by the filter to generate an outcome between 1 and -1. This denotes whether or not a face is present in that region. In the second stage, overlapping detections are merged and arbitrated among multiple networks [10].

Iris images are usually either obtained by cropping face images or captured separately.

4 Face Feature Extraction Techniques

For face feature extraction, the following global and local methods are used. Global methods are those which take the entire image as a whole, for processing and local methods are those which extract features from sub-regions of the image in a repetitive manner across the image.

4.1 Local Binary Pattern Histogram

This is a local method for feature extraction which utilizes 8-bit or 16-bit operator on a fixed area of the image to assign values 0 or 1 with respect to the centre pixel [5]. One method is to consider a circular LBPH with the extended neighbourhood with chi-square distance found between feature vectors of test and train set images for match score.

4.2 Principal Component Analysis

This is one of the global methods which tries to maximize the total scatter of the images in the training dataset. It reduces dimensionality by resulting in an outcome of projecting the training data to a one-dimension eigenface vector space. Test set images are also projected into eigenface vector space. The feature vectors are obtained for both training and test set data available with same dimensions. Euclidean distance between feature vectors of test and train set images are found for match scores [11]. For example, if the face images are of 32×32 size, then the number of dimensions is 1024. If the number of training data is less (hundreds), then it will not be problematic to calculate the Euclidean distance but if the data samples are huge in number, then it is necessary to reduce dimensionality using PCA. The method involves eigenvectors calculation of covariance matrix formed using 1024 dimension feature vectors followed by projection of every feature vector onto the largest eigenvectors in each case. Thus, resulting in a linear consolidate representation of eigenfaces. Largest eigenvectors have maximum variance and have the most information about various areas in the images, so by considerate feature selection, that is, by removing an adequate number of features on order to avoid the curse of dimensionality or loss of information, where it has obtained the required set of features. The feature extraction steps in this technique have: –Centring data: Using scaling (eigenvalues) and rotations (eigenvectors) covariance matrix formation. Eigen decomposition provides the eigenvectors and values. Before PCA application, subtraction of mean from every data point is a necessity. This is the centring process. — Normalizing data: For removal of scale dependency of PCA normalization is performed by division of every feature by the respective standard deviation. — Eigen decomposition calculation: Usually, single value decomposition is utilized for this step to obtain eigenvectors and values. — Data projection: Projecting data over largest eigenvectors.

Some disadvantages of PCA are inability to remove statistical dependency and largest variance may not always correspond to differentiating information.

4.3   Sub-pattern Principal Component Analysis (Sp-PCA)

Sub-pattern PCA works on a set of partitioned sub-patterns of the original pattern. For each partition to extract local sub-features, it then receives a set of projection sub-vectors. After that, there is a need to process them into global features for the following step of classification. Images are partitioned into sub-images. The sub-parts are formulated into row and column matrices. A global feature vector is generated eventually and for face verification, weighted angle distance is used in this algorithm.

4.4   Modular Principal Component Analysis

This is a local method and another variation of PCA in which each image is divided into blocks and PCA is performed on the divided images [13]. Divided images are then transformed into row vectors. The number of images is increased by $N \times d$ where N is the number of blocks and d is the number of training set images. One method could be to find match scores by finding Euclidean distance between feature vectors of test and train set images.

4.5   Linear Discriminant Analysis

This is a global method which also projects the data into a vector space but this method achieves this by modelling the difference between classes. It maximizes inter-class variance and minimizes intra-class variance. One method could be Euclidean distance can be found between feature vectors of train and test set images for match scores [11].

5   Iris Features Extraction Technique

Iris feature extraction is performed on normalized iris images and the outcome must be significant information of the iris pattern in the form of encoding to get a higher level of accuracy when comparisons between templates will be made. The following methods and techniques are used for iris feature extraction.

5.1 Gabor Filter

This method gives an outcome with optimum localization in both space and frequency by modulating sine/cosine wave with Gaussian. Pair of Gabor filters are used for signal decomposition where the specification of a real part and imaginary part is done by a modulated cosine and sine, respectively. Frequency is found by sine/cosine wave frequency and filter bandwidth is found by the width of the Gaussian. Daugman used 2D Gabor filter for getting iris pattern encoding.

5.2 LoG Gabor Filter

This method has the advantage over Gabor filter of producing zero DC components. In this method, Gaussian is on a logarithmic scale and frequency response is given by

$$G(f) = -e^{-\frac{(log(f/f_0))^2}{2(log(\sigma/f0))^2}}$$ (1)

where f_0 = centre frequency, σ = bandwidth of filter.

5.3 Zero Crossing of 1D Wavelets

The zero crossings of dyadic scales of the wavelet filters are used to encode iris features. The formulae for the mother wavelet ψ and wavelet transform for signal $f(x)$ are as below

$$\psi(x) = \frac{d^2\theta(x)}{dx^2}$$ (2)

$$W_s f(x) = f * (s^2 \frac{d^2\theta(x)}{dx^2})x = s^2 \frac{d^2}{dx^2}(f * \theta_s)(x)$$ (3)

where

$$\theta_s = (1/s)\theta(x/s)$$ (4)

5.4 Haar Encoding

In this method, both Gabor transform and Haar wavelet are considered as mother wavelet. A feature vector of 87 dimensions is the outcome of the filtering process. Then, the values ranging between −1 and +1 are mapped to 0 (negative) and 1 (positive) for more compacted templates. This resulted in a little better result than the Gabor transform.

6   Feature Transformation Techniques

Feature transformation is performed for removal of dimensions which are not reliable or required. Feature transformation is also used to create new features from existing original features. This results in a more compact and efficient representation. Few common techniques and algorithms are:

6.1   *Principal Component Analysis (PCA)*

Principal component analysis (PCA) is an algorithm used for reducing feature dimensions [14–16]. Many variants of PCA have been developed namely 2D PCA, 2D HOG representation (using 2D PCA), asymmetric PCA(APCA), WPCA.

6.2   *Linear Discriminant Analysis (LDA)*

Linear discriminant analysis (LDA) is a supervised method which maximizes the ratio of between class and within class scatter. It has been used for transforming features into subspaces in many other works [14, 17].

6.3   *Combination of PCA and LDA*

Combination of PCA and LDA are also used for transformation of features in which there are two steps: Facial descriptors projection into a subspace by PCA followed by linear classifier as an outcome of LDA [6, 12].

7   Fusion Techniques

To create a multimodal verification system, the information has to be fused at one or more levels. This fusion of information can take place using five different fusion techniques namely: signal-level, feature-level fusion, match score-level fusion, decision-level fusion, rank-level fusion.

7.1   *Feature-Level Fusion*

In this fusion technique, the features of the different modalities involved are taken into consideration together to get better information about each entity involved.

Concatenation of feature sets affects verification performance due to an increase in dimensionality and redundancy of data. So, to reduce such effects appropriate feature selection and transformation into a form, where processing becomes simpler are carried out. Feature selection is the process of selecting an optimal number and types of features based on the required task to be carried out. Some of the feature selection methods are genetic algorithm, backtracking search algorithm, particle swarm optimization, and sequential forward floating selection. Some of the feature transformation methods are LDA, PCA, and ICA. So the basic steps carried out for feature-level fusion are: Firstly, feature extraction from the different modalities which in this case are face and iris, then an optional filtering mechanism for richer and more complex features to be extracted, then feature transformation for projection into a different vector space, then feature combination or concatenation of the transformed feature sets, followed by feature selection using a suitable method, and then finally comparison with real information to check if matched or not.

7.2 Match Score-Level Fusion

In this fusion technique, scores are produced by matching or analysing the probability distributions and accuracies obtained from pattern vectors of the different modalities, namely face and iris. There are mainly three types of match score-level fusion techniques: classifier-based, density-based, transformation-based fusion. Classifier-based fusion involves concatenation of scores from various classifiers which together are considered as a feature vector. Every matching score is viewed as an entity of the feature vector. Density-based fusion involves the calculation of score densities separately for imposter scores and actual genuine scores. This estimation although leads to a high level of complexity when implementing. Transformation-based fusion involves normalizing matching scores so that this normalized data falls into a common area of consideration. Some biometric modalities are very incompatible, so for them, a range has to be defined. In face and iris multimodal model, either one iris or both the irises can be taken for match score fusion. Some algorithms for fusing scores are weighted sum rule, simple sum rule, SVM for classification, radial basis function neural network technique, and product rule.

7.3 Decision-Level Fusion

In this fusion technique, based on the received input, every matcher decides on which is the best match and the final decision is made by combining the decision results from the multiple biometric matches. Every decision is taken by comparing scores with a certain threshold value. Some decision-level fusion methods are weighted voting, AND rule, OR rule, majority voting, and behaviour knowledge space. This is generally less studied and applied compared to other fusion techniques because of the higher risks of degradation in performance.

7.4 Rank-Level Fusion

In this fusion technique, certain ranks which are integer numbers are assigned for every possible user identity. This technique combines all such output ranks for verifying the user with a high level of confidence. This technique can be advantageous as they do not require a further transformation into a common domain and hence even have a simpler implementation. When there is lesser number of users or less data, choosing the top rank can be adequate(as in some decision-level fusion techniques), but as the number of users increases, it becomes necessary to even consider secondary options which are used in rank-level fusion.

7.5 Signal-Level Fusion

In this fusion technique, modality samples from multiple sources can be combined to create one more powerful sample. For example, capturing multiple photos of the iris for a better quality of the image(super-resolution).

8 Feature Selection Techniques

8.1 Correlation-Based Feature Selection (CFS)

It is a filter-based algorithm which can be applied to both problems: continuous and discrete. In this algorithm, a basic hypothesis is taken into consideration: Good feature subsets have features correlated to class yet uncorrelated to others [9]. It is a heuristic which takes into consideration individual classifiers for prediction of class labels with a correlation level among them.

8.2 Mutual Information (MI)

This is the information shared between the two variables. Given a variable X, it means how much information is obtained about another variable Y. Here, X is a random variable and takes discrete values. A joint entropy $H(X, Y)$ is used

$$H(X, Y) = - \sum_{x \in X, y \in Y} p(x, y) log2 p(x, y) \tag{5}$$

$$I(X, Y) = H(X) - H(X|Y) \tag{6}$$

$$= H(Y) - H(Y|X) \tag{7}$$

$$= H(Y) - H(Y|X) - \sum_{x \in X, y \in Y} p(x, y) log2 \frac{p(x, y)}{p(x) p(y)} \tag{8}$$

9 Future Scope

- It could be used in multiple applications to improve authentication.
- It could be combined with other biometric authentications to further improve security.
- It could be used in IoT devices for better authentication.

10 Conclusion

This paper has surveyed various techniques and steps involved in the multimodal approach using face and iris modalities. The basic architecture for the multimodal approach using face and iris real-time data (for authentication) will validate it against the stored data in database after pre-processing and template generation. There will be challenges in terms of time constraints and fusing features but this approach will ensure a higher level of security and accuracy.

References

1. Brunelli, R., Poggio, T.: Face recognition: Features versus templates. IEEE Trans. Pattern Anal. Mach. Intell. **15**(10), 1042–1052 (1993)
2. Sirovich, L., Kirby, M.: Low-dimensional procedure for the characterization of human faces. J. Opt. Soc. Am. A, Opt. Image Sci. Vis. **4**(3), 519–524 (1987)
3. Belhumeur, P.N., Hespanha, J.P., Kriegman, D.: Eigenfaces vs. Fisherfaces: Recognition using class specific linear projection. IEEE Trans. Pattern Anal. Mach. Intell. **19**(7), 711–720 (1997)
4. Ma, L., Tan, T., Wang, Y., Zhang, D.: Personal identification based on iris texture analysis. IEEE Pattern Anal. Mach. Intell. **25**, 1519–1533 (2003)

5. Azom, V., Adewumi, A., Tapamo, J.-R.: Face and Iris biometrics person identification using hybrid fusion at feature and score-level: Pattern Recognition Association of South Africa and Robotics and Mechatronics International Conference (PRASA-RobMech) Port Elizabeth. South Africa **26–27**, 2015 (2015)

6. Shan, S., Zhang, W., Su, Y., Chen, X., Gao, W.: Ensemble of piecewise FDA based on spatial histograms of local (Gabor) binary patterns for face recognition. In: Proceedings of the 18th International Conference on Pattern Recognition (ICPR), vol. 4, pp. 606–609 (2006)

7. Estévez, P.A., Tesmer, M., Perez, C.A., Zurada, J.M.: Normalized mutual information feature selection. IEEE T. Neural Networ. **20**(2), 189–201 (2009)

8. Viola, P., Jones, M.: Rapid Object Detection using a Boosted Cascade of Simple Features (2001)

9. Vijaya Kumar, N., Irfan Ahmed, M.S.: Investigation of feature selection techniques for face recognition using feature fusion model. Res. J. Appl. Sci. Eng. Technol. **11**(1), 40–47 (2015)

10. Gupta, V., Sharma, D.: A study of various face detection methods. Int. J. Adv. Res. Comput. Commun. Eng. **3**(5) (2014)

11. Eskandari, M., Onsen, T., Hasan, D.: A new approach for face- iris multimodal biometric recognition using score fusion. Int. J. Pattern Recognit. Artif. Intell. **27**(03), 1356004 (2013)

12. Zhang, W., Wang, X., Tang, X.: Coupled information-theoretic encoding for face photo-sketch recognition. In: Proceedings of the IEEE Conference on Computing Vision Pattern Recognition (CVPR), 2011, pp. 513–520

13. Wang, Y., Tieniu, T., Jain, A.K.: Combining Face and Iris Biometrics for Identity Verification. Audio-and Video-Based Biometric Person Authentication. Springer, Berlin (2003)

14. Déniz, O., Bueno, G., Salido, J., de la Torre, F.: Face recognition using histograms of oriented gradients. Pattern Recognit. Lett. **32**(12), 1598–1603 (2011)

15. Yi, D., Lei, Z., Li, S.Z.: Towards pose robust face recognition. In: Proceedings of the IEEE Conference Computing Vision Pattern Recognition (CVPR), 2013, pp. 3539–3545

16. Cao, Z., Yin, Q., Tang, X., Sun, J.: Face recognition with learning based descriptor. In: Proceedings of the IEEE Conference on Computing Vision Pattern Recognition (CVPR), 2010, pp. 2707–2714

17. Chan, C.-H., Kittler, J., Messer, K.: Multi-scale local binary pattern histograms for face recognition. In: Proceedings of the International Conference Biometrics (ICB), 2007, pp. 809–818

Analysis of Business Behavior in the Australian Market Under an Approach of Statistical Techniques and Economic Dimensions for Sustainable Business: A Case Study

Delio R. Patiño-Alarcón, Fernando A. Patiño-Alarcón, Leandro L. Lorente-Leyva, and Diego H. Peluffo-Ordóñez

Abstract This paper provides a current and future business analysis of small food services and products company. Analysis methods include tabulation of the dataset, as well as hypothesis testing by comparison between directly proportional variables such as price/quality and recommendations/customer loyalty. Other calculations include key economic dimensions, decisions influenced by the JobKeeper payment scheme, data on capital expenditure expectations, and future business conditions. The results of the analyzed data show that customer and company behavior is in parallel with global trade. In particular, the growth of the digital market and customer loyalty where they find a product that meets their quality needs. The analysis finds that the company's prospects in its current position are positive. Of the four variables identified, two will be reinforced by the economic and market strategies to be implemented. The work also investigates the fact that the analysis carried out has possible limitations. Some of the limitations are that not all the company was able to register in the JobKeeper payment scheme and the lack of use of key tools for sustainable marketing.

Keywords Business analytics · Economic dimensions · Hypothesis testing · JobKeeper payment scheme · Sustainable marketing

D. R. Patiño-Alarcón · F. A. Patiño-Alarcón
Academia del Conocimiento, Ibarra, Ecuador
e-mail: deliricard@hotmail.com

L. L. Lorente-Leyva (✉) · D. H. Peluffo-Ordóñez
SDAS Research Group, Ibarra, Ecuador
e-mail: leandro.lorente@sdas-group.com

D. H. Peluffo-Ordóñez
e-mail: dpeluffo@yachaytech.edu.ec

D. H. Peluffo-Ordóñez
Yachay Tech University, Urcuquí, Ecuador

Coorporación Universitaria Autónoma de Nariño, Pasto, Colombia

S. Shakya et al. (eds.), *Proceedings of International Conference on Sustainable Expert Systems*, Lecture Notes in Networks and Systems 176,
https://doi.org/10.1007/978-981-33-4355-9_44

1 Introduction

This analysis is aimed to generate a critical, statistical, and economic analysis, together with all the qualitative (ordinal) and quantitative (discrete) variables that directly affect the global economy and the country in which the case study company belongs at present. The whole process accompanied by scientific sources and data provided by government sources. Some Australian market research [1] analyses the use of quantifiers and government technologies by policymakers to implement the wheat export market liberalization project. In 2018, Mazzarol et al. [2] provide an overview of the cooperative and mutual business model, representing a unique type of organization that has a dual purpose focused on economic and social objectives. It also [3] analyses the drivers and difficulties in the economic relationship between Australia and the European Union as negotiations for a free trade agreement begin. Other research [4] presents how business idea adjustment affects sustainability and creates opportunities for joint value creation in start-up companies.

Based on the analysis conducted, this research covers data updated up to the present year which has been statistically processed and validated through hypothesis testing, commercial impacts due to the global health emergency, data recorded from companies so far this year and finally a forceful and structured marketing tool to sustain the economic flow of the company. Statistical tools, economic dimensions, and market factors are detailed for the proposed analysis, and consequently, all this development becomes a vital analysis because it is a new contribution to future studies within the company.

The government has set up a system of economic support called JobKeeper, which seeks to sustain the economic flow due to the economic impact of the global pandemic that the population is going through. The monopolies are also affecting the economy by implementing measures such as high-interest rates for economic loans that affect medium and small businesses for their economic reactivation.

In summary, it can be said that the objective pursued is to develop a comprehensive analysis to analyze the market behavior of the company case study, while ensuring proper planning and development of its.

The rest of this manuscript is structured as follows: Sect. 2 describes the variables and techniques used for the case study and the statistical verification. The results are presented in Sect. 3. Finally, the conclusions are shown in Sect. 4.

2 Materials and Methods

2.1 Selection of the Variables and Techniques for the Dataset Statistical Study

The vulnerability of the food sector will increase considerably, therefore, these will have to be reinvented through new strategies to minimize human interaction.

Customer loyalty in such businesses will normally continue and customers will be willing to sit in spaces applying social distancing as the norm. New strategies will be employed, such as live cooking, where price and quality will be key variables that will ensure customer loyalty, as well as their subsequent recommendations to the market [5].

The variables declared are as follows:

Q_1: Price;

Q_2: Quality;

Q_3: Recommendations;

Q_4: Customer loyalty.

Based on the above, four essential variables are identified for the statistical study. Once the data have been tabulated, they are displayed as follows: (Tables 1, 2, 3 and 4).

Table 1 Count of the indicator $Q1$ representing the variable named (price)

Indicator	Likert scale/other options	Meaning	Count
Q_1	1	Strongly disagree	2
	2	Disagree	9
	3	Neither agree nor disagree	41
	4	Agree	135
	5	Strongly agree	97

Table 2 Count of the indicator Q_2 representing the variable named (quality)

Indicator	Likert scale/other options	Meaning	Count
Q_2	1	Strongly disagree	3
	2	Disagree	6
	3	Neither agree nor disagree	44
	4	Agree	111
	5	Strongly agree	120

Table 3 Count of the indicator Q_3 representing the variable named (recommendations)

Indicator	Likert scale/other options	Meaning	Count
Q_3	1	Disagree	160
	2	Agree	124

Table 4 Count of the indicator Q_4 representing the variable named (customer loyalty)

Indicator	Likert scale/other options	Meaning	Count
Q_4	1	Disagree	153
	2	Agree	141

2.2 Statistical Verification Through Hypothesis Testing

The statistical indicators used for the verification of the null or alternative hypothesis are mean, variance, and statistical T framed in the T-test for two samples assuming unequal variances, where the mean is between the maximum and minimum value of the analyzed dataset. Variance, on the other hand, is a measure of dispersion that represents the variability of a data series concerning its mean by calculating the sum of the residues squared divided by the total of observations generated. The statistical indicator T is the test that demonstrates whether the null hypothesis is true, as long as the sampled population follows a normal distribution.

Hypothesis testing emerges acceptance or rejection through the results obtained from the study, which may or may not be significant. To this end, it is essential to correctly identify the values included in this process, which, when examined through the hypothesis test, can determine the change in the experimental situation and the approval or rejection of it [6].

As shown in Table 5, the specifications of the hypothesis and their respective conditions for their elaboration in the analysis are represented.

Once the data have been tabulated and processed by the selected software, the process for hypothesis testing between the price and quality variables is represented in Table 6, as follows:

Once the data have been tabulated and processed by the selected software, the process for hypothesis testing between the variables recommendations and customer loyalty is shown in Table 7.

Feedback from the statistical study, the sample is taken, clearly agrees, and prefers quality over price. Customers will be willing to maintain their loyalty to the restaurant where they buy instead of just recommending it.

Table 5 Hypothesis testing specifications

Hypothesis	Definition	Data
Null hypothesis	$H_0 : \mu = \mu_0$	
Alternative hypothesis	$H_1 : \mu \neq \mu_0$;	$\mu =$ Estimated sample mean
	$H_1 : \mu > \mu_0$;	$\mu =$ Population average
	$H_1 : \mu < \mu_0$	

Table 6 Hypothesis testing between price and quality variables

Question	Is the proportion of people surveyed who strongly agree with high quality higher than the proportion who strongly agree with a good pizza price?	
H_0	There is a difference between the analyzed variables	
H_1	There is no difference between the analyzed variables	
Statistical indicator	Price	Quality
Mean	4.112676056	4.193662
Variance	0.665704474	0.7220798
Observations	284	284
Df	565	
T Stat	−1.158531057	
$P(T <= t)$ two-tail	**0.247136828**	
T critical two-tail	1.964171,551	
Test	**0.24714** > 0.05 (α)	

Given that the statistical P-factor is greater than alpha, the null hypothesis (H_0) cannot be rejected.

Table 7 Hypothesis test between the variables recommendations and customer loyalty

Question	Is the proportion of people surveyed who agree to be loyal to their pizzeria higher than the proportion who agree to recommend where they buy their pizza?	
H_0	There is a difference between the analyzed variables	
H_1	There is no difference between the analyzed variables	
Statistical indicator	Recommendations	Customer loyalty
Mean	1.436619718	1.496478873
Variance	0246852138	0.25087095
Observations	284	284
Df	566	
T Stat	−1.429867707	
$P(T <= t)$ two-tail	**0.153306738**	
T critical two-tail	1.964164101	
Test	**0.15331** > 0.05 (α)	

Given that the statistical P-factor is greater than alpha, the null hypothesis (H_0) cannot be rejected

2.3 Regulatory and Economic Dimensions Within the Organizational Macro-environment

The new reality facing especially in the food sector represents an important global economy for the world, as do the millions of people who work there every day. The current crisis will come with a new model of respect for the great food and hotel industries that are the true engines of the new economy [7].

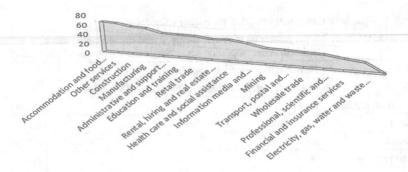

Fig. 1 Employment decisions influenced by the JobKeeper payment scheme, proportion of business by industry

Table 8 Percentages of capital expenditure intentions

	December 2019 (%)	March 2020 (%)
Reduced ability to pay operating expenses	54	22
Decreased cash flow	72	66
Reduced access to a credit or additional funds	25	18

According to Australian Bureau of Statistics [8, 9] says (2020), there is a business preamble, where a survey in April 2020 of a sample size of 2,014 Australian companies with a final response rate of 60% obtains the economic results as shown in Fig. 1 (registration and employment decisions influenced by the JobKeeper payment scheme):

From the business reviews, a 2020–2021 capital expenditure forecast is identified between December 2019 and March 2020, with the capital expenditure expectation data being a major indicator of business confidence. The data shown in Table 8 below are obtained [8]:

2.4 Legal Regulatory Dimensions

See Table 9

2.5 Government Data on Market Factors

As shown by the Australian Bureau of Statistics (2019) [10, 11], a review accompanied by reliable data is essential for the analysis of market factors. The data analyzed have small random adjustments from the government source because of the confidentiality that must be maintained with the data (Fig. 2).

Table 9 Preparing, getting up, and running your business

Preparing yourself for business	What should be done?	Getting up and running	What should be done?
Analyze your business idea	Business potential	Registering your business	Complete all business records
Define your customers and competitors	Market survey	Record keeping	Keep a record of income and expenditure
Develop your business plan	Business plan	Hiring workers	Pay employee benefits
Recognize risk	Emergency management and succession plan	Setting up your business banking	Create a business bank account
Determine your business structure	Define the commercial structure	Setting up your electronic payment systems	Setting up an electronic payment system
Confirm your business name is available	Find your business name		
Apply for an Australian business number (ABN) and register your business name	Request the business number		
Register your Web site domain	Register the Web site domain		

Find advice and support

2.6 Marketing Models for Sustainable Business Results

The pandemic is analyzed as an accelerator of structural change, which presents the consumption and digital transformation in the market. Agents could adopt a digital transformation to the market to recover or even further increase sales after COVID-19, through online advertising, conferencing software that reports 78% growth in the USA, companies are examining in this transition phase decentralized decision-making and new software to make the new digital work culture effective, productive, and fast [12].

As mentioned by [13], to generate an impact on the environment and make a business sustainable, the dominant logic of a company and its marketing is based on focusing on the quality product by orienting its distribution market with marketing strategies that understand the nature and processes of the companies.

The 4P's strategy allows companies to achieve their objectives, through the correct application of these, ensuring that the right product arrives, at the right price, in the right place and that the promotional strategies have a significant influence. Furthermore, with the statistical study and considering the current economic impacts,

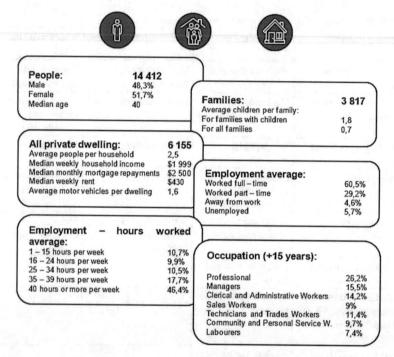

People: **14 412**
Male 48,3%
Female 51,7%
Median age 40

Families: **3 817**
Average children per family:
For families with children 1,8
For all families 0,7

All private dwelling: **6 155**
Average people per household 2,5
Median weekly household income $1 999
Median monthly mortgage repayments $2 500
Median weekly rent $430
Average motor vehicles per dwelling 1,6

Employment average:
Worked full – time 60,5%
Worked part – time 29,2%
Away from work 4,6%
Unemployed 5,7%

Employment – hours worked average:
1 – 15 hours per week 10,7%
16 – 24 hours per week 9,9%
25 – 34 hours per week 10,5%
35 – 39 hours per week 17,7%
40 hours or more per week 46,4%

Occupation (+15 years):

Professional 26,2%
Managers 15,5%
Clerical and Administrative Workers 14,2%
Sales Workers 9%
Technicians and Trades Workers 11,4%
Community and Personal Service W. 9,7%
Labourers 7,4%

Fig. 2 Quick statistics set by Australian bureau of statistics (2016–2019)

there will be a real perception of the customer and its final relationship between affordability and income [14, 15] (Fig. 3).

3 Results and Discussion

The statistical and market study concludes that customers prefer excellent quality at an acceptable price. Customer loyalty is higher than those who could come by advertising or recommendations. More potential customers can be reached through digital channels that dominate the market and will do for a long time. The physical places will have new regulations to fulfill representing a great cost for the company in which the money can be used to potentiate the digitalization of the business. The staff is preserved in a certain way through the financial support of the JobKeeper payment scheme. Finally, all this process will be done in a progressive way to adapt to a change that the world is facing.

By business size, 61% of small, 60% of medium, and 45% of large businesses reported having registered or intending to register for the JobKeeper payment scheme [16–18] (Fig. 4).

Fig. 3 Graphic representation of sustainable marketing models and concepts

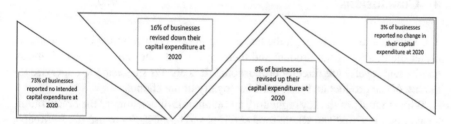

Fig. 4 Business capital expenditure intentions for the 2020–2021 financial year

On the other hand, companies expect restrictions that will affect them within the next few months, and it is estimated that food companies will have a 75% business condition (Table 10).

Table 10 Expected impact of government restrictions in the next two months by the extent of the impact

	Not at all (%)	Small to a moderate extent (%)	To a great extent (%)
Restrictions on trading	37	35	29
Social distancing restrictions	29	46	25
Travel restrictions	50	24	27

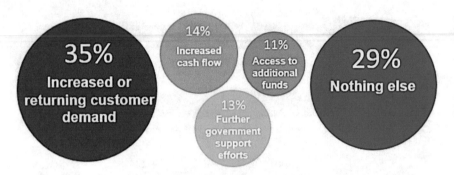

Fig. 5 Requirements to return to post-pandemic business conditions

As additional data, the case study answers the needs of businesses to resume their activities after overcoming to some extent the commercial impact that the global pandemic has generated (Fig. 5).

4　Conclusions

The statistical review through the test of hypothesis is corroborated that the clients prefer the quality of the product and their loyalty with the same one, which means quality and loyalty are directly proportional. It is always suggested to supervise the quality of the product and to reward the loyalty of the clients.

With the study of the key regulatory and economic dimensions of the organization, a panorama is visualized where food services and products try to use the JobKeeper payment scheme in 76%, belonging to an average of 60.5% of the companies categorized between small and medium, also, for the present year (2020), it is already considered a 22% reduction of capacity for operating expenses, 66% reduction of cash flow, and 18% reduced access to some additional credit.

Today's sustainable marketing models are based on a very marked digital structural change, which has grown by approximately 78%. The application of the 4P's tool allows the administration to identify the characteristics required by each strategic zone of this tool, highlighting the value that the product possesses, therefore, it covers the needs and desires of the client with a product, price, correct place, and always with the help of a promotion that truly influences the final purchase of the client.

Acknowledgements The authors are greatly grateful for the support given by the SDAS Research Group (https://sdas-group.com/).

References

1. O'Keeffe, P.: Creating a governable reality: analysing the use of quantification in shaping Australian wheat marketing policy. Agric. Hum. Values **35**(3), 553–567 (2018). https://doi.org/10.1007/s10460-018-9848-6
2. Mazzarol, T., Clark, D., Reboud, S., Limnios, E.M.: Developing a conceptual framework for the co-operative and mutual enterprise business model. J. Manage. Organ. **24**(4), 551–581 (2018). https://doi.org/10.1017/jmo.2018.29
3. Villalta, G.: Drivers and difficulties in the economic relationship between Australia and the European union: from conflict to cooperation. Aust. J. Int. Aff. **72**(3), 240–254 (2018). https://doi.org/10.1080/10357718.2018.1453479
4. Casali, G.L., Perano, M., Tartaglione, A.M., Zolin, R.: How business idea fit affects sustainability and creates opportunities for value co-creation in nascent firm. Sustainability (Switzerland) **10**(1), 189 (2018). https://doi.org/10.3390/su10010189
5. Jain, S.: Effect of COVID-19 on restaurant industry—how to cope with changing demand. SSRN (2020). https://doi.org/10.2139/ssrn.3577764
6. León, A.F.O.: Desarrollo web mediante la metodología híbrida snail y programación estadística para la toma de decisiones por prueba de hipótesis. Machala: Unidad Académica de Ingeniería Civil—Carrera de Ingeniería en Sistemas—Universidad Técnica de Machala (2019)
7. Muller, C.: Restaurant organizations and the power of the new economy: a pandemic, labor value and lessons from the past. En: B. University, ed. Boston Hospital. Rev. Boston, Boston University School of Hospitality Administration (2020)
8. Australian Bureau of Statistics.: Business Response to JobKeeper Payment Scheme (2020)
9. Australian Government.: Australian Government Business (2020). https://www.business.gov.au/planning/new-businesses/preparing-yourself-for-business
10. Australian Bureau of Statistics.: Business Indicators, Business Impacts of COVID-19 (2020). https://www.abs.gov.au/ausstats/abs@.nsf/Latestproducts/5676.0.55.003Main%20Features1April%202020?opendocument&tabname=Summary&prodno=5676.0.55.003&issue=April%202020&num=&view=
11. Australian Bureau of Statistics.: Census QuickStats (2019). https://quickstats.censusdata.abs.gov.au/census_services/getproduct/census/2016/quickstat/POA2111?opendocument
12. Kim, R.Y.: The impact of COVID-19 on consumers: preparing for digital sales. IEEE Eng. Manage. Rev. 1–16 (2020). https://doi.org/10.1109/EMR.2020.2990115
13. Powell, M., Osborne, S.: Social enterprises, marketing, and sustainable public service provision. Int. Rev. Admin. Sci. **86**(1), 62–79 (2020). https://doi.org/10.1177/0020852317751244
14. Karim, R.A., Habiba, W.: Influence of 4P strategy on organisation's performance: a case on Bangladesh RMG sector. Int. J. Entrepreneurial Res. **3**(1), 8–12 (2020). https://doi.org/10.31580/ijer.v3i1.1166
15. Anjana, K., Sreeya, B.: Consumer perception on 4Ps of marketing in malls with special reference to Chennai. Int. J. Innov. Technol. Explor. Eng. **8**(11), 3215–3217 (2019). https://doi.org/10.35940/ijitee.K2522.0981119
16. Australian Bureau of Statistics.: Capital Expenditure Intentions of Businesses (2020)
17. Australian Bureau of Statistics.: Future business impacts and conditions—Expected impact of government restrictions (2020)
18. Australian Government.: Australian Taxation Office (2020). https://www.ato.gov.au/Business/Starting-your-own-business/Getting-up-and-running/

Deep Learning-Based Bluetooth-Controlled Robot for Automated Object Classification

V. Vimal kumar, S. Priya, M. Shanmugapriya, and Aparna George

Abstract This paper presents a pick and place robot which can be monitored and controlled remotely through MATLAB within a range of 10 m. Our system is to retrieve predefined displaced items from a known area. The robot can be remotely controlled by an android phone to move through the area of interest. The robot while moving through its path uses an IP camera to stream images of its current surroundings through Wi-Fi to a deep neural network (DNN) created in MATLAB. The DNN determines if an item of interest is present and triggers appropriate control action to retrieve the item. The three main parts of the machine are, pick and place arm, base with 4 wheels attached to individual DC motors, control and communication system using Arduino, L298N motor drivers and HC05 module. The system works using two Arduino Uno-based controllers and four servo motors that move the arm.

Keywords Pick and place arm · Base · HC05 module · Arduino uno · IP camera · L298N motor drivers · MATLAB · Servo motors

1 Introduction

A remote-controlled robot with deep learning-based computer vision capability has a wide scope for applications. It could be useful in practically any situation, where humans find it dangerous or uncomfortable. Areas with health hazards, tunnels in

V. Vimal kumar (✉) · S. Priya · M. Shanmugapriya · A. George
Applied Electronics and Instrumentation, Rajagiri School of Engineering and Technology, Kochi, India
e-mail: vimalk@rajagiritech.edu.in

S. Priya
e-mail: priyas@rajagiritech.edu.in

M. Shanmugapriya
e-mail: priyams@rajagiritech.edu.in

A. George
e-mail: aparnag@rajagiritech.edu.in

© The Editor(s) (if applicable) and The Author(s), under exclusive license
to Springer Nature Singapore Pte Ltd. 2021
S. Shakya et al. (eds.), *Proceedings of International Conference on Sustainable Expert Systems*, Lecture Notes in Networks and Systems 176,
https://doi.org/10.1007/978-981-33-4355-9_45

607

mica mines, areas of hostile military activity, waste segregation and manipulation of materials located at areas of high radioactivity or temperatures serve as examples. The mechanical design and associated electronics need to be streamlined for a particular application, but intelligent analysis of images to decide on the course of action seems to be a common factor. In this age of increasing computing power, it seems reasonable to act toward the development of a deep neural network capable of identifying objects and situations of all the domains in a holistic way. This may not be achievable by the work of a few experts as there are a multitude of applications and associated constraints related to each of these domains. This could be possible if a method for extensive data collection to develop an expert system takes place along with normal activities of people working in the aforementioned domains. This research work proposed a robotic system with data collection capability and an approach to integrate the collected data to make an intelligent decision in an easy to implement manner using transfer learning. This research work proposes a prototype for simple picking machine as a prototype which is capable of being moved around for data collection by any person knowledgeable enough to use a mobile phone. The data collected in this manner is used to train the last layers of a pre-trained GoogLeNet to create a DNN suitable to implement classification for any domain of interest. This paper presents the system with two Arduino Unos, one for controlling motion by the operator and the other controlling particulars of robotic arm motion in response to signals on Bluetooth communicated by the DNN.

2   Related Work

The papers related to robotic pick and place usually emphasis on a solution for a particular issue like improving upon the precision for a particular application and a robotic application making use of popular devices like android phone, addressing delay encountered between control signal initiation and final element actuation, implementation of an algorithm for pose determination, relieving operators from minute details involved in telecontrol, innovations in sensors, mechanical design or control. The emphasis of this paper is in developing a prototype with a methodology to collect data to bring in the benefits of powerful DNNs in all remote robot applications, irrespective of the aforementioned domains.

Precision and delay reduction is addressed by Manoharan and Ponraj [1] and also the approach of facilitating a user to execute a single task without having to bother about subtasks is discussed. Proposed work integrated subtasks such as motion planning, grasping an object and avoid collision in remote system through DNN in a simple manner. The picking mechanisms could even be integrated with electronic skin. Various research works are available on design, implementation, testing and manufacturing of e-skins [2, 22]. Paper cutting art—kirigami is reported in Lu and Nanshu [3], Huang et al. [4] research models to develop a stretchable transparent electrode. Developments of this electrodes are widely adopted in implantable, biodegradable devices and soft robots. A cost-effective method is reported in Gong et al. [5] research model to fabricate ultra-thin strain sensors. Proposed model uses gold nano-wires to obtain better elasticity and sensitivity in the design. This paper does not discuss the application of advanced sensors, but the data collection step proposed could be improved to take in non-visual sensory inputs to serve as inputs to a DNN expert system. Force feedback and displays generally provide better virtual feeling to the operator [6]. 'Charles Lindbergh operation' is considered as the first telesurgery was performed in 2001. Using ZEUS robot, this operation is performed over 45 min. The robot was preinstalled with laparoscopic cholecystectomy [6] and the surgeon located in New York treated a patient in France. Movement-based robots replicate surgeon movements. In that, Da Vinci robotic system and video-assisted surgery are considered as remarkable telesurgery system [7]. Control system acts as interface between the robot and surgeon and replicated the movement into robot operations [8]. Generally, the levels of autonomy (LOA) [9] are used to define the semiautonomous abilities of a robot [10] and it suggests use of stereocamera-based pick and place robot for high temperature application. This paper though limited to use simple IP camera application, the methodology is scalable to advanced applications. Voice command-based object pick and place robot design are reported in [11] which helps the disable persons. The work in this paper could help disabled persons through mobile apps.

Neural network-based dynamic object recognition system is reported in [12] which performs object grasping by pointing a random object and places the object in bin. Pose estimation is reported in research work [13] which uses vision-based object manipulation to grasp an object. Similarly, finding the object coordinates is reported in literature [14] which converts the pixel coordinates using 2D transformation through its location and object oriented-based robotic arm control strategy. Vision sensor is used to define the fundamental problem in pick the object in this prototype to simplify the coordinate issues. Similar visual perception-based flexible object pick and place model is reported in research work [15] which uses visual data for objection manipulation [16]. Hakani discusses the robot controlled by cell phone to be moved in the desired direction by a touchpad [17]. Discusses a waste segregation strategy by a robot controlled by android application to avoid direct human contact with harmful waste materials. Another pick and place robot system is reported in literature [18] which uses degree of freedom to pick and place process. This paper though discusses a particular system, the methodology used is intended for a general system using a 22 layers deep network called GoogLeNet and uses 1X1 convolution to reduce dimension, increase depth and width without a significant performance penalty [19]. Lin et al. [20] proposed a modified network approach to improve the representation of neural network in object classification and detection process. Recently, deep learning and convolutional neural network-based models increases the detection process in a better fashion. Girshick et al. [21] proposes an approach called Regions with Convolutional Neural Networks (R-CNN) for object classification and detection, which is more effective than merely applying bigger networks.

3 Proposed Work

The robot consists of a base made of multi-wood of dimension 20×20. It supports the arm assembly with four servos starting from servo1 at the bottom and ending with servo4 at the tip for actuating a gripper acting as an end effector. It has four wheels coupled to a separate DC motor. The front and the rear pair of wheels are driven by two L298N motor drivers. The motor drivers are controlled by an Arduino Uno. This Arduino is programmed to communicate with the operator via Bluetooth using an android phone. Thus, the movement of the wheels of a robot can be remotely controlled by the operator using a common android phone. The operator is not supposed to be proficient in programming rather the operator function is limited to positioning the robotic system at a suitable location by using the touchpad of the phone to move the robot back and forth and left and write. It is usually the picking mechanism that needs consideration. In this paper, the emphasis is on developing an easy to implement a deep learning strategy that could encourage as many people as possible to develop datasets suitable for applications to their domain.

Figure 1 shows the black colored robotic arm with four servos mounted on a white base with four wheels connected to separate DC motors and mounted with a stand to fix android phone serving as IP camera.

The base has four DC motors (12 V 500RPM) are operated by two motor drives (L298N). The change in direction and speed of rotation of the DC motor is controlled by the driver. The drivers are in turn controlled by Arduino UNO controller receiving signals on Bluetooth from the android phone of an operator. Figure 2 shows the associated block diagram.

Figure 3 shows the schematic of Bluetooth control of base wheel rotation using the Android app.

Fig. 1 Robot picture with arm, base, IP camera and gripper

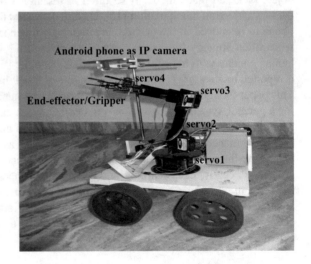

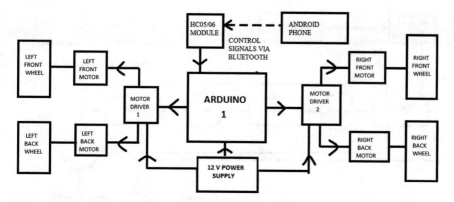

Fig. 2 Block diagram of a control scheme for wheel movement

Fig. 3 Bluetooth control of base wheel rotation using an android app

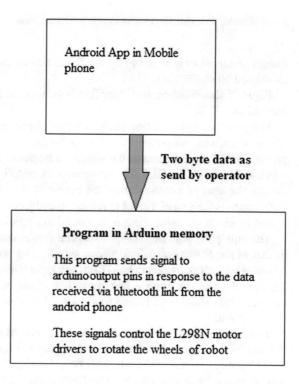

The robotic arm mounted on the base has four servo motors. The end effector of the pick and place robotic arm is a dexterous gripper. The arm has three 180° rotatable MG995 gear servo motors and one SG90 mini servo motor. Another Arduino controlled by another Arduino Uno receiving signals on Bluetooth from a DNN running on MATLAB on a PC. The DNN signals the Arduino in response to the

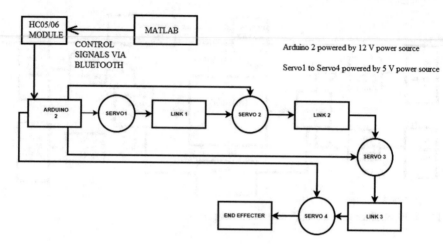

Fig. 4 Control flow associated with servos in the robotic arm

images streamed to it on Wi-Fi by the IP camera on the robot. Figure 4 shows the associated block diagram.

Figure 5 shows the control algorithm implemented in MATLAB to control arm movement.

Figure 5 shows the functional schematic for the working of the robot. An android phone controls the movement of a robot through an area within a 10 m range of the phone. The IP camera streams the images in its field of vision (FOV) to a preview window in MATLAB GUI for monitoring. A MATLAB program automatically detects the item of interest during the course of travel using a DNN and accordingly controls the arm of a robot to retrieve the object of interest. Figure 6 shows the region in which an object needs to be placed to be picked by the robot.

The laptop/PC with MATLAB should be connected to the same Wi-Fi network as that of the IP camera/android phone serving as a server with an IP address. An IP camera object is created in MATLAB workspace with the URL of the IP camera. This can be used to provide live streaming of the position of the robot with its surroundings. This image is also used by a DNN in MATLAB program for object classification.

The Arduino programmed to communicate with MATLAB through a Bluetooth link established by an HC05 module. The program for servo control is present in the PC/laptop running MATLAB. This program signals Arduino to generate output signals as per the objects classified by DNN loaded to MATLAB workspace.

The MATLAB program uses a DNN/CNN to classify objects. Transfer learning was carried out on GoogLeNet to obtain a required classification for five objects. The final layers were replaced appropriately and trained on fresh training data. The final layers of network with modification are shown in Fig. 7.

The network was trained on 124 images collected by using the GUI in Fig. 8.

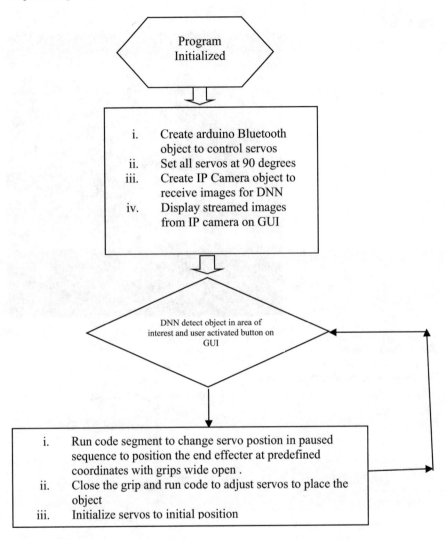

Fig. 5 General arrangement and control flow

The GUI is capable of collecting images for training from the robot moving around its supposed area of operation from the images streamed on Wi-Fi by the IP camera mounted on the robot.

The collected images were separated into five categories.

The categories are as explained below

i.   Bottle: Images with bottle found in the region of interest
ii. BottleOutOfRange: Images with bottle found outside the region of interest
iii. Small_Item: Images with small items like battery, tape and eraser within the region of interest.

Fig. 6 ROI

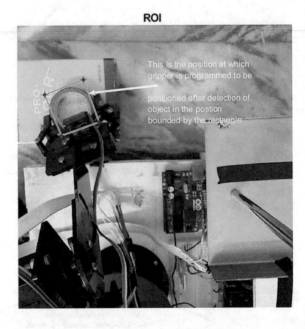

Fig. 7 Final layers of
GoogLeNet replaced with
new_fc and new_classoutput

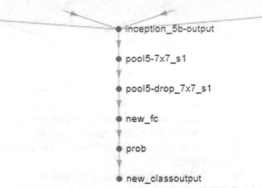

iv. ItemOutOfRange: Images with small items outside the region of interest
v. Unknown: The images encountered in the routine movement of the robot which is not covered in the above four clauses.

4 Results

The DNN was trained on a small set of 124 images. The training was done with 80% of data for training and 20% for validation and data augmentation techniques were used on training data. The network seemed to have stopped training before much

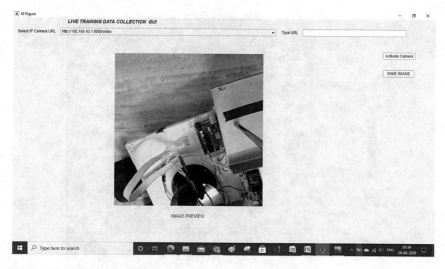

Fig. 8 GUI for live data collection

overfitting as it classified all the validation samples correctly but missed out on one training image. It classified one out of range item as unknown.

Figures 9 and 10 shows the confusion matrices for validation and training data.

Figure 11 shows the network performance on six random samples.

A MATLAB-based GUI was created and was capable of displaying objects classified by DNN and provide an option for the user to trigger an action or take automatic action as per program.

An IP camera positioned at a suitable location stream the video with the current position of the robot. Figures 12, 13, 14, 15 and 16 show the result obtained on the GUI for different scenarios.

5 Conclusion

The paper discussed a simple application of deep learning-based object classification method for robotic pick and place in daily life. The particular application discussed may be implemented by explicit program-oriented methods, but the emphasis of this paper is to develop a methodology with limited computational resources available with individuals in academia to collect domain-specific data and apply that data to keep enhancing a proven deep network with immense capability. Real-life application of DNNs in novel fields of importance requires collective effort within academia and industry to breakdown the tedious task of creation of extensive training data to address unexpected pitfalls in the implementation of a powerful DNN. This paper addressed the problem of freeing users from details of controls involved at lower levels by limiting the picking position at one particular location and leaving the

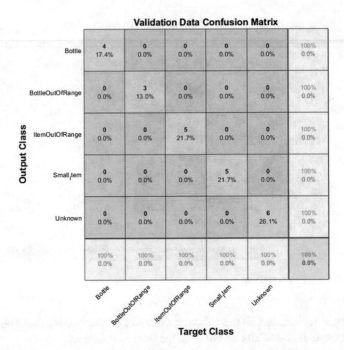

Fig. 9 Validation data confusion matrix

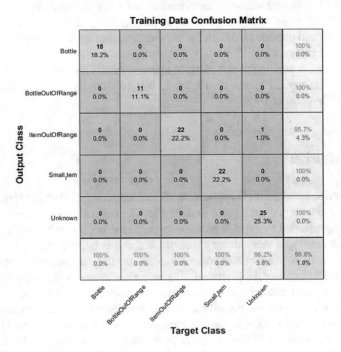

Fig. 10 Training data confusion matrix

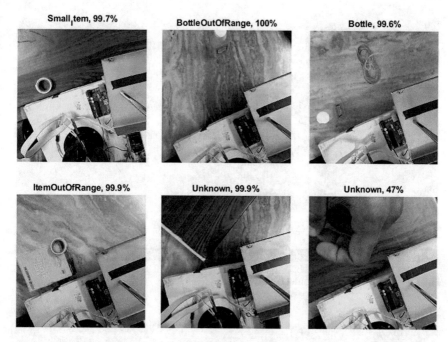

Fig. 11 Six random samples of images classified by the DNN

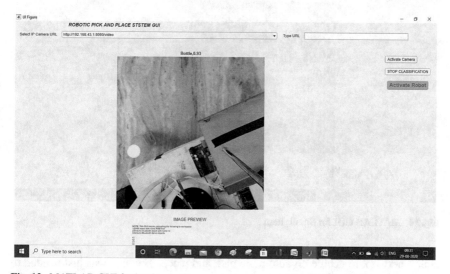

Fig. 12 MATLAB GUI for bottle

Fig. 13 MATLAB GUI for BottleOutOfRange

Fig. 14 MATLAB GUI for Small_Item

neural network to decide on automatic pick and place. The user was just needed to permit to pick in the GUI and the rest was taken care by the code sequence, but the decision on if the position of the robot was conducive for proper performance was taken by the DNN. The mature application of this methodology to combine shared databases will entail considerable time and computational power but should serve to open new frontiers in the application of AI.

Fig. 15 MATLAB GUI for ItemOutOfRange

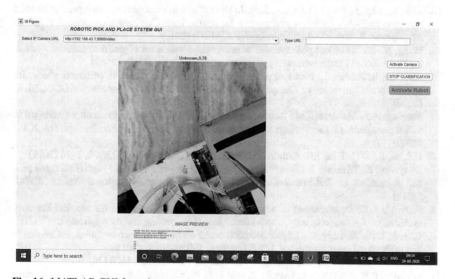

Fig. 16 MATLAB GUI for unknown

References

1. Manoharan, S., Ponraj, N.: Precision improvement and delay reduction in surgical telerobotics. J. Artif. Intell. **1**(01), 28–36 (2019)
2. Vijayakumar, T.: Flexible robotic electronic skin with high sensitivity sensor arrays. J. Electron. **1**(01), 43–51 (2019)
3. Lu, N., Kim, D.-H.: Flexible and stretchable electronics paving the way for soft robotics. Soft Robot. **1**(1), 53–62 (2014)

4. Huang, S., Liu, Y., Zhao, Y., Ren, Z., Guo, C.F.: Flexible electronics: Stretchable electrodes and their future. Adv. Funct. Mater. **29**(6), 1805924 (2019)
5. Gong, S., Lai, D.T.H., Su, B., Si, K.J., Ma, Z., Yap, L.W., Guo, P., Cheng, W.: Highly stretchy black gold E-skin nanopatches as highly sensitive wearable biomedical sensors. Adv. Electron. Mater. **1**(4), 1400063 (2015)
6. Jacques, M., Smith, M.K., Fölscher, D., Jamali, F., Malassagne, B., Leroy, J.: Telerobotic laparoscopic cholecystectomy: initial clinical experience with 25 patients. Ann. Surg. **234**(1), 1 (2001)
7. Khan, A., Anwar, Y.: Robots in healthcare: a survey. In: Science and Information Conference, pp. 280–292. Springer, Cham (2019)
8. Zhou, T., Cabrera, M.E., Wachs, J.P., Low, T., Sundaram, C.: A comparative study for telerobotic surgery using free hand gestures. J. Human-Robot Interact. **5**(2), 1–28 (2016)
9. Sebastian, M., Stückler, J., Behnke, S.: Adjustable autonomy for mobile teleoperation of personal service robots. In 2012 IEEE RO-MAN: The 21st IEEE International Symposium on Robot and Human Interactive Communication, pp. 933–940. IEEE (2012)
10. Kotthauser, T., Mauer, G.F.: Vision-based autonomous robot control for pick and place operations. In: IEEE/ASME International Conference on Advanced Intelligent Mechatronics (2009)
11. Kit, W.S., Venkatratnam, C.: Pick and place mobile robot for the disabled through voice commands. In: 2nd IEEE International Symposium on Robotics and Manufacturing Automation (ROMA) (2016)
12. Kim, K., Cho, J., Pyo, J., Kang, S., Kim, J.: Dynamic object recognition using precise location detection and ANN for robot manipulator. In: International Conference on Control, Artificial Intelligence, Robotics and Optimization (ICCAIRO) (2017)
13. Wang, M.-S.: 3D object pose estimation using stereo vision for object manipulation system. In: International Conference on Applied System Innovation (ICASI) (2017)
14. Andhare, P., Rawat, S.: Pick and place industrial robot controller with computer vision. In: International Conference on Computing Communication Control and automation (ICCUBEA) (2016)
15. Sbnchez, A.J., Martinez, J.M.: Robot-arm pick and place behavior programming system using visual perception. In: Proceedings 15th International Conference on Pattern Recognition. ICPR (2000)
16. Hakani, R.: DTMF based controlled robot vehicle. Int. J. Scient. Res. Dev. **2**, 1–14 (2015)
17. Dhayalini, K., Mukesh, R.: Deterioration and non-deterioration wastes separation using pick and place robot. In: 2nd International Conference on Inventive Systems and Control (ICISC) (2018)
18. Aravind, R.T.: Development of semi-automatic pick and place robot for material handling systems. In: 5th International Colloquium on Signal Processing and Its Applications (2009)
19. Chu, J., Guo, Z., Leng, L.: Object detection based on multi-layer convolution feature fusion and online hard example mining. IEEE Access **6**, 19959–19967 (2018)
20. Lin, M., Chen, Q., Yan, S.: Network in network. CoRR, abs/1312.4400 (2013)
21. Girshick, R.B., Donahue, J., Darrell, T., Malik, J.: Rich feature hierarchies for accurate object detection and semantic segmentation. In: IEEE Conference on Computer Vision and Pattern Recognition, 2014. CVPR 2014 (2014)
22. Saadatzi, M.N., Baptist, J.R., Yang, Z., Popa, D.O.: Modeling and fabrication of scalable tactile sensor arrays for flexible robot skins. IEEE Sens. J. (2019)
23. Szegedy, C., Liu, W., Jia, Y., Sermanet, P., Reed, S., Anguelov, D., Erhan, D., Vanhoucke, V., Rabinovich, A.: Going deeper with convolutions. In: IEEE Conference on Computer Vision and Pattern Recognition, 2015. CVPR 2015 (2015)

Stacked Generalization Ensemble Method to Classify Bangla Handwritten Character

Mir Moynuddin Ahmed Shibly, Tahmina Akter Tisha,
and Shamim H. Ripon

Abstract Image classification is one of the fundamental tasks in the field of machine learning. This study focuses on Bangla handwritten character recognition (BHCR). Recognizing handwritten characters has numerous applications like optical character recognition, post office automation, signboard recognition, number plate recognition, etc. A system that can classify Bangla handwriting on the character level efficiently on a large scale would surely be a primary step to achieve those autonomous user experiences. This study has developed a stacked generalization ensemble system that consists of six convolutional neural networks (CNN). The ensembled CNN model of this study has been able to perform more accurately than the individual CNN models. It can recognize 122 different Bangla handwritten characters with a test accuracy of 96.72% and an $f1$-score of 96.70%. The system has also outperformed most of the existing works by achieving better performance and by recognizing more handwritten characters.

Keywords Ensemble learning · Stacked generalization · Convolutional neural network · Image classification · Computer vision · Bangla handwritten character recognition · Autonomous system

1 Introduction

Approximately, 230 million people all over the world speak Bangla as a native language, and about 37 million people use it as a second language for speaking and writing purposes in their day to day life. Therefore, Bangla is not only the fifth

M. M. A. Shibly (✉) · T. A. Tisha · S. H. Ripon
Department of Computer Science and Engineering, East West University, Dhaka, Bangladesh
e-mail: shiblygnr@gmail.com

T. A. Tisha
e-mail: tahminatish001@gmail.com

S. H. Ripon
e-mail: dshr@ewubd.edu

© The Editor(s) (if applicable) and The Author(s), under exclusive license
to Springer Nature Singapore Pte Ltd. 2021
S. Shakya et al. (eds.), *Proceedings of International Conference on Sustainable Expert Systems*, Lecture Notes in Networks and Systems 176,
https://doi.org/10.1007/978-981-33-4355-9_46

most-spoken native language but also the seventh most-spoken language by the total number of speakers worldwide. For being such a prestigious language, Bangla handwritten character recognition is getting more importance among the researchers. It has a lot of prominent applications such as national ID number recognition, post office automation, automatic plate recognition system for vehicles, online banking, parking lot management system [1], and many more. Even for signboard translation, keyword spotting, digital character conversation, meaning translation, text-to-speech [2], scene image analysis, and most importantly, for Bangla optical character recognition (OCR) [3], the recognition system can play a significant role. But having such a system with optimal performance for Bangla is more challenging than any other language such as English that has a rather simple handwriting pattern.

Bangla script has a basic set of characters, i.e., 11 vowels and 39 consonants, and many conjunct-consonant characters joined by two or multiple basic characters, and these characters resemble each other pretty closely. This morphological complexity, complex-shape cursive characters, the enormous variety of handwriting styles of different people, and also the limited availability of the complete Bangla dataset have created more adversity in recognizing Bangla character. To overcome these challenges, researchers have come out with many articulate architectures such as CNN with transfer learning [4], DCNNs models [1], and ensemble learning [5]. In this study, the stacked generalization method of ensemble learning has been proposed in a new way to achieve a promising performance.

1.1 Background and Problem Statement

Image classification is one of the fundamental tasks in the field of machine learning and computer vision. There are many techniques and methods to categorize images into some classes. Bangla handwritten character recognition is a challenging task as the characters are enriched with various complex patterns. There are numerous methods that researchers have followed in Bangla handwriting related works. Bhattacharya et al. [6] have used a multi-layer perceptron neural network to classify Bangla handwriting images into 50 classes. In another work, hidden Markov models (HMM) have been used for classification [7]. Traditional machine learning algorithms have also been used in this area of research. For example, the support vector machine (SVM) model has been used for handwriting recognition [8]. The majority of recent works in this field have used a convolutional neural network (CNN)-based architecture [9–12]. Saha et al. have created a CNN architecture based on the divide and merge technique [13]. The recurrent neural network has also been used in Bangla handwriting character recognition [14]. On the other hand, transfer learning has shown good prospects in this particular classification task. In [4], a model has been created to classify 50 simple Bangla characters and the pre-trained weights of this model have been used to create another model to classify compound characters. ResNet50 architecture has been adopted for character recognition too [12].

Ensemble learning is a technique where a classifier is built based on multiple base classifiers for improved and accurate prediction. It has been used for image classification and recognition for a long time with better performance. Many works have explored the prospects of ensemble techniques in this area of image processing. An evaluation concerning the performance of the ensemble methods of convolutional neural networks for image recognition has been carried out by Ju et al. in 2018 [15]. The comparison between different methods has been done based on the achieved accuracy of the experiments and prediction. Stacked generalization—an ensemble technique has also been applied with CNN for the classification of document images with intra-domain transfer learning [16]. Stacked generalization can play a vital role in the medical sector for detecting disease. In [17], Rajaraman et al. have shown how this ensemble learning-based algorithm has improved the accuracy of detecting TB in chest radiographs by creating nonlinear decision-making functions. However, there has not been much work based on ensemble learning in Bangla handwriting recognition. There is a gap of knowledge in how an ensemble model would perform in isolated Bangla handwriting character classification.

This study develops an ensemble learning-based system to classify Bangla handwritten characters. The objectives of this study are—(1) To create a few convolutional neural network models for Bangla handwritten character classification. (2) To combine those CNN models to develop a more robust classifier using stacked generalization ensemble technique and (3) to compare the performances of individual models and combined models.

2   Related Works

For recognizing handwritten Bangla character, author Alom et al. [1] first introduced the state-of-the-art deep convolutional neural networks. A comprehensive analysis had been done on the CMATERdb [18] dataset using deep learning models including VGG Net [19], All-Conv Net [20], DenseNet [21], ResNet [22], FractalNet [23], and NiN [24]. Among all these models, DenseNet showed very promising accuracy like 99.13% on digits, 98.31% for Bangla alphabets, and 98.18% on special character recognition which outperformed other existing classical models for BHCR. In total, 73 different characters were classified. In [12], to solve the problem of BHCR using lesser iteration to train, the authors employed transfer learning on ResNet50, a deep learning convolutional neural network model. To fasten the training, the modified version of one cycle policy and the variation of image sizes techniques had been applied on the Banglalekha-Isolated dataset. They claimed to achieve the state-of-the-art result of 96.12% accuracy in 47 epochs on 84 classes without any use of ensemble learning.

The authors in [10] claimed to beat Rahman et al. [11] by achieving accuracy up to 95% using the deep convolutional neural network which seemed better compared to 85.96% achieved in [11] using only the convolutional neural network on Bengali handwritten alphabets. Both of them had 50 classes in their experimented dataset.

According to the authors [10], a more enriched dataset having different writing styles by both female and male, the normalization and preprocessing of images, and the sequential CNNs with max-pooling techniques had distinguished their methodology from the previous work of similar research. S. Bhattacharya et al. [25] achieved a 94.3% character level accuracy on test data consisting of 33,453-word samples by an end-to-end online Bangla handwriting recognition approach. They used SVM as the classifier, and their model was able to predict about 152 different characters from continuous text data by applying the segmentation of lines, words, and stroke having good accuracy. Moreover, the "Bengali Handwritten Keyboard Application" was a part of their proposed system.

In another study [9], the authors experimented on both Ekush and CMATERdb datasets using a convolutional neural network. Their proposed EkushNet model was mainly trained and validated with the Ekush dataset having 122 classes, which achieved a satisfactory accuracy of 96.90% on the training set and 97.73% on the validation set after 50 epochs. Besides, 95.01% accuracy was gained by the model in cross-validation applied on CMATERdb. The authors claimed that the performance of the proposed model was the best than the previous work done before them on Bangla handwritten recognition. Rahaman Mamun et al. [5] had taken an approach to recognize Bangla handwritten digit with the help of an ensemble of deep residual networks. The authors had applied heavy augmentation in the training set of NumtaDB dataset adding dropout layers to avoid overfitting. Finally, they used Xception, an ensemble learning approach models, on a hidden test set to improve the prediction accuracy, which had provided a notable performance of 96.69% accuracy.

3 Dataset

Few datasets consist of Bangla handwritten characters and numerals like BanglaLekha Isolated [26], ISI [27], NumtaDB [28], CMATERdb [18], and Ekush [29]. In these datasets, NumtaDB consists of only numerals and the ISI dataset has Bangla basic characters and numerals of 60 classes. The other datasets CMATERdb and BanglaLekha Isolated have Bangla basic characters, compound characters, and numerals distributed in 73 and 84 classes, respectively. On the other hand, the Ekush dataset also has Bangla modifiers along with basic characters, compound characters, and numerals, and the whole dataset is distributed to 122 classes. As Ekush is larger than any other Bangla handwritten character dataset, it is chosen for this study and it is a multi-purpose dataset. This contains 367,018 different images. The creators of this dataset have collected handwritten images in a form from 3086 people among which 50% are male and 50% are female. Table 1 shows the details of the Ekush dataset. And Fig. 1 demonstrates one character from each of the four types—the first image being a modifier, the second one being a simple character, and third and fourth images being a compound character and a numeral.

Table 1 Ekush dataset details

Character type	No. of classes	No. of instances
Modifier	10	30,667
Basic character	50	154,824
Compound character	52	150,840
Digit	10	30,687
Total	122	367,018

Fig. 1 Different handwritten characters from Ekush dataset

4 Proposed Methodology

In this study, isolated Bangla handwritten characters have been classified into 122 categories. In this section, the working procedures of the study have been described. The dataset contains images having 28 × 28 pixels grayscale images. As machine learning models can only work with numbers, before feeding these images to a model to train, those need to be preprocessed, i.e., converted to some numbers. First, the pixel values of each image are stored in a NumPy array. The pixel values are ranging from 0 to 255. After that, the numbers representing an image is re-scaled in the range of 0–1 by dividing each pixel value by 255. The reason behind this re-scaling is that the deep learning models work better of the input sequences are in 0–1 range. In this way, each image becomes a 28 × 28 matrix containing a value ranging from 0 to 1 in each cell. This matrix representation works as the input of the neural network. When the images are ready to be fed into a deep convolutional neural network, the models are created and trained with the preprocessed images. An example of the original image and its matrix representation after scaling is shown in Figs. 2 and 3. The reason behind preprocessing the input images in this way is that the images are spatial and contextual data. The same number is present in a matrix representation of an image but its placement is different from the other—it means that these two are two different images. Regarding this, the matrix representation is necessary rather than just squashing the pixel value representation of an image to create an input vector for the neural network.

Initially, a deep convolutional neural model has been developed and six models have been trained with different image augmentation techniques and with different

Fig. 2 Original image

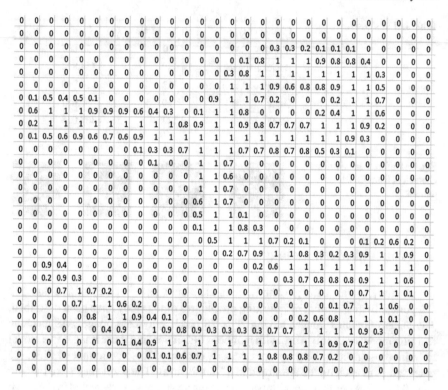

Fig. 3 Scaled matrix representation of the original image from Fig. 2 (input of the neural network)

batch sizes. After that, using those six models, a stacked generalization ensemble model has been created to leverage all the individual learned weights and to develop a more accurate and precise system for Bangla handwritten characters classification. The workflow of this study is given in Fig. 4.

4.1 Convolutional Neural Network

Convolutional neural network (CNN) is a type of deep neural networks that are widely used in image classification. A CNN architecture typically has some hidden convolutional layers that are inspired by the idea of mimicking the working mechanism of the brain. A convolutional layer convolves based on the inputs that are fed to the network. It has few attributes—convolutional kernel, input and output channels, size of filters, and an activation function. The activation function for the convolutional layer is generally rectifier linear unit (RELU). RELU activation function limits the fed numbers to be only positive. It converts the negative numbers to zero and leaves the positive numbers as-is. If x is the input and $f(x)$ is the transformed output, the transformation follows the following equation.

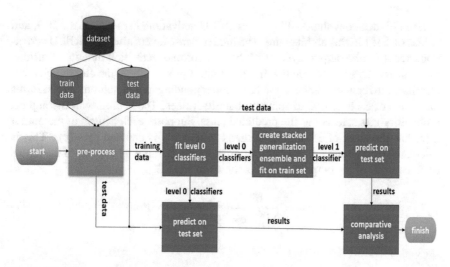

Fig. 4 Block diagram of the workflow of the study

$$f(x) = x^+ = \max(0, x)(\text{relu activation}) \tag{1}$$

Just like other neural networks, it must have an input and an output layer. Apart from these layers, a CNN architecture has max-pooling layers too. The outputs of multiple neuron groups are added by the max-pooling layer, and in this way, the dimensions of data are reduced. A CNN architecture may have some dropout layers too. And, at the top part of the architecture contains some fully connected dense layers. In this study, a 16-layers CNN with 14 hidden layers have been used for the classification task. After the input layer, there are two convolutional layers followed by a max-pooling layer. The two convolution layers have a filter size of 50 and 75, respectively. After the max-pooling layer, another convolutional layer is added with 125 filters and again a max-pooling layer is added. This happens for another two times with the convolutional layers having 175 and 225 filters. After the final max-pooling layer, there is a flatten layer and that is the start of fully connected layers in this architecture. Then, there is a dense layer followed by a 40% dropout layer and another dense layer followed by a 30% dropout layer. And finally, the output layer has 122 neurons indicating the number of classes in this work. Figure 5 shows the complete CNN architecture.

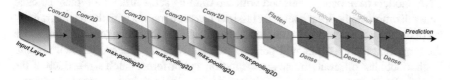

Fig. 5 Proposed CNN architecture

Each hidden convolutional layer has RELU activation, kernel size of 3×3, and strides of 1×1 in this architecture. The hidden dense layers also have RELU activation except for the output layer which has the softmax activation function. Softmax function activates the neuron that has the highest probability in the output layer and classifies the input sequence to the class corresponding to that neuron. It transforms the inputs of each neuron to some probability values. The neuron with the highest probability gets selected as the predicted class. Suppose, each neuron of the output layer has the input of e^{z_i} and the softmax probability is denoted by $\sigma(z)_i$. This is calculated by dividing the input of the corresponding neuron by the sum of the inputs of all neurons of the output layer. It is shown in Eq. (2).

$$\sigma(z)_i = \frac{e^{z_i}}{\sum_{j=1}^{K} e^{z_j}} \text{(softmax activation)} \tag{2}$$

4.2 Stacked Generalization Ensemble CNN

Ensemble learning is a technique used in machine learning fields where more than one model is trained for the same task as opposed to a typical machine learning technique where a single model is developed for solving a particular task. In ensemble learning, hypotheses from different models are combined to create a more generalized, accurate, and robust model to solve a specific problem. There are few ensemble techniques that can develop a generalized model from multiple models. Regarding this, most commonly used ensemble techniques are majority voting, weighted majority voting, Borda count, bagging, boosting, stacked generalization, etc. [30]. In this proposed work, the stacked generalization technique has been adopted to create a system from multiple CNN models for better performance.

In a typical stacked generalization, the original training dataset is divided into n sets and a model for each set is created. Each of the developed models is cross-validated with the same cross-validation set. These models are known as level 0 classifiers. In the next phase, the developed level 0 classifiers are combined to create a level 1 classifier. This level 1 model is known as a meta-model or meta-classifier. The outputs of level 0 models serve as the training data of the meta-classifier. And finally, the meta-model is tested against the original test set of the dataset.

In this study, a modified version of stacked generalization has been used. Six deep CNN models have been developed with the training set. Unlike the typical stacked generalization method, each level 0 classifier has been trained with the whole training set of the dataset but with different image augmentation for better generalization with the hope that the classifier will perform efficiently in the real-world environment. Each model has different image augmentation and different batch sizes. Each of the level 0 classifiers has seen the different versions of the original dataset for different image data augmentation. The architecture described in the previous section has been

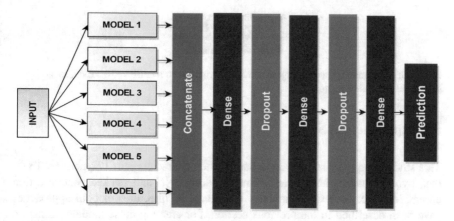

Fig. 6 Proposed stacked generalization ensemble architecture

used to train these level 0 classifiers. Each model has been saved and combined. The output of each model is concatenated and few layers are added on top of that. The layers are a dense layer, a 40% dropout layer, another dense layer, another 30% dropout layer, and the final output layer having 122 neurons representing 122 classes of the Bangla handwritten characters. The final level 1 meta-classifier has been trained with the training set of the dataset without any image augmentation. Figure 6 shows the outlined stacked generalized ensemble system.

4.3 Image Data Augmentation

It has been proven that image data augmentation helps a classifier to recognize images more accurately [31]. In here, while creating the individual CNN models, different types of data augmentation have been applied to the training images. The images have been rotated from 7° to 15°. The height and width shift range also has been adjusted from 0.1 to 0.2 and the zoom range is varied from 0.1 to 0.2. These types of augmentation provide more generalization to the classifiers. Here are some examples of image data augmentation (Figs. 7 and 8).

Fig. 7 Original image (left) and augmented image (right) with rotation = 10°

Fig. 8 Original image (left) and augmented image (right) with rotation = 10°, zoom range = 0.1, height and width shift range = 0.1

5 Experimental Setup

This study aims to classify Bangla handwritten characters into 122 classes. To do that, two approaches have been followed—deep CNN and stacked generalization ensemble of six deep CNN models. The model architectures of both approaches have been described in the previous section. For creating those models, Keras on top of TensorFlow—a Python library has been used. For the stacked generalization model, a scikit-learn library of Python along with Keras has been utilized. The models have been trained on a computer having Ryzen 5, 1600 CPU, 8 GB ram, and Nvidia GeForce 1050 TI GPU with Linux Mint 19.3 operating system.

5.1 Training, Testing, and Validating

The whole dataset has been divided into three sets—train, validation, and test. The train set has 75%, validation set has 10%, and test set has 15% of the data. 275,264 of 367,018 images in the whole dataset are used for training. And for validation and test, the number of images is 36,701 and 55,053, respectively. For level 0 CNN classifiers, each of them has been trained for 40 epochs and has been validated on every epoch. The weights of the model in the epoch that has the best validation accuracy have been saved to the local disk. For each model, this process has been continued. In this way, the best performing of weights of six individual deep CNN has been stored. After that, each of the models have been loaded and tested with a test set of the dataset in terms of precision, recall, $f1$-score, and accuracy. For the level 1 stacked generalization ensemble classifier, a model has been created with the combination of all level 0 classifiers. In the training phase, the layers individual models that constructed the ensemble models have been frozen because there is no point in updating already learned weights. Instead, only the layers that are specific to the ensemble models are made trainable. This model has been trained for 40 epochs too. During training, like individual CNN models, the weights of the best epoch have been stored. And later, it has been tested against the test set. While training, validating, and testing ensemble models, there is no single input channel. Instead, the same input has to be provided to each of the six level 0 models.

5.2 Optimizers, Loss Function, Batch Size, and Evaluation Metrics

While training the level 0 classifier, the RMSprop optimizer with a 0.001 learning rate has been used. For the level 1 stacked generalization ensemble model, the optimizer is Adam with a learning rate of 0.001. As the models have to accomplish a multi-class classification task, the categorical cross-entropy loss function has been used. Another aspect of this study is that, while training, loading the whole training set into memory has not been possible. That is why the dataset has to be loaded batch-wise progressively. The batch size of 512, 768, 1024, 1536, and 2048 have been used during experiments. Moreover, four evaluation metrics have been used throughout the whole working process of this study—precision, recall, $F1$-score, and accuracy. Precision is the portion of the correct predictions made by the classifier with respect to total predicted classes. And, recall is the portion of the correct predictions made by the classifier regarding the total existing accurate classes. $F1$-score is the harmonic mean of precision and recall, i.e., both precision and recall are given equal importance while evaluating the performance of a classifier. In short, accuracy is nothing but the portion of the classifier being right.

$$\text{precision} = \frac{TP}{TP + FP} \tag{3}$$

$$\text{recall} = \frac{TP}{TP + FN} \tag{4}$$

$$f1\_score = \frac{2 \times \text{precision} \times \text{recall}}{\text{precision} + \text{recall}} \tag{5}$$

$$\text{accuracy} = \frac{TP + TN}{TP + FP + TN + FN} \tag{6}$$

Here, TP, TN, FP, and FN are true positive, true negative, false positive, and false negative, respectively.

6 Results

After completing the test phase of both stacked generalization ensemble models, it has been seen that the ensemble model has performed better than any of the individual CNN models. Among the CNN models, the fifth model has shown the best test accuracy of 95.98% , and it also has a training accuracy of 96.65% and a validation accuracy of 95.86% . On the other hand, the stacked generalization ensemble model has a test accuracy of 96.72% . The ensemble model combining six CNN models have a training accuracy of 98.13% . and validation accuracy of 97.38% . after 40. epochs.

6.1 Result Analysis

performance of six individual deep CNN models and stacked generalization ensemble models along with the data augmentation of those models' input images are given in Table 2. It has been observed that model 4 which has 1024 batch size and image augmentation of rotation = 9 degrees, height and width shift = 0.09, and zoom level = 0.09 has the best performance among level 0 classifiers. Train accuracy versus validation accuracy graph and train loss versus validation loss graph of the best performing CNN model has been shown in Figs. 9 and 10. By looking at these two figures, it is seen that the learning of the best CNN model has improved over the number of epochs. These figures also justify the stoping of model training as at the 40th epoch, the accuracy is highest and the loss is minimal. The increasing number of epoch beyond 40 is not that beneficial in this study as the improvement in accuracy is very little compare to the time taken for further training. The other level 0 models demonstrate a similar pattern in the learning curves. Another chart demonstrating the comparison of precision, recall, $f1$-score, and accuracy of all the models is given in Fig. 11. In the bar chart of Fig. 11, it can be seen that the ensemble model has the highest performance than the level 0 classifiers in terms of all the evaluation metrics. This justifies the use of the stacked generalization ensemble technique for the classification task.

Table 2 Performance of level 0 and level 1 classifiers

Model no	Batch size	Image augmentation	Test accuracy (%)
1	512	Rotation range = 10, height and width shift = 0.1, zoom level = 0.1	95.83
2	512	Rotation range = 11, height and width shift = 0.11, zoom level = 0.11	95.36
3	768	Rotation range = 12, height and width shift = 0.12, zoom level = 0.12	95.64
4	1024	Rotation range = 9, height and width shift = 0.09, zoom level = 0.09	95.98
5	1536	Rotation range = 15, height and width shift = 0.15, zoom level = 0.15	95.84
6	2048	Rotation range = 9, height and width shift = 0.09, zoom level = 0.09	95.97
Stacked	**32**	**No augmentation**	**96.72**

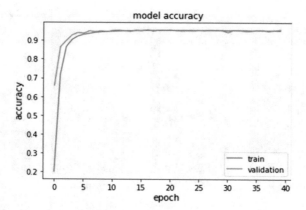

Fig. 9 Best CNN model accuracy (model 4)

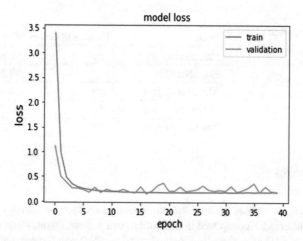

Fig. 10 Best CNN model loss (model 4)

6.2 Result Comparison

There is only one work that has used the ensemble technique for Bangla character recognition and that has only covered the Bangla numerals [5]. This work has obtained 96.69% accuracy. There are few other works, some of them have recognized only simple characters, and some of them have classified simple and numeric characters. There is not much work that has recognized as many characters as this study recognized. A comparison of this work with some existing studies has been presented in Table 3.

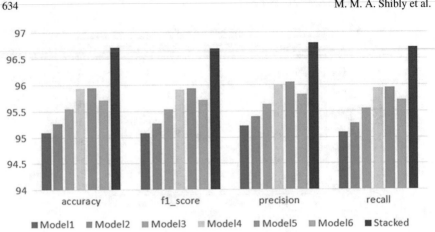

Fig. 11 Level 0 and level 1 classifiers comparison

Table 3 Performance comparison

Work	Number of classes	Test accuracy (%)
Mamun et al. [5]	10	96.69
EkushNet [29]	122	97.73
Ahmed et al. [10]	50	95
Bhattacharya et al. [25]	152	94.3
Proposed method	122	96.72

7   Discussion

T idea of using a stacked generalization ensemble method to classify handwritten Bangla character is first proposed in this study. And it demonstrates that the proposed mhod can be used for the classification task to obtain better and more accurate performance. This method gives higher accuracy than any of the ordinary convolutional neural networks. It also beats the majority of the existing works in terms of performance. In this study, few CNN models have been created to classify the images of the Ekush dataset into 122 classes following the first objective of the study. Those CNN models have acted as the level 0. classifiers for the stacked generalization ensemble classifier. After that, the models have been evaluated based on some evaluation metrics. The ensemble model has the best performance and outperformed most of the existing works in this field of handwriting recognition. It has a training accuracy of 98.13% , validation accuracy of 97.38% , and test accuracy of 96.72% . On the other hand, the level 0 classifiers do not have more than 95.97% test accuracy. Another novel aspect of this study is how the stacked generalization method is adopted. Usually, level 0 classifiers are fitted on different folds of the training set using n-fold cross-validation. Each fold gives a classifier, and those n classifiers are stacked to create the level 1 classifier. This strategy has not been flowed in this study. Each level 0 classifier has been fitted on the full training set, but with different augmentation. By following this strategy, level 0 classifiers see more training data. Additionally, image augmentation helps the classifier to be familiar with different patterns of the handwritten image. If the individual classifiers can handle a variety of images, then the stacked classifier will yield better accuracy.

The comparative analysis of these models' performances shows that the accuracy heavily relies on the batch size of the input instances. The models tend to have higher accuracy if the batch size is larger. Due to the limitation of computational power, it has not been possible to increase the batch size more than 2048. Larger batch size could have yielded better performance. Other reasons behind the errors in the result are that few of the classes are imbalanced, and few of them have wrongly labeled data. For example, the compound characters labeled as 110 and 111 in the dataset have only about 1000. and 400 instances, while the rest of the classes have approximately 3500. instances each. On the other hand, the images labeled as 10. and 11 have a considerable amount of wrongly labeled data. For these reasons, 5–7 classes have poor individual precision and recall scores. And those few classes have affected the overall performance of the classifier. Class-wise outlier detection can be applied to eliminate the wrongly labeled data to make the model more accurate. Adding more data to the imbalanced classes can also boost performance.

Another room for improvement is to work with the CNN architecture of the base classifiers. The architecture used in this work for level 0. classifier is pretty straight-forward. And the same architecture is used for all six base models. So, designing a more complex architecture and using different architectures for the different models could help this study to gain better performance. Moreover, this study does not cover all the existing Bangla characters, rather covers 122 of them. Another limitation is using a dataset that contains only greyscale images. Real-world images are mostly RGB. But the RGB images can be converted to greyscale to be compatible with this work. However, despite all these limitations, this study has shown that ensemble techniques can add another dimension to the research of recognizing handwritten characters. And in this study, it has been proven that the stacked generalization method can obtain better performance.

8 Future Works and Conclusion

Bangla is a very rich language. Though it has only 50 simple characters, there are also many compound characters in Bangla. In the Ekush dataset, there are 52 compound characters. Adding more compound characters to it can help researchers to create more efficient methods to recognize them. There are wrongly labeled data in every publicly available Bangla handwritten character dataset. Cleaning those can help data scientists to build better applications based on handwritten characters. Besides that, only the stacked generalization ensemble technique has been applied in this study. Other methods of ensemble learning can be applied and their efficiency can be measured as well in this case. Another area worth exploring is to understand the prospects of transfer learning in this area by applying various pre-trained models like ResNet50, VGG16, VGG19, Xception, InceptionV3, etc. Those state-of-the-art pre-trained models can add more efficiency to Bangla specific image recognition. Additionally, a systematic literature review on the existing BHCR is missing. This kind of study can lead the researchers in a better direction. Moreover, there are limited existing works that cover the word level and sentence level handwritten image recognition. Working on that can be very helpful to make various Bangla handwriting dependent systems automated.

This study has aimed to explore the prospects of ensemble learning in Bangla handwritten character recognition. And it has proven that ensemble learning techniques can be used successfully, and it can add more efficiency to the model performance. This decision is backed by the implementation of a stacked generalization ensemble method for the classification task as the ensemble classifier performed better than any CNN classifier. This classifier has also performed better than most of the currently available works in handwriting recognition. Such a model can eventually help the software engineers to build sustainable systems for post office automation, number plate recognition, and any other field where handwriting recognition is needed. These autonomous systems can give people more convenient and hassle-free experiences.

References

1. Alom, M.Z., Sidike, P., Hasan, M., Taha, T.M., Asari, V.K.: Handwritten Bangla character recognition using the state-of-the-art deep convolutional neural networks. Comput. Intell. Neurosci. (2018). https://doi.org/10.1155/2018/6747098
2. Manoharan, S.: A smart image processing algorithm for text recognition, information extraction and vocalization for the visually challenged. J. Innov. Image Process. (2019). https://doi.org/10.36548/jiip.2019.1.004
3. Manisha, N., Sreenivasa, E.: Role of offline handwritten character recognition system in various applications. Int. J. Comput. Appl. (2016). https://doi.org/10.5120/ijca2016908349
4. Reza, S., Amin, O.B., Hashem, M.M.A.: Basic to Compound: A Novel Transfer Learning Approach for Bengali Handwritten Character Recognition. Presented at the (2020). https://doi.org/10.1109/icbslp47725.2019.201522
5. Rahaman Mamun, M., Al Nazi, Z., Salah Uddin Yusuf, M.: Bangla handwritten digit recognition approach with an ensemble of deep residual networks. In: 2018 International Conference on Bangla Speech Lang. Process. ICBSLP 2018, pp. 21–22 (2018). https://doi.org/10.1109/ICBSLP.2018.8554674
6. Bhattacharya, U., Gupta, B.K., Parui, S.K.: Direction code based features for recognition of online handwritten characters of Bangla. In: Proceedings of the International Conference on Document Analysis and Recognition, ICDAR (2007). https://doi.org/10.1109/ICDAR.2007.4378675
7. Roy, P.P., Bhunia, A.K., Das, A., Dey, P., Pal, U.: HMM-based Indic handwritten word recognition using zone segmentation. Patt. Recogn. (2016). https://doi.org/10.1016/j.patcog.2016.04.012
8. Mohiuddin, S., Bhattacharya, U., Parui, S.K.: Unconstrained Bangla online handwriting recognition based on MLP and SVM. In: ACM International Conference Proceeding Series (2011). https://doi.org/10.1145/2034617.2034635

9. Azad Rabby, A.K.M.S., Haque, S., Abujar, S., Hossain, S.A.: Ekushnet: Using convolutional neural network for Bangla handwritten recognition. Procedia Comput. Sci. 143, 603–610 (2018). https://doi.org/10.1016/j.procs.2018.10.437
10. Ahmed, S., Tabsun, F., Reyadh, A.S., Shaafi, A.I., Shah, F.M.: Bengali handwritten alphabet recognition using deep convolutional neural network. In: 5th International Conference on Computer, Communication, Chemical, Materials and Electronic Engineering, IC4ME2 2019 (2019). https://doi.org/10.1109/IC4ME247184.2019.9036572
11. Rahman, M.M., Akhand, M., Islam, S., Chandra Shill, P., Hafizur Rahman, M.M.: Bangla handwritten character recognition using convolutional neural network. Int. J. Image. Graph. Sign. Process 7, 42–49 (2015). https://doi.org/10.5815/ijigsp.2015.08.05
12. Chatterjee, S., Dutta, R.K., Ganguly, D., Chatterjee, K., Roy, S.: Bengali handwritten character classification using transfer learning on deep convolutional network. In: Lecture Notes in Computer Science (including subseries Lecture Notes in Artificial Intelligence and Lecture Notes in Bioinformatics) (2020). https://doi.org/10.1007/978-3-030-44689-5_13
13. Saha, S., Saha, N.: A lightning fast approach to classify bangla handwritten characters and numerals using newly structured deep neural network. Proced. Comput. Sci. 132, 1760–1770 (2018). https://doi.org/10.1016/j.procs.2018.05.151
14. Chakraborty, B., Mukherjee, P.S., Bhattacharya, U.: Bangla online handwriting recognition using recurrent neural network architecture. In: ACM International Conference Proceeding Series (2016). https://doi.org/10.1145/3009977.3010072
15. Ju, C., Bibaut, A., van der Laan, M.: The relative performance of ensemble methods with deep convolutional neural networks for image classification. J. Appl. Stat. (2018). https://doi.org/10.1080/02664763.2018.1441383
16. Das, A., Roy, S., Bhattacharya, U., Parui, S.K.: Document image classification with intra-domain transfer learning and stacked generalization of deep convolutional neural networks. In: Proceedings—International Conference on Pattern Recognition (2018). https://doi.org/10.1109/ICPR.2018.8545630
17. Rajaraman, S., Candemir, S., Xue, Z., Alderson, P.O., Kohli, M., Abuya, J., Thoma, G.R., Antani, S.: A novel stacked generalization of models for improved TB detection in chest radiographs. In: Proceedings of the Annual International Conference of the IEEE Engineering in Medicine and Biology Society, EMBS (2018). https://doi.org/10.1109/EMBC.2018.8512337
18. Das, N., Sarkar, R., Basu, S., Kundu, M., Nasipuri, M., Basu, D.K.: A genetic algorithm based region sampling for selection of local features in handwritten digit recognition application. Appl. Soft Comput. J. (2012). https://doi.org/10.1016/j.asoc.2011.11.030
19. Simonyan, K., Zisserman, A.: Very deep convolutional networks for large-scale image recognition. In: 3rd International Conference on Learning Representations, ICLR 2015—Conference Track Proceedings (2015)
20. Springenberg, J.T., Dosovitskiy, A., Brox, T., Riedmiller, M.: Striving for simplicity: the all convolutional net. In: 3rd International Conference on Learning Representations, ICLR 2015—Workshop Track Proceedings (2015)
21. Huang, G., Liu, Z., Van Der Maaten, L., Weinberger, K.Q.: Densely connected convolutional networks. In: Proceedings—30th IEEE Conference on Computer Vision and Pattern Recognition, CVPR 2017 (2017). https://doi.org/10.1109/CVPR.2017.243
22. He, K., Zhang, X., Ren, S., Sun, J.: Deep residual learning for image recognition. In: Proceedings of the IEEE Computer Society Conference on Computer Vision and Pattern Recognition (2016). https://doi.org/10.1109/CVPR.2016.90
23. Larsson, G., Maire, M., Shakhnarovich, G.: FractalNet: ultra-deep neural networks without residuals. In: 5th International Conference on Learning Representations, ICLR 2017—Conference Track Proceedings (2019)
24. Lin, M., Chen, Q., Yan, S.: Network in network. In: 2nd International Conference on Learning Representations, ICLR 2014—Conference Track Proceedings (2014)
25. Bhattacharya, S., Maitra, D. Sen, Bhattacharya, U., Parui, S.K.: An end-to-end system for Bangla online handwriting recognition. In: Proceedings of International Conference on Frontiers in Handwriting Recognition, ICFHR (2016). https://doi.org/10.1109/ICFHR.2016.0076

26. Biswas, M., Islam, R., Shom, G.K., Shopon, M., Mohammed, N., Momen, S., Abedin, A.: BanglaLekha-Isolated: a multi-purpose comprehensive dataset of handwritten Bangla isolated characters. Data Br. (2017). https://doi.org/10.1016/j.dib.2017.03.035
27. Chaudhuri, B.B.: A complete handwritten numeral database of Bangla-A major Indic script. In: 10th International Workshop on Frontiers of Handwriting Recognition (IWFHR), La Baule, France (2006)
28. Alam, S., Reasat, T., Doha, R., Humayun, A.I.: Numta—Assembled Bangla Handwritten Digits, pp. 1–4
29. Rabby, A.K.M.S.A., Haque, S., Islam, M.S., Abujar, S., Hossain, S.A.: Ekush: A multipurpose and multitype comprehensive database for online off-line Bangla handwritten characters. In: Communications in Computer and Information Science (2019). https://doi.org/10.1007/978-981-13-9187-3_14
30. Polikar, R.: Ensemble. Mach. Learn. (2012). https://doi.org/10.1007/978-1-4419-9326-7
31. Perez, L., Wang, J.: The effectiveness of data augmentation in image classification using deep learning (2017)

Extracting Forensic Data from a Device Using Bulk Extractor

Shahana Bano, P. Pavan Kalyan, Y. Lakshmi Pranthi, and K. Priyanka

Abstract Generally, a data or an information of a device will play a key role to find the owner and the things were done in that device from starting day up to the day it was tested were shown in the information which we get from the device. And getting that data was done by data extractors and the problem with the extractor is, for every type of data it needs a different type of extractors to overcome this problem, and this bulk extractor gets all types of data of a device without using any other extractor. This bulk extractor saves the time of testing and gathering information and gets all types of data easily.

Keyword Extractor · Information · Data and device

1 Introduction

Information is just a normal word which used by a human or a machine, but in another way, a small information is more valuable than the whole money in the entire world. Getting an information from a human is not a simple task and getting information from a system or any other device is harder than getting information from a human. We can get information from a human being with help of psychological way, flip flops, logical way, etc. There are many ways to get the information from a human being, but these are not worked on an electronic device. To get data from a device,

S. Bano · P. Pavan Kalyan · Y. Lakshmi Pranthi (✉) · K. Priyanka
Department of Computer Science and Engineering, Koneru Lakshmaiah Education Foundation,
Vaddeswaram, Andhra Pradesh, India
e-mail: pranathi.yr@gmail.com

S. Bano
e-mail: shahanabano@icloud.com

P. Pavan Kalyan
e-mail: pavankalyanpala@gmail.com

K. Priyanka
e-mail: priyankakondenti1999@gmail.com

© The Editor(s) (if applicable) and The Author(s), under exclusive license
to Springer Nature Singapore Pte Ltd. 2021
S. Shakya et al. (eds.), *Proceedings of International Conference on Sustainable Expert
Systems*, Lecture Notes in Networks and Systems 176,
https://doi.org/10.1007/978-981-33-4355-9_47

we need software and that software are called as extractors or data extractors And these extractors are used to get the information from any device easily but those devices take time, and for every different type of data, there is a separate type of data extractor and it will extract the data only based on the categorized type only none of those extractors will never do the work of other extractors work.

2 Introduction to Extractors

Extractor or data extractor is a software which is used to get or recover information from a device, or an electronic gadget and that operation is called as data extraction. Before going to learn about data extractor, user must learn what is data extraction and how it will help the user to learn clearer about data extractor.

Data extraction means it is a process of getting data from a selected device by using the primary and default procedure and techniques of a selected data extraction software. And that software will get all the visible and invisible data of the device which was testing to get the data.

For data extraction, there are plenty of software and those are called as data extractors. The data extractors do only one thing that is the extraction of data for every type of data like mobile data, system data etc., there is a separate data extractor to do the data extraction. And these all extractors work properly and get data from the selected devices and those extractors will only get only particular categorized data from the device only and does not try to extract other categorized data from a fixed categorized data. And all of these software's are not free to use, each and every one of the extractors is a paid software and we get normally a trailer version of that extractor software and in that also we does not get all the operations to work on it Fig. 1.

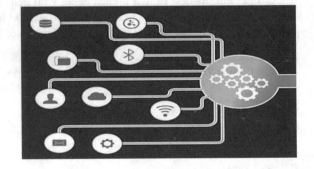

Fig. 1 Working of a data extractor

3 Exiting System

Generally, extractors work only to get data from a connected device only that tool will only grab some kind of data only not entire data. And one of the data extractor process and working was explained in this paper and the faults are also mentioned properly. To start this extractor, we have to set the path of that software stored location and its file name is needed to start that extractor and it is shown in the following image (Fig. 2).

And we can locate the storage of a local disk in terminal and it looks just like as the following image and we can see many details in the result of that command (Fig. 3).

The reason to show the storage location, we need all of the details which are mentioned above to start the extractor. And the command to start this extractor needs to include the storage location and the path and the command is shown in the following image (Fig. 4).

And the same command is used to start extract the data from the device with a small change and the change is in the place of disk name. After executing this command, it will generate an output file for the selected device and that too in HTML file and it is stored in another location of the device and it is locked with a password and there is a command to open those locked files, and the command to open those locked files is as shown in the following image (Fig. 5).

And we can see the locked output file in the location and it will be shown like the following image (Fig. 6).

The problem with these extractors is all the files are in HTML files and all the files are in encrypted form, and these problems are overcome with our proposed system.

Fig. 2 Normal extractor starting

```
·ubuntu:~/Downloads/bulk_extractor-1.3.1$ man bulk_extractor
·ubuntu:~/Downloads/bulk_extractor-1.3.1$ cd ~/Desktop/
·ubuntu:~/Desktop$ sudo fdisk -l
```

Fig. 3 Location of extractor in disk

```
Device Boot    Start      End      Blocks   Id  System
/dev/sdb1   *    2048  125827071  62912512   7  HPFS/NTFS/exFAT
lecturesnippets@lecturesnippets-ubuntu:~/Desktop$
```

Fig. 4 Command to start the extractor

```
·ubuntu:~/Desktop$ sudo bulk_extractor -o output /dev/sdb1
```

Fig. 5 Command to open locks for file

```
·ubuntu:~/Desktop$ sudo chmod -R 777 output/
```

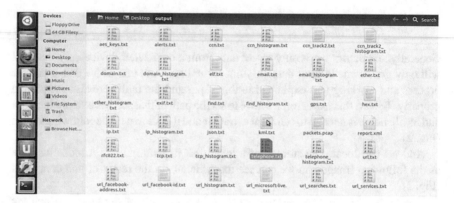

Fig. 6 Output files inside the locked folder

4 Proposed System

As per we explained in the existing process to get data from a device, we can easily identify the fault in the process and the data it will gram through the existing software.

Our proposed procedure and software will help to overcome the problem on existing system. And name of that software is bulk extractor. And to start the bulk extractor, we have to open a terminal in Linux and the command to start is as follows (Fig. 7).

After starting bulk extractor, we can see all the commands in the manual of this software and to see that manual we have to use a command and that command is a follows (Fig. 8).

In that command -h means help that will show each and every command, we can able to perform with bulk extractor and the list of those commands is as follows [5] (Fig. 9).

In the above picture, we can see some of the commands we can perform in the bulk extractor and there are many more commands in bulk extractor to perform. To get the data from a device from this software, we have to insert or connect to the system.

To perform this, I connected an USB drive to a system and then started the bulk extractor. After staring bulk extractor, we have to execute the following command to read the connected USB [6] (Fig. 10).

Fig. 7 Command to start bulk extractor

Fig. 8 Command to open manual of bulk extractor

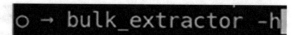

Fig. 9 Some of the
commands we can use in
bulk extractor

```
These scanners disabled by default; enable with -e:
   -e base16 - enable scanner base16
   -e facebook - enable scanner facebook
   -e outlook - enable scanner outlook
   -e sceadan - enable scanner sceadan
   -e wordlist - enable scanner wordlist
   -e xor - enable scanner xor
These scanners enabled by default; disable with -x:
   -x accts - disable scanner accts
   -x aes - disable scanner aes
   -x base64 - disable scanner base64
   -x elf - disable scanner elf
   -x email - disable scanner email
   -x exif - disable scanner exif
   -x find - disable scanner find
   -x gps - disable scanner gps
   -x gzip - disable scanner gzip
   -x hiberfile - disable scanner hiberfile
   -x httplogs - disable scanner httplogs
   -x json - disable scanner json
   -x kml - disable scanner kml
   -x msxml - disable scanner msxml
   -x net - disable scanner net
   -x ntfsusn - disable scanner ntfsusn
   -x pdf - disable scanner pdf
   -x rar - disable scanner rar
   -x sqlite - disable scanner sqlite
   -x vcard - disable scanner vcard
   -x windirs - disable scanner windirs
   -x winlnk - disable scanner winlnk
   -x winpe - disable scanner winpe
   -x winprefetch - disable scanner winprefetch
   -x zip - disable scanner zip
```

Fig. 10 Command to test
device

The result of the command in the previous image will create a folder which consists of all types of the data inside the connected device. And the folder is as shown in the following image (Fig. 11).

And in terminal, we can see the time and other details of operations performed at the time of data gathering from the connected device. And the following image shows the details of how much of time and other operations details (Fig. 12).

After some time, the scanning and extracting data from the device show complete details like time taken to perform operation, number of phases has worked on the device, etc. And the following image will show complete details of the scanning of the connected device [7] (Fig. 13).

In that report, we can see details about the output folder, name assigned by bulk extractor to the output, and it also shown some division of data based on some categories.

After scanning, all the data can be seen in the both file manager and terminal.

To see the output files of bulk extractor, there is a command and that will show each and every file which were inside the folder and the command is shown in the following image (Fig. 14).

After executing that command in terminal where bulk extractor is started, or another terminal will show the result in a very detailed manner and that list of files is shown in the following image (Fig. 15).

Fig. 11 Connected device
and output of bulk extractor

Fig. 12 List of operations
performed in the process

```
bulk_extractor version: 1.6.0
Hostname: kali
Input file: terry-work-usb-2009-12-11.E01
Output directory: bulk_output
Disk Size: 2097152000
Threads: 4
2:28:01 Offset 67MB (3.20%) Done in 0:01:49 at 02:29:50
2:28:02 Offset 150MB (7.20%) Done in 0:00:55 at 02:28:57
2:28:02 Offset 234MB (11.20%) Done in 0:00:39 at 02:28:41
2:28:03 Offset 318MB (15.20%) Done in 0:00:31 at 02:28:34
2:28:04 Offset 402MB (19.20%) Done in 0:00:26 at 02:28:30
2:28:04 Offset 486MB (23.20%) Done in 0:00:22 at 02:28:26
2:28:05 Offset 570MB (27.20%) Done in 0:00:19 at 02:28:24
2:28:05 Offset 654MB (31.20%) Done in 0:00:16 at 02:28:21
```

To see any individual file in the above list again, we have to another command and that command is extension of previous command, and the difference is this time, we have to give the file name which we wanted to see particularly (Fig. 16).

And the result of the above command for the given particular file will show the content inside the selected file which was given in the command shows just like the following image (Fig. 17).

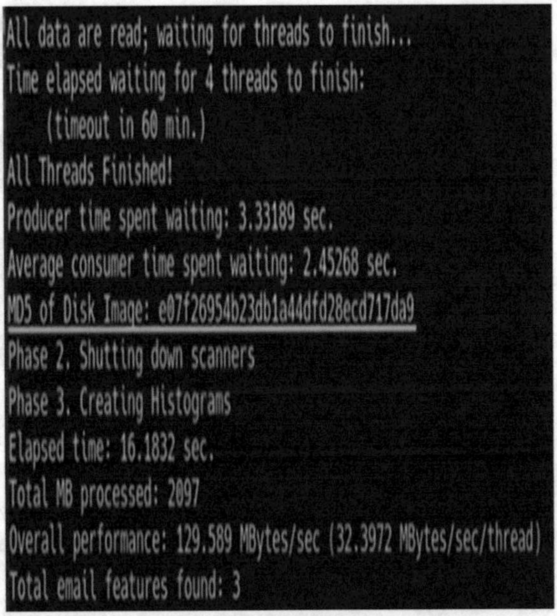

Fig. 13 Complete details of the scanning report

```
All data are read; waiting for threads to finish...
Time elapsed waiting for 4 threads to finish:
    (timeout in 60 min.)
All Threads Finished!
Producer time spent waiting: 3.33189 sec.
Average consumer time spent waiting: 2.45268 sec.
MD5 of Disk Image: e07f26954b23db1a44dfd28ecd717da9
Phase 2. Shutting down scanners
Phase 3. Creating Histograms
Elapsed time: 16.1832 sec.
Total MB processed: 2097
Overall performance: 129.589 MBytes/sec (32.3972 MBytes/sec/thread)
Total email features found: 3
```

Fig. 14 Command to see the output files in terminal

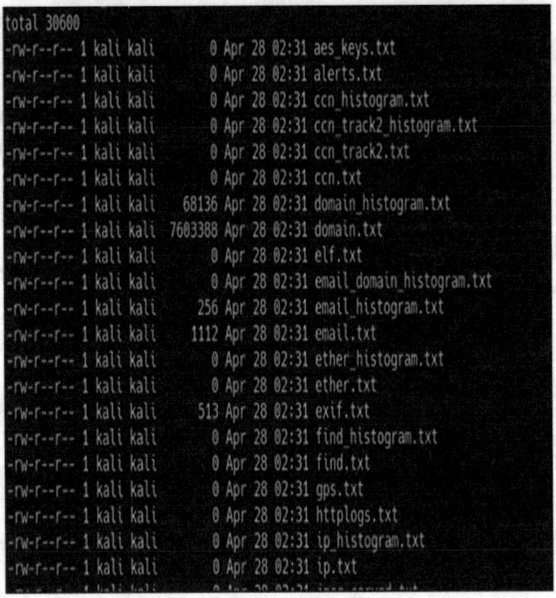

```
○ → ls -l bulk_output
```

Fig. 15 List of gathered files from the device

```
total 30600
-rw-r--r-- 1 kali kali        0 Apr 28 02:31 aes_keys.txt
-rw-r--r-- 1 kali kali        0 Apr 28 02:31 alerts.txt
-rw-r--r-- 1 kali kali        0 Apr 28 02:31 ccn_histogram.txt
-rw-r--r-- 1 kali kali        0 Apr 28 02:31 ccn_track2_histogram.txt
-rw-r--r-- 1 kali kali        0 Apr 28 02:31 ccn_track2.txt
-rw-r--r-- 1 kali kali        0 Apr 28 02:31 ccn.txt
-rw-r--r-- 1 kali kali    68136 Apr 28 02:31 domain_histogram.txt
-rw-r--r-- 1 kali kali  7603388 Apr 28 02:31 domain.txt
-rw-r--r-- 1 kali kali        0 Apr 28 02:31 elf.txt
-rw-r--r-- 1 kali kali        0 Apr 28 02:31 email_domain_histogram.txt
-rw-r--r-- 1 kali kali      256 Apr 28 02:31 email_histogram.txt
-rw-r--r-- 1 kali kali     1112 Apr 28 02:31 email.txt
-rw-r--r-- 1 kali kali        0 Apr 28 02:31 ether_histogram.txt
-rw-r--r-- 1 kali kali        0 Apr 28 02:31 ether.txt
-rw-r--r-- 1 kali kali      513 Apr 28 02:31 exif.txt
-rw-r--r-- 1 kali kali        0 Apr 28 02:31 find_histogram.txt
-rw-r--r-- 1 kali kali        0 Apr 28 02:31 find.txt
-rw-r--r-- 1 kali kali        0 Apr 28 02:31 gps.txt
-rw-r--r-- 1 kali kali        0 Apr 28 02:31 httplogs.txt
-rw-r--r-- 1 kali kali        0 Apr 28 02:31 ip_histogram.txt
-rw-r--r-- 1 kali kali        0 Apr 28 02:31 ip.txt
```

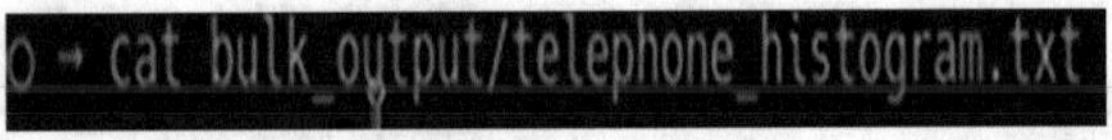
cat bulk_output/telephone_histogram.txt

Fig. 16 Command to see a particular file

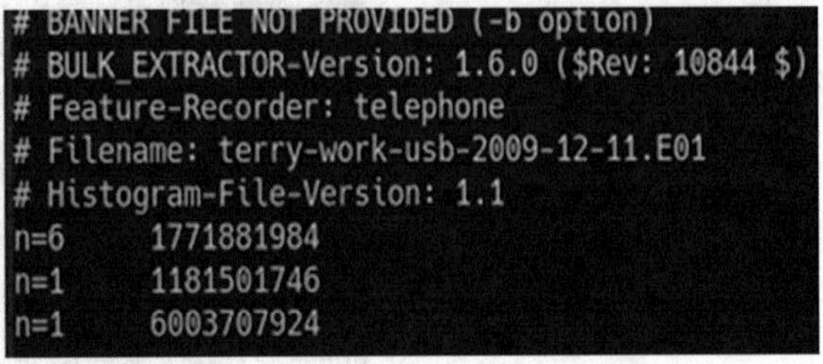
# BANNER FILE NOT PROVIDED (-b option)
# BULK_EXTRACTOR-Version: 1.6.0 ($Rev: 10844 $)
# Feature-Recorder: telephone
# Filename: terry-work-usb-2009-12-11.E01
# Histogram-File-Version: 1.1
n=6 1771881984
n=1 1181501746
n=1 6003707924

Fig. 17 Details which were stored inside selected file

5 Comparison Between Extractors and Bulk Extractor

Data extractor will extract the data from the selected device, or an electronic gadget, and the data was either visible or invisible in some time. And that time to get the data from the device is nearly up to ten minutes. Some of those extractors take some extra time (Table 1).

And, these extractors do not get the data of other categorized. And for every type of data, it will take too much of time to gather the data which was not categorized in the present type of data extractor for that we have to use another type of data extractor and that also take same time of more time.

Instead of using all of those extractors to get the data from the selected device, we better has to use the bulk extractor and this will get almost every data of the selected device and it will get more data than the normal extractors and this bulk extractor works a way faster than the normal extractors.

Bulk extractor is extracts the data better than the other extractors, it extract data from like mobile data, PC data, computer data, hard disk, floppy disk, etc. and reduces the time taken to get the data from the selected device.

Table 1 Comparison of extractor and bulk extractor

S. No.	Gets data	Get All types of data	Data gathering time
Extractor	Yes	No	More than 10 min
Bulk extractor	Yes	Yes	Less than 10 min

6   Result

We learn about how we can find sensitive data from digital evidence files using bulk extractor. Bulk extractor is a rich feature tool that can extract useful information like credit card numbers, domain names, IP addresses, emails, phone numbers, and URLS from evidence hard drives or files found during forensics investigation and it helps analyze image or malware also helps in cyber investigation and password cracking.

7   Conclusion

Bulk extractor is a tool which is used to extract huge amount of information from any kind of electronic device like mobile, laptop, computer, tablet, etc., and this extractor will extract the information very deeply and it gets as much amount of data as possible and it will get all types of data and it works a far better than all types of data extractors. And this extractor will work faster than the other extractors and it will get all types of sensitive data at a time of extracting the normal data and the data it will extractor get is a quite useful and it will help to catch a criminal with the data we get through the extractor.

This extractor not only gets single type of data it will get all types of data and we do not need to set the range or category to get the data and the data was definitely is a useful data and we can implement the filter in the bulk extractor and that will help to get only the needed information.

This bulk extractor will get revolution in the field of digital forensic by its performance and way of extracting the data from the devices.

References

1. MacDermott, A., Baker, T.: IoT forensics: challenges for the Ioa Era. In: IEEE International Conference on New Technologies, Mobility and Security (NTMS), 26–28 Feb 2018
2. Adwan, S., Salamah, F.: A manual mobile phone forensic approach towards the analysis of Whatsapp seven-minute delete feature. In: Saudi Computer Society National Computer Conference (NCC) 25–26 Apr 2018
3. Chung, H., Park, J., Lee, S.: Digital forenisc approaches for Amazon Alexa ecosystem. Digital Invest. **22**, 15–35 (2017)
4. Baig, Z.A., et al.: Future challenges for smart cities: cyber-security and digital forensics. Digital Invest. **22**, 3–13 (2017)
5. Khan, S.: The role of forensics in the internet of things: motivations and requirements. IEEE Internet Initiative eNewsletter (July 2017)
6. Burrows, C., Zadeh, P.B.: A mobile forensic investigation into steganography. In: IEEE International Conference On Cyber Security And Protection Of Digital Services (Cyber Security) 13–14 June 2016

7. Faheem, M., Le-Khac, N.-A., Kechadi, T.: International Conference on Innovative Computing Technology (INTECH) 24–26 Aug 2016
8. Mambodza, W.T., Nagoormeeran, A.R.: Android mobile forensic analyzer for stegno data. In: 2015 International Conference on Circuit Power and Computing Technologies (ICCPCT), pp. 1–8 (2015)
9. Baker, T., Mackay, M., Shaheed, A., Aldawsari, B.: Security-oriented cloud platform for SOA-based SCADA. In: The proceeding of the 15th IEEE/ACM international conference on Cluster Cloud and Grid Computing (CCGrid) (2015)
10. Botta, W.D., Perisco, V. et al.: On the integration of cloud computing and internet of things. In: International Conference on Future Internet of Things and Cloud (FiCloud), pp. 23–30 (2014)
11. Hegarty, R.C., Lamb, D.J., Attwood, A.: Digital evidence challenges in the internet of things. In: Proceeding of the Tenth International Conference (INC 2014), pp. 163–172 (2014)
12. Dykstra, J.: Acquiring forensic evidence from infrastructure-as-a-service cloud computing: exploring and evaluating tools trust and techniques. Digital Foren. Res. Workshop (DFRWS) **43**(12), 12–19 (2012)
13. Glisson, T.: Calm before the storm the challenges of cloud computing in digital forensics. Int. J. Digital Crime Foren. **4**(2), 28–48 (2012)
14. Grispos, G., Storer, T., Glisson, W.: Calm before the storm: the challenges of cloud computing in digital forensics. Int. J. Digital Crime Foren. (IJDCF) (2012)
15. Ibrahim, R., Kuan, T.S.: Steganography algorithm to hide secret message inside an image. Comput. Technol. Appl. **2**, 102–108 (2011)

Low Resource English to Nepali Sentence Translation Using RNN—Long Short-Term Memory with Attention

Kriti Nemkul and Subarna Shakya

Abstract In natural language processing, machine translation is an automated system that analyzes text from a source language (SL), applies some computation on that input, and generates equivalent sentences in the desired destination language (DL) without any human engagement. Recurrent neural network (RNN), a deep learning paradigm has been proved as an effective technique in applications like natural language processing and computer vision. This research work aims to develop a model for the English to Nepali sentence translation using long short-term memory (LSTM) cells in its encoder and decoder with attention. Bilingual evaluation understudy (BLEU) score method is adapted to evaluate the efficacy of the model. The model is trained and tested with different parameters such that the neural network layer considered are 2 and 4, and the hidden units considered are 128, 256. The performance and effectiveness of the model were compared with different parameters, the LSTM cells in encoder and decoder with attention with two layers of neural network and 256 hidden units found to be better in translating English to Nepali sentence with the highest BLEU score 8.9.

Keywords Machine translation · Recurrent neural network (RNN) · Source language · Destination language · Long short-term memory (LSTM) · Bilingual evaluation understudy (BLEU)

K. Nemkul
Central Department of Computer Science and Information Technology, Tribhuvan University, Kirtipur, Nepal
e-mail: kriti.nemkul@gmail.com

S. Shakya (✉)
Institute of Engineering, Tribhuvan University, Kirtipur, Nepal
e-mail: drss@ioe.edu.np

S. Shakya et al. (eds.), *Proceedings of International Conference on Sustainable Expert Systems*, Lecture Notes in Networks and Systems 176,
https://doi.org/10.1007/978-981-33-4355-9_48

649

1 Introduction

Machine translation is the task of deciphering the content in the input language to a destination language, automatically. Machine translation can be considered as a territory of applied research that pulls ideas and techniques from linguistics, computer science, AI, translation theory, and statistics. The recurrent neural network is one of the deep learning paradigms that can be efficiently used for the machine translation task. Deep learning enables computational models built from many processing layers to find outdata representation with various abstraction layers [1]. The encoder–decoder recurrent neural network with long short-term memory (LSTM) can be used to achieve close to state-of-the-art output in machine translation. In this methodology, the encoder neural network takes a source sentence as input and encrypts it into a vector of fixed length. A translation is then rendered by a decoder which decodes the fixed-length vector into a variable-length sentence [2]. This approach of encrypting the variable length input sentence into a vector of fixed length is problematic for translating long sentences [3]. To overcome this limitation, neural machine translation with attention can be adapted. In an attention-based neural machine translation approach, encoding the source sentence into a fixed length vector is avoided. An attention model generates a context vector which is filtered particularly for every output time phase. When the model produces a word during a translation, it scans for a set of places in an input text where the most significant data is focused. The target word is then predicted by the model on the basis of context vector interrelated with these input positions and all target words previously created [3, 4].

2 Literature Review

Machine translation has been an active research topic since 1950s. Greenstein et al. has shown that RNN search model shows the better result with BLEU score 0.73 in the small corpus of size 150,000 sentence-pair in the task of Japanese to English machine translation [2]. Bojar et al. present HindEnCorp, Hindi and English parallel corpus, and HindMonoCorp, Hindi monolingual corpus for the training of statistical machine translation systems [3]. Agrawal and Sharma have used sequence-to-sequence models using RNN for English–Hindi machine translation task. They experiment with the different architecture of RNNs out of which bi-directional LSTM units when complemented with the attention mechanism perform best, especially on compound sentences [4]. Hermanto et al. has shown machine translation based on neural network (RNN based machine translation) shows better result than that of the statistical-based language model on as English to Indonesian translation task. But, the language model based on RNN takes longer to train the model than that of statistical-based language model due to its algorithm complexity [5]. Sustskever et al. had used two recurrent neural networks to encode a source sentence of variable length into a fixed-length vector and to decode that fixed-length vector into a target sentence of variable length. Sustskever et at. had also shown the ability of long short-term memory (LSTM) in RNN based neural machine translation using LSTM to correctly translate very long sentences on an English to French translation task [6]. Cho et al. has proposed a neural network architecture, RNN encoder–decoder that has the ability to learn the mapping from an arbitrary length sequence to another sequence, probably from a divergent set, of random length. The model proposed was capable to either rank a pair of sequences in terms of a conditional probability or produce a destination sequence provided an input sequence [7]. Dobhase English to Nepali machine translation project was developed by Madan Puraskar Pustakalaya (MPP, http://www.mpp.org.np) in collaboration with the Kathmandu University (http://ku.edu.np) which was a rule-based system [8]. Sanat and Bista has proposed interlingua machine translation approach, especially the Nepali generator component of interlingua-based machine translation in which universal networking language (UNL) is used as the interlingua. UNL is an interlingua proposed by United Nations University/Institute of Advanced Studies, Tokyo, Japan, to overcome the language barrier and a digital gap in the World Wide Web in which techniques of syntax design strategies and morphology generation have been used for a translation task [9].

3   Data Collection

During this study, secondary datasets were used. The dataset contains English–Nepali parallel corpus in which one file will contain a set of English sentences and another file will contain the corresponding Nepali sentences aligned parallel. The English–Nepali parallel corpus was collected from different sources: the Kathmandu University, NLP Lab research center, "English–Nepali Parallel Corpus," ELRA catalog [10], and Opus dataset [11]. The collected data are merged and preprocessed to remove the dirty data.

4   RNN Architecture

4.1   Long Short-Term Memory (LSTM)

LSTM, a type of RNN introduced by Hochreiter and Schmidhuber in 1997 to solve the problem of exploding gradients. The model is similar to a standard RNN with a hidden layer in which memory cells are used in place of each ordinary nodes in the hidden layer. Each memory cell is composed of a node with an independent repeating node with fixed weight one which ensures that the gradient will communicate over repetitive steps without vanishing the ability to learn long-term dependencies [12, 13]. All recurrent neural networks in standard RRNs resembles the chain structure of repeated neural network modules with a quite simple structure, like a single layer of

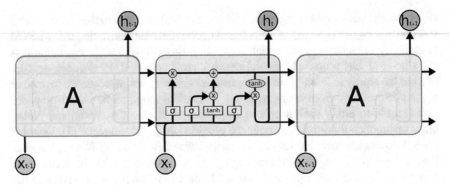

Fig. 1 Repeating module in an LSTM containing four interacting layers [14]

tanh. LSTM also has this chain-like structure with different structures of repeating modules. There are four neural network in LSTM interacting in a very specific way, rather than having a single layer of neural network. The recurrent module in an LSTM containing four interacting layers as shown below [14] (Fig. 1).

The notations in the above diagram are shown below.

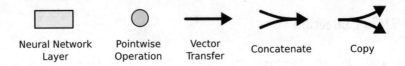

In the diagram above, the entire vector is carried by each line, from one node output to the inputs of others. The pink circles represents pointwise operations, such as vector addition, and the yellow rectangular boxes show the trained neural network layers. Concatenation is represented by the merging lines, while a line forking is used to show its content is replicated, and the copies are sent to various locations [14].

5   RNN Encoder–Attention–Decoder Using LSTM

5.1   *Encoder*

The bidirectional RNN is composed of forward RNN and backward RNN's. The source sequence ordered (from x_1 to x_{T_x}) is scanned by forward RNN $\vec{f}$ and forward hidden state sequences $(\vec{h}_1, ..., \vec{h}_{T_x})$ are calculated. The backward RNN $\overleftarrow{f}$ inputs the series in the reverse order form (from x_{T_x} to x_1) which results in a series of backward hidden states $(\overleftarrow{h}_1, ..., \overleftarrow{h}_{T_x})$.

The model uses of 1—of—K coded word vectors of source sentence as its input

$$x = (x_1, \ldots, x_{T_x}), x_i \in \mathbb{R}^{K_x} \tag{1}$$

and gives outputs as 1—of—K coded word vectors of the translated sentence

$$y = (y_1, \ldots, y_{T_y}), y_i \in \mathbb{R}^{K_y} \tag{2}$$

where K_x and K_y represents the vocabulary size of the source and destination languages, respectively, and T_x and T_y denotes the length of source and destination sentences, respectively.

Initially, the forward states of the RNN are calculated as,

$$\overrightarrow{h_i} = \begin{cases} o_t \tanh(c_t), & \text{if } i > 0 \\ 0, & \text{if } i = 0 \end{cases} \tag{3}$$

where

$$\text{Input gate, } I_i = \sigma\left(W_i E x_{i-1} + U_i h_{i-1} + Z_i \hat{z}_i + b_i\right) \tag{4}$$

$$\text{Output gate, } O_i = \sigma\left(W_0 E x_{i-1} + U_0 h_{i-1} + Z_0 \hat{z}_i + b_0\right) \tag{5}$$

$$\text{Memory, } c_i = f_i c_{i-1} + I_i \tanh\left(W_c E x_{i-1} + U_c h_{i-1} + Z_c \hat{z}_i + b_c\right) \tag{6}$$

$$\text{Forget gate, } f_i = \sigma\left(W_f E x_{i-1} + U_f h_{i-1} + Z_f \hat{z}_i + b_f\right) \tag{7}$$

$E \in \mathbb{R}^{m \times K_x}$ is the word embedding matrix. $U_0, W_0, Z_0, b_0 \in \mathbb{R}^{n \times m}$ are learned weight matrices and biases. The word embedding dimension and number of hidden units are denoted by m and n, respectively. $\sigma(.)$ are logistic sigmoid function.

Similarly, the backward states $\overleftarrow{h_1}, \ldots, \overleftarrow{h_{T_x}}$ are also determined where unlike the weight matrices, the word embedding matrix E is shared between forward and backward RNNs.

Annotations $(h_1, \ldots, h_{T_x})$ are obtained by concatenation of the forward and backward states

where

$$h_i = \begin{bmatrix} \overrightarrow{h_i} \\ \overleftarrow{h_i} \end{bmatrix}. \tag{7}$$

5.2 Decoder

The hidden state s_t of a decoder provided encoder's annotation is calculated as,

$$\tilde{s}_t = O_t \tanh(c_t) \tag{8}$$

where

$$\text{Proposal hidden gate, } i_t = \sigma(W_i E y_{t-1} + U_i s_{t-1} + Z_i \tilde{z}_t + b_i) \tag{9}$$

$$\text{Output gate, } O_t = \sigma(W_0 E y_{t-1} + U_0 s_{t-1} + Z_0 \tilde{z}_t + b_0) \tag{10}$$

$$\text{Memory, } c_t = f_t c_{t-1} + i_t \tanh(W_c E y_{t-1} + U_c s_{t-1} + Z_c \tilde{z}_t + b_c) \tag{11}$$

$$\text{Forget gate, } f_t = \sigma\left(W_f E y_{t-1} + U_f s_{t-1} + Z_f \tilde{z}_t + b_f\right) \tag{12}$$

$E \in \mathbb{R}^{m \times k}$ is the word embedding matrix of the target language. W_0, U_0, Z_0, b_0 are learned weight matrices and biases. c_t will be determined by computing the weighted sum using both previous cell state and current information provided by the cell. Let m and n denote the embedding and LSTM dimensionality, respectively, and σ be the logistic sigmoid activation. The initial hidden state s_0 is computed as,

$$s_0 = \tanh\left(W_s \overleftarrow{h}_1\right), \text{ where } W_s \in \mathbb{R}^{n \times n} \tag{13}$$

The context vector c_i are recalculated by the alignment model at every step as,

$$c_i = \sum_{j=1}^{T_x} \alpha_{ij} h_j \tag{14}$$

where

$$\alpha_{ij} = \frac{\exp(e_{ij})}{\sum_{k=1}^{T_x} \exp(e_{ik})} \tag{15}$$

$$e_{ij} = v_a^{\mathrm{T}} \tanh\left(W_a s_{i-1} + U_a h_j\right) \tag{16}$$

and h_j represents the j-th annotation in the source sentence. $v_a \in \mathbb{R}^{n'}$, $W_a \in \mathbb{R}^{n' \times n}$ and $U_a \in \mathbb{R}^{n' \times 2n}$ weight matrices.

The probability of a target word y_i with the decoder state s_{i-1}, context vector c_i and the last generated word y_{i-1} is,

$$p(y_i|s_i, y_{i-1}, c_i) \propto \exp\left(y_i^T W_0 t_i\right) \tag{17}$$

where

$$t_i = \left[\max\left\{\tilde{t}_{i,2j-1}, \tilde{t}_{i,2j}\right\}\right]_{j=1,\ldots,l}^{T}. \tag{18}$$

and k-th element of a vector $\tilde{t}_i$ ($\tilde{t}_{i,k}$) is calculated as,

$$\tilde{t}_i = U_0 s_{i-1} + V_0 E y_{i-1} + C_0 c_i \tag{19}$$

$W_0 \in \mathbb{R}^{K_y \times l}, U_0 \in \mathbb{R}^{2l \times n}, V_0 \in \mathbb{R}^{2l \times m}, C_0 \in \mathbb{R}^{2l \times 2m}$ are weight matrices [7, 15, 16].

6 Results

6.1 Bilingual Evaluation Understudy (BLEU) Score

BLEU score is a standard for determining the quality of machine-translated sentences from one natural language to another. The core idea behind BLEU is "the closer a machine translation is to a professional human translation, the better it is" [17]. Scores are calculated for individual segments that have been translated, generally sentences. The score is measured by matching such individually translated sentences against a collection of reference translations of high-quality reference. The performance of BLEU is often a numerical value between 0 and 1 which specifies how identical the input text is to the reference texts, with the value closest to 1 indicating more matching texts. In this study, the Moses muiti-bleu score has been used as the evaluation parameter in which the score between 0 and 1 will be converted to the range of 1–100 [18]. A perfect match gives the score 100, while score 0 indicates a perfect mismatch. The BLEU score is calculated a

$$BP = \begin{cases} 1 & \text{if } c > r \\ e^{(1-r/c)} & \text{if } c \leq r \end{cases} \tag{20}$$

Then,

$$BLEU = BP \cdot \exp\left(\sum_{n=1}^{N} w_n \log p_n\right) \tag{21}$$

The ranking behavior is more promptly clear in the log domain,

Table 1 BLEU score of test data with different layers and hidden units

S. No.	Number of hidden units	Number of layers	BLEU score
1	128	2	6.5
2	256	2	8.9
3	128	4	2.6
4	256	4	3.8

$$\log \text{ BLEU} = \min\left(1 - \frac{r}{c}, 0\right) + \sum_{n=1}^{N} w_n \log p_n. \tag{22}$$

where

r: the average no. of words in a reference translation, average overall reference translation

c: total length of input translation corpus [17].

The BLEU score for the sample test dataset is (Table 1),

The best BLEU score of the test data obtained is 8.9 with 256 hidden layers and two layers of the neural network. During translation, tag <unk> has been used for the unknown words.

7 Conclusion

English to Nepali sentence translation is the task of translating the source English language sentence into the target Nepali language sentence. A number of approaches like rule-based, statistical, knowledge passed, phrase-based, and so on have been discovered for the machine translation task. Recurrent neural network encoder–attention–decoder with LSTM cells has been implemented, and analysis of the system has been carried out in this study. The corpus has been collected from the various source from which 80% of data has been used to train the system, 10% of data has been used as a development set, and the remaining 10% of data has been used for testing purpose. The system has been trained with a learning rate of 0.001, and the best BLEU score for the low resource English–Nepali language obtained is 8.9 with 256 hidden layers and two layers of neural network.

The challenge left for the future is to increase the dataset and train the model with more hidden layers and neural layers as well as to better handle the unknown and rare words.

References

1. LeCun, Y., Bengio, Y., Hinton, G.: Deep learning. Nature **521**, 436–444 (2015)

2. Greenstein, E., Penner, D.: Japanese-to-English Machine Translation Using Recurrent Neural Networks, pp. 1–7 (2015)
3. Bojar, O., Diatka, V., Rychlý, P., Straňák, P., Suchomel, V., Tamchyna, A., Zeman, D.: HindEnCorp—Hindi-English and Hindi-only corpus for machine translation, pp. 3550–3555 (2014)
4. Agrawal, R., Sharma, D.M.: Experiments on different recurrent neural networks for English-Hindi machine translation. In: Computer Science and Information Technology (CS & IT), pp. 63–74 (2017)
5. Hermanto, A., Adji, T.B., Setiawan, N.A.: Recurrent neural network language model for English-Indonesian machine translation: experimental study. In: 2015 International Conference on Science in Information Technology (ICSITech). IEEE (2015)
6. Sutskever, I., Vinyals, O., Le, Q. V.: Sequence to sequence learning with neural networks. arXiv vol. 3 [cs.CL] (2014)
7. Cho, K., van Merrienboer, B., Gulcehre, C., Bahdanau, D., Bougares, F., Schwenk, H., Bengio, Y.: Learning Phrase Representations using RNN Encoder-Decoder for Statistical Machine Translation. arXiv vol. 3 [cs.CL] (2014)
8. Bal, B., Shrestha, P.: A Morphological analyzer and a stemmer for Nepali. Working Papers (2004–2007)
9. Keshari, B, Bista, S.K.: UNL Nepali Deconverter, In: 3rd International CALIBER, pp. 70–76 Ahmedabad (2005)
10. European Language Resources Association(ELRA) catalogue. http://catalog.elra.info/
11. Tiedemann, J.: Open parallel corpus. Parallel Data, Tools Interfaces OPUS. http://opus.nlpl.eu/
12. Lipton, Z.C., Berkowitz, J., Elkan, C.: A Critical Review of Recurrent Neural Networks for Sequence Learning. arXiv vol. 4 [cs.LG], pp. 1–38 (2015)
13. Hochreiter, S., Schmidhuber, J.: Long Short—Term Memory. In: Neural Computation. Massachusetts Institute of Technology, pp. 1735–1780 (1997)
14. Colah: Understanding LSTM Networks. http://colah.github.io/posts/2015-08-Understanding-LSTMs/
15. Bahdanau, D., Cho, K.H., Bengio, Y.: Neural machine translation by jointly learning to align and translate. In: 3rd Int. Conf. Learn. Represent. ICLR 2015—Conf. arXiv vol. 7 [cs.CL] pp. 1–15 (2015)
16. Xu, K., Ba, J.L., Kiros, R. et al.: Show attend and tell: a neural image caption generation with visual attention. In: Proceedings of the 32nd International Conference on Machine Learning. vol 37 Farnce (2015)
17. Papineni, K., Roukos, S., Ward, T., Zhu, W.-J.: BLEU: a method for automatic evaluation of machine translation. In: Proceedings of the 40th Annual Meeting on Association for Computational Linguistics—(ACL). Association for Computational Linguistics pp. 311–318 Philadelphia (2002)
18. Koehn, P., Hieu, H. et al.: Moses: open source toolkit for statistical machine translation. In: Proceedings of the ACL. pp. 177–180 Prague (2007)

Portfolio Optimization: A Study of Nepal Stock Exchange

Prakash Chandra Prasad, Anku Jaiswal, Subarna Shakya, and Sailesh Singh

Abstract Portfolio optimization is the process of reallocating the funds available to the best portfolio from the given set of portfolios based on the objectives such as maximizing the expected returns and minimizing the financial risk involved. The mean-variance framework also called as modern portfolio theory proposed by Harry M. Markowitz is the most widely used approach for portfolio optimization. This paper uses the assets of Nepal Stock Exchange (NEPSE) for the portfolio construction and uses the modern portfolio theory-based mean-variance framework to find the optimal weight to be assigned to each asset in the portfolio. The optimal weight of each asset in the portfolio is calculated using the Monte Carlo simulation as well as solving the quadratic equation of the given optimization problem. This paper also tries to explore the limitation and shortcoming of modern portfolio theory, which will form the base for future research work to solve the optimization problem using machine learning techniques.

Keywords Portfolio optimization · Modern portfolio theory · Expected returns · NEPSE · Monte carlo simulation · Quadratic programming · Knowledge engineering

P. C. Prasad (✉) · A. Jaiswal · S. Shakya · S. Singh
Department of Electronics and Computer Engineering, IOE, Pulchowk Campus, Patan, India
e-mail: prakash.chandra@pcampus.edu.np

A. Jaiswal
e-mail: anku.jaiswal@pcampus.edu.np

S. Shakya
e-mail: drss@ioe.edu.np

S. Singh
e-mail: cmcsailesh@gmail.com

S. Shakya et al. (eds.), *Proceedings of International Conference on Sustainable Expert Systems*, Lecture Notes in Networks and Systems 176,
https://doi.org/10.1007/978-981-33-4355-9_49

659

1 Introduction

The financial market, where different types of securities and derivatives are traded, acts as the backbone of the modern economy. It functions as a bridge between the borrower and the savers in that they receive accumulated fund from the individual and firms and then allow other firms or individual to borrow the funds, thus allocating the resources and keeping the balance in the economic cycle. Apart from acting as a tool for the smooth functioning of the business and economy, it also provides the investment opportunity to individuals and organizations. The profit on the investment can be earned, if the price of the given assets can be forecasted. These attractive returns on investment have led to the establishment of many financial institutions, whose sole objective is to maximize returns by minimizing the risks.

One of the major concerns for investors around the world is to choose the best investment opportunities to maximize the returns on their investment. Given investment option, deciding to choose from them, is one of the most complex and challenging problems, especially for the large financial institutions such as banks, insurance, investing, brokerage, and commercial institutions as well as public enterprises who seek to have high return at the cost of moderate or negligible risks. By the process called as diversification, investors of all time try to spread the risk involved, by considering various options of investment such as commodity market (gold, silver, platinum, etc.), securities, bonds, derivatives. A diversified portfolio has a smoother risk behavior that is less variation in expected return. Risk is arising from the uncertainty of data like future investment return which is forecasted. Well, diversification helps to decrease the volatility of portfolio performance since, assuming that a portfolio asset is normally distributed, then the price of all assets does not change in the same direction at the same time and at the same rate otherwise, the portfolio won't be well-diversified.

There are two sides for an investment, namely risk and return. As a general rule in the economy, one who seeks more return must expect more risk too and vice versa. An investor can be classified in one of the three categories; risk-averse is someone who avoids taking risk; thus so conservative, risk-taker, on the contrary, is ready to take more risk hoping to gain more return and risk-indifferent who is neither risk-averse non-risk-taker.

Portfolio optimization is one of the complex and challenging engineering and finance problems. It is the technique of continuously reallocating the funds to the numbers of financial products and instruments, by making sequential decisions with the main aim of maximizing the return on investment while constraining the risk involved.

The mean-variance framework also called modern portfolio theory is one of the most popular approaches to solve the portfolio optimization problem. The optimization problem is formulated based on the mean-variance framework, which tries to maximize the expected return of the portfolio, which is the weighted sum of assets in the portfolio and minimizes the risk involved with the portfolio, which is the standard deviation of the assets in the portfolio.

This paper uses the mean-variance framework to calculate the optimal weight of the portfolio by using the Monte Carlo simulation and quadratic programming. It also tries to explore the shortcoming of the MV framework and its potential to be solved by the machine learning approach.

2 Literature Survey

Jothimani et al. [1] discuss two main issues of mean-variance (MV) framework: (a) estimation error of the mean-variance model (b) instability of the covariance model. For portfolio optimization, the paper presents two-risk parity model: (a) Historical correlation-based hierarchical risk parity model (HRP-HC) (b) Gerber statistics-based hierarchical risk parity model (HRP-GS). The authors have analyzed and tested the result using TSX complete index stock data of 10 years (2007–2016). Since HRP-GS model integrates the advantages of risk parity model and robust statistics, the HRP-GS models outperform the HRP-HC model. According to Hegde et al. [2], the task of portfolio construction is a two-step process (a) determining the risk and reward of the portfolio's instrument using the predictive technique (b) finding the optimal allocation by solving quadratic portfolio optimization problem that maximizes the measure of portfolio performance. As mentioned in this paper, the deep reinforcement learning (DRL) algorithm eliminates the need of the above-mentioned two-step process for finding the optimal allocation, as the DRL algorithm autonomously adjusts to change in the environment, unlike the traditional machine learning algorithm. As the portfolio construction problem has continuous action space, the DRL algorithm suffers from the problem of instability. To solve the problem of instability due to the continuous action space, the paper has suggested the use of deep deterministic policy gradient (DDPG) for the portfolio construction. This paper has evaluated the use of DDPG algorithm for risk-aware portfolio construction and has used the rate of return and sorting ratio as a measure of portfolio performance. The results presented in this paper have suggested the effectiveness of DDPG algorithm for the risk-aware portfolio construction. Hu and Lin [3] have proposed the use of deep reinforcement learning (DRL) for portfolio management and optimization. According to this paper, the DRL algorithm uses its deep learning part for finding the optimal policy and other parts, i.e., reinforcement learning for goal-oriented self-learning without human intervention. The major research issue discussed in the paper is related to policy optimization. This paper has explored one of the deep recurrent neural network (RNN), GRU, to study the influence of previous state and action on policy optimization in the non-Markov decision process. This paper has also presented the risk-adjusted reward function to evaluate the total expected reward for the policy. The DL and RL are integrated, in this paper for DRL approach to harvesting their respective capabilities to find the optimal policy for portfolio optimization.

The capital assets pricing model (CAPM)-based new investment strategy proposed by Gu et al. [4] uses deep learning market prediction approach to optimize the portfolio of the mutual fund. As mentioned in this paper, the authors have evaluated their proposed investment strategy using 22 years of data of S&P500 and mutual fund. According to this paper, the accuracy of their predictive model can reach 84.3% and the rate of return is 13.87%. The authors have claimed that their proposed model is more accurate and profitable than another algorithm so far. The deep learning solution to the portfolio management problem is presented in [5] by Jiang et al. using the financial-model-free reinforcement learning framework. The proposed framework is composed of Ensemble of Identical Independent Evaluators (EIIE) topology, a portfolio-vector memory (PVM), an online stochastic batch learning (OSBL) scheme, and a fully exploiting and explicit reward function. The framework presented in this paper is realized and evaluated using the convolutional neural network (CNN), a basic recurrent neural network (RNN), and a long short-term memory (LSTM). All these three methods along with the other published portfolio selection strategies are examined and tested in the crypto-currency market with a trading period of 30-min. This paper has claimed that all these three mentioned instances of the proposed framework have monopolized the experiment. As per the paper, although with the high commission rate of 0.5% the proposed framework can achieve a return of fourfold in 30 days. The work presented by Chaouki et al. [6] has tested the use of reinforcement learning algorithm in the trading environment for portfolio optimization. According to this paper, the reinforcement learning agent of deep deterministic policy gradient algorithm can successfully learn the optimal trading strategy and can find the optimal solution for the portfolio allocation. The research work of Soleymani and Paquet [7] presents a portfolio management framework called DeepBreath based on the deep reinforcement learning algorithm. The proposed framework is composed of restricted stacked auto-encoder for dimensionality reduction and feature selection and convolutional neural network (CNN) to learn the optimal policy to increase the expected return on the investment by fund reallocation. The proposed framework uses the offline strategy to train the CNN algorithm and the online strategy to handle the drifts in the data distribution resulting from the unforeseen circumstances. The paper has also presented the use of blockchain technology to handle the settlement risk associated with the portfolio. The framework presented in this paper is tested against the three distinct investment periods and has claimed that the performance of the framework has outperformed other optimization strategies and has minimized the risk as well.

3   Related Work

3.1   Mean-Variance Theorem

Harry M. Markowitz's paper titled "Portfolio Selection" which earned him a Nobel Prize in the economy became the foundation of modern portfolio theory. This theory introduced the concept of diversification, according to which it is less risky to own portfolio of assets from different classes than owing the similar classes of assets in a portfolio. The modern portfolio theory also called as mean-variance theory is one of the most popular approaches for the portfolio selection and has led to the quantitative formulation of the portfolio optimization problem. This theory assumes that the return of assets in the portfolio follows the normal distribution.

The portfolio is composed of multiple assets. The return of individual assets is calculated from its historical return series. The daily return of the assets i, denoted by Ri in the kth day, is given by:

$$R_{ki} = \frac{(k\text{th day closing price} - (k-1)\text{th day closing price})}{(k-1)\text{th day closing price}} \tag{1}$$

where R_{ki} is the daily return of the assets i in the Kth day. The expected return for the assets i denoted by μ_i is given by:

$$\mu_i = \sum_{k=1}^{n} \frac{R_{ki}}{n} \tag{2}$$

where μ_i is the expected return for the asset i, R_{ki} is the daily return of the assets i in the Kth day, and n is the number of days considered.

The covariance between the assets i and j denoted by, C_{ij} among n-assets is given by:

$$C_{ij} = \frac{\sum_{k=1}^{n} \left[(R_{ki} - \mu_i)(R_{kj} - \mu_j) \right]}{n} \tag{3}$$

where C_{ij} is the covariance between assets i and j, R_{ki} and R_{kj} are the daily return of the assets i and j for the kth day, and n is the number of days considered.

The positive covariance between the assets denotes that the assets are positively related and have similar behavior under the same market condition, and there is a need for diversification of assets in the portfolio. The negative covariance between the assets denotes that the assets are negatively related and behave differently in the same market condition, and in such case, no diversification is required.

For the portfolio with n-assets, the expected return of assets i is given by μi, where $i = 1, 2, 3, \ldots n$. Let pi denotes the fraction of capital invested in the assets i. The expected return of the portfolio, denoted by μp, is given by:

$$\mu_p = \sum_{i=1}^{n} [(P_i)(\mu_i)] \tag{4}$$

As Eq. (4) suggests, the return of the portfolio is the weighted sum of the weight of individual assets and it is corresponding expected returns. The variance of the portfolio, which denotes that variation in the expected return of the assets in the portfolio, is given by:

$$\sigma_p^2 = \sum_{i=1}^{n} \sum_{j=1}^{n} \left[(P_i)(c_{ij})(p_j) \right] \tag{5}$$

Equation (5) denotes the variance of the portfolio, where p_i denotes the fraction of capital invested in the assets i, pj denotes the fraction of capital invested in the assets j, and C_{ij} is the covariance between assets i and j.

Using the equations, defined above Markowitz's standard mean-variance approach can be defined as follows:

$$\text{minimize } \sum_{i=1}^{n} \sum_{j=1}^{n} \left[(P_i)(c_{ij})(p_j) \right]$$

$$\text{Subject to } \sum_{i=1}^{n} \left[(P_i)(\mu_j) \right] = \mu_p$$

$$\sum_{i=1}^{n} P_i = 1 \text{ where } 0 \leq P \leq 1 \tag{6}$$

The objective of Eq. (6) is to minimize the risk, i.e., the variance associated with the portfolio. The set of constraints for the equation is:

(a) The expected return of the portfolio is denoted by μp
(b) The total sum of capital invested should be equal to 1.
(c) The proportion of capital invested in each asset should lie between 0 and 1.

The solution to Eq. (6) can be used to represent the optimal portfolio, and the solution is a graph of the non-decreasing curve with risk, i.e., standard deviation plotted in X-axis and expected return in Y-axis. The portfolio that lies on the curve can be termed as an optimal portfolio with a higher expected return at a given level of risk. The graph of efficient frontier is as follows (Fig. 1).

3.2 Limitation of Mean-Variance Theorem

Following are the limitation of mean-variance (MV) approach for portfolio optimization:

- MV framework assumes that the expected returns of the assets to be in the normal distribution, which is not the case in real-world trading scenarios.
- MV framework does not consider practical trading constraints such as cardinality constraints (i.e., limiting number of assets in the portfolio), transaction costs, and regulatory constraints and guidelines constraints.
- The portfolio managers expect the expected return and covariance of the assets to be considered for the portfolio to be provided while the calculation of expected return and covariance may result in estimation error.

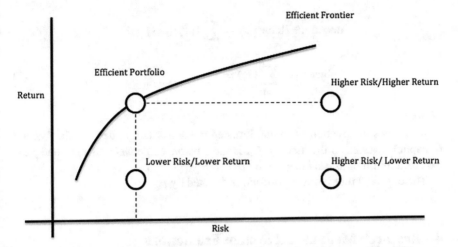

Fig. 1 Efficient frontier

- The estimation error may impact the calculation of the optimal weight of each asset in the portfolio.
- The inversion of the covariance matrix while quadratic programming may lead to erroneous results.

3.3 Portfolio Optimization

The portfolio is defined as a collection of financial instruments such as stocks, bonds, commodity assets, crypto-currencies owned by an individual or the organization. The assets in the portfolio exhibit the time-series property and are affected by the various external factors like political landscape and global economy making them very complex in nature and behavior.

Portfolio optimization is the process of continuously reallocating the available funds/capital to the assets in the portfolio with the main aim of maximizing the expected return and minimizing the risk involved with the portfolio. Thus, the process of portfolio optimization deals with finding the optimal weight that can be assigned to each asset in the portfolio. Due to the complex nature and behavior of the financial assets, determining the optimal weight associated with them has been come very challenging engineering problem.

As per the definition of portfolio optimization, its mathematical formulation becomes bi-objective function and is based on Eq. (6) discussed in the Sect. 3.1. Thus, the bi-objective portfolio optimization function can be formulated as follows:

$$\text{minimize: Risk } \sigma_p^2 = \sum_{i=1}^{n} \sum_{j=1}^{n} \left[(P_i)(c_{ij})(p_j) \right]$$

$$\text{maximize Return } \sigma_p = \sum_{i=1}^{n} [(P_i)(\mu_i)]$$

$$\text{Subject to: } \sum_{i=1}^{n} (P_i) = 1 \qquad (7)$$

where

p_i denotes the fraction of capital invested in the assets i, p_j denotes the fraction of capital invested in the assets j, C_{ij} is the covariance between assets i and j the expected return of assets i is given by μ_i.

The expected return of the portfolio is denoted by μ_p.

4 Research Method and System Framework

4.1 Data Collection

The price history of four companies (NTC, NABIL, NLIC, SCB) was collected from the NEPSE Web site (http://www.nepalstock.com/). The collected data was from 2018-11-21 to 2020-07-29 for approximately 2 years. The python module was created to scrape the data from the mentioned source. The attributes of collected data comprised of Date, No. of Transaction, Max Price, Min Price, Closing Price, Traded Shares, Total Amount, Prev. Close, Difference, % Change. For the experimentation purpose, only closing price was used as input.

4.2 Data Preprocessing

The python scripts were written to remove the null values from the data records and to create a single data frame with the date as an index and four columns labeled by the company's name with closing price as its corresponding values.

4.3 Methodology

The flow of the experiment follows the steps as shown in Fig. 2. The first step consists of determining the number of assets, i.e., stocks to be chosen for the portfolio construction; once the number of assets to be used is determined, the process of relevant data collection starts as mentioned in Sects. 3.1 and 3.2. The collected data is formatted as required by the experiment. The attributes of the final formatted data consist of: date and closing price. The second step consists of determining

Fig. 2 Methodology of proposed work

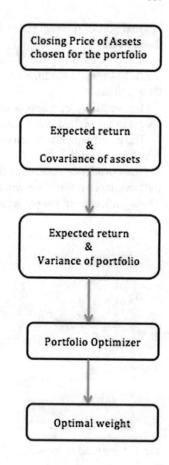

the expected return and covariance of the individual assets in the portfolio using Eqs. (2) and (3). Using Eqs. (4) and (5), the expected return and risk, i.e., variance of the portfolio, is calculated in the third step. The portfolio optimization problem is formulated as per Eqs. (6) and (7) and is passed as input to the portfolio optimizer in the fourth step. The portfolio optimizer after the number of iteration calculates the optimal weight associated with the assets in the portfolio according to the objectives defined in Eqs. (6) and (7).

5 Experimental Result and Discussion

This section discusses the result of the experiment and performs the required analysis and comparison. For the experiment, the portfolio to be constructed consists of four assets, namely NTC, NABIL, NLIC, and SCB. The experiment aims at finding the optimal portfolio composed of these four mentioned assets using the mean-variance framework.

In Fig. 3, closing price of the assets in the y-axis is plotted against the date in the x-axis to generate the time-series plots which show the price trends of each asset in the portfolio.

The volatility of assets is calculated as % change in the daily stock price. The spikes in the plotted graph show the volatility of the mentioned stocks (Fig. 4).

The mean expected return of each asset in the portfolio is calculated using Eq. (2) which is shown in Table 1:

The covariance matrix for the assets in Table 2 is calculated using Eq. (3), and the covariance matrix is essential to understand the relationship between each asset. The significance of the covariance matrix is explained in Sect. 3.1.

Fig. 3 Price trend of the assets

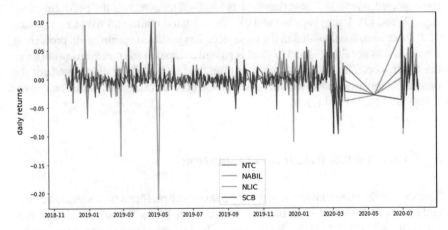

Fig. 4 Volatility of the assets

Table 1 Mean expected
returns of the 4 assets

NTC	−0.000037
NABIL	0.000732
NLIC	0.001193
SCB	0.000854

Table 2 Covariance matrix of the four assets

	NTC	NABIL	NLIC	SCB
NTC	0.000225	0.000135	0.000214	0.000154
NABIL	0.000138	0.000474	0.000320	0.000425
NLIC	0.000214	0.000320	0.000746	0.000377
SCB	0.000154	0.000425	0.000377	0.000603

5.1 Monte Carlo Simulation

Monte Carlo simulation (MCS) is one of the most popular tools to visualize the effect of multiple correlated variables, studying the functional relationship between the variables, testing the strategies, etc. The experiment aims to maximize the Sharpe ratio of randomly generated portfolios to calculate the optimal weight of the assets in the portfolio. The formula for the Sharpe ratio is given below:

$$\text{sharp ratio} = \frac{(R_p - R_f)}{\mu_p} \tag{8}$$

where
R_p = The expected return of the portfolio
R_f = The risk-free return of the portfolio.
Usually, Sharpe ratio greater than 1 is considered good for the investor.

For the experiment, 100,000 portfolios have been simulated with a varying combination of assets weight to find the optimal weight allocation for each asset. For each of the iteration, the Sharpe ratio for each randomly generated portfolio was calculated and recorded, and finally, the portfolio with the maximum Sharpe ratio was extracted as the result of MCS simulation.

The simulation must follow the following constraints:

- The random weight generated for each asset must lie between zero and one.
- The sum of the weight of the asset must equal to one, which is 100% of the total capital.

Figure 5 is to visualize the result of randomly simulated portfolios with annualized expected return (y-axis) and annualized volatility (x-axis). The graph has also annotated two stars to denote two interesting portfolios generated by the simulations.

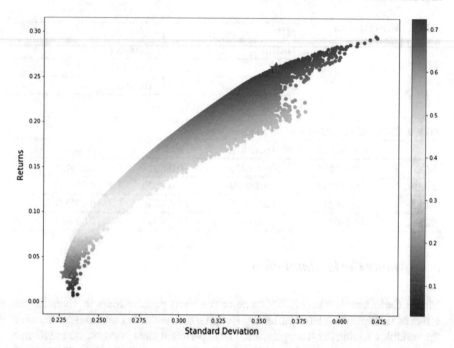

Fig. 5 Efficient frontier generated by MCS

The star at the top denotes the portfolio with maximum variance, and the bottom star denotes the portfolio with minimum variance.

The result of the simulated portfolio with the maximum Sharpe ratio is shown in Table 3. The mentioned table also shows the expected return and risk (i.e., standard deviation) associated with the portfolio.

The result of a simulated portfolio with minimum standard deviation (i.e., risk) is shown in Table 3. The mentioned table also shows the expected return and Sharpe ratio associated with the portfolio (Table 4).

Table 3 Portfolio with a maximum Sharpe ratio

Expected return	Standard deviation	Sharpe Ratio	NTC	NABIL	NLIC	SCB
0.26053	0.35970	0.72431	0.00001	0.19377	0.59971	0.20648
8	5		7	5	9	9

Table 4 Portfolio with a minimum standard deviation

Expected return	Standard deviation	Sharpe Ratio	NTC	NABIL	NLIC	SCB
0.031184	0.228616	0.136401	0.797615	0.171828	0.003942	0.026615

Table 5 Optimal weight generated by SLSQP for the 4 assets	NTC	NABIL	NLIC	SCB
	0.0	0.18	0.63	0.19

5.2 Optimization Using Sequential Least Squares Programming (SLSQP)

The experiment in this section uses SLSQP to calculate the optimal weight allocation for each asset by maximizing the value of the Sharpe ratio for the portfolio and subject to the following constraints:

- The random weight generated for each asset must lie between zero and one.
- The sum of the weight of the asset must equal to one, which is 100% of the total capital.

Table 5 shows the optimal weight generated by SLSQP for the 4-assets in the portfolio, with a maximum weight of 0.63 assigned to the assets NLIC the weight 0.19 to SCB and 0.18 to NABIL. The sum of total weight assigned is equal to 1 which is the constraint for portfolio optimization mentioned in Eqs. (6) and (7).

6 Conclusion

The experiment was conducted using mean-variance theorem to find the optimal weight of the four assets: NTC, NABIL, NLIC, SCB considered for portfolio construction. The result of the experiment shown in Table 5 shows the optimal weight associated with the four assets in the portfolio, whereas the result in Tables 3 and 4 shows the expected return, the risk and optimal weight associated with the portfolio. The conducted experiment was also used to analyze the limitation of MV framework mentioned in Sect. 3.2 and has shown that the MV framework produces erroneous result when considering a large amount of historical data (price history of more than 2 years) for the portfolio construction. So, considering the limitation of MV framework, dynamic, and complex behavior of the financial instrument and market, the task of portfolio optimization is best suited for the machine learning task. The future research work includes exploring the portfolio optimization problem using the deep learning algorithm and using reinforcement learning for an automated trading agent.

References

1. Jothimani, D., Bener, A.: Risk parity models for portfolio optimization: a study of the Toronto stock exchange. In: 2019 International Conference on Deep Learning and Machine Learning in Emerging Applications (DeepML). https://ieeexplore.ieee.org/abstract/document/8876902

2. Hegde, S., Kumar, V., Sing, A.: Risk aware portfolio construction using deep deterministic policy gradients. In: 2018 IEEE Symposium Series on Computational Intelligence (SSCI)
3. Hu, Y.-J., Lin, S.-J.: Deep reinforcement learning for optimizing finance portfolio management. In: 2019 Amity International Conference on Artificial Intelligence (AICAI)
4. Gu, C.-S., Hsieh, H.P., Wu, C.-S., Chang, R.-I., Ho, J.-M.: A fund selection robo-advisor with deep-learning driven market prediction. In: 2019 IEEE International Conference on Systems, Man and Cybernetics (SMC), October 2019. Doi https://doi.org/10.1109/smc.2019.8914183k. Elissa
5. Jiang, Z., Xu, D., Liang, J.: A Deep Reinforcement Learning Framework for the Financial Portfolio Management Problem. Deep Portfolio Management. https://arxiv.org/abs/1706.100 59v2Y
6. Chaouki, A., Hardiman, S., Schmidt, C., Sérié, E., de Lataillade, J.: Deep Deterministic Portfolio Optimization. arXiv:2003.06497v2 [q-fin.MF] 9 Apr 2020
7. Soleymani, F., Paquet, E.: Financial Portfolio Optimization with online deep reinforcement learning and restricted stacked autoencoder—Deep Breath. Expert Systems with Applications. https://www.researchgate.net/deref/https%3A%2F%2Fdoi.org%2F10.1016%2Fj.eswa.2020.113456
8. Zhang, W., Zhou, C.: Deep learning algorithm to solve Portfolio management with proportional transaction cost. In: 2019 IEEE Conference on Computational Intelligence for Financial Engineering and Economics (CIFEr), May 2019. https://doi.org/10.1109/cifer.2019.8759056
9. Usha, B.A., Manjunath, T.N., Mudunuri, T.: Commodity and Forex trade automation using deep reinforcement learning. In: Conference: 2019 International Conference on Advanced Technologies in Intelligent Control, Environment, Computing and Communication Engineering (ICATIECE), March 2019 https://doi.org/10.1109/icatiece45860.2019.9063807
10. Wu, Q., Zhang, Z., Pizzoferrato, A., Cucuringu, M., Liu, Z.: A Deep Learning Framework for Pricing Financial Instrument. arXiv.org>cs>:1909.04497
11. Min-Yuh Day, Jian-Ting Lin: Artificial Intelligence for ETF Market Prediction and Portfolio Optimization,2019 IEEE/ACM International Conference on Advances in Social Networks Analysis and Mining
12. Ma, Y., Han, R., Wang, W.: Prediction-based portfolio optimization models using deep neural networks. https://doi.org/10.1109/access.2020.3003819. IEEE Access
13. Day, M.-D., Cheng, T.-K., Li, J.-G.: AI Robo-advisor with big data analytics for financial services. In: 2018 IEEE/ACM International Conference on Advances in Social Networks Analysis and Mining (ASONAM)
14. Estalayo, I., Del Ser, J., Osaba, E., Nekane Bilbao, M., Muhammad, K., Andrés Iglesias, G.: Return, diversification and risk in cryptocurrency portfolios using deep recurrent neural networks and multi-objective evolutionary algorithms. In: 2019 IEEE Congress on Evolutionary Computation (CEC 2019), 978-1-7281-2153-6/19/$31.00 c 2019 IEEE
15. Pai, N., Ilango, V.: A comparative study on machine learning techniques in assessment of financial portfolios. In: Proceedings of the Fifth International Conference on Communication and Electronics Systems (ICCES 2020) IEEE Conference, Record # 48766; IEEE Xplore ISBN: 978-1-7281-5371-1
16. Fontoura, A., Haddad, D.B., Bezerra, E.: A deep reinforcement learning approach to asset-liability management. In: Reinforcement Learning Approach to Asset-Liability Management, 2019 8th Brazilian Conference on Intelligent Systems (BRACIS)

Machine Learning-Based Real-Time Traffic Accident Detection

Chandrashekhar Choudhary, Prattyush Sinha, Sahil Ratra, and Prakhar Tiwari

Abstract The number of automobiles on the road is growing day by day, and hence, the instances of accidents are constantly on the rise. There is a need to develop a system that detects the accidents and notifies the emergency services to ensure a quick response. This paper presents a real-time traffic accident detection system that makes use of the supervised machine learning algorithm and long short-term memory (LSTM) on CCTV footage, distinguishing accident cases from normal cases. The significance of the result analysis depicts the performance of 87.5%.

Keywords Machine learning · Recurrent neural networks (RNN) · Long short-term memory (LSTM) · Real-time accident detection · Deep learning

1 Introduction

Vehicular traffic, both in India and abroad, has been growing at an alarming rate. The cases of accidents are caused due to road rage [1], mechanical failure or plain human error. The total number of registered vehicles in the National Capital Region, which forms one of the world's largest urban clusters, increased from 1.42 million in 1988 to 11.2 million in 2016.

C. Choudhary · P. Sinha
Department of Computer Science Engineering and Information Technology, Jaypee University of Information Technology, Waknaghat, India
e-mail: karan89009@gmail.com

P. Sinha
e-mail: sinha.prattyush@gmail.com

S. Ratra (✉)
Cognizant Technology Solutions India, Pune, Maharashtra, India
e-mail: sahil07ratra@gmail.com

P. Tiwari
Ingenious E-Brain Solutions, Gurugram, Haryana, India
e-mail: prakhartiwari.1711@gmail.com

© The Editor(s) (if applicable) and The Author(s), under exclusive license
to Springer Nature Singapore Pte Ltd. 2021
S. Shakya et al. (eds.), *Proceedings of International Conference on Sustainable Expert Systems*, Lecture Notes in Networks and Systems 176,
https://doi.org/10.1007/978-981-33-4355-9_50

673

Road accidents [2] are a neglected threat to humans, as they are the cause for thousands of deaths and damage to property. They demand immediate action to cut down their growing stake in human casualties and personal and/or public losses in life or property. According to a report issued by World Health Organization in 2016, an estimated 1.35 million people die every year due to accidents that happen on the world's roads.

There has been no concrete system of reporting the occurred road accidents to the concerned authorities with immediate effect. As per the present scenario, the highway patrolling vehicles or the passers-by have to report the matter or call an ambulance for action. This entire process is elaborated to an extent that it does and has cost several lives, especially in countries where the emergency services' response has not been quick. An information retrieval system is the need of the present scenario which can analyze the collision [3] and produce information which can detect the occurrence of an accident, thereby triggering an immediate response which in turn will alert the local emergency services. This would help the authorities in reaching the affected spots timely for help.

This paper describes the working of a model that can automatically detect accidents occurring on the road through CCTV footage, the result of which can then be used to notify emergency services to take the quickest possible action.

The main features of this paper are:

- This paper demonstrates the usefulness of machine learning [4] in real-time environments
- This paper puts forward an efficient model for real-time traffic [5] accident detection.

The remainder of the paper is organized as follows. The literature survey is discussed in Sect. 2. The methodology is presented in Sect. 3 in which the approach used in the paper is discussed along with the detailed description of the dataset, preprocessing, training and testing, recurrent neural networks and LSTM algorithm used in the paper. Results obtained are discussed in Sect. 4. And the conclusion of the paper is discussed in Sect. 5.

2 Literature Survey

YanjieDuan, YishengLv and Fei-Yue Wang in [6] explored deep learning models and demonstrated a way to apply long short-term memory (LSTM) neural network model to traffic data. Through their application, they were able to develop a model which had a low prediction error of 7% on the test set for traffic series prediction, leaving a further scope for improvement in accuracy for future.

Weicong Kong, Zhao Yang Dong, YouweiJia, David J., Yan Xuand Yuan Zhang in [7] proposed an LSTM recurrent neural network framework for the extremely difficult task of forecasting residential load for an individual. Their research was able to address the potency of the LSTM model through comprehensive testing and comparing to various benchmarks to solve the problem at hand. They were able to assert that many load predicting methods that are successful for substation or grid load prediction find it difficult in the single-meter load prediction problems. The suggested LSTM framework was able to achieve generally the best prediction performance in the dataset.

ShunsukeKamijo, Yasuyuki Matsushita, Katsushi Ikeuchi and Masao Sakauchi in [2] present how traffic accident detection systems can be developed through the use of algorithms. They derived an algorithm, which they call as the spatio-temporal Markov random field model. Through this model, they can track the vehicles on the road and intersections through the use of grayscale images with a success rate of about 93–96%.

Zehang Sun, George Bebis and Ronald Miller in [1] review the recent vehicle detection systems that work on the vision-based on-road guidelines. In Various systems, they reviewed used methods like stereo-based solutions, edge-based solutions, appearance-based solutions and hardware-based solutions. They concluded that vision-based methods in vehicle detection systems are the way forward, with much research happening in the current times, and even more scope in the future.

MantasLukoševičius, HerbertJaeger in [8] provide insights as to how to improve the performance of recurrent neural networks through the use of reservoir computing. They surveyed, systematically, both current and upcoming ways of training different varieties of readouts. They aimed to provide the best possible performance enhancement for recurrent neural networks.

Bradski, and Kaehler in [9] give a comprehensive study about the techniques and methods to learn and gain knowledge about the OpenCV library. It is a very useful resource for information regarding the documentation of various aspects of the OpenCV library, which was used extensively during the research work for writing this paper. The resource is excellent for novices and experts alike.

Andrea Vedaldi and Brian Fulkerson in [10] provide the research community with an open and portable library that consists of computer vision algorithms. They introduced VLFEAT, which can be easily used for fast prototyping for research in the field of computer vision. The common building blocks like k-means clustering, feature extractors, feature detectors and more are all rigorously implemented in the library.

Koresh and Deva in [11] put forward a framework that can sense (detect and recognize) real-time traffic sign, providing an innovatively safe way to plan paths and drive. The method proposed by them makes use of capsules neural networks, which works much better than convolutional neural network by way of getting rid of the manual effort that is necessary. In comparison with CNN and RNN, capsule network showed a 15% higher accuracy with Indian traffic dataset.

Manoharan, Samuel in [12] put forward measures to improve safety probability for vehicles with self-driving capabilities with processors that are enabled with artificial intelligence. Real-time data is used to ensure competence in the proposed measures, and performance evaluation is done using them.

Karlik and Olgac in [13] present a case for various activation functions, their usefulness in different scenarios and their limitations. Bi-polar sigmoid, conic section, uni-polar sigmoid, tanh and radial basis function (RBF) were the activation functions used for experimental purposes.

3 Methodology

To implement the solution for the problem at hand, a proper analysis-based methodology had to be followed (Fig. 1).

The methodology involves three major steps, i.e., making the dataset, preprocessing of the dataset and then finally training and testing the model. The model used was long short-term memory (LSTM).

3.1 Dataset

The first thing we had to do was to work on the collection of data for our dataset [14]. The primary task involves finding the real footages of accidents occurred which can be used to work upon. The dataset will include footage from CCTV cameras across the globe, borrowed through the video library on the social networking video sharing site, YouTube. Other than that, also collected videos [15] of accidents from a Web site called VS lab, where we found accident videos of about 4–5 s in length. These videos were optimal in terms of quality and length for our dataset. On a whole, we

Fig. 1 Flowchart of the methodology followed

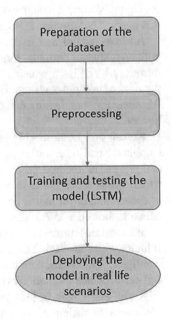

Fig. 2 Snapshot of a video from the dataset

created a sizeable dataset containing both positive and negative videos (i.e., accident occurring and not occurring). The videos were about 978 in number (Fig. 2).

3.2 Preprocessing

Videos are a collection of image frames, and images are an array of pixels with different numbers denoting different colors [9]. In this phase, frames of images are extracted from the videos, converted RGB to grayscale and downgraded them by five using OpenCV and stored it in multidimensional Numpy array. The reason why RGB images are converted to grayscale is that dealing with images in grayscale is easier and faster than RGB. RGB contains three channels, i.e., red, green and blue, whereas grayscale contains only one channel. This impacts the preprocessing of the dataset. Since, in our case with video, datasets are being dealt with, the processing is quite heavy and time consuming, so converting RGB to grayscale saves the processing time. If seen numerically, it means:

RGB color image requires $8 + 8 + 8 = 24$ bits to store a single pixel of the image, whereas grayscale requires 8 bit only.

Now, being ready to perform operations on Numpy array denoting the images, and hence the video. Encoded one hot created a label matrix of size 1×2 (1 row and 2 column). The columns denoted accident and not an accident. Initially, both column values are set to 0. If the video contains an accident, then index 0 is set to 1; else index 1 is set to 1. Figure 3 shows the Numpy array of image frames extracted from the video.

Fig. 3 Numpy array of
image frames of video

array([[[[2.68245882e-01, 2.60559608e-01, 2.53385961e-01, ...,
          9.53695686e-01, 9.52911373e-01, 9.52911373e-01],
         [2.15098118e-01, 2.37058902e-01, 2.50811373e-01, ...,
          9.52127059e-01, 9.51342745e-01, 9.51342745e-01],
         [2.11690196e-01, 2.26592157e-01, 2.54350259e-01, ...,
          9.51342745e-01, 9.48989804e-01, 9.48989804e-01],
         ...,
         [9.65729412e-02, 9.65729412e-02, 9.61249412e-02, ...,
          9.30436863e-02, 8.86617725e-02, 8.60185098e-02],
         [1.38298431e-01, 1.18690588e-01, 1.03340627e-01, ...,
          7.99223216e-02, 8.50320627e-02, 8.63535686e-02],
         [2.46533725e-01, 2.29749412e-01, 1.73615137e-01, ...,
          8.18710588e-02, 8.32390588e-02, 8.36658824e-02]],

        [[2.54378290e-01, 2.63464565e-01, 4.73365757e-01, ...,
          9.52911373e-01, 9.55264314e-01, 9.56832941e-01],
         [2.59324424e-01, 2.85052580e-01, 2.68990761e-01, ...,
          9.52911373e-01, 9.54323137e-01, 9.55264314e-01],
         [2.40569584e-01, 2.67127576e-01, 3.07665098e-01, ...,
          9.52911373e-01, 9.52911373e-01, 9.52911373e-01],

3.3   Training and Testing

Dataset is spliced into training and testing sets separately, taking a ratio of 70–30
(70% data is associated with training and the rest 30% for testing). The main objective
was to use a model such that the time dependencies are maintained while training it.
Hence, the LSTM model is used and implemented using the Keras library in Python.
The results of the same are elaborated in further sections (Fig. 4).

3.4   Recurrent Neural Network

Neural networks are a combination of nonlinear elements interconnected through
weights adjustable by nature. The basic property of a recurrent neural network [16]
is that it contains at minimum one feedback association to facilitate the flow of
activations around the loop. This technique helps the recurrent network to perform
processing of the temporal kind and realizes sequences, for example, performing
reproduction of sequence or recognition or forecasting and temporal association.

Learning in architectures which are simple and predestined activation functions
can be attained with the help of similar origin procedures to those leading to feed-
forward networks or back-propagation algorithm.

A multilayer perceptron with the previous set of hidden unit activations feeding
back into the network along with the inputs is the simplest form of fully recurrent
neural network [8] (Fig. 5).

Problem with recurrent neural networks—vanishing gradient descent [17]: As
the numbers of layers are increased using some activation functions in RNNs, the
gradient of the loss function tends to zero, and this makes hard to train the model
because our inputs and its output are dependent on one another.

To tackle this problem, LSTM is used. This paper demonstrates how LSTMs are
used to overcome the shortcomings of RNN and apply it to the problem at hand.

Fig. 4 Training of the model under process

```
12/12 [==============================] - 4s 343ms/step

Epoch 00009: val_acc did not improve from 0.58333
Epoch 10/20
12/12 [==============================] - 4s 329ms/step

Epoch 00010: val_acc did not improve from 0.58333
Epoch 11/20
12/12 [==============================] - 4s 337ms/step

Epoch 00011: val_acc improved from 0.58333 to 0.66667,
Epoch 12/20
12/12 [==============================] - 4s 338ms/step

Epoch 00012: val_acc did not improve from 0.66667
Epoch 13/20
12/12 [==============================] - 4s 313ms/step

Epoch 00013: val_acc did not improve from 0.66667
Epoch 14/20
12/12 [==============================] - 4s 310ms/step

Epoch 00014: val_acc did not improve from 0.66667
Epoch 15/20
12/12 [==============================] - 4s 307ms/step

Epoch 00015: val_acc did not improve from 0.66667
Epoch 16/20
12/12 [==============================] - 4s 310ms/step

Epoch 00016: val_acc improved from 0.66667 to 0.75000,
Epoch 17/20
12/12 [==============================] - 4s 306ms/step
```

Fig. 5 A simple RNN perceptron

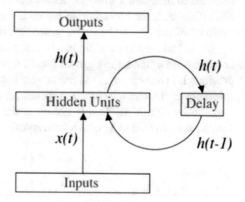

3.5 Long Short-Term Memory

LSTM or long short-term memory networks [6] are helpful when our neural network needs to switch between recollecting fresh stuff and stuff from the past time. It maintains two memories; one is long-term memory and the other is short-term memory,

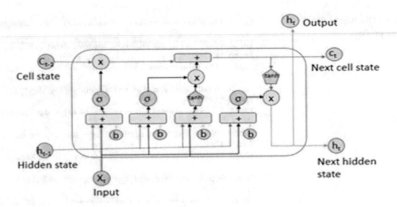

Fig. 6 Diagrammatic representation of LSTM

and these memories help to retain patterns. LSTM deals with the disadvantage of RNNs, i.e., vanishing gradient descent (Fig. 6).

A general LSTM network [7] consists of memory blocks known as cells. Two states get moved to the following cell, the cell state and the hidden state. The state called the cell state is the principal tie of the stream of data that permits the data to move ahead basically unaltered. Be that as it may, linear transformations of some sort may happen. The data can get included in or maybe expelled by means of sigmoid the gates from the cell state. A gate can be defined as a layer or a progression of matrix operations that consist of individual weights that are distinctive. LSTMs are intended to maintain a strategic distance from the issue of dependency (long-term), since it controls the procedure of remembering by using gates.

To construct an LSTM network, we first need to identify the information that is not needed and has to be excluded, in that move, from the cell. For this identification and omission of data, the sigmoid function comes into play. It takes the yield of the previous LSTM unit (p_{t-1}) at time $t - 1$ and that of the present input (Y_t) at time t. Moreover, the sigmoid function makes sure which portion of the previous output needs to be discarded. This is known as the forget gate (f_t); f_t is a vector whose values range from 0 to 1, which corresponds to every number in the cell state, C_{t-1}.

$$f_t = \sigma\left(T_f\left[p_{t-1}, Y_t\right] + a_f\right) \tag{1}$$

Here, the sigmoid function [18] is σ, and a_f and T_f are the bias and weight matrices of the forget gate, respectively.

The next move is choosing and saving data taken from the latest input (Y_t) in the cell state and also the cell state updation. This move comprises two things, first the sigmoid layer followed by the tanh layer [13]. Firstly, the sigmoid layer makes sure if the new info needs to be amended or avoided (0 or 1), and secondly, the function tanh assigns weight to the values that came through it, establishing their level of significance (-1 to 1). The values thus obtained, through multiplication, are then used for the upcoming cell state update. The resulting new memory is further concatenated to the previous memory C_{t-1}, leading to C_t.

$$i_t = \sigma\left(T_i\left[p_{t-1}, Y_t\right] + a_i\right) \tag{2}$$

$$M_t = \tanh\left(T_n\left[p_{t-1}, Y_t\right] + a_n\right) \tag{3}$$

$$C_t = C_{t-1}f_t + M t i t \tag{4}$$

Here, C_t and C_{t-1} are the cell states at time t and $t - 1$, and meanwhile a and T are the bias and matrices of weight of the cell state, respectively.

In the last move, the output values (p_t) are dependent on the output cell state (O_t) which is a version that is filtered. Firstly, a sigmoid layer makes sure what portions of the cell state get through to the output. After that, the new values that tanh produces from the cell state (Ct) are multiplied with the sigmoid gate output (O_t) value of which ranges from -1 to 1.

$$O_t = \sigma\left(T_o\left[p_{t-1}, Y_t\right] + a_o\right) \tag{5}$$

$$p_t = O_t \tan h(C_t) \tag{6}$$

Here, a_o and T_o are the bias and the weight matrices of the output gate, respectively. This is how LSTM helps us to maintain time dependencies.

4 Results

After training our LSTM model with data from our dataset, its performance is analyzed against a test dataset. The proposed model was successful in achieving an accuracy of 87.5%.

To further analyze the results in detail, the following parameters are made use of:

$$\text{Accuracy} = \left(T_p + T_n\right)/\left(T_p + T_n + F_p + F_n\right) \tag{7}$$

$$\text{Sensitivity} = T_p/(T_p + F_n) \tag{8}$$

$$\text{Precision} = T_p/(T_p + F_p) \tag{9}$$

where the true positive outcomes are denoted by T_p, the true negative outcomes are denoted by T_n, the false positive outcomes are denoted by F_p and the false negative outcomes are denoted by F_n.

Fig. 7 Confusion matrix

```
confusion matrix
[[7 3]
[ 1 14]]
True negative: 7
False positive: 3
False negative: 1
True positive: 14
```

Table 1 Summary of results

Parameters	Result value
Accuracy	0.875
Sensitivity	0.933
Precision	0.823

Fig. 8 Input video to the model

From the confusion matrix shown below, the values of accuracy, sensitivity and precision can be calculated (Fig. 7).

Table 1 summarizes the results of the accuracy, sensitivity and precision the model was able to achieve.

Shown below are snapshots of the videos fed to the model, and the model making the prediction on them (Figs. 8 and 9).

5   Conclusion

This paper has successfully demonstrated the working of a model for real-time traffic accident detection through the implementation of long short-term memory (LSTM)

Fig. 9 Model correctly predicting the occurrence of an accident

networks. The evaluation of the results from the model's performance affirms that the LSTM model successfully detected and identified accidents through the use of CCTV footage, with an accuracy of 87.5%. Furthermore, future implementations can facilitate that once an accident gets detected through the proposed method, an alert can be sent to the emergency services so that they can provide relief as soon as possible, thereby reducing the risk of loss of lives and property.

Moreover, results from the model can also be sent to users through the Internet, warning them about the accident, so that they can take evasive actions well ahead of time.

References

1. Sun, Z., Bebis, G., Miller, R.: On-road vehicle detection: a review. IEEE Trans. Patt. Anal. Mach. Intell. **28**(5), 694–711 (2006)
2. Kamijo, S., Matsushita, Y., Ikeuchi, K., Sakauchi, M.: Traffic monitoring and accident detection at intersections. IEEE Trans. Intell. Transp. Syst. **1**(2), 108–118 (2000)
3. Teschner, M., Heidelberger, B., Müller, M., Pomerantes, D., Gross, M.H.: Optimized spatial hashing for collision detection of deformable objects. In *Vmv* (Vol. 3, pp. 47–54) (2003, November)
4. Singh, A., Thakur, N., Sharma, A.: A review of supervised machine learning algorithms. In: 2016 3rd International Conference on Computing for Sustainable Global Development (INDIACom) (pp. 1310–1315). IEEE (2016, March)
5. Cucchiara, R., Piccardi, M., Mello, P.: Image analysis and rule-based reasoning for a traffic monitoring system. IEEE Trans. Intell. Transp. Syst. **1**(2), 119–130 (2000)
6. [Duan, Y., Lv, Y., Wang, F.Y.: Travel time prediction with LSTM neural network. In: 2016 IEEE 19th International Conference on Intelligent Transportation Systems (ITSC) (pp. 1053–1058). IEEE (2016, November)

7. Kong, W., Dong, Z.Y., Jia, Y., Hill, D.J., Xu, Y., Zhang, Y.: Short-term residential load forecasting based on LSTM recurrent neural network. IEEE Trans. Smart Grid **10**(1), 841–851 (2017)

8. Lukoševičius, M., Jaeger, H.: Reservoir computing approaches to recurrent neural network training. Comput. Sci. Rev. **3**(3), 127–149 (2009)

9. Bradski, G., &Kaehler, A. (2008). Learning OpenCV: Computer vision with the OpenCV library. "O'Reilly Media, Inc."

10. Vedaldi, A., Fulkerson, B.: VLFeat: an open and portable library of computer vision algorithms. In: Proceedings of the 18th ACM international conference on Multimedia (pp. 1469–1472) (2010, October)

11. Koresh, M.H.J.D., Deva, J.: Computer vision based traffic sign sensing for smarttransport. J. Innov. Image Process. (JIIP) **1**(01), 11–19 (2019)

12. Manoharan, Samuel: An improved safety algorithm for artificial intelligence enabled processors in self-driving cars. J. Artif. Intell. **1**(02), 95–104 (2019)

13. Karlik, B., Olgac, A.V.: Performance analysis of various activation functions in generalized MLP architectures of neural networks. Int. J. Artif. Intell. Expert Syst. **1**(4), 111–122 (2011)

14. Xiao, H., Rasul, K., &Vollgraf, R. (2017). Fashion-mnist: a novel image dataset for benchmarking machine learning algorithms. arXiv preprint arXiv:1708.07747

15. Tao, A., Takeuchi, S., Ueno, S., Komori, Y.: U.S. Patent No. 6,441,832. Washington, DC: U.S. Patent and Trademark Office (2002)

16. Zaremba, W., Sutskever, I., Vinyals, O.: Recurrent neural network regularization (2014). *arXiv preprint* arXiv:1409.2329

17. Hochreiter, S.: Recurrent neural net learning and vanishing gradient. Int. J. Uncert. Fuzz. Knowl. Based Syst. **6**(2), 107–116 (1998)

18. Ito, Y.: Representation of functions by superpositions of a step or sigmoid function and their applications to neural network theory. Neur. Netw. **4**(3), 385–394 (1991)

Clustering-Based Pattern Analysis on COVID-19 Using K-Means Algorithm to Predict the Death Cases in Bangladesh

Mohammad Ashraful Hoque, Thouhidul Islam, Al Amin, and Tanvir Ahmed

Abstract In December 2019, the COVID-19 rapidly infected virus disease was an outbreak in China. Woefully, the number of infected and death cases is increasing day after day. Still, there is no effective vaccine and antibiotic. As a result, our fighting starts with an invisible enemy. Up to July 31, 2020 the total number of death cases is 3111 (Male: 2446 and Female: 665), infected cases are 237,661, and recovered cases are 135,136 in Bangladesh. The most infected city is Dhaka with 47.83% death cases. Among our overall population, senior citizens and infants are more valuable assets for the nation. In this research work, the K-means algorithm was implemented to discover a pattern from collected data with the help of a data mining technique to predict the age level for death cases on COVID-19. Data were collected from 8 March 2020 to 31 July 2020 from authentic recognized sources. In this research work, R language is used for clustering on the K-means algorithm. The result analysis of this research shows that the most infected age range is above 60 years in this period, and they have a high possibility of death cases in the near future.

Keywords Coronavirus disease · COVID-19 · Data mining · Clustering · K-means algorithm · R language · RStudio · R^2 · IEDCR · DGHS · Bangladesh

M. A. Hoque (✉) · T. Islam · A. Amin
Department of Computer Science and Engineering, Southeast University, Dhaka, Bangladesh
e-mail: ashraful@seu.edu.bd

T. Islam
e-mail: bappythouhidrx15@gmail.com

A. Amin
e-mail: alaminwww07@gmail.com

T. Ahmed
Department of Computer Science and Engineering, Ahsanullah University of Science and Technology (AUST), Dhaka, Bangladesh
e-mail: tanvir.cse@aust.edu

© The Editor(s) (if applicable) and The Author(s), under exclusive license
to Springer Nature Singapore Pte Ltd. 2021
S. Shakya et al. (eds.), *Proceedings of International Conference on Sustainable Expert Systems*, Lecture Notes in Networks and Systems 176,
https://doi.org/10.1007/978-981-33-4355-9_51

685

1 Introduction

The novel coronavirus disease 2019 is named as COVID-19. COVID-19 is an infectious disease. In Wuhan, capital of Hubei province in China, COVID-19 was out broken. After that, it spreads globally. And today, it is a global pandemic situation according to WHO's statement. COVID-19 is a direct threat to human health. It keeps its impact on the global economy, environment, and education. The characteristics of this virus are changing geographical environments all around the world. Since, for COVID-19, there is no antibiotic or vaccine, people followed a few precautionary steps such as social distancing, frequent hand washing, avoid public gathering, and self-control. Finally, it is very much needed to increase the awareness and quarantine themself if anybody senses any symptom of COVID-19 virus [1, 2].

Data mining is a technique, which is used to extract the hidden pattern from the given data. To handle the very large data set, it is required to precise our data. Without data mining, it is not possible because a traditional database management system (DBMS) is not enough at all. Data mining has its technique for extracting the pattern; the amount of them clustering is a machine learning technique. K-means algorithm is an unsupervised clustering algorithm. It is a simple and robust algorithm among all clustering algorithms. For analyzing big data, researchers are using a statistical programming language and analyzing tool "RStudio" [3].

In this research work, a pattern is developed with the help of the K-means clustering algorithm to prophesy the age levels from a data set that is in chancy to the death cases due to COVID-19 in near future. The data is collected from 8 March 2020 to 31 July from WHO, Bangladesh IEDCR, and DGHS Web portals.

2 Literature Study

Habibzadeh et al. [4] present an overview of the coronavirus in detail with the pathogenesis, epidemiology, clinical, presentation, and diagnosis with the treatment methodology. Coronavirus is a human-to-human transmittable virus. It is a type of respiratory and enteric infection for both humans and animals. Worldwide, peoples are frequently infected with four types of coronavirus such as 229E, OC43, NL63, and HKU1. Its symptoms are as similar to upper respiratory tract infection. Up to 03 February 2020, approximately total infected and death cases were 17,496 and 362 respectively around the globe. Due to Chinese New Year, the peak travel reason played a significant rote to spread the infection. Thailand was the first infected country after China. The proper precautionary steps should be taken to prevent the transmission of this virus. There are clinical similarities between COVID 19 and SARS-CoV for infection causes. Generally, for the human coronavirus, the incubation period is 2–4 days, but for SARS-CoV, itis 4–6 days. A non-specific symptom of COVID 19 is similar to SARS-CoV such as fever, malaise, and dry cough in the prodromal phase. A study conducted at Wuhan Hospital shown that among 41 confirmed cases of COVID-19 about 98% had a fever, 76% had a cough, 44% had fatigue and 55% had dyspnea as a common symptom. Few have upper respiratory tract symptoms like sore throat and rhinorrhea. In general, the median age range (49–61 age) and male individuals are more frequently infected than others. As a treatment, there are no vaccines or antivirus for COVID-19. In the first case, the USA used remdesivir as an anti-viral drug, which was developed for the treatment of the Ebola virus. World Health Organization (WHO) recommends hospital admission for confirmed cases on requirement basis. However, isolation and oxygen therapy with high flow canella and fluid management will be the best way to prevent the COVID-19.

Azarafza et al. [5] present the analysis of COVID-19 infection spread analysis in Iran. To prepare this pattern, they utilized the clustering algorithm and geographical information system. Geographical information system (GIS) was implemented from the starting point to other points in Iran to determine the possibilities of spreading. They took Qom city as a starting point. The major task of this research work was to design a visual view of GIS mapping with a few interactive queries. Finally, they present all operations results in COVID-19 spread pattern map. They showed the infection situation from 19 February to 22 March of this year in Iran. They used Gephi software to draw the figure. This research is a combination of clustering and GIS mapping. These clusters were determined with the help of K-means clustering. Based on their research observation, 11, 27 and 31 provinces were affected according to first, second, and third week with the help of the transportation system.

Isabella et al. [3] present the implementation of the K-means algorithm to find the clusters using a language "*R*". To cluster the data values, they used normalized and actual values. They observed three data sets, Iris, Lenses, and Soybean (Large), with different variables to produce various accuracy rates with the clustering methodology.

3 Coronavirus Disease 2019 (COVD-19)

The 2019 novel coronavirus disease (COVID-19) initially upsurged in China (Wuhan). Later on, the following countries Thailand, Japan, South Korea, and the USA informed about the coronavirus confirmed cases. To discover the drug, surveillance system, and the prevention method from this epidemic, its origin was determined based on the phylogenomic analysis. It shows that COVID-19 is almost similar to Severe Acute Respiratory Syndrome (SARS-2). That is why it is also like SARS-Cov-2. Firstly, this new coronavirus was isolated from a seafood market in Wuhan, China. After that, this virus was spread to other provinces also in other countries [6].

3.1   COVID-19 in Bangladesh

Bangladesh declared their first three COVID-19 positive cases on 8 March 2020. After 9 days, Bangladesh reported its first death case on 18th March 2020. He is a 70-year-old senior citizen [2].

According to the WHO COVID-19 situation report [7] on 27 July 2020, 254,274 are COVID-19 new confirmed cases. 5490 deaths are in new cases. And total COVID-19 confirmed cases are 16,114,449 and total death cases are 646,641. The World Health Organization (WHO) declared COVID-19 as a global pandemic for its rapid spreading of the infection all around the world. According to the WHO Bangladesh COVID-19 situation report [8] on 27 July 2020, 2772 are COVID-19 new confirmed cases and 37 are new deaths. And the total COVID-19 confirmed cases are 226,225 and total death cases are 2965 where recovered cases are 125,683. From a total of 81 COVID-19 laboratories all over the country, 83,259 samples were collected in the last seven days. Overall, 60.8% was tested inside the capital of Bangladesh, Dhaka. From that 60.8% tested samples, 20.1% were positive.

There was a study on 700 young adult participants from two different states in Bangladesh to find precautionary behavior adoption. Among 700, 350 participants were from Dhaka. The study found that younger adults who have completed their post graduation are more adapted to precautionary behaviors with a mean value of 26.2. And higher secondary completed younger people are less adapted to precautionary behavior. The mean value was 23.7. This type of evaluation was played among 350 young adults in Tangail, nearest district of Dhaka. In this study, it was found that the adoption of precautionary behavior mean value for graduate young adults was slightly better than higher secondary young adults. This study confirmed that precautionary behaviors are higher for postgraduate's participants to prevent COVID-19 [2].

4   Data Mining

Data mining is a method by which a pattern may be found, which was unknown previously. The process for extracting useful knowledge and information from a high amount of data is called data mining. For mining data from a data warehouse, it follows three steps. These are exploration, pattern identification, and deployment. In exploration stage, data becomes clean and gets ready to shift in another form. After exploring the data in the second stage, data makes its pattern identification to perform the best prediction. In the final stage, patterns are deployed as an output. Data mining is a part of knowledge discovery in data [9]. Knowledge discovery process in data is shown in Fig. 1.

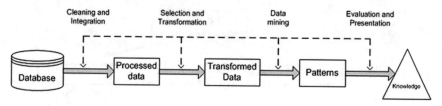

Fig. 1 Knowledge discovery process

4.1 Data Mining Technique

For mining data from a database, different technique algorithms are used such as classification, clustering, regression, neural networks, association rules, decision trees, genetic algorithm, and nearest neighbor method. Data mining techniques are shown in Fig. 2.

4.1.1 Classification and Clustering

The most commonly used data mining technique is classification. It works with the help of a pre-classified data set. This technique is commonly used with credit card fraud detection and risk management application systems. The types of classification methods are neural classification, support vector machines, classification by decision tree, Bayesian classification, etc. [9].

Clustering is similar to classification as the data is grouped. To accomplish clustering, determine the similarity among the collected data on the predefined attributes. The most similar data is grouped in different clusters. Using clustering methodology, researchers can discover the pattern and correlations among the data. The types of clustering methods are the density-based method, grid-based method, model-based method, partitioning method, etc. [9]. Clustering methods are shown in Fig. 3.

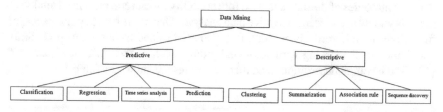

Fig. 2 Data mining techniques

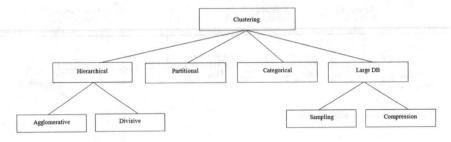

Fig. 3 Clustering methods

4.1.2 Regression

The regression method is used to predict anything from the given data. It is also used to find the relationship between the independent and dependent variable or variables. The types of regression methods are linear regression, nonlinear regression, multivariate linear regression, multivariate nonlinear regression [9].

From large data sets, frequent itemsets can be found using association rules. It helps to make a particular decision in a business system like customer shopping behavior analysis, catalog design, etc. The types of association rules are multidimensional association rule, multi-level association rule, and quantitative association rule [9].

5 *R* Language

R is an open-source statistical programming language. It is an extension of S language. It was developed in 1992 by Ross Ihaka and Robert Gentleman. *R* language initial version was released in 1995. R is an interactive programming language with built-in statistical functions which are used for detailed statistics. It gives a synopsis of the main attributes of the data and used for variability measurement, central tendency, etc. R is widely used for investigative data analysis. One of the best popular packages for visualisation libraries is ggplot2. It is also used to find a continuous probability distribution. It is working with command prompt such as ">", "#", "<−", and "$" [10]. Some command prompts with their meanings are shown in Table 1.

Table 1 Meaning of command prompts	Command prompt	Meaning
	>	Starts a program
	#	Mention comment line
	<−	Assignment operator
	$	String type data

A sample *R* language program for two vectors addition is given below,

```
> x       # first vector
[1] 7 6 5
> y       # second vector
[1] 9 6 3
> x + y   # doing addition
[1]  16 12  8
```

6   Research Methodology

In this research work, the K-means algorithm is utilized for clustering among the data of death cases due to the novel coronavirus. *R*-statistical programming platform is used to cluster the collected data. The workflow of this research work is given in Fig. 4.

6.1   Data Source

After detecting the first COVID-19 positive patient from 8 March 2020, Institute of Epidemiology, Disease Control, and Research (IEDCR) continues the press briefing every day. IEDCR and DGHS have dedicated an online platform to show the complete data on tested, infected, recovered, and death [11]. From the given data, by summarizing the death cases from 18 March to 31 July 2020, a pattern is extracted to predict the death range according to the age level. After forecasting the death age ranges, each and everyone's life may be saved by taking precautionary behavior.

Table 2 shows the total test, infection, recovery, and death cases from 8 March 2020 to 31 July 2020.

6.2   K-Means Algorithm

K-means is a standard prototype-based clustering algorithm in data mining. It was proposed in 1967 firstly by Mac Queen. This clustering technique is simple and very fast. It is an unsupervised, numerical, non-deterministic, and iterative algorithm. For creating globular clusters, it is very suitable. For partitioning methodology, the clusters made from the K-means algorithm are independent and compact [12]. K-means algorithm is given in Fig. 5.

Fig. 4 Flowchart of research
work

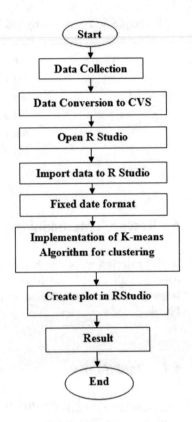

Table 2 Data of total test, infection, recovery, and death cases

Month	Test cases	Infection cases	Recovery cases	Death cases
March	1,602 51 25			5
April	63,064	7616	135	163
May	244,264	39,486	9621	482
June	457,530	98,330	49,843	1197
July	410,349	92,178	75,512	1264

The time complexity of the K-means algorithm is $O(n * k * i)$, where n is total elements and k is the number of cluster iterations. Generally, the K-means algorithm is required $k \ll n$, $t \ll n$. The K-means algorithm may miss the global optimum. Because the K-means algorithm does not work on categorical data, it works on the previously defined mean. The cluster mean of $k_i = \{t_{i1}, t_{i2}, \ldots t_{im}\}$ is defined as, $m_i = \frac{1}{m} \sum_{j=1}^{m} t_{ij}$ [12].

A similarity measure can be described as the distance among different data points. In reality, the performance of many algorithms depends upon choosing a good distance function over a data set such as Euclidean distance, cosine distance, Jaccard

Fig. 5 K-means algorithm

Input:

$D = \{t_1, t_2, \ldots\ldots\ldots, t_n\}$ // Set of elements

k // Number of desired clusters

Output:

K // Set of clusters

K-means algorithm:

assign initial values for means $m_1, m_2, \ldots\ldots\ldots, m_k$;

repeat

     assign each item t_i to the cluster which has the closest mean;

     calculate new mean for each cluster;

until

     convergence criteria is met;

distance, and Manhattan distance. In this paper, the K-means clustering algorithm uses the Euclidean distance to identify the similarities between the objects. Euclidean distance computes the root of the square difference between coordinates of pairs of objects. To calculate the Euclidean distance between two points x and y in a line segment $dis(x, y) = \sqrt{\sum_{i=1}^{n}(x_i - y_i)^2}$, the formula is used.

7 Result Analysis and Discussion

In this section, the analyzed data is taken from different data sources between 8 March 2020 and 31 July 2020. According to the data of the Institute of Epidemiology, Disease Control and Research (IEDCR) [13], and Directorate General of Health Services (DGHS) [14], data was prepared and clustered with the K-means algorithm in RStudio. These clustered results are given in Figs. 6, 8, 9, 10, and 11. And Figs. 7, 12, and 13 show the R^2 value. As have known, the R squared method is widely used

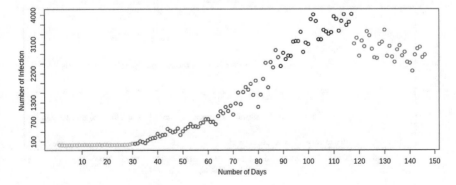

Fig. 6 Infection cases clustering

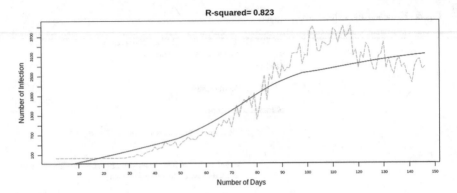

Fig. 7 R^2 value of infection cases

to elaborate the future outcomes and also used to predict a model. This method helps us to measure the accuracy of a model as well as to eradicate the prediction error. $r = \dfrac{n(\sum xy) - (\sum x)(\sum y)}{\sqrt{[n\sum x^2 - (\sum x)^2][n\sum y^2 - (\sum y)^2]}}$ equation is followed to find the value of R^2.

Figure 6 shows the data representation of infection cases between 8 March 2020 and 31 July 2020. The neon green, red, blue, black, and green color, respectively, mentioned the March, April, May, June, and July months. It is observed that the infection is rapidly accumulating constantly. It is also seen that the infection case in July was higher than the other months. Comparatively in the March, infection case was minimal among the other months.

Figure 7 shows the data representation of the R^2 value of infection cases between 8 March 2020 and 31 July 2020. For infection cases, the R^2 value is 0.823 where the referred value range is 0–1.

Figures 8, 9, 10 and 11 show the data representation of death cases between 18 March 2020 and 31 July 2020. The actual death toll from COVID-19 is much higher

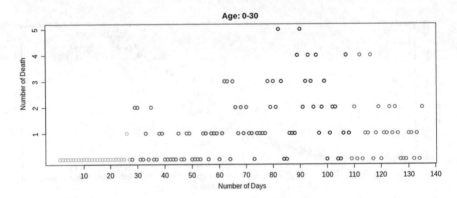

Fig. 8 Death cases clustering (0–30)

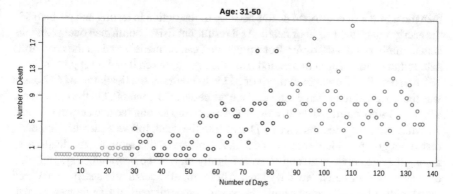

Fig. 9 Death cases clustering (31–50)

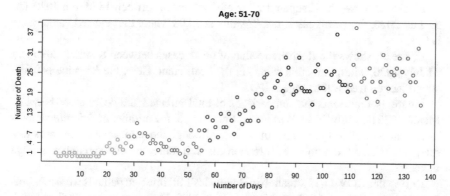

Fig. 10 Death cases clustering (51–70)

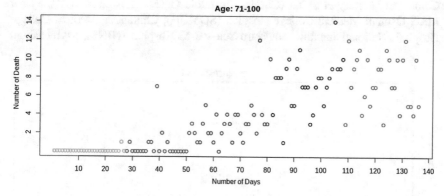

Fig. 11 Death cases clustering (71–100)

than the number of confirmed deaths because of restriction in testing and problems in the ascription of the origin of death. As a result, published confirmed deaths and the actual number of deaths differ. But a prediction can be made based on the death tolls data found so far. Within this period from the ages between 0 and 30 approximately, 135 persons died from a total death of 3111. In contrast, the death toll of 642 cases was recorded from the same total of the ages between 31 and 50. On the other hand, a total of 1809 confirmed death cases were registered among the age between 51 and 70. Moreover, between the ages of 71 and 100, the death toll was surprisingly lower than other groups which were around 525. It is observed that although the death cases among the age between 20 and 40 were slightly high at the beginning days, this rate dramatically declined a few days later. After a month, death rate sharply increased in the midst of middle-aged people. However, senior citizens of Bangladesh were in the least suffering position from COVID-19.

Figure 12 shows the R squared value of death cases between 18 March 2020 and 31 July 2020. For death cases, the R^2 value is 0.808 where the referred value range is 0–1.

Figure 13 shows the R squared value of death cases between 8 March 2020 and 31 July 2020, where infection cases are the coefficient. Here, the R^2 value is 0.883 where the referred value range is 0–1.

Figure 14 represents the comparisons of total infected and death cases between March 2020 and July 2020. Within this time period, the number of day infections is increasing as well as the number of deaths. Although the number of death cases was often lower than the number of infection cases, it increased as the number of days increased.

During this COVID-19 situation in Bangladesh, all infected patients were rehabilitated in 10 public hospitals from March to May 2020. These are Bangabandhu Sheikh Mujib Medical University (BSMMU), Dhaka Medical College Hospital (DMCH), Combined Military Hospital (CMH), Kurmitola General hospital (KGH), Sylhet MAG Osmani Medical College Hospital (SOMCH), Chittagong Medical College Hospital, National Institute of Neuro-Science & Hospital (NINS), Mymensingh

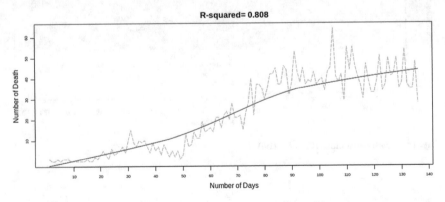

Fig. 12 R^2 value of death cases

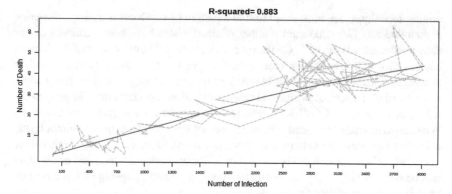

Fig. 13 R^2 value of death cases against infection cases

Fig. 14 Comparisons between infection and death cases

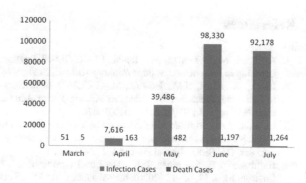

Medical College Hospital (MMCH), and Centre for the Rehabilitation of the Paralysed (CRP). During lockdown time, the Outdoor Patients Department (OPD) was completely closed. But some doctors practiced using the virtual consulting platform over the Internet with the patients [15].

8 Limitations and Conclusion

From this clustering analysis, a pattern can be discovered for death cases on COVID-19. It shows that the maximum death occurred within 51–70 years age limits, which are 69.36, almost one-third of the total death cases. This research work has few limitations such as; data is collected from several online sources and the affected areas could not be verified from authentic sources. Human-to-human virus transmission methods could be changed during this long-term evolution of mutation rate.

This research study forecasts the death cases for COVID-19 in Bangladesh from 18 March 2020 to 31 July 2020 (136 days) based on the recognized platforms. As the methodology, R language is used for clustering to get the prediction result. And mining techniques are introduced to find a pattern of death cases during this period in Bangladesh. The maximum number of death cases have been observed of age above sixty years is 46.19%. On the other side, the minimum number of death cases has recorded of age below ten years is 0.58%. The most infected city in Bangladesh is Dhaka, whereas Chittagong (24.40%) as the second-largest city in Bangladesh also recorded the second position in case of infection and death rate. To prevent the rate of death cases of COVID-19, people have to follow some guidelines as said by WHO such as maintain social distances, contact with others, avoid traveling, trying to increase the immune system, and also many more. In the future, this work will be extended for a minimum of six months or 287 days collected data, and death cases will be compared globally for Southeast Asian countries to upsurge the predication level from clustered data for Bangladesh.

References

1. Azarafza, M., Azarafza, M., Tanha, J.: COVID-19 Infection Forecasting based on Deep Learning in Iran. medRxiv, https://doi.org/10.1101/2020.05.16.20104182, pp. 1–7. (2020)
2. Imtiaz, A., Khan, N.M., Hossain, M.A.: COVID-19 in Bangladesh: Measuring Differences in Individual Precautionary Behaviors Among Young Adults, in medRxiv, pp. 1–24 (2020). https://doi.org/10.1101/2020.05.21.20108704
3. Isabella, S.A., Srinivasan, S.: Analysis of k-means algorithm for big data analytics using R language. In: 4th International Conference on Cyber Security (ICCS) 2018. Int. J. Adv. Stud. Sci. Res. (IJASSR), Special Issue, pp. 232–236 (2018)
4. Habibzadeh, P., Stoneman, E.K.: The novel coronavirus: a bird's eye view. Int. J. Occup. Environ. Med. **11**, 65–71 (2020). https://doi.org/10.15171/ijoem.2020.1921
5. Azarafza, M., Azarafza, M., Akgün, H.: Clustering method for spread pattern analysis of corona-virus (COVID-19) infection in Iran, in medRxiv, doi: medRxiv preprint, pp. 1–6 (2020). https://doi.org/10.1101/2020.05.22.20109942
6. Zhang, L., Chen, F., Lin, Z.: Origin and evolution of the 2019 novel coronavirus. Clinical Infectious Diseases February 2020, The Infectious Diseases Society of America (IDSA), doi: medRxiv preprint doi: 10.1093/cid/ciaa112, pp. 1–2 (2020)
7. The World Health Organization (WHO): Coronavirus disease (COVID-19) weekly epidemiological update and weekly operational update, Situation reports July 2020, Situation report—189," 27 July 2020, pp. 1–18 (2020)
8. The World Health Organization (WHO): South–East Asia, Bangladesh "Coronavirus disease (COVID-2019) Bangladesh situation reports, Situation reports July 2020, Situation report—22," 27 July 2020, pp. 1–17 (2020)
9. Ramageri, B.M.: Data mining techniques and applications. Indian J. Comput. Sci. Eng. (IJCSE) **1**(4), 301–236305 (2010)
10. Venables, W.N., Smith, D.M., R Core Team: An introduction to R, notes on R: a programming environment for data analysis and graphics, Version 4.0.2, June 22 2020
11. Bangladesh Corona Information Portal. Available https://corona.gov.bd/
12. Na, S., Xumin, L., Yong, G.: Research on k-means clustering algorithm—an improved k-means clustering algorithm. In: 3rd International Symposium on Intelligent Information Technology and Security Informatics (IITSI), IEEE, doi: medRxiv preprint https://doi.org/10.1109/iitsi.2010.74, pp. 63–67 (2010)
13. Directorate Gene Institute of Epidemiology, Disease Control and Research (IEDCR). Available: https://iedcr.gov.bd/
14. Directorate General of Health Services (DGHS). Available: https://www.dghs.gov.bd

15. Uddin, T., Islam, M.T., Rahim, H.R., Islam, M.J., Hossıan, M.S., Hassan, M.I., Mamun, M.A.A., Hossaın, M.S., Rahman, M.A., Chobı, F.K.: Rehabilitation perspectives of COVID-19 pandemic in Bangladesh. J. Bangladesh College Phys. Surg. **38**, COVID-19 (Supplement Issue), July-2020, pp. 76–81 (2020). https://doi.org/10.3329/jbcps.v38i0.47345

A Secure Smart Shopping Cart Using RFID Tag in IoT

Naresh Babu Muppalaneni and Ch. Prathima

Abstract The Internet of things is evolving in everyday life by partner conventional articles into gathering. For example, in a general store, all things can be related to each other, modeling a splendid shopping system. In such an Internet of things system, an efficient radio frequency identification tag can be associated with each thing which, when put into a shopping cart, can be thus examined by a truck outfitted with a radio frequency identification for every client. In like manner, charging can be coordinated from the shopping bushel itself, shielding customers from holding up in a long queue at the checkout. Moreover, sharp racking can be incorporated into this system, outfitted with radio frequency identification per clients, and can screen stock, perhaps in like manner invigorating a central server. Another favorable position of this kind of structure is that stock organization ends up being significantly less complex, as all things can be normally scrutinized by a radio frequency identification for every client as opposed to physically analyzed by a specialist.

Keywords Internet of things · Smart shopping cart · Security

1 Introduction

Web of things [1, 2] is a collaboration among physical articles that have transformed into a reality. Regular articles would presently have the option to be equipped with figuring power and correspondence functionalities, allowing objects any place to be related. This has acquired another upset mechanical, money related, and ecological frameworks and activated extraordinary difficulties in information the board, remote interchanges, and ongoing basic leadership. Also, different security and affirmation issues have been made and lightweight cryptographic frameworks are in over the

N. B. Muppalaneni (✉)
National Institute of Technology, Silchar, India
e-mail: nareshmuppalaneni@gmail.com

Ch. Prathima
Sree Vidyanikethan Engineering College, Tirupati, India
e-mail: prathima.ch@vidyanikethan.edu

S. Shakya et al. (eds.), *Proceedings of International Conference on Sustainable Expert Systems*, Lecture Notes in Networks and Systems 176,
https://doi.org/10.1007/978-981-33-4355-9_52

top energy to fit with the Internet of things applications. There has been a great deal of Internet of things look into different applications, for instance, shrewd houses, e-thriving structures, skillful gadgets, and so forth. In this review paper, an astute shopping framework dependent on radio frequency identification advancement is rotated around, which has not been well-thought about as of now. In such a structure, everything open to be acquired is related to a radio frequency identification tag [3], so they can be trailed by any contraption equipped with a radio frequency Identification for every customer in the store. Normally, this carries the going with inclinations: (1) Items put into a sharp shopping canister (with radio frequency identification looking at the point of confinement) can be therefore investigated and the charging data can in like way be made on the clever truck. As requirements are, clients do not have to hold up in long lines at checkout. (2) Smart leaves that are additionally outfitted with radio frequency identification per customer can screen every single stacked thing and send things to notice to the server. Exactly when things become sold out, the server can urge workers to restock. (3) It winds up being clear for the store to do stock association as all things can be consequently investigated and sufficiently logged. In this developing system, each smart truck is outfitted with an ultra high-frequency radio frequency identification per client, a microcontroller, a liquid crystal display contact screen, Zigbee connector, and weight sensor. To the keen truck can consequently peruse the things put into a truck by means of the radio frequency identification per user. All together for the keen truck to talk with the server, having picked Zigbee [4] advancement as it has low control and is sensible. All of us additionally have an excess weight scanner introduced on typically the shrewd truck for calculating the weight of things. At the point any time client completes typically the process of shopping, they will pay at the peruse point utilizing the developed charging data on the particular brilliant truck. RFID is set to each cart to check whether the things in the particular truck have been purchased. Protection issues additionally can be found in such a construction that the contender of the store may get very simple access to the scattering of wares for cash-related methodology, and consumer inclinations could be surmised by simply gathering the product data in customers' buying baskets. As it is an IoT application, the electric power utilization must be very low. With respect to the particular customer server correspondence: At this progression, elliptic curve cryptography-based cryptosystems are used, as the key size is a lot littler contrasted with different cryptosystems, for example, RSA algorithm. As the things are put on, the costs will get put on a complete tab. Individuals have reliably created and developed an advancement to help their needs as far back as the beginning of mankind. The central explanation behind the movement in development has been in restricting assignments and making standard tasks less complex and faster, free of various zones open. An imperative task on which individuals are found contributing a broad proportion of vitality is shopping. As indicated by a survey, roughly a large portion of the people goes through 1.5 h day by day on shopping. An enormous number of clients will constantly in general leave a line if the line is especially long.

2 Literature Survey

In balanced advertising [3], showcasing material is focused on and tweaked for a specific client, and it along these lines considers their specific individual needs. The fundamental motivation behind this examination depends on an organization of two early advances: remote individual cell phones (e.g., PDAs or individual computerized collaborators) and radio recurrence ID (RFID). The last innovation empowers ongoing following of each thing by joining a little tag or transponder to each thing. Under the main situation, it accepted that customers enter a retail location with an individual cell phone preloaded with the shopping list. The rundown is later transmitted to the focal mall server, which at that point registers a shopping way or course and imparts it back to the client. The clever thought is to incorporate with the course areas with advanced things. These areas are processed dependent on the shopping list or potentially dependent on the chronicled obtaining propensities for the client (e.g., if a devotion program is as a result). In the resulting situation, which requires thing level stamping and shrewd racks, it is recognized that an investigator, which is a touch of sharp racks, examines the starting at now acquired things of the client by investigating the labels connected to these things in the shopping basket. A significant way to deal with the model client's discrete decisions is the MNL model. The two primary situations are (1) the mall does not have a radio frequency identification organization, or (2) the mall has a radio frequency identification sending at the thing. In the subsequent situation, the mall is radio frequency identification conveyed with savvy racks [5] and thing level labeling is utilized. If a customer frequents the store, the store's data framework might foresee her needs. In a savvy rack framework, items are labeled by RFID transponders to give them a one of a kind recognizable proof. Chosen tracks are furnished with a reception apparatus framework and questioner unit, which are associated with a data framework. The client has just bought chosen things, and these can be distinguished by brilliant racks. A faithfulness program is normally utilized in retailing to upgrade the general offer and to improve client unwaveringness. It should propel purchasers to make next buys through limits and potential quicker administration. The shopping basket can figure consequently and show the all-out costs of how many numbers of items inside it. This makes it simple for the client to know the cost of items that the person in question needs to pay while shopping and not at the checkout counter. Along these lines, the client can get quicker administration at the checkout. This innovation is progressively encouraging to the degree of a potential substitution of the standardized identification framework as new minimal effort radio frequency identification label producing systems have risen. Another Zigbee module working at a similar recurrence can without much of a stretch catch the transmitted information. This issue should be settled explicitly regarding charging to advance purchaser certainty. Further, a progressively refined miniaturized scale controller and bigger showcase framework can be utilized to give better purchaser experience [6].

An inventive thing with the social assertion is the one that helps the solace, comfort, and gainfulness in standard everyday presence. Securing and shopping at gigantic strip malls are ending up each day movement in metro urban zones. A huge flood can be seen at these malls on rests and parts of the bargains. Individuals buy various things and kept the products in the basket. After fulfillment of gets, one needs to go to charging station for portions. At charging counter, the agent sets up the bill using institutionalized distinguishing proof per client which is a repetitive system [7] and results in a long queue at charging station. In this paper, it is talked about an item 'Insightful Shopping Cart [8]'' being helpful to an individual in regular shopping to decrease the time spent when purchasing the things. The essential objective of the proposed system is to give an advancement arranged, negligible exertion, viably flexible, and extreme structure for helping shopping up close and personal. The utilization of RFIDs in this framework fathom advantages, for example, expanding security and the subsequent decrease in item misfortune, diminished human mediation and mistake, expanded speed in included procedures, one of a kind distinguishing proof of items with extra data and accessibility of ongoing data, among others. There will be snappy updates of the truck's position at whatever point it is moved by the client. To make this possible, a shopping container arranging advancement is required in a couple of supermarket regions with the goal that the lasting observing empowers continuous truck position refreshes. The decision of a situating innovation to our answer is especially troublesome on account of various qualities every innovation presents. Angles, for example, run, vitality utilization, well-being, accuracy, among others, are significant for our answer. This venture report surveys and abuses the current improvements and different kinds of radio recurrence ID innovations which are utilized for item distinguishing proof, charging, and so forth.

The guideline objective of the proposed structure is to give an improvement arranged, negligible exertion, reasonably versatile, and extreme framework for helping shopping very close. The server communication segment [9] sets up and keeps up the relationship of the shopping holder with the basic server. User interface and show portion give the user interface, and automatic charging fragment handles the accusing in a relationship of the server communication section. Savvy shopping baskets [10] with electronic showcases, in correspondence with a PC framework, can show a general portrayal with cost subtleties related to a shopping list databases [11]. Shrewd truck, likewise outfitted with RFID labels, can likewise check the purchase of the things as they are put in the truck and, whenever wanted, speak with a charging framework to consequently charge the customer for the buys. Radio frequency identification names, or basically "names", are little transponders that react to demand from each customer by remotely transmitting a back to the back number or near identifier. The attractive sign is transmitted by the circle receiving wire associated alongside this circuit which is utilized to examine the radio frequency identification card number. In this undertaking, radio frequency identification card is utilized as a security access card. So everything has the individual radio frequency identification card which tends to the thing name. Radio frequency identification per customer is interfaced with a microcontroller. Here the microcontroller is the blast sort reprogrammable microcontroller [5] which as of late changed with a card number. The made thing is not hard to use, negligible exertion, and need not bother with any unique preparation. This undertaking report audits and endeavors the current improvements and different kinds of radio recurrence ID advances which are utilized for item distinguishing proof, charging, and so on. They have likewise taken in the engineering. This venture audits and adventures the current advancements and different kinds of radio recurrence distinguishing proof advances which are utilized for item recognizable proof, charging, and so on. Therefore, the review paper studies and assesses look into understanding in RFID frameworks.

3 System Analysis

3.1 Proposed System

In this system, the cost of the items has simply appeared on the liquid crystal display with the help of the radio frequency identification reader [12] and infrared sensor. If the user has completed his shopping, the total is in a like manner decided and appeared on the liquid crystal display itself.

Disservices:

1. No billing is given.
2. Difficult to put the product in precisely in the middle of the sensors to distinguish.
3. Hard to distinguish the item.

3.2 Designed System

In this system, the structure has been completed viably to move the data successfully to the billing session. In this system, radio frequency identification reader and Zigbee to data transferring are being used (Fig. 1).

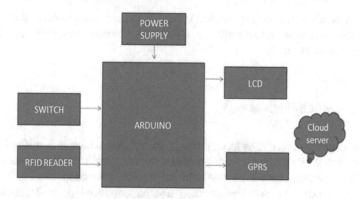

Fig. 1 Block diagram

RFID per user is utilized to peruse the information present in the RFID tag [13]. RFID perusers or collectors are made out of an RFID module, a control unit, and a reception apparatus to examine electronic labels by means of radio recurrence correspondence. Numerous additionally incorporate an interface that speaks with an application. Perusers can be handheld or mounted in vital areas to guarantee they can peruse the labels as the labels go through a "cross-examination zone." A radio frequency identification per user's capacity is to question radio frequency identification labels. The LCD displays the item name, cost, terminate date, and an aggregate sum. Complete postings of the items alongside their cost on LCD are shown. The power supply is that an off-the-rack Arduino connector must be a DC connector (e.g., it needs to put out DC, not AC). It ought to be somewhere in the range of 9 and 12 V DC (see note underneath). It must be appraised for at least 250 mA current yield, even though you will probably need something progressively like 500 mA or 1 A yield, as it gives you the present important to control a servo or twenty LEDs on the off chance that you need to. The power supply must have a 2.1 mm power plug on the Arduino end, and the fitting must be "center positive", that is, the center stack of the attachment must be the positive association. A cloud server is a sound server that is produced, encouraged, and passed on through a conveyed figuring stage over the Internet. Cloud servers have and show similar capacities, and value to a normal server anyway is gotten too remotely from a cloud master association. A cloud server may in like manner be known as a virtual server or virtual private cutoff. General Packet Radio Service is a group-based remote correspondence organization that ensures data rates from 56 up to 114 Kbps and a steady relationship with the Internet for wireless and PC customers. The higher data rates empower customers to partake in video gatherings and interface with sight and sound Web regions and amount applications using flexible handheld devices PCs similar to scratch cushion.

3.3 Questioning Statement

The essential objective of the proposed system is to give an advancement arranged, insignificant exertion, adequately versatile, and extreme structure for helping shopping face to face.

3.4 Project Objectives

The fundamental objective is to assemble a savvy card which is furnished with a radio frequency identification reader module, microcontroller, and a power supply. To reduce the waiting time in the queue for billing uding the checkout. The essential objective of the structure is to give advancement that organized insignificant exertion, adequately adaptable, and intense system for helping shopping up close and personal.

3.5 Feasibility Study

This investigation incorporates system resource requirements like expense and advantages, labor required, and venture length. Cost and advantages of the expense of this undertaking are around 4000–5000. The advantages are low cost, more solid, time sparing, manpower, and span manpower required.

4 Results

4.1 Home

Figure 2 is the model of the cart, and initially, it welcomes the customers to do secure shopping.

Initial count will be displayed as 0 (zero) as the customer did not select an item so the total bill is also 0.

LCD shows that the customer shopped items for example the rice with a quantity of 25 kg and the cost of it is Rs. 1000. This process continues for all the items.

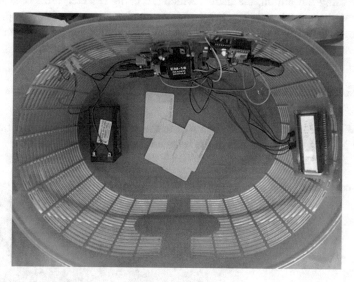

Fig. 2 Model of the cart

Fig. 3 Uploading the data

4.2 Uploading the Data

After selecting the required items into the basket, the customers need to press the button which is used to upload the details of the selected items automatically to the server based on the basket ID (Fig. 3).

4.3 Data Is Sending to the Server

After pressing the button, the details which are being selected are stored into the server based on the basket ID. For customer confirmation, the LCD displays the message as "Data sending… to server" to know whether the details are stored or not.

4.4 Confirmation for the Customer

Here the LCDs the message as "data sent to the server" for customer confirmation purpose to know that the details of the selected items are stored (Fig. 4).

Fig. 4 Confirmation for the customer

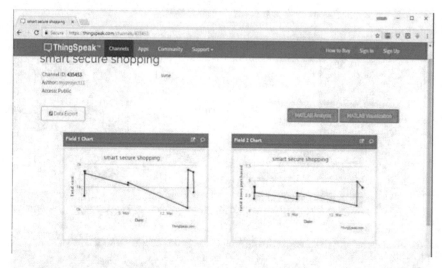

Fig. 5 Details of the basket per day

4.5 Message Displayed

After storing the details of the selected items into the server correctly, then LCD displays the message as "Thank You Visit Again".

4.6 Details of the Basket Per Day

In this portal, it displays the plot which consists of the total cost of selected items in that basket on a particular day, and also in another plot it displays many items purchased on a particular day (Fig. 5)..

4.7 Data Export

After displaying the plot, the data is needed to export to generate the bill by using the.csv files such as total cost and total items purchased in the particular basket (Fig. 6)..

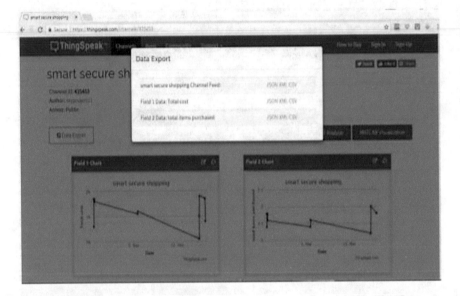

Fig. 6 Data export

4.8 Displaying the Stored Details

See Fig. 7.

CREATED_AT	ENTRY_ID	TOTAL
2018-02-26 07:46:56 UTC	1	600
2018-02-26 07:48:34 UTC	2	1700
2018-02-26 10:32:05 UTC	3	1700
2018-02-26 10:35:29 UTC	4	1600
2018-03-05 10:34:16 UTC	5	1100
2018-03-05 11:30:57 UTC	6	1200
2018-03-15 06:26:04 UTC	7	100
2018-03-15 06:31:06 UTC	8	1000
2018-03-15 07:30:28 UTC	9	1800

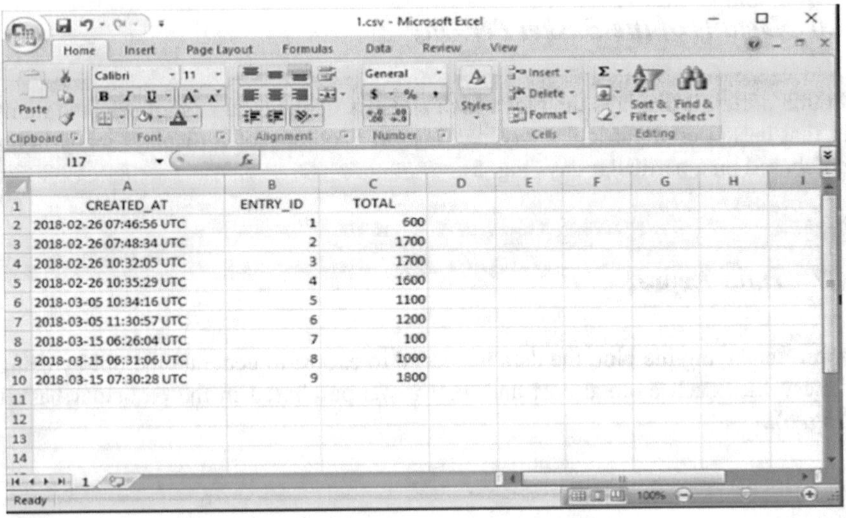

Fig. 7 Displaying the stored details

The details of the selected items which are stored in the server are displayed in the system with the format of the .csv file with the details such as time and date at which the basket is selected, the basket id, and the basket total cost.

Project execution time: It takes a few seconds for the execution. The project execution also depends on system performance. System performance is based on microcontroller, system hardware, and space available in the system.

5 Conclusion

A safe brilliant shopping framework uses RFID innovation. This is the main event when that UHF RFID is used in improving shopping experiences and security issues are examined with regard to a keen shopping framework. The plan of a total framework is detailed and constructs a model to test its capacities. A protected correspondence convention is additionally structured and presented security examination and execution assessments.

With radio frequency identification advancement, what is more, our appraisal is the main one in the improvement of a sharp shopping framework. Our future research will concentrate on improving the to and for development framework, for instance, by decreasing the computational overhead at the shrewd truck side for higher proficiency, and how to improve the correspondence capacity while saving security properties.

References

1. Xia, F., Yang, L.T., Wang, L., Vinel, A.: Web of things. Int. J. Commun. Syst. **25**(9), 1101 (2012)
2. Gubbi, J., Buyya, R., Marusic, S., Palaniswami, S.: Internet of Things (IoT): a vision, architectural elements, and future directions. IEEE (2011). https://doi.org/10.1109/i-smac.2017.805 8399
3. Klabjan, D., Pei, J.: In-store coordinated showcasing. J. Retail. Consum. Serv. **18**(1), 64–73 (2011)
4. Ali, Z., Sonkusare, R.: Rfid based brilliant shopping and charging. Univ. J. Adv. Res. Comput. Commun. Eng. **2**(12), 4696–4699 (2013)
5. Ali, Z., Sonkusare, R.: RFID based savvy shopping and charging. Global J. Adv. Res. Comput. Commun. Eng. **2**(12), 4696–4699 (2013)
6. Sanghi, K., Singh, R., Raman, N.: The smart cart—an enhanced shopping experience. TA, Justine Fortier Team 41 (2012)
7. Gangwal, U., Roy, S., Bapat, J.: Smart shopping cart for automated billing purpose using wireless sensor networks. IEEE (2013). https://doi.org/10.1109/icices.2014.703399
8. Kumar, R., Gopalakrishna, K., Ramesha, K.: Intelligent shopping cart. Int. J. Eng. Sci. Innov. Tech. (IJESIT) **2**(4) July 2013
9. Kamble, S., Meshram, S., Thokal, R., Gakre, R.: Building up a multitasking shopping trolley dependent on RFID technology. January 2014 Int. J. Soft Comput. Eng. (IJSCE)

10. Yewatkar, A., Inamdar, F., Singh, R., Bandal, A., et al.: Savvy truck with programmed charging, item data, item proposal usingrfid and zigbee with antitheft. Proced. Comput. Sci. **79**, 793–800 (2016)
11. Mitton, N., Papavassiliou, S., Puliafito, A., Trivedi, K.S.: Consolidating cloud and sensors in a brilliant city condition. EURASIP Diary Wirel. Commun. Netw. **2012**(1), 1 (2012)
12. Yathisha, L., Abhishek, A., Harshith, R., Darshan Koundinya, S.R., Srinidhi, K.: Automation of shopping cart to ease queue in malls by using RFID (2015). https://doi.org/10.1109/icices. 2014.7033996
13. Ahsan, K., Shah, H., Kingston, P.: RFID applications: an introductory and exploratory study. IJCSI Int. J. Comput. Sci. **7**(1), 3 (2010)

Author Index

Printed in the United States
by Baker & Taylor Publisher Services